COMPOSITE
AIRFRAME STRUCTURES

"FASTENERLESS COMPOSITES"

The part shown is a full scale demonstration article of the forward segment of a composite canopy frame. This innovative design consists of an upper skin, seven internal bulkheads and a lower closeout skin which are assembled by co-consolidation without using a single fastener — see Section 12.4 in Chapter 12 for more details.

(Thanks to both Lockheed Aeronautical Systems Co. and Composites Horizons, Inc.)

COMPOSITE
AIRFRAME STRUCTURES

PRACTICAL DESIGN INFORMATION AND DATA

Michael Chun-Yung Niu

HONG KONG CONMILIT PRESS LTD.

About the Author

Michael C. Y. Niu is the president of AD Airframe Consulting Co. and is a metallic and composite airframe structural consultant. He was a Senior Research and Development Engineer, Composites Development Center (CDC). Lockheed Aeronautical Systems Company, Burbank, California, USA. Mr. Niu has acquired over 25 years experience in aerospace and airframe structural analysis and design. He has been the acting Department Manager in charge of various structural programs, including development of innovative structural design concepts for both metallic and composite materials that are applicable to current and/or future advanced tactical fighters and advanced transports. He was lead engineer responsible for the L-1011 wide body derivative aircraft wing and empennage stress analysis.

During 1966 and 1968, he served as stress engineer to B727, B747 etc. at The Boeing Company.

Mr. Niu received a B.S. in Hydraulic Engineering from the Chungyuan University, Taiwan, in 1962 and a M.S. in Civil Engineering from the University of Wyoming in 1966.

He is author of a highly acclaimed book **AIRFRAME STRUCTURAL DESIGN** (1988) and teaches a popular UCLA Engineering Short Course on the same subject. He has also written Lockheed's "**Composite Design Guide**" and "**Composites Drafting Handbook**".

He received the Lockheed Award of Achievement and Product Excellence in 1973 and 1986 respectively, and is listed in *Who's Who in Aviation* 1973.

© 1992 Conmilit Press Ltd. Hong Kong All rights reserved.
No part of this book may be reproduced in any form or
by any electronic or mechanical means, including
information storage and retrieval devices or systems,
without prior written permission from the publisher

First edition – Jan., 1992
Second edition – Sept., 1993
Second published – Nov., 1996

All inquiries should be directed to: HONG KONG **CONMILIT** PRESS LIMITED
11/F Kwong Ah Building   or   P.O.Box 23250
114 Thomson Road   Wanchai Post Office
Wanchai   Hong Kong
Hong Kong   Tel: (852) 2838 7608
                              Fax: (852) 2834 2156

U.S. order/inquiry: TECHNICAL BOOK COMPANY
2056 Westwood Blvd
Los Angeles, CA 90025
U.S.A
Tel: (310) 475-5711

Please forward any suggestions or comments to:
Prof. Michael C. Y. Niu
P.O. Box 3552
Granada Hills, CA 91394
Fax: (818) 701-0298

DISCLAIMER: The author and publisher do not guarantee the accuracy of the information provided in this book, and it should not be referenced as an authoritative source. Information and data used in this book does not constitute official endorsement, either expressed or implied, by the manufacturers or the publisher. In case of conflict between the material herein and the manufacturer's specification, the specification shall prevail.

ISBN 962-7128-06-6   *Printed in Hong Kong*

PREFACE

In the past decades considerable progress in advanced composite technology has been made. However, the full potential in the design, manufacturing and especially the application of composites has not been realized. The use of composites in heavily loaded primary structures has been limited, mainly due to lack of hands-on experience and confidence. This book is intended to advance the technical understanding and practical knowledge of advanced composites, emphasizing the design and manufacture of airframe structures. All aspects at composite design will be discussed in a thorough and rigorous fashion which includes guidelines, observations, design factors, pros and cons of design cases, and troubleshooting techiques. However, neither the basic chemistry of materials nor laminate strength (or stress) analysis will be discussed in detail. Such information can be found in numerous composite books and published papers.

This book (which may be used as a handbook) is divided into twelve chapters, with emphasis on itemized write-ups, tables, graphs and illustrations which focus on points of interest. While some of the data and information will be superseded as technology advances, the basic technical information and data will hold true. However, some modification of the design concepts and manufacturing processes described in this book may be required in due course. One purpose of this book is to give composite designers and engineers a practical design tool, containing broad data and information gained from past experience and lessons learned in design and fabrication of composite components, that can be used to design low cost and weight efficient composite structures with high structural integrity.

Composite structures are not just an extension of their metal counterparts and should not be considered only as piece-by-piece replacements (the BLACK ALUMINUM approach) that merely save structural weight by their lower material density advantage. The designer's ingenuity and resourcefulness is needed to develop innovative concepts which will reach the ultimate goal of composite structures that meet the requirements of durability, damage tolerance, maintainability, repairability, crashworthiness, low weight and cost effectiveness.

Early interface and support from producibility at the predesign level is critical in composite design to insure that cost-effective producibility features. Engineering design should also seek interface and criteria from tool design, production, manufacturing, industrial engineering, and quality assurance. It is recommended that during the composite design process on-board support guidelines are combined with previous counterpart metal experience. It must be remembered that composite airframe structural design encompasses almost all the engineering disciplines, and engineers who go to the computer workstation or drawing board today need hands-on information and data that are technically sound and emphasize rational design and technical analysis, and this book was written with that need in mind. It is not practical, however, to cover all the information and data within this handy compact book. Selected relevant references are presented at the back of each chapter so that reader can explore his own personal interests in greater detail.

In preparing this book, it was necessary to collect vast amounts of information and data from many sources. Sincere appreciation and thanks is given to those who contributed to this book, to my previous colleagues at Lockheed Aeronautical Systems Company, and to other specialists from various companies for their gracious help. Special thanks to **Dr. John T. Quinlivan**, Manager of B777 Empennage Structures, The Boeing Company, for his valuable comments and review of the entire draft of this book.

While some of the material presented in this book may be controversial, the author has tried to the best of his knowledge to make this book of practical use in the airframe structures arena. Any constructive suggestions and comments for improvement and future revision would be greatly appreciated by the author.

Michael Chun-yung Niu
（牛春匀）
L.A., California, U.S.A.

Jan., 1992

CONTENTS

PREFACE			v
GLOSSARY			xi
Chapter	**1.0**	**INTRODUCTION**	1
	1.1	Introduction	1
	1.2	Characteristics of Composites	11
	1.3	Composites Vs. Metals (Aluminum Alloys)	14
	1.4	Design for Low Cost Production	20
	1.5	Criteria	27
	1.6	Certification of Composite Airframe Structures	29
	1.7	Laminate Orientation and Symbol Notations	35
Chapter	**2.0**	**MATERIALS**	41
	2.1	Introduction	41
	2.2	Organic Matrices	47
	2.3	Metal, Carbon and Ceramic Matrix Composites	57
	2.4	Reinforcements (Fibers)	72
	2.5	Material Forms	85
	2.6	Sandwich Core Materials	111
Chapter	**3.0**	**TOOLING**	126
	3.1	Introduction	126
	3.2	Tool Design Considerations	138
	3.3	Metallic Tooling	142
	3.4	Non-metallic Tooling	147
	3.5	Chamber Systems	159
	3.6	Self-contained Tooling System	165
	3.7	Elastomeric Tooling System	168
	3.8	Matched Die Tooling	173
Chapter	**4.0**	**MANUFACTURING**	176
	4.1	Introduction	176
	4.2	Manufacturing Guidelines and Practices	186
	4.3	Placements	200

	4.4	Bagging System	217
	4.5	Mold Forming	228
	4.6	Thermoforming (Platen Press)	238
	4.7	Molding	251
	4.8	Pultrusion	257
	4.9	Sandwich Structures	265
	4.10	Layup-over-foam Method	270
	4.11	Hybridized Fabrications	272
	4.12	Cutting and Machining	276
Chapter	**5.0**	**JOINING**	**285**
	5.1	Introduction	285
	5.2	Mechanical Fastening	290
	5.3	Bonding	330
	5.4	Welding	349
	5.5	Sewing and Wet-crush-rivet	353
Chapter	**6.0**	**ENVIRONMENTS**	**357**
	6.1	Introduction	357
	6.2	Weathering Effects	357
	6.3	Lightning Strikes	361
	6.4	Galvanic Corrosion	379
	6.5	Other Environmental Concerns	380
Chapter	**7.0**	**LAMINATE DESIGN PRACTICES**	**383**
	7.1	Introduction	383
	7.2	Material Allowable Determination	393
	7.3	Laminate Strength Analysis	397
	7.4	Fatigue and Impact Damage	422
	7.5	Design Practices	433
	7.6	Preliminary Sizing	446
Chapter	**8.0**	**STRUCTURAL TESTING**	**453**
	8.1	Introduction	453
	8.2	Durability and Damage Tolerance Tests	460
	8.3	Coupon Tests	464
	8.4	Element and Components Tests	486
	8.5	Verification Full-scale Tests	490
Chapter	**9.0**	**QUALITY ASSURANCE**	**500**
	9.1	Introduction	500
	9.2	Material Qualification	504
	9.3	Types of Defects (Flaws)	506
	9.4	NDI Methods	511
	9.5	Difficult-to-inspect Designs	529

Chapter	**10.0**	**REPAIRS**	533
	10.1	Introduction	533
	10.2	Bolted Repairs	541
	10.3	Bonded Repairs	544
	10.4	Honeycomb Repairs	554
	10.5	Injection Repairs	557
Chapter	**11.0**	**COMPOSITE APPLICATIONS**	560
	11.1	Introduction	560
	11.2	Commercial Transport Aircraft	562
	11.3	Military Aircraft	576
	11.4	All-composite Utility Aircraft	590
Chapter	**12.0**	**INNOVATIVE DESIGN APPROACHES**	621
	12.1	Introduction	621
	12.2	Fiber Orientation Concept	623
	12.3	Modular Concept	630
	12.4	Monolithic (Integral) Concept	632
	12.5	Assembly Concept	636
	12.6	Other Concepts	637

Appendix A Commonly Used Conversion Factors (English units Vs. SI units)   647

Appendix B Composite Engineering Drafting Practice   648

Appendix C A List of Schematic Drawings Related to Composite Structures   655

Index   657

GLOSSARY

A

ADHEREND: The material being bonded by an adhesive.

ADHESION: The property denoting the ability of a material to resist delamination or separation into two or more layers.

ADHESIVE: A substance capable of holding two materials together by surface attachment. In the composites, the term is used specifically to designate structural adhesives, those which produce attachments capable of transmitting significant structural loads.

ADHESIVE FAILURE: Failure at the adhesive/adherend interface.

AGING: The effect on materials of exposure to an air environment for an interval of time.

AMORPHOUS: A material, such as a liquid, which has no crystallinity; i.e., there is no order or pattern to the distribution of the molecules in the material.

ANELASTICITY: A characteristic exhibited by certain materials in which strain is a function of both stress and time, such that while no permanent deformations are involved, a finite time is required to establish equilibrium between stress and strain in both the loading and unloading directions.

ANISOTROPIC: Not isotropic; having mechanical and/or physical properties which vary with direction relative to natural reference axes inherent in the material.

ANNEALING: In plastic, heating to a temperature where the molecules have significant mobility, permitting them to reorient to a configuration having less residual stress.

ARAMID: A type of highly oriented organic material derived from polyamide (nylon) but incorporating aromatic ring structure.

AREAL WEIGHT: The weight of fiber per unit area (width × length) of tape or fabric.

ARTIFICIAL WEATHERING: Exposure to laboratory conditions which may be cyclic, involving changes in temperature, relative humidity, radiant energy, and any other elements found in the atmosphere in various geographical areas.

AUTOCLAVE: A closed vessel which applies pressure to objects inside, such as a bagged laminate. The pressurizing medium is a gas (usually nitrogen or carbon dioxide). "Hydroclave" is an autoclave, except that water is used as the pressurizing medium.

AUTOCLAVE MOLDING: A process in which the layup is covered by a pressure bag, and the entire assembly is placed in an autoclave capable of providing heat and pressure for curing the part. The pressure bag is normally vented to the outside.

B

B-STAGE: An intermediate stage in the reaction of a resin; that is, partial cure.

BAGGING: The process of applying an impermeable layer of film over a part and sealing the edges so that a vacuum can be drawn. The bag permits a pressure differential to exist between the pressurizing medium (usually the working fluid of the autoclave or hydroclave) and the part, thereby applying pressure to the part.

BALANCED LAMINATE: A composite laminate in which all laminate at angles other than 0° and 90° occur only in pairs (not necessarily adjacent), and is symmetrical about a centerline. See Symmetrical Laminate.

BARRIER FILM: The layer of film used to permit removal of air and volatiles from a composite layup during cure while minimizing resin loss.

BATCH (or LOT): In general, a quantity of material formed during the same process and having identical characteristics throughout. As applied to the composites, a batch of prepreg tape is defined as a quantity of tape which is produced from a single batch of matrix material. The prepreg tape batch is not necessarily produced at one time, but all sub-batches are produced in the same equipment under identical conditions. The filaments included in a batch, in this context, are restricted only to being acceptable within the requirements of applicable filament procurement specifications.

BINDER: A bonding resin used to hold strands together in a mat or preform during manufacture of a molded object.

BISMALEIMIDES (BMI): Lower temperature capability addition polyimide resin systems which have epoxy-like processability but have improved elevated temperature properties.

BLEEDER CLOTH: Material, such as fiberglass, used in the manufacture of composite parts to allow the escape of excess gas and resin during cure. The bleeder cloth is removed after the curing process and is not part of the final composite.

BOND: The adhesion of one surface to another, with or without the use of an adhesive as a bonding agent.

BRAIDING: Weaving of fibers into a tubular shape instead of flat fabric. This technique is currently used in the fabrication of graphite fiber reinforced golf shafts. It is being considered for aerospace tubular structures.

BREATHER CLOTH: A layer or layers of open weave cloth used to enable the vacuum to reach the area over the laminate being cured, such that volatiles and air can be removed and also causing the pressure differential that results in application of pressure to the part being cured.

BRIDGING: Special techniques must be used so that the fibers will move into radii and corners; otherwise, they "bridge" the gap, resulting in dimensional control problems and voids. Care must also be taken to prevent bridging of separators, bleeders, barrier films, venting layers, and bagging.

BRITTLE: Most of the high performance resin systems are brittle; namely, they exhibit very low elongation and almost no plastic deformation prior to failure. In composites, the mismatch in thermal coefficients of expansion between the fibers and the resin matrix, in combination with a brittle resin, often results in formation of microcracks.

BROADGOODS: A term loosely applied to prepreg material greater than about 12 inches in width, usually furnished by suppliers in continuous rolls. The term is currently used to designate both collimated uniaxial tape and woven fabric prepregs.

BUCKLING (COMPOSITE): Buckling is a mode of failure characterized generally by an unstable lateral deflection caused by compressive or shear action on the structural element involved. In composites, buckling may take the form not only of conventional general instability and local instability, but also a microinstability of individual fibers.

C

C-SCAN: The back and forth scanning of a specimen with ultrasonics. An NDT (nondestructive test) technique for finding voids, delaminations, defects in fiber distribution, etc.

CARBON: The element which provides the backbone for all organic polymers. Graphite is a more ordered form of carbon; diamond is the densest crystalline form of carbon. Most of the high strength "graphite" fibers are actually carbon fibers; with higher temperature processing, the same organic precursor fiber can be converted to the higher modulus graphitized form.

CARBON FIBERS: Fibers made from a precurser by oxidation and carbonization, and not having a graphitic structure.

CATALYST: A chemical which promotes a chemical reaction without becoming a part of the molecular structure of the product. In resin systems, catalysts (and accelerators) lower the temperature at which significant amounts of reaction occur, affecting reaction rate and changing the characteristics of the cure cycle.

CAUL PLATES: Smooth plates, free of surface defects, used during the curing process to provide a controlled surface on the finished laminate.

CELANESE COMPRESSION: A specialized compression test using a fixture designed by Celanese Corp. Now a standard ASTM test.

CERAMIC TOOLING: Use of a castable ceramic to make a tool shape. Ceramic tooling is seldom used unless a very large number of complex parts are to be made; otherwise, tooling such as graphite tooling is more cost-effective.

CHROMATOGRAM: A plot of detector response against peak volume of solution (eluate) emerging from the system for each of the constituents which have been separated.

CLOTH: A woven product made from continuous yarns or tows of fiber. "Cloth" and "fabric" are usually used interchangeably.

COCURING: The act of curing a composite laminate and simultaneously bonding it to some other prepared surface during the same cure cycle.

COEFFICIENT OF THERMAL EXPANSION: The change in a dimension of a specimen per unit change in temperature, expressed as a ratio. Typical units are microinches/inch/°F and microinches/inch/°C. High modulus graphite fibers have a negative coefficient of thermal expansion (CTE) in the axial direction; they shrink when heated. The CTE only describes the reversible portion of the dimensional change.

COEFFICIENT OF VARIATION: The standard deviation divided by the mean.

COHESIVE FAILURE: Describing the failure surface of an adhesive joint where the failure occurred primarily in the adhesive layer.

COMPLEX CURVATURE: Describing a surface which curves in more than one direction, such as a saddle or spherical shape. In other words, the surface has both concave and convex areas.

COMPOSITE: As used generally in this book, composite describes a matrix material reinforced with continuous filaments.

COMPOUND: An intimate mixture of polymer or polymers with all the materials necessary for the finished product.

COMPRESSION MOLDING: Putting a reinforced resin into a mold cavity, closing the mold, and applying pressure and heat in order to force the material to completely fill the mold cavity and to cure the material.

CONDITIONING: Maintaining the test specimens in a controlled environment for a specific length of time prior to testing.

CONSOLIDATION: In metal matrix or thermplastic composites, the diffusion bonding operation in which an oriented stack of preplies is transformed into a finished composite laminate.

CONTINUOUS FILAMENT YARN: Yarn formed by combining two or more continuous filaments into a single, continuous strand.

CRAZING: The development of a multitude of very fine cracks in the matrix material.

CREEL: The rack used to hold fiber spools so that a large number of fiber tows can be used simultaneously, as in making tape prepreg or for the warp fiber tows of a weaving loom.

CROSSLINKING: Chemical reaction between molecules resulting in the formation of a three-dimensional network of molecules. Crosslinking requires that at least one of the molecules involved in the reaction have three or more reactive groups; otherwise, the reaction only results in forming a longer molecule (chain extension).

CROSS-PLIED: A laminate with plies in different directions. Laminates made from fabric automatically have fibers in two directions, but are not considered as "cross-plied" unless the different layers (plies) are oriented in different directions.

CRYSTALLINITY: Polymers such as nylon form localized area of crystallinity (highly ordered sections) formed by alignment of sections of a polymer chain (by folding, etc.) or of adjacent molecules. The localized areas of crystallinity change the physical behavior of the polymer.

CURE: To change the properties of a thermosetting resin irreversibly by chemical reaction. Cure may be accomplished by addition of curing agents, with or without catalyst, and with or without heat and pressure.

CURE CYCLE: The time/temperature/pressure cycle used to cure a composite resin system or prepreg.

CURE MONITORING: Use of electrical techniques to detect changes in the electrical properties and/or mobility of the resin molecules during cure.

CURE STRESS: A residual internal stress produced during the cure cycle of composites containing reinforcements and/or resins with different thermal coefficients of expansion.

D

DAM: Boundary support used to prevent excessive edge bleeding of a laminate and to prevent crowning of the bag.

DEBOND: A deliberate separation of a bonded joint or interface, usually for repair or rework purposes.

DEBULKING: Compacting the thickness of a layer of prepreg or of a prepreg layup by using pressure and/or vaccum to remove most of the air.

DELAMINATION: The separation of one or more layers of a laminate.

DENIER: A textile term for the weight, in grams, of 9000 meters of fiber tow.

DESORPTION: A process in which an absorbed material is released from another material. Desorption is the reverse of absorption, adsorption, or both.

DIELECTROMETRY: Use of electrical techniques to measure the changes in loss factor (dissipation) and in capacitance during cure of the resin in a laminate.

DIFFERENTIAL SCANNING CALORIMETRY (DSC): Measurement of the energy absorbed (endotherm) or produced (exotherm) as a resin system is cured. Also detects loss of solvents and other volatiles.

DIFFERENTIAL THERMAL ANALYSIS (DTA): An experimental analysis technique where a specimen and a control are heated simultaneously and the difference in their temperatures is monitored. The difference in temperature provides information on relative heat capacities, presence of solvents, changes in structure (i.e., phase changes such as melting of one component in a resin system), and chemical reactions. Also see Differential Scanning Calorimetry.

DISBOND: A lack of proper adhesion in a bonded joint. This may be local or may cover a majority of the bond area. It may occur at any time in the cure or subsequent life of the bond area and may arise from a wide variety of causes.

DISSIPATION FACTOR: A measure of the lag in phase angle caused by presence of a material between the plates of a condensor (capacitor). Also called power factor, loss tangent, and tangent δ. In dynamic dielectric analysis of a resin undergoing cure, it is a measure of the energy used in aligning (or attempting to align) the dipoles present in the resin system with the constantly reversing (AC) electrical field. When the resin voscosity is low, it takes very little energy to align the dipoles; the dissipation factor increases as resin viscosity increases, but decreases when the resin is cured (when cured, the dipoles are immobilized, and very little power is lost).

DISTORTION: In fabric, the displacement of fill fiber from the 90° angle (right angle) relative to the warp fiber. In a laminate, the displacement of the fibers (especially in radii), relative to their idealized location, due to motion during layup and cure.

DRAPE: The ability of a prepreg to conform to a contoured surface. If the resin becomes hard, due to loss of solvent or due to staging, the prepreg becomes stiff ("boardy") and loses its drape characteristics.

DRYING TOWER: In prepregging via a solvent process, a conveyor belt carries the prepreg through a drying section which uses heated air to remove excess solvent from the prepreg. Usually this heated section is vertical, due to space limitations.

E

EDGE BLEED: Removal of volatiles and excess resin through the edge of the laminate, as in matched die molding of a laminate. In autoclaved parts, edge bleeding is discouraged, since excess resin will only be removed from the area near an edge; resulting in uneven resin distribution.

ELASTOMERIC TOOLING: A tooling system utilizing the thermal expansion of rubber materials to form composite hardware during cure.

ENVELOPE BAG: A vacuum bag which encloses the laminate and the tool.

EXPANDABLE TOOLING: Use of hollow rubber mandrel which can be pressurized to form composite hardware during cure.

F

FABRIC: A material constructed of interlaced yarns, fibers, or filaments, usually arranged in a planar structure. Nonwovens are sometimes included in this classification.

FAYING SURFACE: The portion of a component's surface that, upon assembly, will be pressed against another component and hence must be pre-cleaned and otherwise treated for bonding.

FIBER: A single homogeneous strand of material, essentially one-dimensional in the macrobehavior sense, used as a principal constituent in composites because of its high axial strength and modulus.

FIBER CONTENT: The amount of fiber present in a composite. This is usually expressed as a percentage volume fraction or weight fraction of a cured composite.

FIBER DIRECTION: The orientation or alignment of the longitudinal axis of the fiber relative to a stated reference axis.

FIBER FINISH: A material applied to the surface of fibers to improve the bond of the fiber to the resin matrix and/or to protect the fiber against abrasion damage during operations such as weaving.

FIBER TOW: A loose, untwisted bundle of continuous fibers. In composite technology, "tow" is often used interchangeably with "yarn", the twisted version.

FIBER VOLUME: The volume percent of fiber in a composite.

FIBERGLASS: The generic name for glass fibers and for composites using glass fibers for reinforcement.

FILAMENT: Fibers characterized by extreme length, such that there are normally no filament ends within a part except at geometric discontinuities. Filaments are used in filamentary composites and are also used in filament winding processes which require long continuous strands.

FILAMENT WINDING: An automated process in which continuous filament (or tape) is treated with resin and wound on a removable mandrel in a prescribed pattern.

FILAMENT WOUND: Pertaining to an object created by the filament winding method of fabrication.

FILL: Yarn oriented at right angles to the warp in a woven fabric.

FILLER: A second material added to a basic material to alter its physical, mechanical, thermal, or electrical properties. Sometimes used specifically to mean particulate additives.

FILLER PLY: Partial plies, usually located on sandwich edgebands, which do not extend onto any portion of the honeycomb surface.

FILM ADHESIVE: Adhesive supplied in the form of a film.

FINISH: A material, with which filaments are treated, which contains a coupling agent to improve the bond between the filament surface and the resin matrix in a composite material. In addition, finishes often contain ingredients which provide lubricity to the filament surface, preventing abrasive damage during handling, and a binder which promotes strand integrity and facilitates packing of the filaments.

FLAME-SPRAYED TAPE: A form of metal matrix preply in which the fiber system is held in place on a foil sheet of matrix alloy by a metallic flame-spray deposit. Each flame-sprayed preply is usually combined in the layup stack with a metal cover foil and/or additional metal powder to ensure complete encapsulation of the fibers. During consolidation, all the metallic constituents are coalesced into a homogeneous matrix.

FLASH: Excess resin which forms at the parting line of a mold or die, or which is extruded from a closed mold.

FLEXURAL STRENGTH: A mechanical test to measure the strength of a laminate in bending. The strength reported is a calculated value which assumes that the material is isotropic in the thickness direction, which is approximately true for unidirectional specimens but is definitely not true for cross-plied laminates. In addition, failure modes can be in compression, in tension, in interlaminar shear, or a combination thereof. Therefore, the test is only suitable as a comparison or quality control test.

G

GEL PERMEATION CHROMATOGRAPHY (GPC): A form of liquid chromatography in which the polymer molecules are separated by their ability or inability to penetrate the material in the separation column.

GEL TEMPERATURE: In a cure cycle, the temperature at which the viscosity of a thermosetting resin becomes so high that no further dimensional change occurs. The temperature at which the resin gels can be changed by changing the cure cycle (heatup rate, hold times, etc.). The laminate dimensions become fixed at the gel temperature (for all practical purposes); hence, the tool dimensions at the gel temperature control the dimensions of the cured laminate.

GEL TIME: The amount of time required before a resin sample advances to the gelation point, when held at a pre-defined constant temperature.

GELATION: In composite technology, referring to the point in a resin cure where the resin viscosity has increased to the point where it barely moves when probed with a sharp instrument.

GELCOAT: A resin applied to the mold to provide an improved surface for the composite.

GLASS: An inorganic product of fusion which has cooled to a rigid condition without crystallizing. In the composites, all reference to glass will be to the fibrous form as used in filaments, woven fabric, yarns, mats, chopped fibers, etc.

GLASS CLOTH: Woven glass fiber material (see also SCRIM).

GLASS TRANSITION TEMPERATURE (T_g): One method of describing the temperature at which increased molecular mobility results in significant changes in the properties of a cured resin system. The glass transition temperature (T_g) can be defined as the inflection point on a plot of modulus vs. temperature. T_g is defined as the inflection point — properties can have decreased significantly before T_g is reached.

GRAPHITE: The crystalline, allotropic form of carbon. In bulk form, used for advanced composite tooling and for such items as the "lead" in pencils. See Graphite Fibers.

GRAPHITE FIBERS: Technically, a highly oriented form of graphite. However, in common usage it also includes highly oriented carbon fibers which have only a small amount of graphite content.

GRAPHITIZATION: Conversion of carbon to its crystalline allotropic form by use of very high temperatures. Diamond is also crystalline allotropic form of carbon, but requires extremely high pressures (over one million psi) in addition to very high temperature in order to be formed.

H

HAND LAYUP: A process in which components are applied either to the mold or on a working surface, and the successive plies are built up and worked by hand.

HARDENER: The compound which reacts with a resin to form the crosslinked (thermoset) plastic.

HARDNESS: Resistance to deformation; usually measured by indentation. Types of standard tests include Brinell, Barcol, and Rockwell.

HARNESS SATIN: Describes a set of weaving patterns which produce a fabric having a satin appearance. "8HS" describes a harness satin weave where the warp fiber tows go over seven fill tows and then under one fill tow, for a repeating total of 8. By itself, "8HS" is not a complete description, because there are many possible patterns of where the crossover points of adjacent tows are located.

HOT MELT PROCESSING: Refers to the process of heating a resin to reduce its viscosity, using a doctor blade arrangement to spread a controlled thickness layer onto transfer paper, and subsequently forcing the hot resin (briefly heated to a higher temperature to reduce the viscosity) into collimated fibers. The process can also be used to prepreg fabric, but care has to be taken to avoid crushing the fibers where they cross over.

HYBRID: A composite laminate or prepreg contianing two or more types of composite systems. Usually only the fibers differ since cocuring two different resins is difficult.

I

INCLUSION: A physical and mechanical discontinuity occurring within a material or part, usually consisting of solid, encapsulated foreign material. Inclusions are often capable of transmitting some structural stresses and energy fields, but in a noticeably different degree from the parent material.

INTEGRAL COMPOSITE STRUCTURE: Composite structure in which several structural elements, which would conventionally be assembled together by bonding or mechanical fasteners after separate fabrication, are instead laid up and cured as a single, complex, continuous structure; e.g., spars, ribs, and one stiffened cover of a wing box fabricated as a single integral part. The term is sometimes applied more loosely to any composite structure not assembled by mechanical fasteners.

INTEGRALLY HEATED: Referring to tooling which is self-heating, through use of electrical heaters such as cal rods. Most hydroclave tooling is integrally heated; some autoclave tooling is integrally heated to compensate for thick sections, to provide higher heatup rates, or to permit processing at a higher temperature than the capability of the autoclave.

INTERFACE: The boundary between the individual, physically distinguishable constituents of a composite.

INTERLAMINAR SHEAR (ILS): Ideally, test methods used to measure ILS apply a pure shear load to the interface between two plies of a composite. The short beam shear (SBS) test does not apply a pure shear load; SBS strength values do not relate directly to ILS strength, but are suitable for quality control purposes.

INTERPLY: Two or more different reinforcements are combined in discrete layers and fibers are not mixed within a layer.

INTRAPLY: When reinforcements are mixed within a layer such as alternating strands in a fabric.

ISOTROPIC: Having uniform properties in all directions. The measured properties of an isotropic material are independent of the axis of testing.

K

KEVLAR: An organic polymer composed of aromatic polyamides having a pare-type orientation. (Parallel chain extending bonds from each aromatic nucleus.)

L

LAMINA: A single ply or layer in a laminate made of a series of layers.

LAMINAE: Plural of lamina.

LAMINATE: A produce made by bonding together two or more layers or laminae of material or materials.

LAMINATE ORIENTATION: The configuration of a crossplied composite laminate with regard to the angles of crossplying, the number of laminae at each angle, and the exact sequence of the lamina layup.

LAYUP: A process of fabrication involving the assembly of successive layers of resin impregnated material.

M

MACRO: In relation to composites, denotes the gross properties of a composite as a structural element but does not consider the individual properties or itentity of the constituents.

MANDREL: A form fixture or male mold used for the base in the production of a part by layup or filament winding.

MAT: A fibrous material consisting of randomly oriented chopped or swirled filaments loosely held together with a binder.

MATCHED DIE: A mold, in two or more pieces, which is capable of producing parts with two or more dimensionally controlled surfaces.

MATRIX: The essentially homogeneous material in which the fiber system of a composite is imbedded.

MELTING RANGE: Thermoplastics whose makeup includes a distribution of molecular weights will not have a well-defined melting point, but have a melting range.

MICROCRACKING: Microcracks are formed in composites when thermal stresses locally exceed the strength of the matrix. Since most microcracks do not penetrate the reinforcing fibers, microcracks in a cross-plied tape laminate or in a laminate made from cloth prepreg are usually limited to the thickness of a single ply.

MOISTURE CONTENT: The amount of moisture in a material determined under prescribed conditions and expressed as a percentage of the mass of the moist specimen, i.e., the mass of the dry substance plus the moisture present.

MOISTURE EQUILIBRIUM: The condition reached by a sample when it no longer takes up moisture from, or gives up moisture to, the surrounding environment.

MOLD RELEASE AGENT: A lubricant applied to mold surface to facilitate release of the molded article.

MOLD SURFACE: For an autoclave or hydroclaved laminate, the mold surface is the side of the laminate which faced the mold (tool) during cure.

MOLDED EDGE: An edge which is not physically altered after molding for use in final form, and particularly one which does not have fiber ends along its length.

MOLDED NET: Description of a molded part which requires no additional processing to meet dimensional requirements.

MODLING: The forming of a composite into a prescribed shape by the application of pressure during the cure cycle of the matrix.

N

NDI: Nondestructive inspection. A process or procedure for determining the quality or characteristics of a material, part, or assembly without permanently altering the subject or its properties.

NDT: Nondestructive testing. Broadly considered synonymous with NDI.

NOVOLAC: A phenolic-aldehyde resin which remains permanently thermmoplastic unless a source of methylene or other groups are added.

O

ORTHOTROPIC: Having three mutually perpendicular planes of elastic symmetry.

OUT TIME: The time a prepreg is exposed to ambient temperature; namely, the cumulative amount of time prepreg is out of the freezer. The main effects of out time are to decrease drape and tack of the prepreg while also allowing it to absorb mositure from the air.

OVEN DRY: The condition of a material that has been heated under prescribed conditions of temperature and humidity until there is no further significant change in its mass.

OXIDATION: In carbon/graphite fiber processing, the step of reacting the precursor polymer (rayon, PAN, or pitch) with oxygen, resulting in stabilization of the structure for the hot stretching operation. In general usage, oxidation refers to any chemical reaction in which electrons are transferred.

P

PAN: Polyacrylonitrile, used in fiber form as a percursor for making carbon/graphite fibers.

PAS: Polyarysulfone.

PEEL PLY: A layer of open-weave material, usually fiberglass or heat-set nylon, applied directly to the surface of a prepreg layup. The peel ply is removed from the cured laminate immediately before bonding operations, leaving a clean, resin-rich surface which needs no further preparation for bonding, other than application of a primer where one is required.

PEEK: Polyetheretherketone.

PES: Polyethersulfone.

PHENOLIC: Any of several types of synthetic thermosetting resin obtained by the condensation of phenol or substituted phenols with aldehydes such as formaldehyde.

PI: Polyimide.

PITCH: High molecular weight material left as a residue after processing of petroleum (crude oil). After further purification, can be processed into fiber form, useful as a precursor for production of carbon/graphite fibers.

PITCH FIBER: Fibers derived from a special petroleum pitch.

PLAIN WEAVE: A weaving pattern where the warp and fill fibers alternate; i.e., the repeat pattern is warp/fill/warp...Both faces of a plain weave are identical. Properties are significantly reduced relative to a weaving pattern with fewer crossovers.

PLASTIC: A general term for the mixture of a polymer and ingredients such as hardeners, fillers, reinforcing fibers, plasticizers, etc. After processing, thermoplastics can be resoftened to their original condition by heat, while the thermosetting plastics cannot.

PLASTICIZER: A material of lower molecular weight added to a polymer to separate the molecular chains. This results in a depression of the glass-transition temperatures, reduced stiffness and brittleness, and improved processability.

PLY: A single layer of prepreg. Used synonymously with "Lamina."

PLY WRINKLE: A condition where one or more of the plies are permanently formed into a ridge, depression, or fold.

POISSON'S RATIO: The ratio of transverse strain to the corresponding axial strain below the proportional limit caused by a uniformly distributed axial stress.

POLYAMIDEIMIDE: A polymer containing both amide ("nylon") and imide (as in polyimide) groups; properties combine the benefits and disadvantages of both.

POLYARYLSULFONE (PAS): A high temperature thermoplastic previously marketed under the trade name of Astrel 360. "Polyarylsulfone" is also occasionally used to describe the family of resins which includes polysulfone and polyethersulfone.

POLYIMIDE (PI): Generic name for a family of high temperature resins. Both thermoplastic and thermosetting versions are available.

POLYMER: An organic material composed of long molecular chains consisting of repeating chemical units.

POLYMERIZATION: A chemical reaction in which the molecules of nonomers are linked together to form polymers.

POLYPHENYLENE SULFIDE (PPS): A high temperature thermoplastic useful primarily as a molding compound. Optimum properties depend on slightly cross-linking the resin. Best known under the trade name of Ryton.

POLYPHENYLSULFONE: A relatively new thermoplastic having properties similar to polyethersulfone but with increased resistance to water and some solvents. Available under the trade name of Radel.

POLYSULFONE: A thermoplastic polymer with the sulfone linkage with a T_g of 375°F (190°C).

POROSITY: A condition of trapped pockets of air, gas, or voids within a cured laminate, usually expressed as a percentage of the total nonsolid volume to the total volume (solid + nonsolid) of a unit quantity of material. See Voids..

POSTCURE: Completing the cure cycle of a laminate in an oven instead of tying up the equipment used for the initial cure.

POT LIFE: The period of time during which a reacting thermosetting composition remains suitable for its intended processing after mixing with a reaction-initiating agent.

PPS: Polyphenylene Sulfide. Better known under the trade name of Ryton.

PRECURSOR: In carbon/graphite fiber technology, the organic fiber which is the starting point for making carbon or graphite fibers. In resin technology, sometimes used to describe the polymers present at an intermediate stage in the formulation of a cured resin.

PREFIT: A process to check the fit of mating detail parts in an assembly prior to adhesive bonding to ensure proper bond lines. Mechanically fastened structures are also prefit sometimes to establish shimming requirements.

PREMOLDING: The layup and partial cure at an intermediate cure temperature of a laminated or chopped fiber detail part to stabilize its configuration for handling and assembly with other parts for final cure.

PREPLY: A composite material lamina in the raw material stage ready to be fabricated into a finished laminate. The lamina is usually combined with other raw laminae prior to fabrication. A preply includes all of the fiber system placed in position relative to all or part of the required matrix material that together will comprise the finished lamina. An organic matrix preply is called a prepreg. (Metal matrix preplies include green tape, flame-sprayed tape, and consolidated monolayers.)

PREPREG: Ready to mold or cure material in sheet form which may be fiber cloth, or mat, impregnated with resin and stored for use. The resin is partially cured to a "B" stage and supplied to the fabricator for layup and cure.

PRESS CLAVE: A simulated autoclave made by using the platens of a press to seal the ends of an open chamber, providing both the force required to prevent loss of the pressurizing medium and also providing the heat to cure the laminate inside.

PRESSURE INTENSIFIER: A layer of flexible material (usually a high temperature rubber) used to assure that sufficient pressure is applied to a location, such as a radius, in a layup being cured.

PROCESS CONTROL: During laminate cure, the use of electrical technique to monitor the cure cycle. Also refers to the overall procedure of recorging cure cycle temperatures (in or near the part), vacuum, and pressure, plus mechanical testing of a "process control" panel cured along with the part.

PULTRUSION: A process to continuously process structural shapes or flat sheet by drawing prepreg materials through forming dies to produce the desired constant cross-sectional shape and simultaneously curing the resin.

Q

QUASI-ISOTROPIC: A layup sequence of the 0°, +45°, −45°, 90° family, with equal amounts of fiber in each direction. With the fiber axes in four directions, laminate properties in the plane of the fibers are nearly isotropic.

R

REINFORCEMENT: A relatively high strength or stiffness material inbedded in a matrix to improve their mechanical properties.

RELEASE AGENT: See Mold Release Agent.

RELEASE FILM: An impermeable layer of film which does not bond to the resin being cured. See Separator.

RESIN: A polymer (or polymers) and their associated hardeners, catalysts, acclerators, etc., which can be converted to a solid by application of energy, normally in the form of an elevated temperature.

RESIN CONTENT: The amoung of matrix present in a composite by percent weight.

RESIN RICHNESS: An area of excess resin, usually occurring at radii, steps, and the chambered edge of core.

RESIN STARVED: An area deficient in resin usually characterized by excess voids and/or loose fibers.

ROVING: A number of strands, tows or ends collected into a parallel bundle with little or no twist.

RUBBER: Crosslinked polymers whose glass transition temperature is below room temperature and which exhibit highly elastic deformation and have high elongation.

S

SANDWICH CONSTRUCTION: A structural panel consisting in its simplest form of two relatively thin, parallel sheets of structural material bonded to and separated by a relatively thick, lightweight core.

SCRIM: (also called Glass Cloth, Carrier): A reinforcing fabric woven into an open mesh construction, used in the processing of tape or other B-stage material to facilitate handling.

SECONDARY BONDING: The joining together, by the process of adhesive bonding, of two or more already cured composite parts.

SEMICRYSTALLINE: In plastics, refers to materials which exhibit localized crystallinity. See Crystallinity.

SEPARATOR: A permeable layer which also acts as a release film. Porous Teflon coated fiberglass is an example. Often placed between layup and bleeder to facilitate bleeder system removal from laminate after cure.

SHELF LIFE: The length of time a material, substance, product, or reagent can be stored under specified environmental conditions and continue to meet all applicable specification requirements and/or remain suitable for its intended function.

SHELL TOOLING: A mold or bonding fixture consisting of a contoured surface shell supported by a substructure to provide dimensional stability.

SHORT BEAM SHEAR: A flexural test of a specimen having a low test span to thickness ratio (e.g., 4/1) such that failure is primarily in shear.

SIZING: Material applied as a very thin coating on fibers to improve their processability and/or to increase the fiber/matrix bond strength in composites. See Fiber Finish.

SPECIFIC GRAVITY: The weight of a specimen compared to the weight of the volume of water it displaces.

STAGING: Heating a pre-mixed resin system, such as in a prepreg, until the chemical reaction (curing) starts, but stopping the reaction before the gel point is reached. Stating is often used to reduce resin flow in subsequent press molding operations.

STOPS: Metal pieces inserted between die halves; used to control the thickness of a press molded part. Not a recommended practice, since the resin will end up with less pressure on it, and voids can result.

STRESS, RESIDUAL: The stress existing in a body at rest, in equilibrium, at uniform temperature an not subjected to external forces.

SYMMETRICAL LAMINATE: A composite laminate in which the ply orientation is symmetrical about the laminate midplane.

T

TACK: Stickiness of a prepreg.

TACKING: To locally join together layers of thermoplastics, by localized melting of the resin.

TAPE: In composites technology, tape is the prepreg form consisting of collimated fibers and resin, supported on a layer of release paper. Almost all graphite tape is made to have an 0.005″ thickness, as cured.

TEFLON: DuPont trade name for both TFE and FEP flourocarbon polymers.

TFE: Tetraflouroethylene, or polytetraflourethylene (PTEE), the "Teflon" with the highest elevated temperature resistance. However, yield strength is very low; Teflon is known for having "cold flow" problems.

THERMAL CONDUCTIVITY: The capability of a substance to "conduct" heat from a hot area to a cooler area. Graphite fibers have good conductivity (both thermal and electrical) in the axial direction, but relatively poor conductivity in the radial direction.

THERMOFORMING: Forming a thermoplastic material after heating it to the point where it is soft enough to be formed without cracking or breaking reinforcing fibers.

THERMOPLASTIC: A plastic that repeatedly can be softened by heating and hardened by cooling through a temperature range characteristic of the plastic, and that in the softened stage can be shaped by flow into articles by molding or extrusion.

THERMOSET: A plastic that is substantially infusible and insoluble after having been cured by heat or other means.

TOUGHNESS: Describes a material which has both high elongation to failure and good strength, such that the area under the stress/strain curve (a measure of the energy required to deform the material) is very high.

TOW: An untwisted bundle of continuous filaments. Commonly used in referring to man-made fibers, particularly carbon and graphite fibers in the composites industry.

TRACER: A fiber or tow or yarn added to a prepreg for verifying fiber alignment and, in the case of woven materials, distinguishing warp fibers from fill fibers.

V

VACUUM BAG: The plastic or rubber layer used to cover the part so that a vacuum can be drawn.

VACUUM BAG MOLDING: A process in which the layup is cured under pressure generated by drawing a vacuum in the space between the layup and a flexible sheet placed over it and sealed at the edges.

VENT CLOTH: A layer or layers of open weave cloth used to provide a path for vacuum to "reach" the area over a laminate being cured, such that volatiles and air can be removed and also causing the pressure differential that results in application of pressure to the part being cured. Often used synonymously with "Breather Cloth".

VENTING: In autoclave curing a part of assembly, venting refers to turning off the vacuum source and venting the vacuum bag to the atmosphere. The pressure on the part then becomes the pressure difference between the pressure in the clave and atmospheric pressure. Venting is usually used to prevent the resin "boiling" that can occur when a resin is heated and simultaneously subjected to reduced pressure (vacuum).

VISCOSITY: The property of resistance to flow exhibited within the body of a material.

VOID: A physical and mechanical discontinuity occurring within a material or part which may be 2-D (e.g., disbonds, delaminations) or 3-D (e.g., vacuum-, air-, or gas-filled pockets). Porosity is an aggregation of micro-voids. Voids are essentially incapable of transmitting structural stresses or nonradiative energy fields.

VOLATILES: Refers to gaseous materials leaving a laminate that is being cured, and which were liquids or solids before the cure cycle started. Volatiles produced usually include residual solvents and absorbed or adsorbed water. Many materials also produce volatiles as by-products of the curing reactions.

W

WARP: The longitudinally oriented yarn in a woven fabric (see Fill); a group of yarns in long lengths and approximately parallel.

WATER JET: Water emitted from a nozzle under very high pressure. Useful for cutting materials.

WET LAYUP: A method of making a reinforced product by applying the resin system as a liquid as the reinforcement is put in place.

WET STRENGTH: The strength of a composite measured after conditioning the test specimen in water or water vapor.

WET WINDING: A method of making filament in which the fiber reinforcement is coated with the resin system as a liquid just prior to wrapping on a mandrel.

WETOUT: The process of wetting a fiber bundle with a resin; that is, wetout is achieved when all of the fiber surface area is in intimate contact with the matrix resin.

WHISKER: A short single fiber or filament. Whisker diameters range from 1 to 25 microns with length-to-diameter ratios between 100 and 15,000.

WORK LIFE: The period during which a compound, after mixing with a catalyst, solvent, or other compounding ingredients, remains suitable for its intended use.

Y

YARN: Generic term for strands of fibers or filaments in a form suitable for weaving or otherwise intertwining to form a fabric.

Z

ZERO BLEED: A laminate fabrication procedure which does not allow loss of resin during cure. Also describes prepreg made with the amount of resin desired in the final part, such that no resin has to be removed during cure.

Chapter 1.0

INTRODUCTION

1.1. INTRODUCTION

Combinations of different materials which result in superior products started in antiquity and has been in continuous use down to the present day. In early history mud bricks were reinforced with straw to build houses; more recently man-made stone was reinforced with steel bars (reinforced concrete) to build modern buildings, and bridges, etc., and now composites of matrix reinforced with fibers are used to build airframe structures.

Modern composites owe much to glass fiber-polyester composites developed since the 1940's, to wood working over the past centuries, and to nature over millions of years. Numerous examples of composites exist in nature, such as bamboo which is a filamentary composite. Through the years, wood has been a common used natural composite whose properties with and against the grain vary significantly. Such directional or anisotropic properties have been mastered by design approaches which take advantage of the superior properties while suppressing the undesirable ones through the use of lamination. Plywoods, for example, are made with an old number of laminae. Such a stacking arrangement is necessary in order to prevent warping. In the language of modern composites, this is referred to as the symmetric lay-up or zero extension-flexure coupling (orthotropic).

(1) PROGRESS IN COMPOSITES

The emergence of boron filaments gave birth to a new generation of composites in the early 1960s. The composites that employ high modulus continuous filaments, like boron and carbon (or graphite), are referred to as advanced composites. This remarkable class of materials is cited as a most promising development that has profoundly impacted today's and future technologies of airframe design. The term composites or advanced composite material is defined as a material consisting of small-diameter (around 6 to 10 microns), high-strength, high-modulus (stiffness) fibers embedded in an essentially homogeneous matrix as shown in Fig. 1.1.1. This results in a material that is anisotropic (it has mechanical and physical properties that vary with direction). Terms commonly used in describing advanced composites are provided in the glossary of this book.

Chapter 1.0

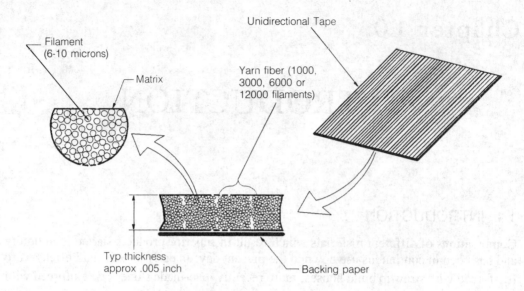

Fig. 1.1.1 Contents of Composite Materials

In the airframe industry the matrix (or resin) binds together high-strength fibers like carbon, glass, Kevlar or ceramic to create lighter and more efficient airframes which result in light-weight commercial transports which are able to economize on fuel. Weight savings also enable combat aircraft to achieve greater speed and range, extending mission capabilities. Future fighters will undoubtedly have airframes with a high percentage of advanced composites as shown in Fig. 1.1.2.

By courtesy of Lockheed Aeronautical System Co.

(a) YF-22 (Lockheed, Boeing and General Dynamics)

Fig. 1.1.2 U.S. Advanced Tactical Fighters (ATF)

(b) YF-23 (Northrop and Madonneell-Douglas)

Fig. 1.1.2 U.S. Advanced Tactical Fighters (ATF) (cont'd)

Serious development work with advanced composite materials started in the mid'60s with the boron fibers embedded in an epoxy resin matrix. Since that time, a host of new materials has been added, including three types of carbon and graphite fibers, organic material fibres such as Kevlar, and new matrix materials which include polyimides, thermoplastics, and even metals such as aluminum, titanium, and nickel. Due to the remarkable specific properties of composite materials, component weight savings of up to 30% have been achieved. However, the resulting structures are generally much more expensive than the metal counterpart. High raw material costs, extensive processing and quality assurance procedures and the fact that the major emphasis is on maximum weight savings have led to these high costs. To accomplish the objective of cost and weight savings, design approaches should emphasize structural simplification, reduced part count, and elimination of costly design features, as illustrated by the examples in Fig. 1.1.3.

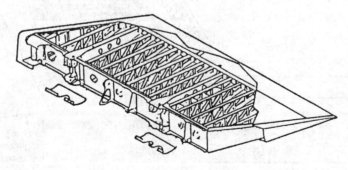

(i) Aluminium

Fig. 1.1.3 Comparison of Composite Aileron and Vertical Fin to Aluminum Counterpart — L-1011

Chapter 1.0

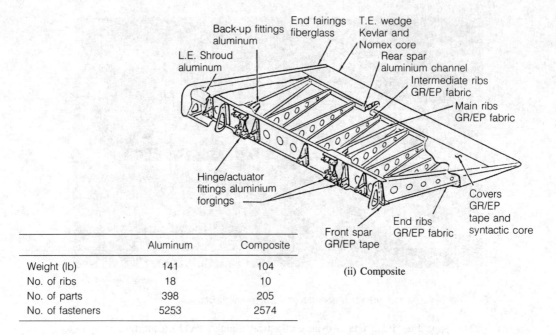

	Aluminum	Composite
Weight (lb)	141	104
No. of ribs	18	10
No. of parts	398	205
No. of fasteners	5253	2574

(a) Inboard aileron

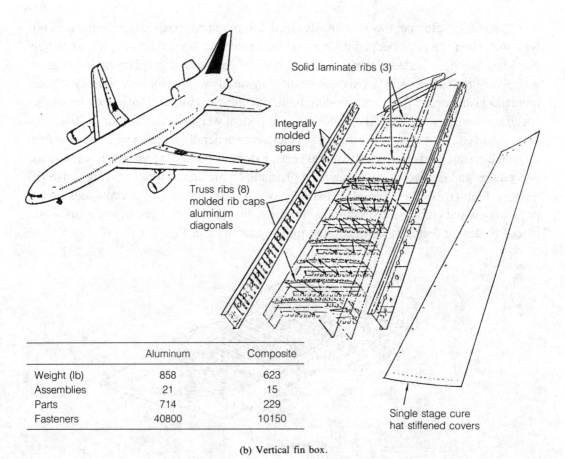

	Aluminum	Composite
Weight (lb)	858	623
Assemblies	21	15
Parts	714	229
Fasteners	40800	10150

(b) Vertical fin box.

Fig. 1.1.3 Comparison of Composite Aileron and Vertical Fin to Aluminum Counterparts — L-1011 (cont'd)

Ten years ago, airframe manufacturers had very modest experience with the use of continuous graphite/carbon fiber reinforced thermosets, and that only in secondary structures. In recent years, a substantial amount of airframe research has been directed at developing advanced composites for use as heavily-loaded primary structures such as wing, fuselage and empennage components for both commercial and military aircraft. Both thermoset and thermoplastic matrices are being developed for subsonic transports, advanced tactical fightest and future supersonic/hypersonic transports (see Fig. 1.1.4).

By courtesy of NASA

(a) U.S. proposed X-30 hypersonic airplane (17,000 MPH in orbital speed) by the year 2010 or 2020

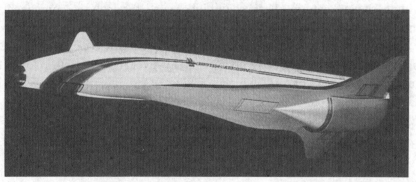

By courtesy of Birtish Aerospace PLC

(b) British aerospace's HOTOL (Horizontal takeoff and landing)

Fig. 1.1.4 Hypersonic Experimental Planes

(2) IMPORTANCE OF COMPOSITES IN AIRFRAME DESIGN

Composite materials are ideal for structural applications where high strength-to-weight and stiffness-to-weight ratios are required. Aircraft and spacecraft are typically weight-sensitive structures (see Fig. 1.1.5), in which composite materials can be cost-effective. When the full advantages of composite materials are utilized, both aircraft and spacecraft will be designed in a manner much different from the present, as shown in Fig. 1.1.6.

The study of composites actually involves many topics, such as manufacturing processes, anisotropic elasticity, strength of anisotropic materials, and micromechanics. Truly, no one individual can claim a complete understanding of all these areas. In this book, the emphasis is hence on practical design rather than on theroretical analysis. Adequate references to the latter found in either composite publications or references listed in this book.

Over the past decades, a variety of composite materials have been developed which offer mechanical properties that are competitive with common aluminum and steel but at fractions of their wieght. Fig. 1.1.7 gives a comparison of the properties of several different composites with conventional metallic materials. It is possible for the designer to locate and orient the reinforcement in sufficient quantity and in the proper direction, even in very localized areas to withstand the anticipated loads. With composite materials it is possible can be arranged in such a way to create structures such as the forward swept wing of the X-29A fighter, as shown in Fig. 1.1.8., which have aerodynamics characteristics that would not be possible if the structrures were composed metal. Historically, aluminum materials have been the primary materials for aircraft and spacecraft construction. Today, structural weight and stiffness requirements have exceeded the capability of conventional aluminum, and high-performance payloads have demanded extreme thermo-elastic stability in the aircraft design environment. To achieve the best composite structure design, composite designers should be trained to obtain basic knowledge as well as experience about metal structures. As matter of fact, composite designers should not consider composite materials to be a panacea, because in some areas of airframe structure the use of metal material is still the most cost effective choice. As mentioned previously, composite material costs are high compared with common airframe metals. Design costs are also higher with these new materials because of higher costs for analysis, components testing, certification and documentation testing. Furthermore, production and prototype tooling costs are higher than with conventional metals. Quality control, especially non-destructive inspection (NDI), is another high-cost operation. However, it is possible to lower costs by:
- Innovative design concepts which consider of producibility
- Lowering part counts
- Elimination of costly fasteners or the use of fewer fasteners
- Use of automation methods to cutdown manufacturing costs

• Small civil aircraft	$50/lb
• Helicopters	$300/lb
• Advanced fighters	$400/lb
• Commercial transports	$800/lb
• Supersonic and hypersonic transports	$3000/lb
• Orbit satellites	$6000/lb
• Synchronous satellites	$20,000/lb
• Spacecraft	$30,000/lb

Fig. 1.1.5 Value of Weight Savings in Airframe Structures (1990)

> New Freedoms:
> - Ability to engineer the material as well as the structure
> - Superior structural properties
> - Ability to assemble and form in a soft condition and bake to harden
>
> New Problems:
> - New stress methods (anisotropic or orthotropic) needed
> - Computer analysis programs required
> - Designing around material weaknesses
> - No reservoir of service experience

Fig. 1.1.6 Design with Advanced Composites

	Graphite/Epoxy (Unidirectional)		Kevlar/Epoxy (Woven cloth)	Glass/Epoxy (Woven cloth)	Boron/Epoxy	Aluminum	Beryllium	Titanium
	High Strength	High Modulus						
Specific strength, 10^6 in	5.4	2.1	1	0.7	3.3	0.7	1.1	0.8
Specific stiffness, 10^6 in	400	700	80	45	457	100	700	100
Density, lb/in^3	0.056	0.063	0.05	0.065	0.07	0.10	0.07	0.16

Fig. 1.1.7 Composite Vs. Conventional Materials.

By courtesy of Grumman Corp.

Fig. 1.1.8 X-29A Forward Swept Composite Wing

The maturation of composite technology is still in progress, but the basses of understanding has broadened significantly. Fig. 1.1.9 illustrates the diversity of developmental experience obtained and use on advanced commercial transport structures. However, the application of advanced composite materials in civil aircraft has generally lagged behind military usage because:
- Cost is a more important consideration to commercial aircraft manufacturers
- Safety is a more critical concern, both to the airframe manufacturer and government certifying agencies
- A general conservatism due to financial penalties from equipment downtime

Chapter 1.0

The use of composite materials in military fighter aircraft construction has fluctuated in the past and is expected to change further in the future as shown in Fig. 1.1.10. The development of advanced composites in the 1960s resulted in a quantum jump in weight saving potential. This trend will continue with the introduction of new high-strain and high-toughness composite materials such as toughened thermosets, or thermoplastics, which have been in development for many years and have produced results not possible with nontoughened thermosets.

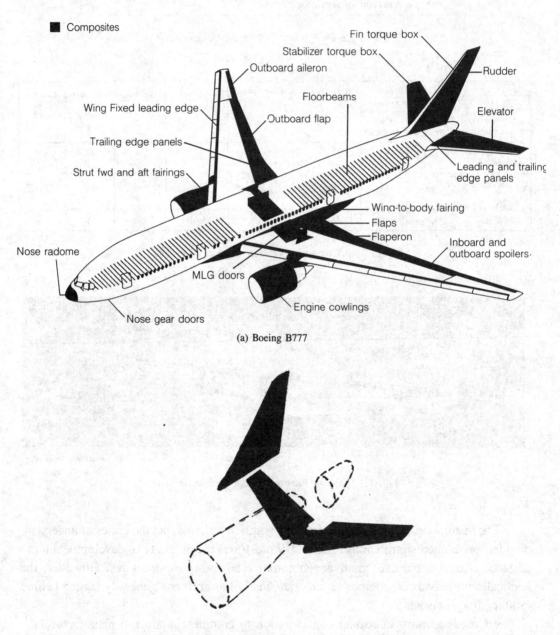

(a) Boeing B777

(b) All composite horizontal and vertical stabilizer- A320

Fig. 1.1.9 Composites Applications on Commercial Transports

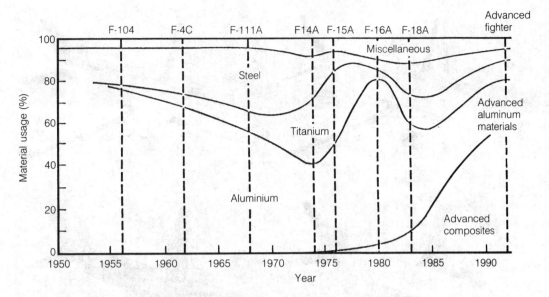

Fig. 1.1.10 Progress in Using Composite Materials on Military Fighter Aircraft

In summary, use of composites is based on demonstration that:
- Significant weight savings can be achieved
- Use of composites can reduce cost, or can be cost effective
- Composite structures have been validated by tests as meeting all structural requirements under aircraft environmental conditions.
- Cost-weight trade studies should conducted as part of design activity to determine appropriatse use of composites versus metals

Structural weight reduction is the key advantage in using composite materials. The relatively high raw material cost of composites can be offset by carefully evaluating design and manufacturing processes to minimize the cost of fabrication, inspection and repair. Obviously, the strongest of materials pound for pound, composites draw most of their strength from their hidden fibers, which come in many types and can be arranged in various patterns, some in three dimensional shapes by braiding or weaving. These complex patterns can produce shaped with enormous strength in all directions. New contributions are being made by specialized industries, as shown in Fig. 1.1.11, that until now have not been involved in the airframe manufacturing business.

- Textile Industry (Braiding and Weaving)
- Injection Mold Industry (One-piece structures created by injecting matrix into braiding, 2-D weaving or 3-D weaving)
- Sewing Industry (Stitching)
- Rattan Industry (Very complex shape or VCS weaving technique)

Fig. 1.1.11 New Involvement of Industries

Chapter 1.0

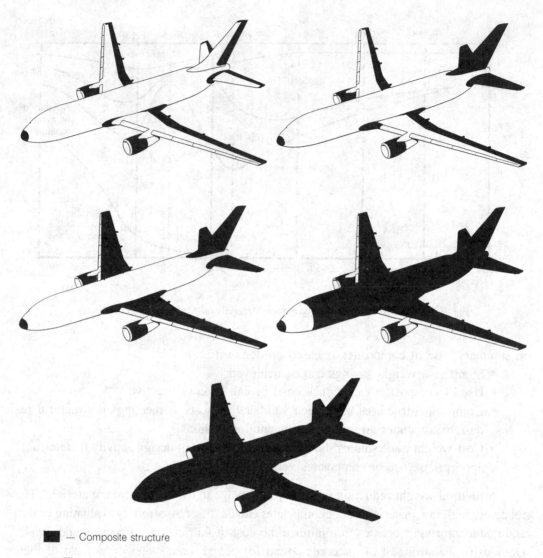

■ — Composite structure

Fig. 1.1.12 The Progressive Use of Composites on Commercial Transport Airframes

That the use of composites is now becoming almost commonplace illustrated by the extensive use of the lighter fiber-reinforced materials in transport airframe construction and their progressively increased use, as shown in Fig. 1.1.12. Transport airframe manufacturers are extending use from non-critical areas to the more critical areas of secondary structure, including flight control surfaces and primary empennage structures. It is likely that use will soon include wing and fuselage structures where the greatest pay-off from weight and cost savings would be immediately appreciated.

Since the beginning of the 1980's, an all- or mostly-composite airframe has almost become a must in the developing and manufacturing of business aircraft, as shown in Fig. 1.1.13, as well as general aviation aircraft. Design approahes which differ from those of most transport airframes and used to reduce cost and structural weight. These innovative designs and manufacturing techniques are pioneers in composite airframe structure development.

(a) Lear Fan 2100

By courtesy of Beech Aircraft Corp.

(b) Starship

By courtesy of AVTEK Corp.

(c) AVTEK400A

By courtesy of P.T. Cipta Restu Sarana Svaha

(d) AA-200

Fig. 1.1.13 All-or Mostly-Composite Aircrafts

1.2 CHARACTERISTICS OF COMPOSITES

The most commonly used advanced composite fibers are carbon and graphite, Kevlar and boron, Carbon fibers are manufactured by pyrolysis of an organic precursor such as rayon or PAN (Polyacrylonitrile), or petroleum pitch. Generally, as the fiber modulus increases, the tensile strength decreases. Among these fibers, carbon fiber is the most versatile of the advanced reinforcements and the most widely used by the aircraft and aerospace industries. Products are available as collimated, preimpregnated (prepreg) unindirectional tapes or woven cloth. The wide range of products makes it possible to selectively tailor materials and configurations to suit almost any application.

Matrix materials used in advanced composites to interconnect the fibrous reinforcements are as varied as the reinforcements. Resins or plastic materials, metals, and even ceramic materials are used as matrices. Today, epoxy resin is the primary thermoset composite matrix for airframe and aerospace applications. In all thermoset materials, the matrix is cured by means of time, temperature, and pressure into a dense, low-void-content structure in which the reinforcement is aligned in the direction of anticipated loads.

An important element in determining the material behavior is the composition of the matrix that binds the fibers together. The selected matrix formulation determines the cure cycle and affects such properties as creep, compressive and shear strengths, thermal resistance, moisture sensitivity, and ultraviolet sensitivity, all of which affect the composite's long-term stability. Characteristics of a selection of composite matrices include:

(a) Epoxy
  - most widely used
  - best structural characteristics
  - maximum use temperature of 200°F (93°C)
  - easy to process
  - toughened versions now available
(b) Bismaleimide
  - maximum use temperature of 350°F (180°C)
  - easy to process
  - toughened versions becoming available
(c) Polyimide
  - variety of matrix types
  - can be used up to 500 — 600°F (320°C)
  - difficult to process
  - expensive
(d) Polyester
  - relatively poor structural characteristics limit usage to non-structural parts
  - easy to process
(e) Phenolic
  - same limitations as polyesters
  - more difficult to process
  - provide higher use temperature than polyesters and epoxies
  - low smoke generation
(f) Thermoplastics
  - greater improved toughness
  - unique processing capabilities
  - have processing difficulties

The major advantages of the thermoplastic matrix over thermoset are
- high service temperature
- shorter fabrication cycle
- no refrigeration required for storage
- increased toughness
- low moisture sensitivity
- no need for a chemical cure

A detailed discussion which compares thermoplastic and thermoset composites is contained in Chapter 2.2.

The matrix can also be affected by exposure to a water. Since it is the matrix, and not the fiber (except for Kelvar), that exhibits these hydroscopic characteristics, the matrix-sensitive properties are seriously reduced, especially at high temperatures by exposure to moisture. For airframe structures, which experience rapid changes of environment, this loss of mechanical performance due to moisture absorption must be accounted for in design.

Kevlar® (Aramid) is the trade name for a synthetic organic fiber. A density of 0.052 lb/in^3 gives Kevlar a specific tensile strength higher than either boron or most carbon fibers. When compared to other composite materials such as carbon and boron, Kevlar has poor compressive strength. This inherent characteristic of Kevlar results from internal buckling of the filaments. However, Kevlar demonstrates a significant increase in resistance to damage compared to other composite materials. Kevlar fibers are hygroscopic and this fact must be considered in designing with Kevlar.

High performance advanced composites are often used in stiffness-critical applications. Thus, when developing new materials, the tendency is to maximize longitudinal moduli while maintaining acceptable levels of strength, impact resistance, strain-to-failure, and fracture toughness. Tensile properties are fiber-dominated; therefore, the choice of fiber is dictated by the application.

Compressive properties in unidirectional laminates are both fiber- and matrix-dependent. While compressive moduli are related to the fiber, compressive strength is dictated by the neat matrix shear modulus. But for homogeneous, isotropic materials, the neat matrix shear modulus is related to matrix tensile modulus. Therefore, having relatively high matrix strength will prevent or minimize intraply cracking in the composite under impact conditions, and will also insure acceptable tranverse properties. Fracture toughness is important in matrices to minimize the propagation of cracks and defects, especially at crossly interfaces.

Retention of compressive strength and strain after impact is important property in high performance composites. It should be emphasized that, although damage prevention is important, damage containment is even more crucial. Therefore, to prevent impact-generated cracks from propagating and causing excessive delamination, adequate interlaminar fracture toughness is required in composites used for the airframe structures.

Guidelines for the synthesis of improved matrices have evolved primarily from experiential data which highlights weaknesses;
- design criteria which considers the most dangerous threat to performance degradation
- limitations in process technology
- evaluations of the relationships between neat matrix properties and composite properties

These concept reveal that the most desirable matrix and composite properties (ideal goals) are shown as follows:

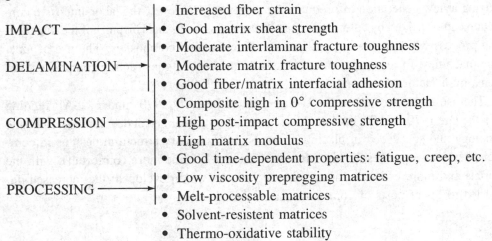

IMPACT
- Increased fiber strain
- Good matrix shear strength

DELAMINATION
- Moderate interlaminar fracture toughness
- Moderate matrix fracture toughness
- Good fiber/matrix interfacial adhesion

COMPRESSION
- Composite high in 0° compressive strength
- High post-impact compressive strength
- High matrix modulus
- Good time-dependent properties: fatigue, creep, etc.

PROCESSING
- Low viscosity prepregging matrices
- Melt-processable matrices
- Solvent-resistent matrices
- Thermo-oxidative stability

Metal/matrix composite (MMC) materials have very high tensile and compressive strength and stiffness compared to most carbon/epoxy materials as shown below:
- Boron/Aluminum — Simple members for high tension or compression load. Beef-up aluminum member for additional strength
- Boron/Titanium — Higher strength structures
- Borsic/Aluminum — Higher strength structures
- Borsic/Titanium — Higher strength and high temperature applications
- Carbon/Aluminum — Aerospace
- Carbon/Magnesium — Aerospace

Much of the MMC development work has been government funded; the major characteristics of MMC's are
- good strength at high temperature
- good structural rigidity
- dimensional stability
- light weight
- and processing flexibility

1.3 COMPOSITES VS. METALS (ALUMINUM ALLOYS)

For some time it seemed as if composite materials would replace aluminum as the material of choice in new aircraft designs. This put pressure on aluminum developers to improve their products. One result was aluminum-lithium (The first aluminum-lithium alloy, called 2020, was actually developed in the 1950s for the U.S. Navy RA-5C Vigilante). One of the main efforts of the developers is to save weight and cost compared to composites because the conventional aluminum manufacturing facilities can be used on aluminum-lithium.

The early demonstrations of the 25-35% weight savings composites offer over aluminum constructions plus a substantial reduction in the number of parts required for each application represents a major attraction of these composites. The obstacles to a wider use today of composite materials are their high acquisition cost compared with aluminum, the labor-intensive construction techniques required and substantial capital costs involved in buying a new generation of production equipment. However, the labor-intensive construction can be solved by automation (see later discussion in this Chapter) of the manufacturing process which is the key technology in developing composites. The use of tape-laying machines, for example, can cut the time and cost of constructing composite components by a factor of ten or more.

The use of composites in the U.S. began in the early 1970s under USAF funding and in the late 1970s NASA instituted a series of programs aimed at developing composite technology and succeeded in placing primary and secondary structural designs in commercial services. As a result, aircraft manufacturers became more comfortable with the materials and more efficient construction techniques were developed; the increased demand led to lower costs of composite materials.

Metals are isotropic, having structural properties which are the same in all directions. Composites are anisotropic (see in Fig. 1.3.1), a single ply having very high strength and stiffness in the axial direction but only marginal properties in the crosswise direction. Cross-plying based on load and function enables composites to meet and surpass the properties of metals. However, composites can be laid up to be quasi-isotropic (having nearly isotropic properties).

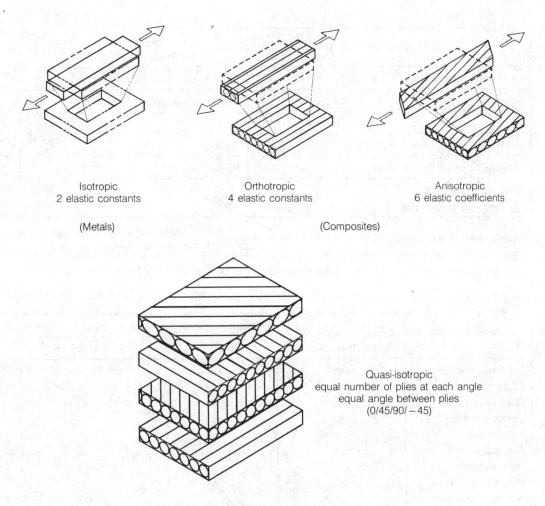

Fig. 1.3.1 Composites VS. Metals Characteristics

(1) COMPOSITES VERSUS METALS

Comparison of composites and aluminum alloy characteristics, and the relative advantages of thermosets, thermoplastics and metal are shown in Fig. 1.3.2 and Fig. 1.3.3 respectively.

Composites differ from metals as their
- Properties are not uniform in all directions
- Strength and stiffness can be tailored to meet loads
- Possess a greater variety of mechanical properties (see Fig. 1.3.4)

- Poor through the thickness (i.e., short transverse) strength
- Composites are usually laid up in essentially two-dimensional form, while metal may be used in billest, bars, forgings, castings, etc.
- Greater sensitivity to environmental heat and moisture
- Greater resistance to fatigue damage (see Fig. 1.3.5)
- Propagation of damage through delamination rather than through-thickness cracks

Condition		Composite Behavior Relative to Metals
Load-strain relationship		More linear strain to failure
Notch-sensitivity	Static	Greater sensitivity
	Fatigue	Less sensitivity
Transverse properties		Weaker
Mechanical properties variability		Higher
Sensitivity to aircraft hydrothermal environment		Greater
Damage growth mechanism		In-plane delamination instead of through-thickness cracks

Fig. 1.3.2 Composites Versus Metals

Material Properties	Relative Advantage		
	Thermoplastics	Thermosets	Metal
Corrosion resistance	xxx	xxx	x
Creep	xxx	xxx	x
Damage resistance	xx	x	xxx
Design flexibility	xxx	xxx	x
Fabrication	xx	xx	x
Fabrication time	xxx	xx	x
Final part cost	xxx	xx	x
Finished part cost	xxx	xx	x
Moisture resistance	xx	x	xxx
Physical properties	xxx	xxx	xxx
Processing cost	xxx	xx	x
Raw material cost	x	xx	xxx
Reusable scrap	xx	—	xxx
Shelf life	xxx	x	xxx
Solvent resistance	xxx	xx	x
Specific strength	xxx	xxx	x
Strength	xxx	xxx	x
Weight saving	xxx	xx	O

Note: xxx-best; xx-good; x-fair; O-not applicable

Fig. 1.3.3 Relative Advantages of Thermoset, Thermoplastics and Metals.

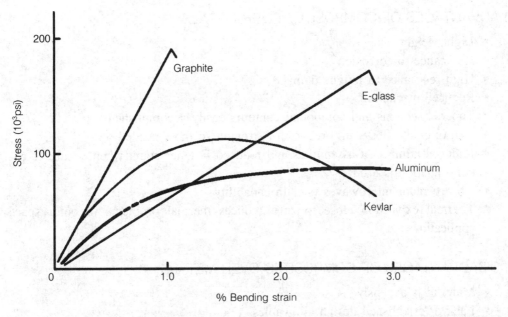

Fig. 1.3.4 Epoxy/Unidirectional Tapes Composites Vs. Metal Stress/Strain Surves Comparison

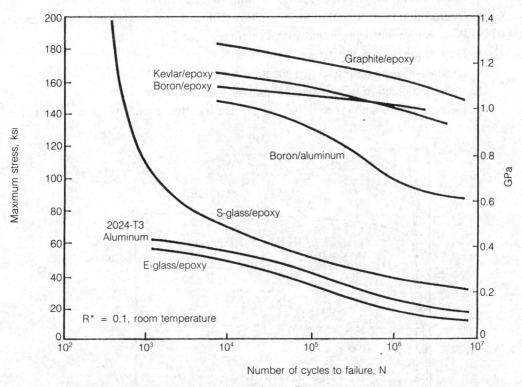

*R = minimum stress/maximum stress for tension-tension cyclic test

Fig. 1.3.5 Fatigue Behavior of Unidirectional Composites and Aluminum

(2) ADVANTAGES OF COMPOSITES OVER METALS

- Light weight
- Resistance to corrosion
- High resistance to fatigue damage
- Reduced machining
- Tapered sections and compound contours easily accomplished
- Can orientate fibers in direction of strength/stiffness needed
- Reduced number of assemblies and reduced fastener count when cocure or co-consolidation is used
- Absorb radar microwaves (stealth capability)
- Thermal expansion close to zero reduces thermal problems in outer space applications

(3) DISADVANTAGES OF COMPOSITES OVER METALS

- Material is expensive
- Lack of established design allowables
- Corrosion problems can result from improper coupling with metals, especially when carbon or graphite is used (sealing is essential)
- Degradation of structural properties under temperature extremes and wet conditions
- Poor energy absorption and impact damage, see Fig. 1.3.6
- May require lightning strike protection
- Expensive and complicated inspection methods
- Reliable detection of substandard bonds is difficult
- Defects can be known to exist but precise location cannot be determined

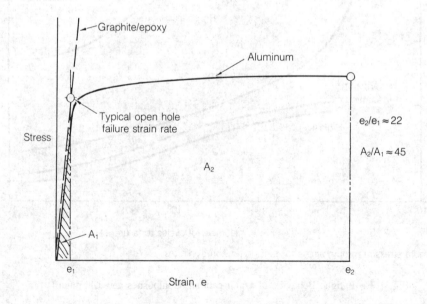

Fig. 1.3.6 Stress/strain Relationship of GR/EP and Aluminum

Material		Units	F^{tu} Ksi	F^{ty*} Ksi	F^{cu} Ksi	F^{cy*} Ksi	E^c Msi	F^{su**} Ksi	G^{**} Msi
7075-T6 Al. sheet, (B), $\varrho = .101$			80	73	—	72	10.5	48	3.9
Boron/Epoxy $V_f = .5$ $\varrho = .0725$	Uni-directional	Absolute ratio	180 .319	164 .319	400 —	364 .142	30 .251	13/67 2.65/.514	.7/7.8 /.359
	Quasi-isotropic	Absolute ratio	55 1.044	50 1.048	146 —	133 .389	10.8 .698	21 1.641	2.5 1.120
High strength C/EP $V_f = .6$ $\varrho = .056$	Uni-directional	Absolute ratio	180 .246	164 .247	180 —	164 .243	21 .277	12/65 2.22/.409	.65/5.5 /.393
	Quasi-isotropic	Absolute ratio	65 .682	59 .686	65 —	59 .677	7.6 .766	22 1.210	1.88 1.150
High Modulus C/EP $V_f = .6$ $\varrho = .058$	Uni-directional	Absolute ratio	110 .418	100 .419	100 —	91 .454	25 .241	9/43 3.06/.641	.65/6.5 /.344
	Quasi-isotropic	Absolute ratio	39 1.178	35 1.198	36 —	33 1.253	8.7 .693	14 1.969	2.1 1.066
Ultra-high Modulus C/EP $V_f = .6$ $\varrho = .061$	Uni-directional	Absolute ratio	84 .575	76 .580	80 —	73 .596	40 .159	6/40 4.83/.725	.6/10.1 /.233
	Quasi-isotropic	Absolute ratio	30 1.611	27 1.633	28 1.739	25 .450	14.1 1.380	21 1.380	5.4 .436
Intermediate Modulus C/EP $V_f = .6$ $\varrho = .055$	Uni-directional	Absolute ratio	160 .272	145 .274	160 —	145 .270	17 .336	10/50 2.61/.523	.65/4.5 /.472
	Quasi-isotropic	Absolute ratio	60 .726	54 .736	60 —	54 .726	63 .908	18 1.452	1.6 1.327

Notes: All material properties are in room temperature.
"Ratio" is the ratio of the specific property of aluminum to that for the composite.
* Composite "yield" strengths are arbitrarily taken as 91% of the ultimate property.
** Shear strength and stiffness of composites are shown for unidirectional (100% 0°)/and (100% ±45°)
(1) Quasi-isotropic = 0°/±45°/90°
(2) Typical calculation of absolute ratio as follow:

$$\left(\frac{80}{0.101}\right) \bigg/ \left(\frac{180}{0.0725}\right) = 0.319$$

Fig. 1.3.7 Comparison of Aluminum and Composite Material Properties

(4) STRUCTURAL EFFICIENCY OF COMPOSITES VS. METALS

In general, composite materials are most effectively utilized when they are preferentially oriented. That is, the laminae are oriented so that the majority of the fibers are placed in the principal load direction and the proportion of the transverse or angle plies is determined by the relative values of the biaxial load components or torsional stiffness requirements. The range of weight reduction potentials of five classes of composite materials, compared to 7075-T6 aluminum sheet (B-values). is shown in Fig. 1.3.7. To quantify the extremes of these directional characteristics, both unidirectional and quasi-isotropic (0/±45/90) properties are shown in the figures. For shear strength or stiffness 100% of ±45 laminate properties are shown, as they represent the upper limit for shear properties. The ratio of the specific strength or stiffness of aluminum to that for the extremes of composite properties is also shown. The average of these is identified as the mean ratio, and the range of ratio is also illustrated. It is postulated that the mean ratio represents the most likely measure of the weight reduction to be expected from the use of composites. It is evident from the values shown that substantial weight savings are possible through the judicious use of composites. It is also evident that the selection of the optimum composite material depends on the application requirements in terms of stiffness and loading type. In general, the high strength Carbon/Epoxy (C/EP) has the best balance of properties. However, the intermediate modulus C/EP is less costly and is being used extensively where modulus is not the determining factor.

1.4 DESIGN FOR LOW COST PRODUCTION

Low cost production starts with designing components so that they can be built using techniques which are feasible and available. Cost has to be controlled as a design parameter when building composite airframe structures in the same way as it was controlled in the design of their metal counterparts. The design definition and drawing stages are vital, but the initial production phases are no less important to cost control.

Novel materials and processes can involve cost pitfalls. Unforeseen problems with production and quality control may be added to the extra outlay for pioneering techniques, so that apparently cost-effective solutions are not necessarily so. Conventional technology, e.g., metal superplastic forming techniques which can be used to form thermoplastic composite laminates, should always be re-examined closely before going on to try some new technology. High-investment, high-waste fabrication methods, such as integral machining, must be evaluated carefully and weighted against the alternatives. The repeatability of quality must be realizable, not just promised, and the intended economies in labor must actually occur.

The benefits of reduction of the number of parts is shown below:

(1) Fewer parts mean less of everything required to manufacture a product:
- Number of tools
- Production planning
- Production floor space
- Engineering time, drawings changes
- Production control records and inventory
- Number of purchase orders, suppliers, etc.
- Number of bins, containers, stock locations, etc.
- Material handling equipment, receiving docks, inspection stations, etc.
- Accounting details and calculations
- Service parts, catalogs, and training
- Production equipment, facilities, training

(2) A part that is eliminated:
- Costs nothing to make, assemble, move, handle, orient, store, purchase, clean, inspect, rework, service
- Fewer jams or automation breakdowns
- Fewer failures, malfunctions, or needed adjustments

Composites, with many of their qualities determined during manufacture, encourage designers to take more account of production engineering; however, cost assessment should be given the same emphasis. Early composite products, such as fiberglass, can still be considered as alternatives to graphite/carbon for minimum-cost structures where there is no critical stress requirement.

Composite construction enables the designer to make extensive use of configurations, such as the sinewave beam, which have always proved expensive in metal. In composite wing skins, tighter manufacturing tolerances can be obtained to meet the requirement of manufacturing assembly, and then checked by robotic scanning — all at great cost savings. The number of fasteners has been reduced which eliminates many problems:

- Fasteners are difficult to feed
- Fasteners tend to jam
- Fasteners require monitoring for presence and torque
- Fasteners are expense (especially installation costs)
- Fasteners allow possible galvanic corrosion when composites are attached to metals

It is no secret that designers need to design composite structures with manufacturing cost effectiveness in mind. Studies in Concurrent Engineering indicate that a large percentage of the final cost of a product is determined in the early phases of the product life cycle. However, advanced composites are expensive to manufacture, to the point that fabrication cost is currently a major issue affecting the ultimate widespread use of these materials. There is a general consensus that several areas can play a key role in the reduction of the high cost of manufacturing composite structures; these include:

- part design
- materials selection
- cost-effective manufacturing processes
- automated systems

In the near term, engineers must show that in many applications, the high performance capabilities of composites justify the high costs. In long term, to make real progress in driving prices lower and also in meeting the ultra-light structural weight expectations, revolutionary manufacturing technology must be developed in parallel with innovative design concepts.

Cost factors which are often given insufficient attention include plant modifications such as;

- air-conditioning ventilation
- safe health-conscious handling of composites
- new tooling
- cleaning of equipment
- protective clothing or operatives
- special training

Production methods vary as to the cost involved due to processing time, consumables and tooling. The types, shapes, and sizes of components for which each method is most suitable must be addressed. Cocuring and reusable self-sealing rubber bags optimize the use of autoclaves, for instance.

All of the factors discussed above obviously impact costs. The designer needs to know which aspects of a design are the cost drivers as the design is handled in the manufacturing, inspection or maintainability areas. At the early stage of a design, decisions are made which will influence 90-95% of the total cost, including operations and maintenance costs. The preliminary design phase gives the designer the maximum opportunity to influence the direction and cost of the entire project. The designer must develop an open approach to other disciplines and be assertive in demanding their involvement and informed and considerate response.

(1) CONCURRENT ENGINEERING

Concurrent Engineering is the buzz word being used today which describes a new communication method that involves all of the disciples shown below:
- Design (focal point)
- Manufacturing
- Tooling
- Materials and Processes
- Maintainability
- Repairability
- Stress Engineering

The mechanism which is described by concurrent engineering is not new and it is rather something that we lost it along the way. Prior to the 1960's, concurrent engineering was the accepted practice and the disciplines of design and manufacturing were co-equals in the definition of a product. In the mid-60's, however, manufacturing was de-emphasized and product innovation highlighted. Simultaneously worldwide competitiors went in the opposite direction and the results are very visible.

Concurrent engineering is the process where all participants in a project are communicating from the start. Project development is performed in sequential fashion, with the designer starting at the beginning with the loads group. The loads group passes the baton to the structural analyst who passes their requirements to the designer. The designer continues to pull the requirements into a cogent package and finally tooling, manufacturing, and other disciplines are given the go-ahead. It is not, however, until the design is sufficiently developed that the designer feels comfortable about discussing it with most other disciplines.

Considering that 70% of the product cost is determined in the first 5% of the design process, concurrent engineering changes the project development from a sequential to a simultaneous involvement of the necessary disciplines. It means that the designer takes materials, structures, tooling, manufacturing, inspection, maintainability, and cost problem into consideration to a greater degree than takes place in the sequential system. This method makes the designer more aware of manufacturing and quality assurance needs during the design phase. No one expects the designer to become thoroughly familiar with all of the other disciplines; this would be an impossible task. What is intended, however, is to bring the other disciplines into the design process (see Fig. 1.4.1) at the beginning so they can evaluate the concepts as they develop.

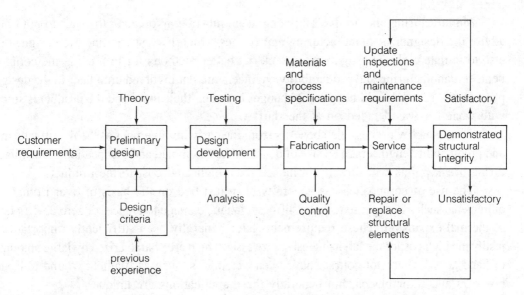

Fig. 1.4.1 Elements of Structural Integrity Planning

The important aspect of this system is that data is reviewed as it is needed and required, not when the generator of the data or design is ready to release it. It is presumed that if the various disciplines have access to data and design concepts at earlier stages, even as it is developing, it will open interdisciplinary dialogue and result in products of higher quality and lower cost. Each discipline is able to measure and evaluate the impact of a change on their area of expertise.

(2) DESIGN FOR PRODUCTION

The maturation of composites technology has increased at an accelerating pace over the last decades. The development has taken the technology from that of a laboratory curiosity into real hardware application in the aerospace, automotive, marine, sports, and medical fields. The structural configurations have ranged from simple flat laminates and tubes to complex stiffened skins and sandwich structures with compound curvatures.

Designing for production is designing for manufacturability. The result is the generation of a design which exploits a manufacturing method and yields a product of high quality and low cost. This occurs because of the interdisciplinary involvement of manufacturing and their various representatives with the designers. The following is describes the goals of the product development cycle:

(a) Design for customer (Quality)
- Design a functionally and visually appealing product
- Provide mechanical reliability for the long haul

(b) Design for manufacturability (Cost)
- Product can be made cost-effectively

(c) Reduce design cycle (Time)
- Do a better job in a shorter length of time
- Minimize liabilities

Manufacturing and producibility engineers must be involved at the start to not only advise the designer of manufacturing approaches which may streamline the design, but also to initiate early investigations to seek out better methods for producing the component. If manufacturing and producibility engineers are not involved until later in the design process, the schedule will not permit changes based on their input and the product is stuck with whatever was decided on at the start.

The assembly process also benefits from the early involvement of the manufacturing and producibility engineers. The size of details to be assembled and the location of manufacturing break points will directly influence the weight and cost to the product.

The use of composites has generally been justified on the basis of over-riding requirements such as weight reduction, stiffness, fatigue damage tolerance, or zero co-efficient of thermal expansion. These requirements have generally been sufficiently important to justify the high cost of the labor-intensive processes which are characterized by large amounts of hand labor. There, of course, have been exceptions, such as filament wound railroad box cars and rotor blades, but generally these applications are unique.

It is time now to step off the threshold and move the technology of advanced composites manufacturing into higher volume production and into the factory environment. To do this it is necessary to reduce highly-labor-intensive procedures, increase the production rate, and decrease the amount of scrap created.

In metal design, tooling cost was all-important and so was analysis of the tooling. The case with composites manufacturing is somewhat different. The actual labor and material cost (see Fig. 1.4.2) of the part itself is usually the greatest single item in its cost, while the amortization of the tooling, engineering, and other considerations is much less important. However, tool durability and the high cost of some composite tooling concepts must be considered in the design cycle.

Composite Fibers	Costs per pound
Boron	> $200
Graphite	
Standard PAN based (33×10^6 modulus)	
1,000 TOW	$135
3,000 TOW	$ 38
6,000 TOW	$ 28
12,000 TOW	$ 24
Intermediate modulus	
12,000 TOW	$ 45
6,000 TOW	$ 85
High modulus	$650
Kevlar	$ 15-20
Fiberglass	$ 3-5
Quartz	$110

Fig. 1.4.2 Comparison of Composite Fibers Cost (Ref. only)

Some concerns have suffered tremendously because of that ingenious instinct of the engineer to create and design something new instead of using an existing part or assembly, which, with reasonable ingenuity, may be made to serve. The natural barrier against using something created by someone else must be broken down. a true mark of a good designer is the ability to see reasons why anything formerly used can be employed again, instead of the tendency to think of reasons why it is not once again functionally satisfactory. Standard types of equipment and standard structural parts must be used. The use of standard parts involves:
- Developing a modular designs
- Designing parts for repeat use
- Using off-the shelf components
- Standardizing and rationalizing

Simplicity in design consists of the following:
- Designing for ease of component fabrication
- Designing for ease of assembly
- Designing for commonality
    — interfacing with other disciplines
    — including sufficient details in the preliminary concept to enable other disciplines to evaluate the design
    — taking tooling into consideration
    — using simple processes and operations

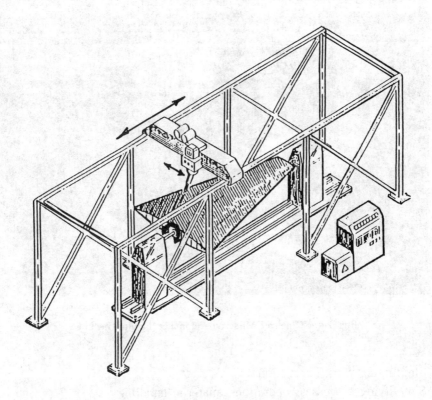

Fig. 1.4.3 Automated Complex Shape Fabrication

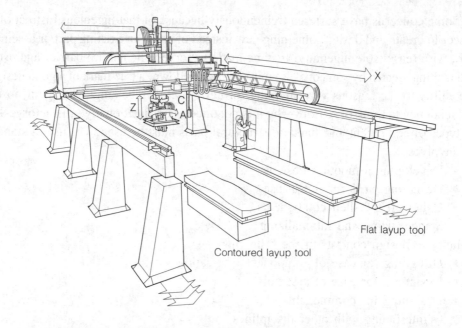

By courtesy of Cincinati Milacron

Fig. 1.4.4 Cincinati Milacron Automated Layup Machine

- Simplifying work instructions
- Simplifying processes, operations, material handling
- Simplifying tooling and fixtures

(3) AUTOMATION

Automation is the key to the practical application of composites technology and it has already been extensively applied to simple structures such as tubes and straight constant-cross-section configurations. Fabricators of these structures have been able to exploit the filament or tape winding (see Fig. 1.4.3), and pultrusion processes. Certainly any design which exploits the braided or woven structures processes can also be said to be automated. This is also true of simple press and injection molding methods.

However, laminate and sandwich structures of the size and complexity normally encountered in the airframe industry have proven difficult to automate. This results in excessively labor-intensive processes with resultant high part acquisition costs. Several companies are currently developing automated tape laying systems, as shown in Fig. 1.4.4. However, all of the systems merely automate the lay up of the laminate. A completely integrated system is needed which will permit a hands-off fabrication of wing covers and fuselage. The cure cycle control is also being automated and will permit the automatic monitoring and control of cure cycles.

The degree of automation which can be contemplated will be determined by the early participation of the manufacturing and producibility engineers. The designer is not expected to be up-to-date on the most current manufacturing systems available. The designer, however, is expected to involve those aware of the new systems in the development of the design.

The type of basic automation that immediately offers high cost savings in composite construction is the use of high-speed tape laying machines. More and more composite materials are being prepared under automatic processing, and filament-wound structures are amenable to it, but accurate timing is crucial if curing requirements are to be met. A relevant axiom is "never underestimate the difficulties of automating the role of a skilled operator, and do not try to imitate his skill directly". After all, the diminishing labor pool and highly intensive processes dictate that for advanced composites technology to grow it must automate.

1.5 CRITERIA

The selection of composite materials for specific applications is generally determined by the physical and mechanical properties of the materials, evaluated for both function and fabrication. The functional considerations include items such as the strength, weight, hardness, and abrasion resistance of the finished part. Fabrication considerations include cure cycle (time, temperature, pressure), quantity of parts, tooling costs, equipment, and availability of facilities.

(1) STANDARDS

There are a number of standards and specifications which are intended to ensure repeatable results by carefully defining either the technical requirements of a material or the specific steps used in the manufacturing process.

(a) Military specifications: These are issued by the Department of Defense (DoD) to define materials, products, or services used only or predominantly by military entities.
(b) Military standards: These provide procedures for design, manufacturing, and testing, rather than giving a particular material description.
(c) Federal specifications and standards: These are similar, except that they have come out of the General Services Administration, and are primarily for federal agencies. However, in the absence of military specifications and standards for a given product, federal specifications and standards are acceptable for use.
(d) Federal Aviation Regulations (FAR): In addition to military specifications, the Federal Aviation Administration (FAA) has advisory circulars for composite materials and part fabrication methods which as acceptable for aircraft. The FAA advisory circulars discuss several areas:
   - Test Plan — A unified program and schedule for tests that verify design allowables. These tests for composites might include a coupon test, static full-scale test for durability, environmental tests, stress analysis, and tests for subcomponents of a major structure.
   - Process Specifications — This includes both a material specification to be used to help select a commercial product and a process specification.
   - Quality Assurance Plan — This details acceptance and repeatability tests for material and process inspection of fabricated parts.
   - Report Submission — Includes final report submission, audit tests, how tests are accomplished, and who witnesses them.
(e) Company specifications — In many cases, companies feel that military standards and specifications do not reflect the most up-to-date materials and processing techniques. So companies develop specifications that will ensure all the requirements for fulfilling the military contracts.
(f) International standards — The International Standards Organization Technical Committee 61 and its subcommittee 13 covers reinforced composites.
   Among the latest efforts from the DoD itself are two materials specifications:
   - MIL-P-46179A for thermoplastic composite, which covers polyamide-imide, PES, PEEK, Polysulfone, PEI, PPS, etc.
   - MIL-P-46187 for high-temperature thermoset composites

To date, the greatest problem is how to get users to agree on test measures and reduce the number of tests, especially since every company has its own proprietary products and has established its own specifications and handbooks. In the composites industries, agreed-upon composites standards are not only a must, but essential for cutting costs.

(2) REQUIREMENTS AND SPECIFICATIONS

When the type and use of the aircraft is defined, reference is made to the requirements of the customer involved, and in the case of the commercial aircraft, the requirements of the licensing agency. The minimum structural requirements for aircraft for the various agencies are presented in the following documents:

(a) Civil aircraft
- Federal Aviation Regulations (FAR), Volume III, Part 23 — Airworthiness Standards: Normal, Utility, and Aerobatic Category Airplanes
- Federal Aviation Regulations (FAR), Volume III, Part 25 — Airworthiness Standards: Transport Category
- British Civil Airworthiness Requirements, Section D — Aeroplanes
- Joint Airworthiness Requirements (JAR), JAR — 25 Large Aeroplanes
- FAA Advisory Circular 20-107A
- JAA Advisory Circular

(b) Military aircraft (U.S.A.)
Air Force:
- MIL-A-008860A(USAF), Airplane Strength and Rigidity General Specification For, 1971
- AFGS-87221A, General Specification for Aircraft Structures, Air Force Guide Specification, 1990
- MIL-STD-1530A(11), Aircraft Structural Integrity Program, Airplane Requirements, Military Standard, 1975

Navy:
- MIL-A-8860B(AS) Airplane Strength and Rigidity General Specification For, 1987

(3) SPECIAL REQUIREMENTS

General requirements do not always apply to new types of aircraft. Consequently, interpretations and deviations from the requirements are often necessary. These deviations and interpretations are then negotiated with the licensing or procuring agency. In some cases, special requirements may be necessary to cover unusual aircraft configurations.

The manufacturer's requirements are usually the result of experience or an advancement in the "state-of-the-art" by that manufacturer. The trend on commercial aircraft in recent years has been toward the establishment of specially designated "Special Conditions" for each individual aircraft design. The FAA specifies these conditions by negotiation with the airframe manufacturer.

1.6 CERTIFICATION OF COMPOSITE AIRFRAME STRUCTURES

Certification requirements of airframe structures are ultimately identical whatever materials are used. Certification of composites has become more complex than certification of conventional materials (aluminum alloys, etc.) because of special design considerations and increased material variability. Because the use of composites usually requires fabrication from perishable raw materials, more controls are required over them. The attitude of the certificating authorities is that use of new materials should not subject aircraft operators to higher levels of risk than they accept with existing materials. It is the composite designer's responsibility to determine how this assurance is to be provided.

In July 1978, the FAA put out Advisory Circular AC20-107 (the document for complying with specific foreign countries certification requirements is AC21-2) on the certification of composite airframe structures. It is a brief document stating that the evaluation of a composite should be based on achieving a level of safety at least as high as that currently required for metal structures. It also emphasizes the need of testing for the effect of moisture absorption on static strength, fatigue and stiffness properties for the possible material property degradation of static strength after application of repeated loads. Typical test requirements for composite structural certification are shown below:

- 150% design limit load (DLL) test requirement of critical design condition(s)-check civil or military certification requirements
- Fatigue testing — Damage tolerance on primary structures
- Design (Environmental effects dominate)
  — Hot/wet
  — Cold/dry
  — Notched effect

In addition, damage tolerance (with particular reference to the effects of moisture and temperature) and crashworthiness must be addressed. Other data include flammability, lightning protection, weathering, ultraviolet radiation and possible degradation by chemicals and fuel; and also specifications covering quality control, fabrication technique, continuing surveillance and repair.

In 1982, the FAA published special "rules" which were applicable solely to the Lear Fan 2100 aircraft (American first all-composite airframe design), shown in Fig. 1.1.13(a). This airframe is made of advanced composite material and extensive use was made of bonding during assembly. The material and assembly technique is completely different from typical aluminum structures.

The following contains information the U.S. certification of the composite airframe structures (in case of conflict between the material herein and the detail specification and requirements, the certification specification and requirements shall prevail.):

(1) COMMERCIAL AIRFRAME — Federal Aviation Administration (FAA)

Federal regulations require that all civil aircraft operated in the United States should receive an airworthiness certificate. The certification process is administered by the FAA Part 23, 25, 27 and 29 Type Certificate, if the aircraft complies with design and safety regulations and standards. An Airworthiness Certificate is issued when the FAA determines that the particular aircraft was built in accordance with the specifications approved under the Type Certificate. The Airworthiness Certificate remains effective so long as required maintenance and repairs are performed on the aircraft and its equipment in accordance with the FAA regulations.

The certification process, shown in Fig. 1.6.1 and Fig. 1.6.2, is rigorous and generally takes several years to complete. Several of the phases required for certification are briefly described below (for the actual number of phases and requirements consult directly with local FAA representatives):

(a) Engineering Data Package — Preparation of this package includes drawings and specifications; load and structure analyses; structural, ground and flight test proposals and reports; flight and maintenance manuals; and an equipment list and parts catalogue. These reports essentially explain and record the aircraft's performance and constituent parts.

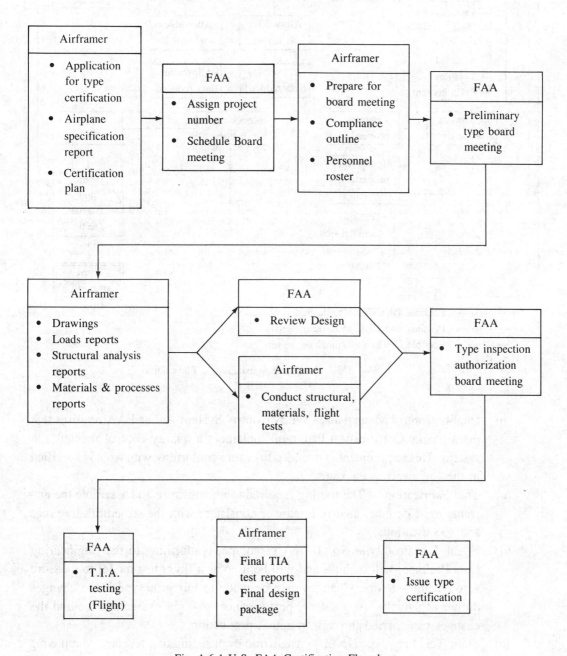

Fig. 1.6.1 U.S. FAA Certification Flowchart

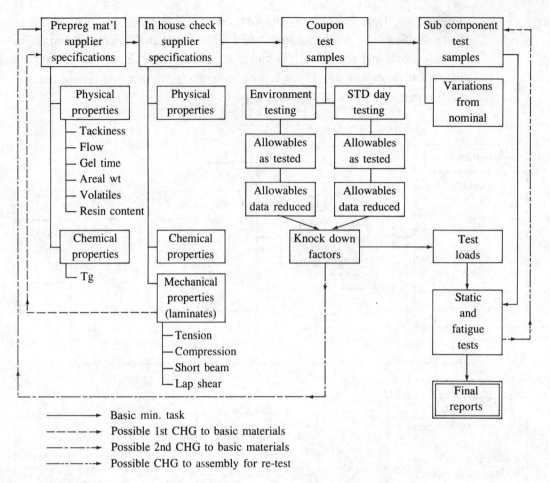

Fig. 1.6.2 Composite Substantiation Flow Chart

(b) Quality Control Manual and Quality Control System — The FAA requires that the airframe Certification Program implement a quality control manual and system. This requirement is intended to ensure conformity with what is specified in the Engineering Package.

(c) Tool Fabrication — The tooling required to manufacture and assemble the airframe must be fabricated in a manner consistent with the structural drawings and specifications.

(d) Flight Test Prototype No. 1 — A prototype, conforming to the Engineering Data Package, will be built and will be used as a flight test model to measure aerodynamic factors. Changes may be made to this prototype if the changes do not negatively affect safety, performance or flight characteristics and the changes can carried through to actual production.

(e) Static Test Prototype No. 2 — An airframe consisting of a fuselage, main wing and tail (or canard or both) will be built and will be used for testing under static load conditions.

(f) Fatigue Test Prototype No. 3 — An airframe will be built to test cyclic loads. The tests will determine the number of safe structure flight hours. The airframe will be intentionally damaged to determine the extent to which it can be damaged and still carry the critical load as defined by the FAA.

(g) Environmental Test Assemblies — Several assemblies will be load tested before, during and after exposure to extreme temperature and humidity far beyond expected in-service conditions.

(h) TIA (Type Inspection Authorization) Inspection and Ground Test — The FAA will issue a TIA inspection after a review of all technical data and upon a finding that the airframe is safe for FAA flight testing.

(i) TIA Flight Test — Upon successful completion of the above phases, the FAA will conduct the flight test to determine that the airframe prototype meets regulatory requirements.

(j) Type Certificate Issuance — After completing the above phases, a final Engineering Data Package containing all engineering data is submitted to the FAA, and a final Type board meeting will be held to ensure that all agenda certification items have been completed and properly demonstrated for compliance. When all items are cleared, the FAA will issue a Type Certificate.

(2) MILITARY AIRFRAME

Certification of composites for aircraft requires meeting the specifications of the following documents;

(a) Military specification MIL-A-8860A (USAF) through MIL-A-8870A (USAF) series
(b) Military specification MIL-A-8860B (AS) through MIL-A-8870B (AS) series
(c) Military specification MIL-A-87221 (USAF) — Aircraft Structures, General Specification For
(d) Military standard MIL-STD-1530A (USAF) — Aircraft Structural Integrity Program (ASIP), Airplane Requirements

Certification of a military airframe is generally a process negotiated the user and the airframe manufacturer based on the applicable specifications:

- U.S. Air Force — use the above mentioned item (a), (c) and (d)
- U.S. Navy Air Force — use the above mentioned item (b) plus a standard document similar to (d) which is now under study

For example, when certifying Air Force airframes, the above mentioned USAF Aircraft Structural Integrity Program (ASIP) is primarily used for full scale development of metal and composite structures. Fig. 1.6.3 summarizes the total five tasks include in item (d). When composite structures are designed and built there is some shifting of emphasis in Task II and Task III. A brief description of materials and joint allowables, stress analysis and design development tests, and full scale testing for these two Tasks are shown in Fig. 1.6.4 and Fig. 1.6.5 respectively.

Chapter 1.0

Task I	Task II	Task III	Task IV	Task V
Design information	Design analyses and development tests	Full scale tests	Force management data package	Force management
ASIP master plan Structural design criteria Damage tolerance and durability control plans Selection of mat'ls, processes, and joining methods Design service life and design usage	Materials and joint allowables Loads analysis Design service loads spectra Design chemical/ thermal environment spectra Stress analysis Damage tolerance analysis Durability analysis Sonic analysis Vibration analysis Flutter analysis Nuclear weapons effects analysis Non-nuclear weapons effects analysis Design development tests	Static tests Durability tests Damage tolerance tests Flight and ground operations tests Sonic tests Flight vibration tests Flutter tests Interpretation and evaluation of test results	Final anaylsis Strength summary Force structural maintenance plan Loads/environment spectra survey Individual airplane tracking program	Loads/environment spectra survey Individual airplane tracking data Individual airplane maintenance times Structural maintenance records

Fig. 1.6.3 USAF Aircraft Structural Integrity program (ASIP)

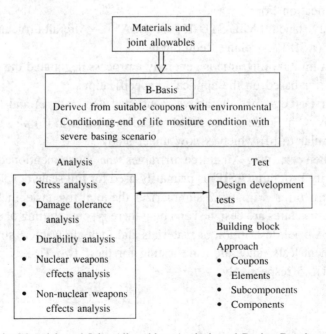

Fig. 1.6.4 Composite Materials and Joint Allowables, Analysis and Design Development Tests (Task II)

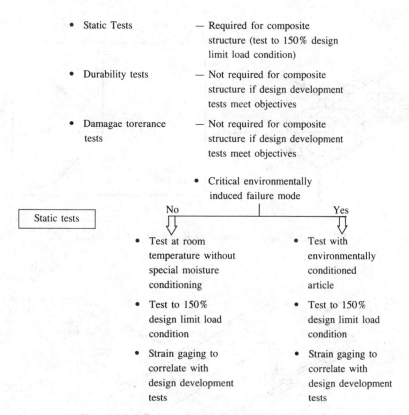

Fig. 1.6.5 Full Scale Testing of Composite Structure

1.7 LAMINATE ORIENTATION AND SYMBOL NOTATIONS

No universal or standard sign conventions and symbol notations for laminate composites have been recognized or accepted by the worldwide composite industry. However, convenient laminate sign conventions and symbol notations are defined below and will be used throughout this book.

(1) SIGN CONVENTION

The four standardly used ply orientations are $0°$, $45°$, $-45°$ and $90°$ (although composites are not limited to these orientations) and the sign convention shown in Fig. 1.7.1 is used in this book because it has been widely used by most of the composite industry and also academic organizations. Appendix B describes how ply orientations and stacking sequences are presented on engineering drawings for composites.

(2) SYMBOL NOTATIONS

Since composites are more complicated compared to metals of homogenous material, and involve layers of plies and oriented fibers, a set of basic and simple symbols and notations are needed with which to define a laminate in conjunction with the equations of composite laminate strength analysis. Symbols and notations are shown in Fig. 1.7.2 and Fig. 1.7.3 respectively.

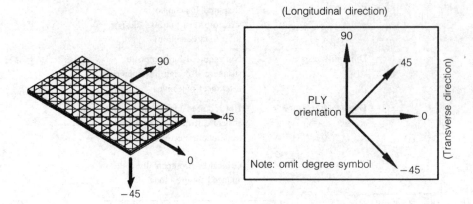

Fig. 1.7.1 Typical Laminate Structure Sign Convention

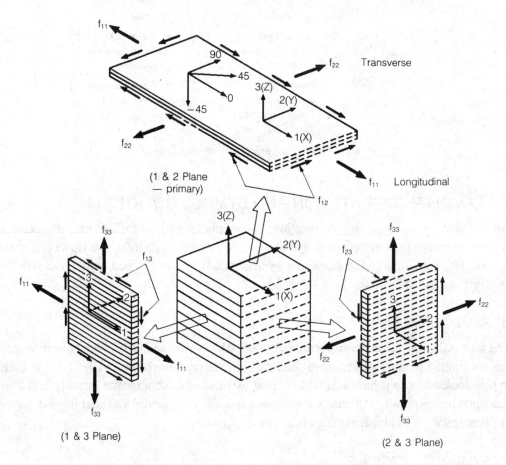

Fig. 1.7.2 Typical Laminate Composite Notations

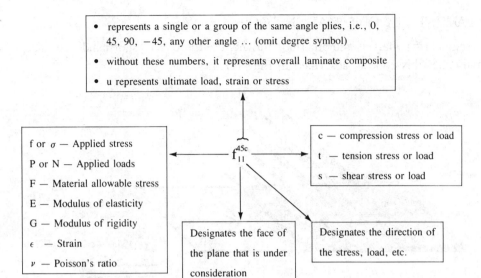

- Generally, 1, 2 & 3 represented lamina (ply) or laminae (plies) coordinates; x, y & z represente laminate coordinates (see Fig. 7.3.4).
- 1 or x — Longitudinal direction or 0°
- 2 or y — Transverse direction or 90°
- 3 or z — Perpendicular to the plane of 1 and 2 (laminae) or the plane of x and y (laminate) overall thickness direction)

Fig. 1.7.3 Lamina and Laminate Description of Notation

EXAMPLE I (Lamina or laminae in 1, 2 & 3 coordinates):

f_{11}^c
- Lamina or laminae applied compression stress
- The face of the plane is in "1" direction (or 0°)
- Direction of the stress is "1" direction (or 0°)

f_{22}^{45t}
- 45° lamina or laminae is in applied tension stress
- The face of the plane is in "2" (or 90°)
- Direction of the stress is in "2" (or 90°)

f_{12}^s
- Lamina or laminae is in plane applied shear stress
- The face of the plane is between "1" (or 0°) and "2" (or 90°)
- The direction of the shear stress is in the plane between "1" (or 0°) and "2" (or 90°)

EXAMPLE II (Laminate in x, y & z coordinates):

σ_x^{UC} or σ_{xx}^{UC}
- Laminate applied ultimate compression stress
- The face of the plane is in "x" direction (or 0°)
- Direction of the stress is "x" direction (or 0°)

N_{xy}
- Laminatae is in plane applied shear load
- The face of the plane is between "x" (or 0°) and "y" (or 90°)
- The direction of the shear load is in the plane between "x" (or 0°) and "y" (or 90°)

(3) LAMIATE PLY ORIENTATION CODE

For engineering analysis and other nondrawing use there is a "shorthand" laminate orientation code; this code is described in Fig. 1.7.4.

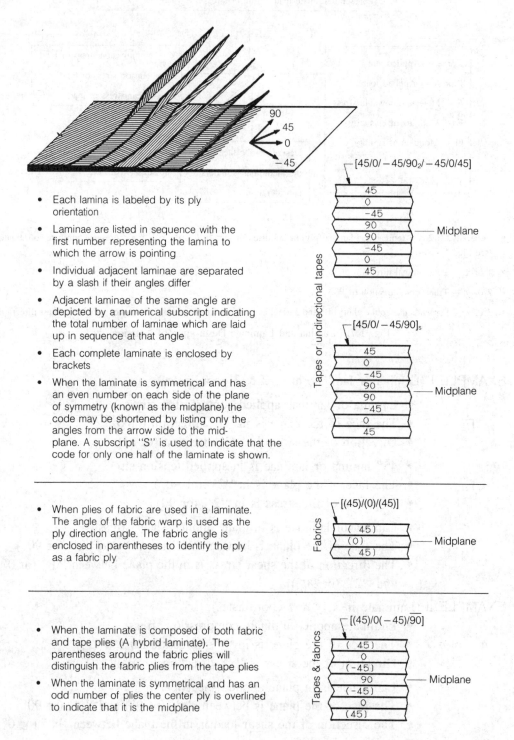

- Each lamina is labeled by its ply orientation
- Laminae are listed in sequence with the first number representing the lamina to which the arrow is pointing
- Individual adjacent laminae are separated by a slash if their angles differ
- Adjacent laminae of the same angle are depicted by a numerical subscript indicating the total number of laminae which are laid up in sequence at that angle
- Each complete laminate is enclosed by brackets
- When the laminate is symmetrical and has an even number on each side of the plane of symmetry (known as the midplane) the code may be shortened by listing only the angles from the arrow side to the midplane. A subscript "S" is used to indicate that the code for only one half of the laminate is shown.

- When plies of fabric are used in a laminate. The angle of the fabric warp is used as the ply direction angle. The fabric angle is enclosed in parentheses to identify the ply as a fabric ply

- When the laminate is composed of both fabric and tape plies (A hybrid laminate). The parentheses around the fabric plies will distinguish the fabric plies from the tape plies
- When the laminate is symmetrical and has an odd number of plies the center ply is overlined to indicate that it is the midplane

Fig. 1.7.4 Shorthand Laminate Orientation Code

References

1.1 Anon., "Composite Certification for Commercial Aircraft", AEROSPACE ENGINEERING, April, 1990. pp. 29-32.

1.2 Anon., "Safe Handling of Advanced Composite Materials Components: Health Information", Suppliers of Advanced Composite Materials Association. 1600 Wilson Blvd., Arlington, VA 22209. 1989.

1.3 Niu, C. Y., "AIRFRAME STRUCTURAL DESIGN", Conmilit Press Ltd., 101 King's Road, North Point, Hong Kong. 1988.

1.4 Lenoe, E. amd etc., "FIBROUS COMPOSITES IN STRUCTURAL DESIGN", Plenum Press, 227 W. 17th St., New York, NT 10011. 1980.

1.5 Anon., "ADVANCED COMPOSITES", Edgell Communications, Inc., 7500 Old Oak Blvd., Cleveland, OH 44130.

1.6 MIL-A-8860A (USAF) — General Specification for Airplane Strength and Rigidity.

1.7 MIL-A-8860B (AS) — General Specification for Airplane Strength and Rigidity.

1.8 MIL-A-87221 (USAF) — Aircraft Structures, General Specification for.

1.9 MIL-A-1530A (USAF) — Aircraft Structural Integrity Program (ASIP), Airplane Requirements.

1.10 Federal Aviation Regulations (FAR), Vol. III, Part 23 — Airworthiness Standards: Normal, Utility, and Aerobatic Category Airplanes.

1.11 Federal Aviation Regulations (FAR), Vol. III, Part 25 — Airworthiness Standards: Transport Category.

1.12 Anon., "MODERN PLASTICS ENCYCLOPEDIA"

1.13 Anon., "INTERNATIONAL ENCYCLOPEDIA OF COMPOSITES", VCH Publishers, Inc., 220 E. 23rd St., New York, NY 10010.

1.14 Anon., "ENGINEERING MATERIALS HANDBOOK, Vol 1 — Composites", ASM International, The ASM Composite Materials Collection, Metals Park, OH 44073.

1.15 Herakovich, C.T. and Tarnopolskii, Y.M., "HANDBOOK OF COMPOSITES, Vol 2 — Structure and Design", Elsevier Science Publishers, B.V., 1989.

1.16 MIDDLETON, D.H. "COMPOSITE MATERIALS IN AIRCRAFT STRUCTURES", Longman Scientific and Technical, Harlow, Essex, U.K. 1990.

1.17 Niu, C. Y., "COMPOSITES DRAFTING MANUAL", Internal Publication of Lockheed Aeronautic Systems Co. 1988.

1.18 Niu, C. Y., "COMPOSITES DESIGN GUIDE", Internal Publication of Lockheed Aeronautic Systems Co. 1989.

1.19 MIL-HDBK-17B, "Military Handbook — Polymer Matrix Composites, Vol. 1, Guidelines", Feb. 29, 1988.

1.20 Anon., "How US Companies are Attacking Production Costs", INTERAVIA, April, 1986. pp. 419-423.

1.21 English, L. K., "Issues and Answers in Industrial Composites", MATERIALS ENGINEERING — Composite Materials, April, 1989. pp. 37-41.

1.22 Middleton, D., "A Technology Revolution — Composite Materials", AIRCRAFT ENGINEERING, July, 1989. pp. 14-15.

1.23 Anon., "ADVANCED COMPOSITES III: Expanding and Technology", Proceedings of 1987 Conference, Engineering Society of Detroit and ASM Conference book.

1.24 Shook, G., "REINFORCED PLASTICS FOR COMMERCIAL COMPOSITES", an ASM source book. Metals Park, OH 44073. 1986.

1.25 Anon., "ADVANCED COMPOSITES: The Latest Developments", Proceedings of 1986 Conference, an Engineering Society of Detroit and ASM Conference Book.

1.26 Lubin, G. "HANDBOOK OF COMPOSITES", published by Van Nostrand Reinhold Co., New York NY. 1982.

1.27 Broutman, L. J. and Krock, R. H., "MODERN COMPOSITE MATERIALS", published by Addison-Wesley Publishing Co., Reading, Massachusetts, 1967.

1.28 Tsai, S. W., "INTRODUCTION TO COMPOSITE MATERIALS", published by Technomic Publishing Co. Landcaster, PA. 1980.

1.29 Anon., "Advanced Composites Design Guide, Vol. 1 through V", Air Force Materials Laboratory, Wright-patterson AFB, OH. 1983.

Chapter 2.0

MATERIALS

2.1 INTRODUCTION

Composite materials have gained their acceptance among structural engineers during the last decades. The performance of a composite depends upon:
- The composition, orientation, length and shape of the fibers;
- The properties of the material used for the matrix (or resin);
- The quality of the bond between the fibers and the matrix material.

Composite materials consist of any of various fibrous reinforcements coupled with a compatible matrix to achieve superior structural performance. The most important contribution to material strength is that of fiber orientation. Fibers can be unidirectional, crossed ply, or random in their arrangement and, in any one direction, the mechanical properties will be proportional to the amount of fibers oriented in that direction. Reduced properties result from the shear strength of the weak matrix. In fact, both the strength and moduli of a composite in a ply are reduced considerably when the angle of the applied load deviates from the direction of the filaments in the composites. Fig. 2.1.1 shows how a unidirectionally reinforced composite will have a far lower strength in transverse tension than one loaded exactly in the direction of the fibers. Therefore, it is evident that with increasingly random directionality of fibers, mechanical properties in any one direction are lowered. Thus, because of their low mechanical properties normal to the fiber direction, laminate composites will need to be strengthened or stiffened by laying up plies (unidirectional tape or woven fabric) in different directions. Such lamination will be necessary because stresses in a loaded component or panel can vary in both the "X" and "Y" direction.

Laminate properties of various combinations of plies oriented at different angles can be calculated through the use of computer programs to produce the best design. These computer aids are particularly helpful, because of the composite's non-isotropic properties, in calculating the various properties of any combination of oriented plies.

In this Chapter only those materials which are used on airframe structures will be discussed the data given is general and, while it may be used by the designer to do rough sizing, it is not appropriate for stress analysis or final sizing use. No composite material design allowable data is given (usually this data is part of a company's proprietary data) in this chapter because numerous varieties exist and many improved products are available every year.

Chapter 2.0

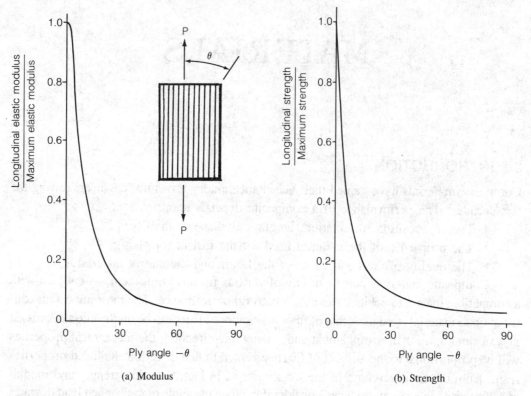

Fig. 2.1.1. Modulus And Strength Of Composites Drop Steeply As The Angle Between The Fibers And The Direction Of Load Increases

(1) MATERIAL SELECTION

Material selection plays a large part in final cost, not only because the raw material itself is expensive but also because the material selected often determines downstream manufacturing costs. The material selection criteria are given below:
- Cost
- Available mechanical and environmental properties database
- Suitability for use in proposed manufacturing processes
- Structural performance
- Ease of processing
- Ease of handling
- Supportability
- Maximization of knowledge base
- Available processing data
- Immediate or near term commercial availability

There are many cases where more than one material can meet the structural and/or weight requirements specified for a given part. Assume there is a choice between a unidirectional material form (which can be used on the automated lay-up machine) and a broadgoods form (fabric or woven) of the same material. Clearly there is a difference in the costs of these raw material forms: unidirectional prepreg generally less expensive because the material supplier has not gone through the added step of weaving the broadgoods fabric. At the same time, it may take more time to fabricate a laminate component from unidirectional tape than form broadgoods. Therefore, there is a tradeoff between actual raw material costs and the downstreams manufacturing costs which are predetermined in choosing a particular raw material.

Obviously, along with cost considerations, any design must carefully match the requisite properties with the candidate composite material. Once the optimum, or best available, material is chosen, the designer must be concerned with any additional limitations that material selection might impose on the capabilities of the design. The common areas of concern include hot/wet properties, notched effect (if fasteners or small cutouts are used in design), in-service temperature, and impact resistance. Critical limitations that must be considered in composite design include the relatively low strength and stiffness in the out-of-plane direction and often poor shear properties. These factors, must be considered to prevent delamination under compressive loading or inadequate out-of-plane load-carrying capabilities.

Various kinds of composite materials with temperature resistant matrices and high-performance reinforcements are currently available or are in advanced stages of development. Their upper (and partially overlapping) service temperatures are shown in Fig. 2.1.2.

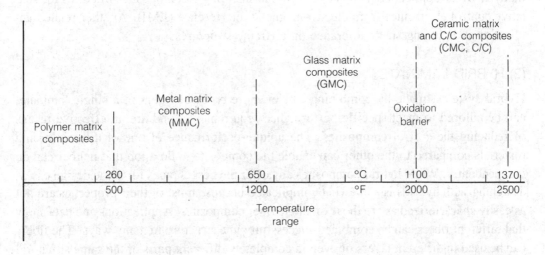

Fig. 2.1.2 Temperature Resistance of Composites

Since the properties of composites depend critically on the processes used to make them, designers must work with prepreg and fiber producers to achieve desired results. The designers should be aware of the weaknesses of various fibers and construction methods and so design around them. The following are material specification requirements:
- Fiber and fabric properties
- Storage and retest requirements
- Uncured prepreg properties
- Cured or co-consolidated prepreg properties
- Mechanical properties
- Environmental testing
- Processability trials
- Chemical characterization
- Non-destructive inspection (NDI)

In general, most of the major reinforcement systems have been well characterized for many years, and performance improvements have occurred in relatively small increments. However, improvements in the matrix resins (both thermoset and thermoplastic) have allowed great strides in composite fabrication, producibility, performance, and stability.

The reinforcing fiber may have a negative thermal expansion coefficient along its axis, a property that makes possible the design of structures with zero or very low linear and planar thermal expansion. Thus, the main support truss for the mirrors of the "Space Telescope" is made of a carbon/epoxy composite to meet extreme close tolerance requirements.

It is worthwhile to note research on organic conducting polymers, which would have many airframe applications, such as to provide shielding on composite structures for sensitive control electronics from electro-magnetic interference (EMI). Another related application is lightning strike tolerance on airframe structures.

(2) HYBRID LAMINATES

Hybrid systems, made by combining two or more types of fibers in a single laminate, can be tailored to meet specific performance requirements, and are an effective means of reducing the cost of composites. The unique performance of one of the reinforcing materials compared to the other can enable the composite to do a job that neither can do independently. While hybrid composites any offer the best choices for some design cases, designing with hybrids is somewhat complicated because most of their properties are not as easily characterized as are those of single fiber composites. Applications and data show that different fibers can be combined successfully in a structure in many ways. The fibers can be used in different layers or even in completely different parts of the same structural element. They also can be blended to form a hybrid tape or woven to form a hybrid fabric.

For hybrid constructions, directional response and failure parameters should be defined for each material. Care must be taken to provide reinforcement for all loading directions. Since carbon or graphite is conductive both thermally and electrically and has a slightly negative coefficient of thermal expansions, it is conceivable for designers to develop hybrid material geometries with structural responses totally different from the existing conventional metal materials.

As the cost/performance tradeoff becomes more critical, hybrids may become the material system of choice for more structural uses, making them materials with a future. Nevertheless, the future of hybrids in the airframe industry appears uncertain and much still needs to be learned about this systems. Issures peculiar to hybrid systems are described below:
  (a) Material more tailored to specific needs than is available with a single fiber-matrix combination is very desirable.
  (b) Different fibers have different:
     • Strains to failure
     • Moduli
     • Coefficients of thermal expansion
     • Coefficients of moisture expansion
  (c) Thermally induced stresses exist in every hybrid lamina (below the laminate level)
     • Caused by different thermal expansion characteristics of constituents
     • Can be large enough to cause failure without mechanical load
  (d) Effectively an infinite variety of hybrids is possible
     • Each new hybrid must have some minimum level of material property qualification
     • The wide variety of possible hybrids (just like the wide variety of possible laminates) must be deliberately restricted on purely practical grounds.

There are typically three methods of hybridization:
  • Interply — Different reinforcements are stacked in separate layers with no mixing within the layers
  • Intraply — Different reinforcements are commingled within a layer either by alternating strands or mixing chopped fibers
  • Selective placement — The laminate is basically composed of one reinforcement, but a different reinforcement is added in certain areas (such as corners, ribs, etc.)

Hybrid reinforcements can be combined in almost all material forms including:
  • Prepregs
  • Fabric
  • Woven roving
  • Chopped fibers

Common hybrids include:
  (a) Carbon/Aramid
      Can be combined without residual thermal stresses since coefficients of thermal expansion are very similar

(b) Carbon/Glass
- Increased impact strength
- Improved fracture toughness
- No galvanic corrosion
- Reduced cost over all carbon fiber laminates

(3) MATERIAL PROPERTIES SUMMARY

A summary of material density, coefficient of thermal expansion and thermal conductivity data is shown in Fig. 2.1.3.

Material	Density lb/in^3	Thermal Expansion in/in/°F $\times 10^{-6}$	Thermal Conductivity (but) (in.) (hr) (ft^2) (°F)
Aluminum	0.1	12.6-13	1300-1400
Beryllium	0.066	6	1121
Magnesium	0.063	14	1092
Steel	0.284	6-6.7	310-460
Titanium	0.160	4.7-5.6	112
Glass	0.09	3.0	5.4
Phenolic	0.048	11-35	1.1
Epoxy	0.0457	20-60	2-6
Polyester	0.046	16-35	2-5
Polyimide	0.052	70	3-5
Silicone rubber	0.045	45-200	2
Kevlar 49	0.052	-3.5	—
Silicon	0.11	1.4	—
Boron	0.098	2.7-3.5	22
Carbide	0.125	2.2	—
Ceramic	0.058-0.072	0.5-5	10-80
Monolithic Graphite	0.063	1-2	800
Boron filament	0.083	2.7	—
Graphite fiber	0.063	-0.05	840
Invar	0.29	3	—
Graphite/epoxy [0]	0.055-0.059	0.24	—
E Glass/epoxy [0]	0.065-0.070	4.8	—
Kevlar 49/epoxy [0]	0.046-0.050	-3.0	—
Boron/epoxy [0]	0.07-0.075	2.3	—
Nickel	0.32	7.4	400

Fig 2.1.3 Summary of Material Density, Coefficient of Thermal Expansion and Thermal Conductivity Data

2.2 ORGANIC MATRICES

The purpose of the matrix is to bind the reinforcement (fiber) together and to transfer load to and between fibers, and to protect the flaw- or notch-sensitive fibers from self-abrasion and externally induced scratches. The matrix also protects the fibers from environmental moisture and chemical corrosion or oxidation, which can lead to embrittlement and premature failure. In addition, the matrix provides many essential functions from an engineering standpoint: the matrix keeps the reinforcing fibers in the proper orientation and position so that they can carry the intended loads, distributes the loads more or less evenly among the fibers, provides resistance to crack propagation and damage, and provides all of the interlaminar shear strength of the composite. The matrix generally determines the overall service temperature limitations of the composite and may also control its environmental resistance.

In summary, the matrix:
- distributes loads through the laminate
- protects fibers from abrasion and impact
- determines:
  - compressive strength
  - transverse mechanical properties
  - interlaminar shear
  - service operating temperature
  - selection of fabrication process and tool design
- contributes to fracture toughness

With any fiber, the material used for the matrix must be chemically compatible with the fibers and should have complementary mechanical properties. Also, for practical reasons, the matrix material should be reasonably easy to process.

The development of high strength and high thermal resistance is frequently accompanied by complex cure procedures or brittleness in thermosets. Overcoming these obstacles has proven the key to developing viable composite matrices, with processing/fabrication constraints of fiber wet-out, prepreg shelf life, tack and drape, cure shrinkage, etc., adding to the complexity.

The organic matrices commonly used are broadly divided into the categories of thermoset and thermoplastic; organic matrices commonly used on airframe structures are given below:

THERMOSET	THERMOPLASTIC
• Expoxy	• Polyethylene
• Polyester	• Polystyrene
• Phenolics	• Polypropylene
• Bismaleimide (BMI)	• Polyetheretherketone (PEEK)
• Polyimides	• Polyetherimide (PEI)
	• Polyethersulfone (PES)
	• Polyphenylene Sulfide
	• Polyamide-imide (PAI)

The relative advantages of thermosets and thermoplastics include:

THERMOSET MATRICES	THERMOPLASTIC MATRICES
(Characteristics)	
• Undergo chemical change when cured	• Non-reacting, no cure required
• Processing is irreversible	• Post-formable, can be reprocessed
• Low viscosity/high flow	• High viscosity/low flow
• Long (2 hours) cure	• Short processing times possible
• Tacky prepreg	• Boardy prepreg
(Advantages)	
• Relatively low processing temperature	• Superior toughness to thermosets
• Good fiber wetting	• Reusable scrap
• Formable into complex shapes	• Rejected parts reformable
• Low viscosity	• Rapid (low cost) processing
	• Infinite shelf life without refrigeration
	• High delamination resistance
(Disadvantages)	
• Long processing time	• Less chemical solvent resistance than thermosets
• Restricted storage	• Requires very high processing temperatures
• Requires refrigeration	• Outgassing contamination
	• Limited processing experience available
	• Less of a database compared to thermoset

Compared to thermoplastics, thermoset matrices offer lower melt viscosities, lower processing temperatures and pressures, are more easily prepregged and are lower cost. On the other hand, thermoplastic matrices offer indefinite shelf life, faster processing cycles, simple fabrication, and generally do not require controlled-environment storage or post curing.

(1) THERMOSET MATRICES

The most prominent matrices are epoxy, polyimides, polyester and phenolics; a matrix comparison is given in Fig. 2.2.1.

Thermoset matrix systems have been dominating the composite industry because of their reactive nature. These matrices allow ready impregnation of fibers, their malleability permits manufacture of complex forms, and they provide a means of achieving high-strength, high-stiffness crosslinked networks in a cured part. Fig. 2.2.2 is a comparison of selected thermoset matrices which are commonly used in airframe design for primary structural applications.

Characteristics	Thermosetting resins				
Property	Polyester	Epoxy	Phenolic	Bismaleimide	Polyimide
Processability	Good	Good	Fair	Good	Fair to difficult
Mechanical properties	Fair	Excellent	Fair	Good	Good
Heat resistance	180°F	200°F	350°F	350°F	500-600°F
Price range	Low — Medium	Low — Medium	Low — Medium	Low — Medium	High
Delamination resistance	Fair	Good	Good	Good	Good
Toughness	Poor	Fair — Good	Poor	Fair	Fair
Remarks	Used in secondary structures, cabin interiors, primarily with fiberglass	Most widely used, best properties for primary structions; principal resin type in current graphite production use	Used in secondary structures, primarily fiberglass good for cabin interiors for low smoke generation	Good structural properties, intermediate temperature resistant alternative to epoxy	Specialty use for high temperature application

Fig. 2.2.1 Comparison of Properties for General Thermoset Matrices

Matrix Type	Tack	Drape	Thermal Stability	Cure Temp	Cure Pressure	Void Content	Cost	Other Problems
Epoxy	Excellent	Excellent	200°F Dry 180°F Wet	350°F (177°C)	100 psi	Low	Low	Low temp Storage High Moisture Pickup Brittle
Toughened Epoxy	Very good	Excellent	180°F Dry 160°F Wet	350°F (177°C)	100 psi	Low	Moderate	Low temp Storage
BMI	Good When heated	Good When heated	350°F to 400°F Dry 300-350°F Wet	350°F Cure 400 — 500°F Post cure	100 psi	Low	Moderate	Microcracks on tempertaure Cycling
Condensation PI	Good	Good	Excellent 500°-600°F	600°-700°F	500 psi or higher	High	Moderate	Complex process cycle Large parts difficult
Acetylene Terminated PI	Poor-boardy	Poor	Very good 500-550°F	500°F Cure 600°F — 650°F post	200 psi	Low	Moderate to high	Small process window
PI (PMR)	Good	Good	400-500°F Higher for brief periods	Complex to 650°F	500 psi	Low to moderate	Moderate	

Fig. 2.2.2 Comparison of Commonly Used Thermoset Matrices

(A) EPOXY

Epoxy systems are the major composite material for low-temperature applications [usually under 200°F (93°C)] and generally provide outstanding chemical resistance, superior adhesion to fibers, superior dimensional stability, good hot/wet performance, and high dielectric properties. Epoxy can be formulated to a wide range of viscosities for different fabrication processes and cure schedules. They are free from void-forming volatiles, have long shelf lives, provide relatively low cure shrinkage, and are available in many thoroughly-characterized standard prepreg forms. They also have good chemical stability and flow properties, and exhibit excellent adherence and water resistance, low shrinkage during cure, freedom from gas formation, and stability under environmental extremes. In addition, on other very important advantage is the wealth of database information available.

The epoxy family is the most widely used matrix system in the advanced composites field. Because it is generally limited to service temperatures, this restricts use in many aerospace applications, where higher service temperatures are required.

The baseline system of epoxy used in the majority of applications includes:
- Superior mechanical properties
- 250°F curing: −65°F to 180°F (−53 to 82°C) service temperatures
- 350°F curing: −65°F to 250°F (−53 to 121°C) short-term or 200°F (93°C) long-term service temperature

Epoxy matrices, the workhorse of the advanced composites industry today, are suitable for use with glass, carbon/graphite, aramid, boron, and other reinforcements and hybrids. Yet greater demands can be met by conventional epoxies are being made for today's parts, so a wide variety of epoxies are being developed to handle the ever-increasing requirements for speed of fabrication, toughness, and higher service temperatures.

Unmodified epoxies are brittle. When subjected to impact from a flying stone, an occasional bump, or a dropped wrench, etc., they can be damaged internally and suffer loss of laminate compressive strength. Epoxies have been modified or improved to increase their damage resistance. The result is "toughened" epoxies.

Epoxies have a tendency to absorb moisture, this absorbed moisture can lead to decrease mechanical properties especially at elevated temperatures. The presence of water decreses the glass transition temperature of the epoxy matrix, hence the term "wet Tg" This effect must be considered in design.

The following environmental hazards have detrimental effects on epoxy matrices:
- Moisture
- Temperature
- Ultraviolet light
- Hydraulic fluid
- Fuel
- Cleaning agents

(B) POLYIMIDES (HIGHER SERVICE TEMPERATURE MATRICES)

Polyimides are thermo-oxidatively stable and retain a high degree of their mechanical properties at temperatures far beyond the degradation temperature of many polymers, often above 600°F (320°C). Several types with superior elevated temperature resistance are listed below:

- Bismaleimides: good to 450°F (230°C), relatively easy to process
- Condensation types: good to 600°F (320°C), very difficult to process
- Addition types: good to 500 — 600°F (260 — 320°C), improved processability compared to condensation types

(a) BMI (Bismaleimides), a special polyimide system, operates around a 350°F to 450°F (177 to 230°C) upper limit. BMI offer good mechanical strength and stiffness, but are generally brittle and may have cure-shrinkage. Other BMIs have significant improvements in toughening which greatly enhances their usefulness. When good hot/wet performance or thermal stability beyond epoxy limits is desired, BMI matrices may be the matrix of choice. BMI characteristics are summarized below:

- BMIs provide increased thermal stability compared to epoxies, with comparable processability
- The major problem with BMIs has been increased brittleness over epoxies — with reduced damage resistance and toughness
- BMI systems with improved toughness are available at the sacrifice mechanical properties

(b) PMR-15 (Polymerization of Monomeric Reactants) is a thermoset addition polyimide which offers higher continuous service temperatures. Originally developed by NASA. Thermo-oxidative stability, relatively low cost, and availability in a variety of forms make PMR-15 one of the candidates for airframe industrial applications where performance from 500 to 600°F (260° — 320°C) is the key material selection criterion. PMR-15 processing is complicated, requiring application of near 600 psi (4.1 Mpa). A heated tool is often necessary to achieve faster heatup rates than are possible with conventional tooling in an autoclave. The room temperature properties of PMR-15 are similar to those of 350°F (177°C) epoxy, but, unlike epoxy, properties do not decrease significantly until temperatures over 500°F (260°C) are reached, even after exposure to moisture. NASA has developed LARC-160, a "PMR" system which provides a significant improvement in processability over the PMR-15 matrix, with only a small loss in elevated temperature properties. However, both NASA's PMR-15 and LARC-160 matrices are still in progress under their continuing development programs.

(C) POLYESTER

Polyesters matrices can be cured at room temperature and atmospheric pressure, or at a temperature up to 350°F (177°C) and under higher pressure. These matrices offer a balance of low cost and ease of handling, along with good mechanical and electrical properties, good chemical resistance properties (especially to acids), and dimensional stability. Polyester combined with fiberglass fibers becomes a very good radar-transparent structural material and polyester is also a relatively inexpensive matrix that offers a compromise between strength and impact resistance for use in aircraft radomes. Low mold-pressure requirements helped promote the manufacture of large polyester composite structures, and this was further helped along by their relatively quick cure characteristics.

Vinyl esters are a subfamily of polyesters, derived from epoxy-matrix backbones, which provide higher tensile elongation, toughness, heat resistance, and chemical resistance than conventional unsaturated polyesters.

(D) PHENOLICS

Phenolics are the oldest of the thermoset matrices, and have excellent insulating properties, resistance to moisture, and good electrical properties (except arc resistance). The chemical resistance of phenolics is good, except to strong acids and alkalis. Phenolics are available as compression-molding compounds, and injection-molding compounds. This material is very useful in military and high-performance aerospace applications where radiation-hardness, dimensional stability at high loads and temperatures, and ability to ablate may be critical to component survival.

Fig. 2.2.3 shows a number of the thermoset resin products which are available for reference.

Matrix resin (Vendor)	Resin type	Available Material Forms	Thermal Stability	Remarks
934 (Fiberate) 5208 (Narmco) 3501-6 (Hercules)	Epoxy	(Carbon, glass Kevlar, boron, & ceramic fibers) Unitape Various fabrics Prepreg tow	180°F Serv temp 250°F Dry 180°F Wet	• Best overall structural characteristics • High compressive strength but poor damage tolerance • Absorbs moisture • Structural properties affected by moisture content • Easy to process • Good chemical resistance • Good environmental resistance • Brittle
3502 (Hercules) 8551-7A (Hercules) 8552 (Hercules) R6376 (Ciba Geigy) 977-2 (Fiberite)	Toughened Epoxy	(Carbon & Glass fibers) Unitape Various fabrics	>200°F Dry 180°F Wet	• Same as epoxy but not as brittle because of improved toughness • Lower hot/wet performance than standard epoxy
F650 (Hexcel) 5250-2 (BASF/Narmco) 5245C (BASF/Narmco)	Bismaleimide (BMI)	(Carbon, glass quartz & ceramic fibers) Unitape Various fabrics prepreg tow	250-450°F Dry 250-300°F Wet	• Easy to process • More brittle than epoxies • Microcracks on temperature cycling with some systems
V398 (U.S. poly) F655 (Hexcel) 5250-4 (BASF/Narmco)	Toughened Bismaleimide	Carbon, glass quartz & ceramic fibers) Unitape various fabrics prepreg tow	250-400°F Dry 200-350°F Wet	• Better damage tolerance • Lower service temperature than other BMI's
PMR-15 (Various)	Polyimide	Unitape fabric powder impregnated tow & tape	400-500°F Higher for brief periods	• Difficult to process

Fig. 2.2.3 Thermoset Prepreg Resin Choices (Epoxy and Polyimide)

(2) THERMOPLASTIC MATRICES

In recent years, thermoplastic matrix systems have been introduced. Their major advantages are
- Service temperatures of up to 540°F (280°C)
- Excellent strain capabilities
- High moisture resistance
- Unlimited shelf-life
- Short processing cycles

Disadvantages are
- High processing temperatures
- As yet marginal processing experience
- Lack of drapeability

Thermoplastic matrices are not new to the airframe industry. They have been used for many years for various components, mainly in aircraft fuselage interiors and for other non-structural parts. The engineering thermoplastic resins have high continuous service temperatures, from 250°F to 400°F (121 to 200°C), high matrix melting temperatures, and high viscosity which leads to higher mold pressure in autoclave operations. Thermoplastic matrices provide better interlaminar fracture toughness combined with acceptable postimpact compression, better resistance to high temperatures and solvents, and have low moisture sensitivity. The major advantage over thermoset matrices is their shorter fabrication cycle and the fact that a chemical cure does not take place, allowing reprocessing or reconsolidation of a flawed part after manufacture.

Thermoplastics offer potential cost reduction by:
- Reforming capability
- Welding capability
- Eliminating cold refrigeration storage and having unlimited shelf life
- Rapid processing cycles times
- Recyclable scrap
- Being less difficult to drill and machine

Product forms are still being developed and the most recently available prepregs are stiff and boardy. They lack the drape and tack needed for handleability (forms that handle well such as commingled fabrics will be discussed in Materal Forms in Section 2.5). Tack and drape in some thermoplastic prepregs are achieved by the presence of solvent, as is the case with some ployimides.

The "Wet" prepregs compared to "Dry" prepregs are described below:

(a) "Wet" Thermoplastic Prepreg Materials
- Wet matrices include KIII, AIX-159, Torlon 696, etc.
- Wet matrices have inherent tack and drape at room temperature
- Wet matrices may react chemically during processing (and in the past have been referred to as "pseudo-thermoplastics")
- Have half life and out time constraints
- Process like thermosets
- Have high volatile content, 12-25% by weight

- Some require post-cure for maximization of Tg
- Require volatile management during in processing

(b) "Dry" Thermoplastic Prepreg Materials
- More difficult to prepreg than "Wet"
- Dry matrices have no inherent tack and drape at room temperature
- Have no shelf life or time constraints
- Melt fusible, no chemical reaction
- Solution or hot melt impregnated
- Amorphous and/or semi-crystalline

One of the most critical factors is lack of an extensive database of performance properties over service time. Military aircraft structural applications are one of the major drivers to develop thermoplastics for use as high-temperature composite matrices and four major requirements are:

- High temperature capabilities (range 350°F (177°C)) under severe hot/wet environmental conditions
- Better damage tolerance in primary structures
- Easily automated in order to drive down manufacturing process costs
- Lower total part-acquisition and lifetime costs (including material, processing, and supportability)

To understand differences between thermoplastic matrices, an overview of properties dependent on their microstructure is shown in Fig. 2.2.4.

Morphology	Processing	Characteristics
Amorphous Thermoplastic 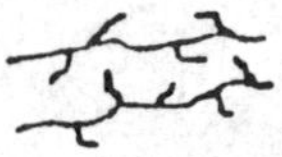	Melt Fusion	Poor solvent resistance Lower temp capability Better formability
Pseudothermoplastic 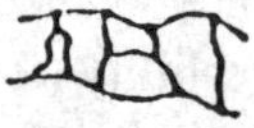	Condensation	Better solvent resistance Higher temp capability Some not reprocessible
Crystalline 	Melt fusion	Good solvent resistance Crystallinity dependent on processing
Liquid crystal 	Melt Fusion	Anisotropic properties Directionality of crystal dependent on processing Injection molding & extrusion

Fig. 2.2.4 Thermoplastics properties Depend on Microstructure

The characteristics of two major divisions (semi-crystalline and amorphous) of thermoplastic matrials are described below:
(a) Semi-crystalline matrices
- Have a definite melting point
- Better resistance to halogenated hydrocarbons and paint strippers
- Gradual loss of properties after Tg (glass transition temperature) is reached
- Density varies slightly depending on degree of crystallinity
- Mechanical properties may vary depending on degree of crystallinity
- Degree of crystallinity dependent on processing

(c) Amorphous Matrices
- No definite melt temperature
- Can be solvated for ease of fabric impregnation
- Free from the problems associated with crystallinity
- More susceptible to methyl chemical paint strippers

Fig. 2.2.5 compares the modulus vs. temperature curves for a typical amorphous polymer and a typical semi-crystalline polymer [note that in both cases the modulus has decreased by over an order of magnitude before the Tg (glass transition temperature) or the Tm (melting temperature) is reached]. Understanding the role of crystallinity in thermoplastics immensely improves the success of the engineer in both the design and manufacturing. Fig. 2.2.6 shows a number of available thermoplastic matrices.

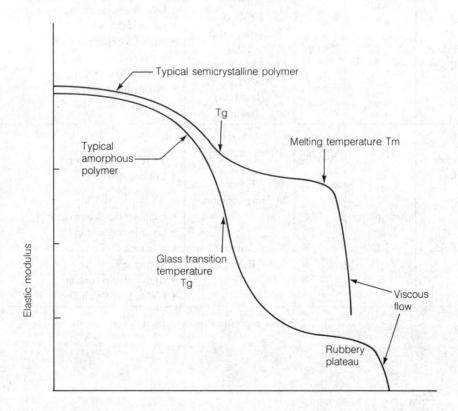

Fig. 2.2.5 Typical Modulus Vs. Temperature Curves for Thermoplastics

Polymer type		Serv temp	Material forms available	Remarks
Polyether-imide (PEI)	Amorphous	300°F	Fabric Film Molding compound Cowoven Commingled	• Good database • Unidirectional tape available, but poor quality • Low processing temperature • Good impact resistance
Polyether-sulfone		350 – 300°F	Unitape, Various fabrics Cowoven Commingled Film Molding compound Neat resin	• Very good processing characteristics • Solvent free prepreg
Polyarylene-sulfide (PAS)		300°F	Unitape Fabric Neat resin Molding compound	• Low processing temperature • Fiber/resin interface not optimal • Solvent free prepreg
Polyphenyl-enesulfide	Semi-crystalline	150 – 200°F	Unitape Fabric Discontinuous	• Very low processing temperature • Low service temperature
Polyether-etherketone (PEEK)		250°F	Unitape Fabric Commingled Film Molding compound Powder slurry Neat resin Towpreg	• Low moisture pickup • Large database • High processing temperatures • Solvent free prepreg • Processing window defined
Polyether-sulfone		350°F	Unitape powder slurry Film Fabric Molding compound	• Solvent free prepreg • Limited database
Polyimide	Amorphous	>300°F	Unitape Fabric	• High service temperature • Long processing cycle • Has tack • Moisture sensitive • High volatile content • Limited reformability

Fig. 2.2.6 Thermoplastic Prepreg Matrices

2.3 METAL, CARBON, AND CERAMIC MATRIX COMPOSITES

During the last decades, metal, carbon, and ceramic matrices have not progressed to the same extent as organic matrices because of high cost has limited applications. Most of these composites are still in the research and development stage and are not widely used on today's airframes.

(1) METAL MATRIX COMPOSITES

Work with metal matrix composites (MMC) has concentrated on boron/aluminum (B/Al), graphite/aluminum (GR/Al), and silicon carbide/aluminum (SC/Al) composites but other types of matrix materials are also being studied, including titanium and magnesium. Metal matrices offer greater strength and stiffness than those provided by polymers. Fracture toughness is superior, and metal matrix composites offer less-pronounced anisotropy and greater temperature capability in oxidizing environments than do their polymeric counterparts. The greatest applications are where stiffness and light weight are needed. The following describes the metal matrix choices:

(a) Aluminum
- Principal metal matrix material
- Greatly improved properties when reinforced
- Lightweight
- Easily processed
- Fig. 2.3.1 demonstrates a typical fabrication scheme to produce Boron/Aluminum composites

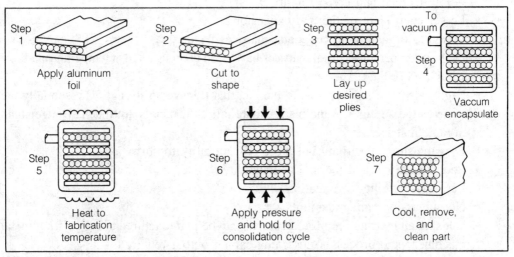

*Similar preforms are available with SiC filaments.

By courtesy of Textron Inc.

Fig. 2.3.1 Flow Chart of Producing Boron/Aluminum Composites

(b) Titanium
- Lightweight
- Good resistance to high temperatures
- Difficult to reinforce
- Expensive

(c) Magnesium
- Good interface with reinforcements
- Poor corrosion resistance
- Lightweight

(d) Copper
- Improved shear strength over aluminum at elevated temperatures
- Heavier than aluminum

From the standpoint of airframe design, the most interesting materials for aircraft components are SiC (Silicon Carbide) reinforced aluminum and titanium; of main interest for space structures are graphite-reinforced aluminum or magnesium.

A variety of reinforcement-matrix combinations are used for metal matrix composites and some representative materials are shown in Fig. 2.3.2. Each class of material can have a broad range of properties, depending upon the specific fiber, matrix, and fiber volume. Advantages of metal matrix composites vs. metals include:

- Higher strength/density ratio
- Higher stiffness/density ratio
- Under highly elevated temperatures metal matrix composites still have better properties:
  — higher strength
  — lower creep and creep rupture
- Lower coefficient of thermal expansion

In comparison to organic matrices metal matrices have:

- High on temperature (metal matrices have been demonstrated at temperatures above 2000°F (1100°C)
- High transverse strength due to the fact that transverse strength is essentially the same as the strength of the matrix material, and, metals are much stronger than organic matrices.
- Less moisture sensitivity but more susceptibility to corrosion
- Better electrical and thermal conductivity
- Less susceptibility to radiation
- No outgassing contamination
- Whisker and particle reinforced metals can be manufactured using existing metal machinery and processes, lessening capital output required.

Fig. 2.3.3 provides a comparison of polymer composites vs. metal matrix composites for design use.

Currently, four methods of production are the most common for reinforced metal matrices:

(a) Diffusion bonding — Diffusion bonding is most often used when the reinforcement is continuous fibers. Strands or mats of fibers are sandwiched between sheets of the matrix metal. The laminate is then sealed in an evacuated can, heated and pressed to full density.

(b) Conventional casting — This process uses a proprietary treatment to promote wetting of the reinforcement by the molten metal.

(c) Power metallurgy (P/M) — Particulate reinforcements are mixed with the metal powder, and the mixture is processed in conventional P/M processes.

Some difficulties remain to be solved for metal matrix composites:
- High cost
- Cannot be extruded or forged
- Lack of machining and joining techniques
- Lack of non-destructive testing techniques
- Need to improve the adhesion of the fibers to the matrix
- Complex and expensive fabrication methods for metal matrix composites with continuous reinforcements

Selected metal matrix reinforced design concepts, as applied to structural components such as airframe floor beams, stiffeners, columns, and rods and tubes, are shown in Fig. 2.3.4. To carry the primary axial loads, Boron reinforcements are selectively introduced into the beam or stiffener flanges in the form of aligned boron filaments threaded completely through lobes or other apertures embodied in the metal structure. This method of reinforcement makes possible a lighter structure while still permitting the retention of traditional metal joining and assembly techniques such as riveting and welding. It is claimed that boron-reinforced structural components are from 25 to 45% lighter than metal counterparts.

Fiber	Matrix	Potential Applications
Carbon	Aluminum	Satellite, missile, and helicopter structures
	Magnesium	Space and satellite structures
	Copper	Electrical contacts and bearings
Boron	Aluminum	Compressor blades and structural supports
	Magnesium	Antenna structures
	Titanium	Jet engine fan blades
Borsic	Aluminum	Jet engine fan blades
	Titanium	High-temperature structures and fan blades
Alumina	Aluminum	Superconductor restraints in fusion power reactors
	Magnesium	Helicopter transmission structures
Silicon Carbide	Aluminum	High-temperature structures
	Titanium	High-temperature structures
	Superalloy (Co-based)	High-temperature engine components
Molybdenum	Superalloy	High-temperature engine components
Tungsten	Superalloy	High-temperature engine components

Fig. 2.3.2 Selected Various Metal-Matrix Composite Materials

	Polymeric Matrix	Metal matrix
Unidirectional Material usage	Not recommended (Handling & general damage tolerance a major concern)	Can be used for components subjected to basically uniaxial loading
Moisture Exposure	A concern in strength critical & dimensionally stable structures	Not critical under most usage conditions
Temperature capability	Poor	Good
Nonlinear effects	Usually not important in typical design considerations	Plasticity of matrix has pronounced effected due to it strong influence on mechanical behavior
Stiffness and strength anisotropy	High E_{11}/E_{22} & E_{11}/G_{12} and associated strength ratios can create design problems	Stiffness anisotropy usually relatively weak but transverse tensile strength may limit applicability of unidirectional material

Fig. 2.3.3 Polymer Matrix Composites Vs. Metal Matrix Composite Design

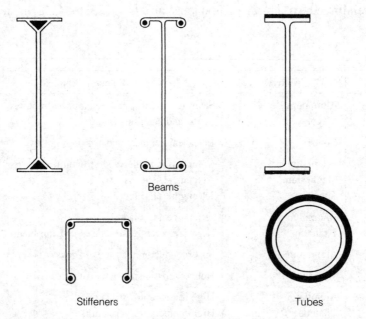

Examples of boron reinforcement (shown in solid black) in metal structures

Fig. 2.3.4 Boron-reinforced Aluminum Design Cases

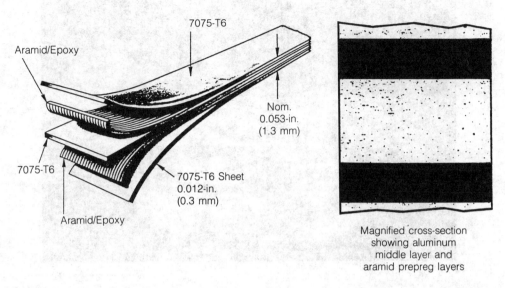

Fig. 2.3.5 ARALL® Aluminum Laminate

ARALL® (ARamid Aluminum Laminate) is made by bonding layers of thin sheets of aluminum with epoxy matrix reinforced with aramid fibers, as shown in Fig. 2.3.5, by prestressing the material with the aramid fibers. ARALL combines the strength and fatigue resistance of polymer composites with aluminum's machineability and formability.

(2) CARBON/CARBON (C/C) MATRIX COMPOSITES

Carbon matrix composites, developed specifically for parts that must operate in extreme temperature ranges, are composed of a carbon matrix reinforced with carbon yarn fabric, 3-D (three-dimensional or three-directional) woven fabric, 3-D braiding, etc. depending upon application (see Section 2.5). They have been used on aircraft brakes, rocket nozzles and nose cones, jet engine turbine wheels, high-speed spacecraft, and other planetary exploration spacecraft. The following describes a few of the current applications:

(a) Aircraft brakes (see Fig. 2.3.6) — The rapid deceleration required for a landing aircraft generates a considerable amount of frictional heat. C/C composite brakes retain strength at high temperatures. Unliked steel brakes, C/C maintain more consistent performance over the life of the part, with no increase in stopping distance. C/C has a high heat absorption capacity, so it can act as a lightweight heat sink and can endure thousands of thermal cycles with little or no fatigue. In addition, its resistance to wear outlasts steel brakes by two fold, which means fewer overhauls and lower maintenance costs.

By courtesy of B. F. Goodrich

Fig. 2.3.6 C/C Composites Brakes

(b) Rocket Nozzles (see Fig. 2.3.7) — Hot gases rush through the nozzles at extremely high velocity, stressing (up to 300 ksi (2.1 Gpa)) and eroding the nozzle walls; C/C composites can resist erosion and abrasion with very little burning away. However, rocket nozzles are not typically reused, so they are not often coated for oxidation protection and are allowed to partially burn. This burning must be taken into account in the design of the nozzle.

(c) Rocket nose cone — Similar to a rocket nozzle, the nose cone and leading edge of the space craft must endure the searing heat and high stressing from the atmosphere during reentry, up to more than 3000°F (1650°C). The cone must also endure solar radiation and the attack of atomic oxygen. C/C is very good at resisting thermal shock which occurs during the rapid transistion from less than $-200°F$ ($-129°C$) in space to more than 3000°F (1650°C) during reentry, and allows the nose cone to remain dimensionally stable over this wide temperature range under high stress.

(d) Jet engine turbine wheel (see Fig. 2.3.8) — Advanced materials for jet engines must withstand static and dynamic loads caused by high and varying centrifugal forces, as high as more than 200,000 g's at high temperatures (at 40,000 RPM and more than 3000°F (1650°C)), and chemically aggressive combustion gases. The higher combustion temperature permits greater efficiency and performance, while dictating engine size, weight, and fuel consumption. To prevent burning and oxidation, a coating (e.g., ceramic), which does not bear structural loads, is critical for C/C composites.

(e) Future National Aerospace Plane (NASP) program (see Fig. 1.1.4 (a) of Chapter 1.0) — The NASP will fly at speeds up to 17,000 MPH (27350 km/h) and hypersonic velocities will produce aerodynamic heating, resulting in surface temperatures higher than most metals can endure.

By courtesy of Hercules INC.

Fig. 2.3.7 C/C Rocket Motors

By courtesy of LTV Missiles and Electronics Group

Fig. 2.3.8 C/C Jet Engine Turbine Wheel (15 inch Dia.)

C/C composites meet applications ranging from aircraft to aerospace because of their ability to maintain and even increase their structural properties at extreme temperatures. The characteristics and advantages of C/C are described below:

- Extremely high temperature resistance [up to 3500 — 5000°F (1930 — 2760°C)], see Fig. 2.3.9
- Strength actually increases at higher temperatures [(up to 3500°F (1930°C)] which would be devastating to metal
- High strength and stiffness
- Unaffected by sharp temperature variations
- Holds dimensional stability at high temperatures
- Good ablative qualities
- Ablates evenly thus maintaining structural properties
- Good mechanical properties
- Good resistance to thermal shock
- Carbon fibers, when used with existing 2-D or 3-D textile technology (see Fig. 2.3.10 and Section 2.5) results in reinforcing materials that are functionally superior.

Typically, C/C composites are fabricated from both 2-D laminates 3-D preforms (see Section 2.5 for details). Although most 3-D preforms are made with dry fibers, one technique is to braid phenolic prepreg. After being braided, the preform is cured. Then the freestanding structure is densified. The following briefly discusses preform constructions:

- 2-D preform — Use of woven fabric or unidirectional tape cross-plied in the X and Y planes. For additional strength, reinforcement is added in the third dimension in the form of fibers in the "Z" plane
- 3-D preform — Like cylindrical shapes, the fibers are interwoven axially, radially, and circumferentially. 3-D preforms are most important in C/C composites because the carbon matrix is inherently brittle and a 3-D preform adds toughness.
- Multi-directional preform — Most multi-directional preforms used for C/C composites are represented by the orthogonal or polar constructions or by some modification of these constructions.

C/C processing can be divided into six primary cycles:

- Layup
- Cure
- Pyrolysis
- Impregnation
- Coating
- Sealing

Manufacturing of C/C composites utilizes special forming processes and equipment and at present manufacturing C/C composites is an extremely time-consuming process, requiring slow pyrolyzing to drive off gases without cracking the matrix.

Two primary C/C processes, liquid impregnation and gaseous infiltration, have been developed to fabricate C/C composites. A manufacturing flow chart for C/C composites is shown in Fig. 2.3.11.

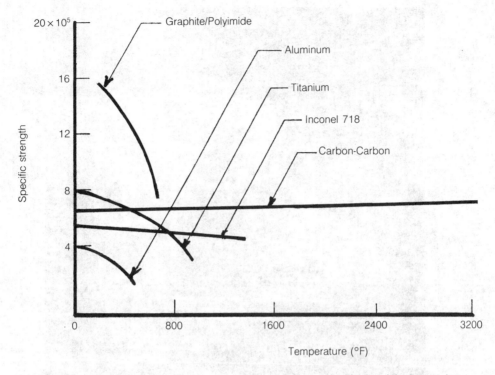

Fig. 2.3.9 Tensile Strength Vs. Temperature Curves of Conventional Materials Composites

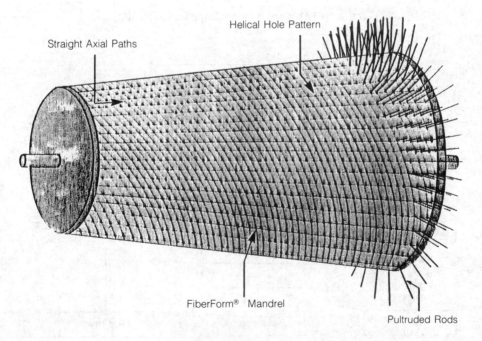

Fig. 2.3.10 3-D Weaves for C/C Application

Chapter 2.0

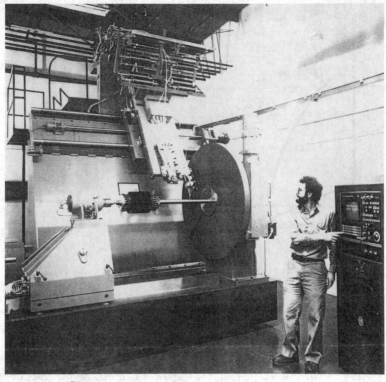

ULTRALOOM™ *By courtesy of Fiber Materials Inc.*

Fig. 2.3.10 3-D Weaves for C/C Application (cont'd)

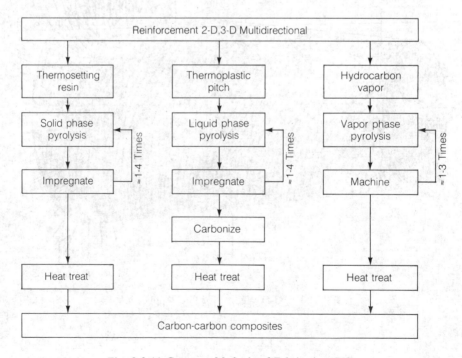

Fig. 2.3.11 Common Methods of Fabricating C/C

C/C composites can be manufactured by one of the following methods:
(a) Liquid impregnation — Carbon fiber prepregs are fabricated into a desired shape or dry fiber is laid up into a preform and then impregnated with a liquid matrix (or resin) or pitch. Densification of fiber preforms is accomplished by impregnation and carbonization, and sometimes graphitization, of thermosetting resins such as phenolics or thermoplastic or petroleum based pitches. The process may be done under various pressures and temperatures depending on the precursor, desired carbon yields, density, and part shape and thickness. Examples of C/C composite manufacturing procedures used to manufacture a 3-D C/C radially pierced nozzle billet are shown in Fig. 2.3.12. An automated weaving system for the C/C exit cone of a solid rocket, prior to densification, is shown in Fig. 2.3.13.
(b) Chemical Vapor Infiltration (CVI) — CVI is used to densify graphite fiber preforms starting with a hydrocarbon gas, such as methane, as a precursor. All of these involve multiple cycles to achieve final densification, see Fig. 2.3.14. High pressure pitch impregnation processes with high final densities are favored for ablative products. The ablative capacity is a direct function of mass, so high density is a desirable physical property for this application. High density is not a necessary prerequisite for high strength and stiffness. In fact, high mechanical properties and relatively low density are achieved with a phenolic matrix impregnation/CVI follow-up process.

Additional work has to be done after C/C densification processing:
(a) Part is machined to final configuration
(b) C/C parts must be coated with silicon carbide and sealed with silicon glass to resist oxidation at high temperatures

Opportunities for broader use of C/C composites are evidenced by the requirements established for a variety of future aerospace vehicles and systems such as the previously mentioned NASP (see Fig. 1.1.4(a) in Chapter 1.0) which is expected to experience surface temperatures approaching 1700°F (930°C) with the nose leading edge temperatures possibly twice as hot. Because of their unique properties, C/C composites should become a leading candidate for high-speed spacecraft and reentry vehicles and other high temperature applications.

(3) CERAMIC MATRIX COMPOSITES

Ceramics in general are characterized by high melting points, high compressive strength, good strength retention at high temperatures, and excellent resistance to oxidation. As the aerospace industry seeks to build stronger, lighter weight, and more fuel-efficient aircraft, designers are turning to a whole host of new materials to fulfill these requirements. Ceramics have been used for the braking systems of commercial and military aircraft, including the F-16, Space Shuttle, B747-400, A320, etc. Some companies have produced preprototype aircraft engine parts, missile guidance fins, and prototype hypersonic fuselage skins made of ceramic composites. Ceramic composites also offer excellent oxidation resistance. In addition, they are made from preforms to net or near-net shape, thus requiring little post-machining.

Chapter 2.0

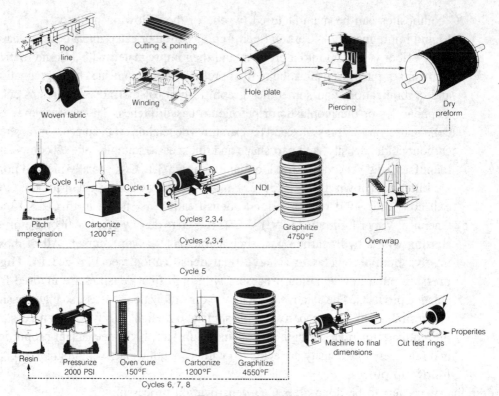

(a) Flow Chart of Fabrication

(b) Finish part of a c/c billet

By courtesy of Textron Inc.

Fig. 2.3.12 3-D C/C Radially Pierced Nozzle Billet Fabrication

(a) Automated system

(b) A cone as it is removed from the system

By courtesy of Textron Inc.

Fig. 2.3.13 Automated Weaving System

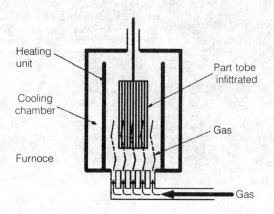

Fig. 2.3.14 C/C Chemical Vapor Infiltration (CVI) Process

Before ceramic composites can be successfully produced for applications, aerospace engineers must learn how to design with these materials. Ceramics cannot be joined with conventional fasteners, or machined as easily as metals. Therefore, engineers have turned to large furnaces in which one-piece shapes can be made to near-net or net shape, eliminating the need for joining and the need for much post-machining. For example, a high temperature integral component of a Hermes leading section (shown in Fig. 2.3.15) was produced from C/SiC, and an engine rotor (shown in Fig. 2.3.16) was produced by briading SiC fibers into 3-D preforms for subsequent infiltration of the SiC-matrix by chemical vapor deposition. The braided fiber/ceramic radome used on Patriot missiles (see Fig. 2.3.17) is also manufactured in one step.

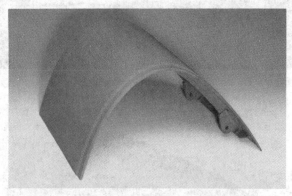

By courtesy of Snecma Group

Fig. 2.3.15 C/SIC Leading Edge. This ½ Scale Hermes (European Spaceplane) Leading Edge Section Has Been Tested at 3200°F.

By courtesy of Snecma Group

Fig. 2.3.16 High Temperature SiC/SiC 3-D Preform Rotor (2370°F or 1300°C)

Materials

By courtesy of Martin Marietta Corp.

Fig. 2.3.17 A Ceramic Radome Combining Heat Resistance with Radar Signal Permeability Is Used on Patriot Missiles

The challenge here will be developing furnaces big enough to handle large parts and improving processes to create the optimum ceramic composites. On a per pound basis, it will be expensive, but the payoff will be better on a systems basis. If ceramic parts that operate at high temperatures are lighter weight and are more efficient than other counterparts, then the production costs will be paid back in the long run.

The dimensional stability shown by ceramics at high temperatures may prove beneficial for aerospace applications. Their dimensional stabiltiy is ever better than that of metal metrix composites, and in the form of reinforced glass ceramics they are very tough. They can be manufactured much like graphite or carbon/epoxy composites. Dimensional stability makes glass ceramic composites particularly applicable for structures used in spacebased optical systems, where accurate operation depends on virtually no change in dimensions under varying temperatures. Another use is in low observables because ceramic composites have lower radar detectability than other composites.

There are a number of manufacturing processes for producing ceramic composites, including:
- chemical vapor deposition
- viscous glass consolidation
- polymer conversion
- powder and hot press techniques
- and gas-liquid metal reaction.

The weaknesses of ceramics include relatively low tensile strength, poor impact resistance, and poor thermal-shock resistance. The addition of a high-strength fiber to a relatively weak ceramic does not always result in composite with a tensile strength greater than that of the ceramic alone. That is, at stress levels sufficient to rupture the ceramic, the elongation of the matrix is insufficient to transfer a significant amount of the load to the reinforcement, and the composite will fail unless the volume percentage of the fiber is extraordinarily high.

Therefore ceramic matrices are usually chosen for their ability to be processed without cracking. This requires a coefficient of thermal expansion that is close to that of the reinforcement. Approach to improving ceramics is to toughen the ceramic using whiskers or chopped fibers to reinforce the matrix. The aim is to retain the thermal strength, hardness and wear resistance which makes ceramics so desirable, while greatly increasing toughness, making failure more predictable and improving producibility.

A number of important technological barriers need to be overcome before advanced ceramics can achieve their full potential. Matrix brittleness, shrinkage associated with sintering, reactivity, and the generation of internal stresses due to thermal mismatch have proven to be only a few of the problems associated with the development of ceramic matrix composites. Difficulties also lie in the areas of production costs, reliability in service, and reproducibility in manufacture. Over all, overcoming brittleness has been the major stumbling block in the development of new ceramic products and applications. With the demand for high-temperature performance products, the future of advanced ceramics is growing and solutions to these challanges are being pursued.

2.4 REINFORCEMENTS (FIBERS)

While composite materials owe their unique balance of properties to the combination of matrix and reinforcement, it is the reinforcement system that is primarily responsible for such structural properties as strength and stiffness. The reinforcement dominates the field in terms of volume, properties, and design versatility. Almost all fibers in use in airframe structures today are solid and have a circular cross section or nearly so. Generally, the smaller the diameter the greater the strength of the fiber. Other potential shapes such as polygonal, hexagonal, rectangular, irregular, and unusual shapes, as shown in Fig. 2.4.1, are under development with possibilities for improved fiber strengths. Hollow fibers have been developed, are commercially available, and show promise for improved mechanical properties of composites, especially in compressive strength.

Composite reinforcement fibers are more expensive than current aluminum materials and represent a high percentage of the recurring cost in composite components. The following list gives a fiber cost comparison:

FIBER	$/lb.
Aluminum (for comparison)	1 — 5
Fiberglass	3 — 5
Kevlar-49®	10 — 20
Quartz	120

Standard PAN based [33 Msi (227 Gpa) modules] fiber
  1000 tow   140
  3000 tow   40
  6000 tow   30
 12000 tow   25
Intermediate modulus fiber, 40-50 Msi (275-345 Gpa) modulus
  6000 tow   85
 12000 tow   45
Ultra high modulus fiber, >70 Msi (483 Gpa) modulus   300 — 700
Boron fiber   more than 200
(Note: Fiber cost comparison — Reference only, 1990 prices)

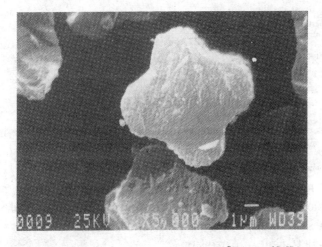

By courtesy of BASF Inc.

Fig. 2.4.1 Unusual Cross-sections Give Superior Compression and Shear Strength

Because fiber materials have borrowed some of the terminology of the textile industry, the following terms are defined:
- Filament — the basic structural fibrous element. It is continuous, or at least very long campared to its average diameter, which is usually 5 — 10 microns.
- Yarn — A small, continuous bundle of filaments, generally fewer than 10,000. The filaments are lightly stranded together so they can be handled as a single unit and may be twisted to enhance bundle integrity.
- Tow — A large bundle of continuous filaments, not twisted. The number of filaments in a bundle is usually 3000, 6000, or 12000 (3K, 6K or 12K tow). 12K tow is the cheapest, 3K tow is the most expensive (see list above). The smaller tow sizes are normally used in weaving, winding, and braiding applications while large tow sizes are used in unidirectional tapes. Very thin tapes are also made from low filament-count tow for satellite applications.

- Graphite
  Most widely used for primary structural applications; best balance of properties, cost, handleability; wide range of fiber types and properties
- Aramid (Kevlar-49)
  Uses limited by low compression strength and poor bond to resins; lighter weight and lower cost than graphite; good impact resistance
- Boron
  Very high cost, difficulty in handling have limited boron to specialty use where its combination of high strength and stiffness is needed
- Fiberglass
  Much lower stiffness has limited use to secondary structures

(a) Fiber selection summary

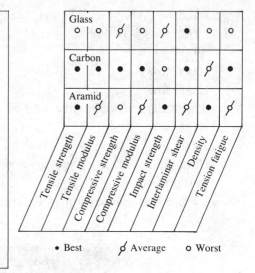

(b) Fiber comparison

Fig. 2.4.2 Composite Fiber Comparison

- Fabric — Fabric is a planar textile structure produced by interlacing yarns, fibers, or filaments.

The function of the reinforcement (continuous fibers) includes:
- principal load-carrying member of the composite
- responsible for tensile, compressive, flexural strength and stiffness of the composite
- determines electrical properties
- thermal coefficient of expansion

Reinforcement fibers are generally one of three types:
- Organic fibers — Organic fibers, like organic matrices, offer high strength and light weight, e.g., glass, aramid, PE, graphite carbon, etc.
- Ceramic fibers — Ceramic fibers can withstand high temperatures and insulate against heat, e.g., quartz, silicon carbide, alumina, etc.
- Metallic fibers — Metallic fibers permit composite to conduct or dissipate heat and electricity

Discontinuous reinforcement fiber materials provide:
- Cost effective production methods
- High dimensional accuracy
- Rapid cycle times
- High tooling cost
- Lower properties than continuous fibers
- Possibility for complex shapes
- Faster manufacturing methods

Fig. 2.4.2 shows the comparison of several common composite reinforcements while detailed descriptions are given.

A fiber properties comparison which includes all fibers discussed above is given in Fig. 2.4.3.

Fig. 2.4.4 shows selected material properties of commonly used fibers in airframe applications.

Fiber	Diameter (microns)	Density (lb/in^3)	Tensile strength (ksi)	Modulus (Msi)	Service temp. (°F)
S-glass	7	.09	500 - 650	13	600 - 700
Aramid	12	.052	400	10 - 25	500
PE	27-38	.035	375 - 430	17 - 25	230
Carbon	7	.06	350 - 450	33 - 55	1000
Quartz	9	.079	500	10	2000
SiC	10 - 20	.083-.094	400	28	2400
Alumina	20	.141	200 - 300	55	1800
Boron	50 - 200	.09	500	58	3500

Fig. 2.4.3 Properties Comparison of Various Fibers

Fiber ident	Fiber type	Fiber modulus (Msi)	Fiber tensile strength (ksi)	Elongation to break	comments
Kevlar 29	Aramid (Kevlar)	9-12	400	3.6	• Exceptionally tough, with good impact, chemical and fatigue resistance
Kevlar 49		17-19	525-600	2.5	• Lighter weight and lower cost than graphite • Use limited by low compression strength and poor bond to resins
Kevlar 149		27	500	2.0	• Very hard to machine and finish with conventional tools
E-glass	Glass	10.5	500	4.8	• High strength and light weight but low stiffness limits use to secondary structures • Good damage resistance
S-glass		12.4	665	5.7	• Comparatively high fiber density
Boron	Boron	58	510	—	• Limited to specialty use where high strength strength and stiffness is needed • Very high cost • Difficult to handle and machine • Poor impact resistance

Fig 2.4.4 Commonly Used Fibers

Fiber ident	Fiber type	Fiber modulus (Msi)	Fiber tensile strength (ksi)	Elongation to break (%)	Comments
P-55	Graphite	55	300	0.5	Widely used for primary strutcutal
P-75	(pitch	75	300	—	applications
P-100	based)	105	325	0.31	
T300		34	450	1.4	Best balance of properties, cost and handleability
T650-42		42	730		Wide range of fiber forms and properties
T650-50	Carbon	50	650		Difference between graphite & carbon fibers:
T650-66	(PAN	60	650		graphite fibers:
30-500	based)	34	550	1.62	• tensile modulus > 50 Msi (345 GPa)
G-40-600		44	620	1.43	• high orientation
G-40-700		44	720	1.66	• processed > 3100°F (1700°C)
G-45		47	658	1.39	carbon fibers:
Apollo IM 53-750		53	830	1.55	• tensile modulus > 50 Msi • lower orientation • processed < 3100°F(1700°C)
Apollo IM 55-550		55	500	0.91	
Apollo IM 55-500		55	435	0.74	
Apollo 43-750		43	820	1.90	
AS4		35	580	1.5	
IM6		40	635	1.5	
IM7		44	—	1.8	
IM8		45	750	1.66	
Magnamite UHM		64	550	0.80	
Hitex 46-8B		46	825	1.7	
Hitex 50-8B		50	840	—	
M-30		43	560	—	

Fig. 2.4.4 Commonly Used Fibers (cont'd)

(1) FIBERGLASS

The most widely used fiber is unquestionably fiberglass, which has gained acceptance because of its low cost, light weight, high strength, and non-metallic characteristics. Fiberglass composites have been widely used for aircraft parts that do not have to carry heavy loads or operate under great stress. They are used principally for fuselage interior parts such as window surrounds and storage compartments, as well as for wing fairings and wing fixed trailing edge panels. Fiberglass is extensively used in primary structures of sport and utility aircraft as well as helicopter rotor blades.

The two most common grades of fiberglass are "E" (for electrical board) and "S" (high strength for structural use). E-glass provides a high strength-to-weight ratio, good fatigue resistance, outstanding dielectric properties, retention of 50% tensile strength to 600°F (320°C), and excellent chemical, corrosion, and environmental resistance. While E-glass has proved highly successful in aircraft secondary structures, some applications require higher properties. To fill these demands, S-glass was developed, which offers up to 25% higher compressive strength, 40% higher tensile strength, 20% higher modulus, and 4% lower density. This glass also has higher resistance to strong acids than E-glass, and more costly.

The use of other glass types such as A-glass, C-glass and even D-glass has been limited because they are lower strength and not suitable for structural purposes.

Designing with fiberglass is much simpler than designing with some other composite systems because of the large volume of empirical data collected over the years and the availability of standard systems with well-documented properties from many manufacturers. Fig. 2.4.5 shows the properties of E and S-glass.

Hollow glass fibers used in certain applications have demonstrated improved structural efficiency where stiffness and compressive strength are the governing criteria. The transverse compressive strength of a hollow fiber is lower than that of a solid fiber. Hollow fibers are quite difficult to handle as they break easily and may absorb moisture.

(2) KEVLAR® (Aramid fiber-Dupont product)

Kevlar fiber has been used for structural applications since the early 1970s. Combining extremely high toughness and energy-absorbing capacity (very good projectile and ballistic protection characteristics has led to use in bullet-proof vests), tensile strength, and stiffness with low density (the lowest in recently developed advanced composite materials), Kevlar fiber offers very high specific tensile properties (see Fig. 2.4.6). Low compressive strength is one of the weaknesses of Kevlar. But where the highest compressive strength is needed, hybrids of Kevlar and carbon fibers are generally used.

As shown in Fig. 2.4.6, Kevlar have very high specific tensile strength. This provides the basis for the claim that Kevlar, on a pound-for-pound basis, is five times as strong as steel. Fig. 2.4.7 illustrates tensile stress/strain curves for tensile loading. Like most other composite materials, Kevlar has a classically brittle response with a tensile strength a little greater than 200 ksi (1.38 Gpa) and tensile modulus of 11 Msi (76 Gpa) for a typical unidirectional composite. When Kevlar is under compression, the behavior is quite different from the tensile response. At a compressive load about 20% of the ultimate tensile load, a deviation from linearity occurs. This is an inherent characteristic of the Kevlar 49 fiber representing an internal buckling of the filament. This unusual characteristic of Kevlar 49 fiber has made fail-safe designs possible because fiber continuity is not lost in a compressive or tensile failure. It has also limited the use of the fiber on major structural applications.

When tensile and compressive loadings are combined in flexural bending, instead of the brittle failure encountered with glass and carbon fibers, the bending failure of Kevlar 49 is similar to what is observed with metals (see Fig. 1.3.4 of Chapter 1.0). This helps to explain the outstanding toughness and impact resistance of composites reinforced with Kevlar.

Another area in which Kevlar excells is vibration damping; Fig. 2.4.8 shows the decay of free vibrations for various materials, Kevlar is less prone to flutter and sonic fatigue problems than most other materials, Kevlar fibers also offer good fatigue, and chemical resistance, and retain their excellent tensile properties to relatively high temperatures. Because of their high specific properties and fewer handling problems, these fibers have replaced glass fibers in many applications. However, relatively low compressive strength has kept them out of many aircraft primary structures. New methods for machining Kevlar are also needed because the fibers are too tough to cut with conventional tools.

(3) POLYETHYLENE (PE)

Polyethylene (PE) fiber resists impact better than glass or carbon. PE/epoxy shows a low dielectric constant: 30% less than that of aramid/epoxy, and half that of glass/epoxy. Like aramid (previously mentioned as Kevlar), PE is resistant to projectile penetration leading to use in ballistic protection for armor, radomes, etc. PE fibers melt at relatively low temperatures up to 230°F (110°C) and they absorb very little moisture. Typical types of PE fibers are Spectra 900 and Spectra 1000.

- Spectra 900   — specific strength is 35% greater than glass or aramid, twice that of carbon
  - — specific modulus is less than carbon that of, but 30% greater than that of aramid and three times that of glass
- Spectra 1000 — It is 15% stronger and 40% stiffer than Spectra 900
- Polyethylene is difficult to "wet" and processing is in the development stage

Property	E-glass	S-glass
Specific Gravity	2.6	2.49
Density (lb/in^3.)	0.092 (2.55 g/cc)	0.09 (2.49 g/cc)
Tensile strength (monofilament) (ksi)	500 (3.45 Gpa)	665 (4.6 Gpa)
Modulus (Msi)	10.5 (72.4 Gpa)	12.6 (87 Gpa)
Elongation (%)	4.8	5.4
Thermal (bulk glass)		
coefficient expansion (80°F-212°F) (10^{-6} in./in./°F)	2.8	3.1
specific heat	0.192	0.176
Conductivity, BTU/ft-hr-°F	0.56	—
Electrical (bulk glass)		
Dielectric constant 10 KHz, 75°F (24°C)	6.13	5.21
Less tangent 10 KHz, 75°F (24°C)	0.0039	0.0068

Fig. 2.4.5 Properties of E versus S-glass

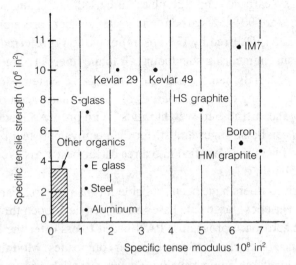

Fig. 2.4.6 Specific Tensile Strength Vs. Specific Modulus of Composite Fibers.

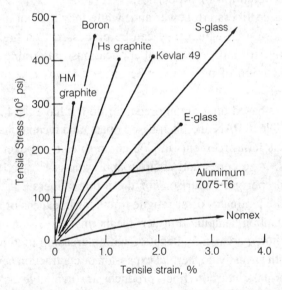

Fig. 2.4.7 Stress and Strain Curves of Various Fibers/epoxy.

Material	Loss fact $\times 10^{-4}$
Cured polyester resin	400
Kevlar 49/Epoxy	180
Fiberglass/Epoxy	30
Graphite/Epoxy	17
Stainless steel	6

Fig. 2.4.8 Loss Factor From Vibration Decay.

(4) CARBON (or Graphite)

Although the names "carbon" and "graphite" are used interchangeably when describing fibers, carbon fibers are 93-95% carbon, and graphite fibers are more than 95% carbon. Use of either graphite (preferred by U.S. users) or carbon (preferred by European users) is acceptable from an engineering standpoint. Graphite fibers are among the strongest and stiffest composite materials being combined with matrix systems for high-performance structures. The outstanding design properties of carbon/matrix composites are their high strength-to-weight and stiffness-to-weight ratios. With proper selection and placement of fibers, composites can be stronger and stiffer than equivalent steel parts at less than half the weight. Carbon fibers are classified into three categories; PAN-polyacrylonitrile, pitch, and rayon-based fibers.

(a) PAN-derived fibers have been available for many years. For several of the lower modulus varieties, large data bases have been developed through their use in aircraft and aerospace programs. PAN-based fibers offer the highest strength and best balance of mechanical properties in composites. Moduli up to 130 Msi (897 Gpa) are available commercially. They are available in standard, intermediate, and high modulus grades. These fibers are generally selected for their high strength and efficient retention of properties.

(b) The pitch-based fibers are newer and, while they are not as strong as the PAN fibers, the ease with which they can be processed to a higher modulus makes them attractive for stiffness-critical applications. Favorable cost projections for volume production of pitch fibers because of lower raw material cost has not been realized.

(c) Carbon fibers based on a rayon precursor do not have the high mechanical properties available in PAN and pitch-based fibers, and recently have been used almost exclusively as reinforcements in C/C composites for rocket nozzle throats, aircraft brakes, nose cones and ablative applications.

Fig. 2.4.9 shows a property comparison for these three fibers.

Carbon composite laminates offer fatigue limits far in excess of aluminum, or steel, along with superior vibration damping. Further, the thermal expansion coefficients of carbon composite fibers become increasingly negative with increasing modulus. This allows the design of structures with virtually no thermal expansion or contraction across ranging thermal cycles. As with fiberglass, carbon fiber products are available as prepreg, molding-compound, and other standardized product forms.

Carbon or graphite is generally available in three forms:
(a) HTS — High tensile strength fiber
  $F = 350$ ksi (2.4 Gpa) and $E = 30$ Msi (207 Gpa)
(b) HM — High modulus fiber
  $F = 200$ ksi (1.35 Gpa) and $E = 50$ Msi (345 Gpa)
(c) UHM— Ultra high modulus fiber
  $F = 150$ ksi (1.03 Gpa) and $E = 70$ Msi (483 Gpa)

Fig. 2.4.10 gives typical carbon (or graphite) fiber properties.

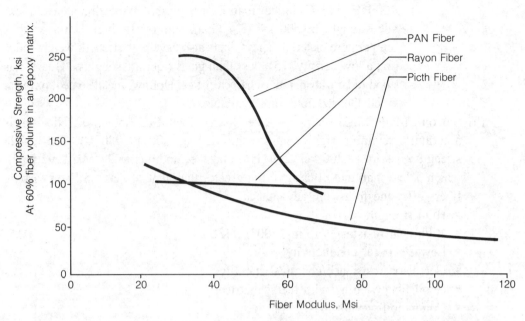

Fig. 2.4.9 Compressive Strength Vs. Modulus Curves of Three Most Common Carbon Fibers

Typical TOw properties	Type							
	AS2	AS4	IM6	IM7	XIM8	HMS4	HMU	XUHM
Filaments/TOW	6K 12K	3K 6K 12K	6K 12K	12K	12K	12K	1K 3K 6K	12K
Tensile Strength, ksi*	400	580	745	785	750	400	400	550
Tensile Modulus, Msi*	33	33	39.5	40	45	48	52	62
Tensile Strain, %*	1.3	1.6	1.75	1.85	1.63	0.8	0.7	.75
Approximate Yield ft./lb.	3260 1630	6800 3400 1700	6380 3190	3190	3360	1700	19936 6645 3323	4300
Density, lb./in^3	0.065	0.065	0.063	0.065	0.065	0.064	0.067	0.068
*Typical tow value, ksi=psi×10^3; Msi=psi×10^6								

Source: Hercules Incorporated

Fig. 2.4.10 Carbon Fiber Properties

(5) CERAMIC FIBERS

When the matrix is metal or ceramic, temperatures associated with component fabrication and use can be extremely high. For these applications, ceramic fibers are typically the material of choice.

(a) Quartz — fibers can be used continuously to over 1900°F (1040°C), more than 1000°F (540°C) higher than E or S-glass. Quartz fibers have a tensile strength of 500 ksi (3.4 Gpa); one of the highest for a high-temperature material. These fibers also have high strength/weight ratios and a low density, 15% less than glass fibers. Like glass, Quartz has good radar transparency characteristics. Hollow, metal coated, or metal filled Quartz fibers are available.

(b) Silicon carbide (SiC) fibers — Like Quartz fiber, SiC fiber (i.e., Nicalon) is a ceramic, retaining significant strength above 1800°F (980°C). The tensile strength goes up to 400 ksi (2.7 Gpa) and the modulus to 28 Msi (192 Gpa), which is less than that of carbon, but greater than that of glass. Silicon carbide fibers offer the following advantages:
- High strength
- High heat resistance up to 2200°F (1200°C)
- Low electrical conductivity
- Corrosion resistance/chemical stability
- Good fiber wetting with metal matrices
- Compatibility with plastic matrices

The excellent wetting properties and oxidative resistance of SiC fibers allow them to reinforce polymers, polyimides, ceramics, metals (i.e., aluminum, titanium, magnesium, etc.), and even carbon matrices.

(c) Alumina — Alumina fiber has modulus of 55 Msi (380 Gpa); 40% greater than that of SiC fibers. Its service temperature is over 1800°F (980°C). Pure alumina is relatively brittle, but it is well suited for reinforcing aluminum matrices.

(6) BORON

Boron fibers (fiber diameter from 50, 100, 140 and 200 microns) are formed by chemical vapor deposition (CVD) of boron gas onto a tungsten filament (see Fig. 2.4.11). Boron fiber is stronger than carbon, with a tensile strength of 500 ksi (3.4 Gpa). Its modulus is 58 Msi (400 Gpa); greater than that of carbon, and five times that of glass. It can be mixed with organic and metal matrices to form lightweight materials. Boron fibers have found limited use although their high stiffness has made possible the early use of composites in primary aircraft structures. So far, the relatively high cost and large fiber diameter of boron have kept this fiber from high volume usage.

A major drawback of boron fibers is the difficulty in handling the material. The fibers are extremely stiff and brittle, difficult to work with, and limit the minium radius around which they can be wrapped. Another drawback involves ply thickness, which is determined by the filament diameter and desired fiber volume fraction. Most airframe structures require thin skins and need thin plies to tailor the ply orientation to optimize the light weight. But boron filaments are an order of magnitude larger than carbon fibers. The thickness of individual plies will always exceed fiber diameter in unidirectional laminates unless the fiber volume fraction is sufficiently low to permit nesting.

Boron epoxy has also become very valuable in the field of aircraft maintenance. It is commonly used as a patch to extend fatigue life and prevent crack propagation in many aircraft repairs.

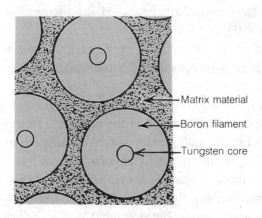

Fig. 2.4.11 Typical Boron Fibers

(7) PARTICLES

Particles can constitute either a major or a minor part of the composition of a composite. The structure of particles can range from irregular masses to precise geometrical forms such as spheres, polyhedrons, or short fibers. The following functions can be accomplished by using particles:
- To change the overall properties of the composite
- To help strengthen the composite
- To reduce weight
- To reduce the quantity of matrix used
- To change properties, such as electrical conductivity

(8) HOLLOW MICROSPHERES

The primary use of hollow microspheres (microballoons) is to reduce the weight of the matrix system. Commercial products are available in densities from

0.09 g/cc (.0032 lb/in^3) to .22 g/cc (.008 lb/in^3) — silicate-based; diameter from 65 to 75 microns

0.15 g/cc (0.0054 lb/in^3) to .38 g/cc (.0137 lb/in^3) — glass

0.5 g/cc (.018 lb/in^3) to 0.7 g/cc (.0253 lb/in^3) — ceramic aluminosilicate

When hollow microspheres were first introduced, their major application areas were limited to thermosetting matrix systems. Primarily, this was a function of the low molding pressures required for thermosets. However, stronger microspheres, such as ceramic aluminosilicate, are available which are up to five times stronger than the standard grades and these may be used with thermoplastic matrix systems.

(9) WHISKERS

Whiskers are short single crystal fibers or filaments used as a reinforcement in a matrix. Diameters range from 1-25 microns with an aspect ratio between 100 and 1500. Whisker composites resist crack growth better than monolithics, but not as well as continuous fibers. Most whiskers have a small diameter and short length, so they are randomly oriented. Therefore, it is difficult for the composite to achieve the full strength because the load must be transferred from whisker to whisker within the matrix. The strength of whisker-reinforced composites exceeds that of particle reinforcement, but is less than that of continuous fibers.

As long as the whiskers are parallel, whisker composites can provide uniform mechanical properties in the plane of the whiskers. However, whiskers can be difficult to line up parallel to one another in a matrix, resulting in uneven strength. Although properties approaching isotropic can be obtained in continuous fiber composites, angle plying is required to produce quasi-isotropic properties. Whisker-reinforced composites can be fabricated into complex shapes and processed like monolithic materials to save cost, while still maintaining reasonable strength. Ceramic and carbon are the most commonly used whisker materials. Other common whisker materials are alumina, sapphire, silicon, silicon nitride, and silicon carbide.

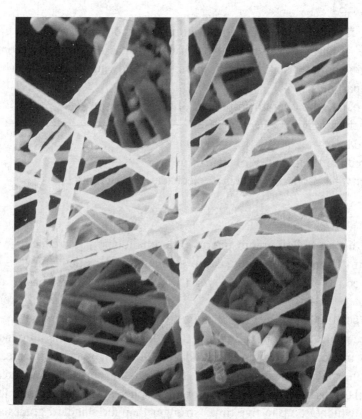

(American Matrix Whisker, Photo Courtesy of Dr. Sam C. Weaver of Third Millennium Technologies, Inc.)

Fig. 2.4.12 SiC Whisker (0.5+03 microns in diameter)

Submicron whiskers have diameter from 0.002 to 0.05 microns with an aspect ratio of 50 to 300 which permits processing at higher speeds and greater injection pressure. These whiskers can be dispersed more uniformly, leading to more isotropic mechanical properties and more isotropic mold shrinkage. The following are the features of silicon carbide whiskers (see Fig. 2.4.12):
- High modulus and high strength
- Good wetting properties with metal matrices such as aluminum and magnesium alloys
- High temperature stability
- Chemically resistance

2.5 MATERIAL FORMS

In this chapter the end products used directly on airframe structures are described rather than the description of how to produce these preforms, which is beyond the scope of this chapter. A general classification of fiber reinforced composites is shown in Fig. 2.5.1. Common use of material forms and a general classification of the composite forms is shown in Fig. 2.5.2.

Fiber/matrix combinations are available in a wide variety of reinforcement forms which allow for great flexibility in design and manufacturing. Most forms can be obtained either "dry" or "preimpregnated" with the desired matrix. Dry forms require some method of applying the matrix during the lay-up process, while the more common "prepreg" forms required no extra matrix application.

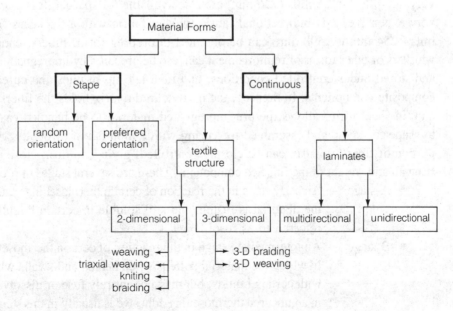

Fig. 2.5.1 General Classification of Fiber Reinforcements

Forms	Advantages	Disadvantages
Uni-tape	Maximum structural properties, design flexibility	Poor drapability, possible fiber misalignment
Woven fabrics	Good drapability, reduced lay-up costs	Some loss of properties due to fiber crimp, width limitations, less design flexibility
Unidirectional Fabrics	Improved drapability, fiber alignment, minimal reductions in fiber strength	Slight weight penalty
Pre-plied fabrics stitched fabrics, pre-forms	Reduces lay-up costs Provides exceptional fiber stability need for pultrusion, resin injection molding; some forms provide "Z" direction strength	Loss of design flexibility Weight penalty, loss of design flexibility, increased cost

Fig. 2.5.2 Common Use of Composite Material Forms

Prepregs are available in a variety of types. Thin plies of continuous fibers, unidirectional tape generally has a cured ply thickness of 5 to 10 mils (.127 mm to .254 mm); very thin plies of 2.5 mils (.06 mm) are also available for spacecraft applications but are very expensive. 2-D bidirectional fabric (woven fabric with a thickness from 10 to 20 mils (.254 mm to .508 mm) can be obtained in prepreg form. Fibers, chopped strands, whiskers or other forms of reinforcement can also be prepared by impregnation with matrix, and stored under controlled conditions, and then laid up in plies and cured into typical composite components. In prepreg, the partly cured matrix holds the fibers in alignment and, in sheet form, allows the pre-impregnated material to be handled easily, to be cut to shape for layup and assembled for curing with a minimal of difficulty. Also, the proportion of fiber to matrix can be closely controlled, giving, in turn, close control of the strength and weight of the finished component. There are several stages of prepreg "cure":

- A-stage — An early stage in the reaction of certain thermosetting matrices in which the material is fusible and still soluble in certain liquids. Sometimes referred to as resol.
- B-stage — An intermediate stage in the reaction of certain thermosetting matrices in which the material softens when heated and swells when in contact with certain liquids, but may not entirely fuse or dissolve. The matrix in an uncured thermosetting adhesive is usually in this stage. Sometimes referred to as resitol.

- C-stage — The final stage in the reaction of certain thermosetting matrices in which the material is relatively insoluble and infusible. Certain thermosetting matrices in a fully cured adhesive layer are in this stage. Sometimes referred to as resite.

Some common forms of prepregs are:

(a) Tapes (unitapes or unidirectional tapes) — Tapes usually come impregnated with the matrix (thermoset or thermoplastic) as a prepreg to hold the longitudinal continuous fibers together to form a continuous sheet (see Fig. 2.5.2) and are ready for layup. Therefore, the properties of a composite transverse to the fiber direction depend largely upon the matrix material but, in any case, are very weak compared to the longitudinal properties. Consequently, in the design of most structures that are subject to both longitudinal and transverse loadings, the fibers must be oriented in specific directions to withstand these loads. Design data are available from many sources that dictate layup patterns required for combining layers of tape to achieve desired direction properties.

Currently the most common tape thickness is from 5 to 10 mils (.127 mm to .254 mm) but thicker tape has been considered because utilizing 10-mil (.254 mm) tape for automated tape layup can save up to 40% of the layup time and the mechanical properties of such laminates are comparable to laminates produced from 5-mil (.127 mm) tape.

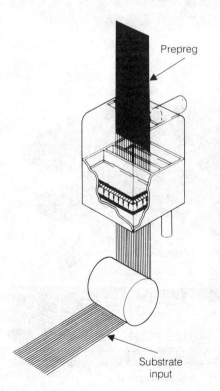

By courtesy of Electrostatic Technology, Inc.

(a) Prepreg method

APC-2 product forms. Left to right, single tow tape, 9 in (216 mm) wide prepreg, preconsolidated laminate and 6 in (140 mm) wide prepreg.

By courtesy of ICI Fiberit

(b) Prepreg products

Fig. 2.5.2 Typical prepreg Tape (continue Fibers)

(b) Discontinuous fiber tapes have been developed mainly for thermoplastic composites consisting of a thermoplastic matrix and highly aligned discontinuous fibers, as shown in Fig. 2.5.3. They possess the post-formability that is an inherent benefit of these systems. Discontinuous fiber tapes have the following characteristic and benefits:
- Average fiber length is 1 — 6 inches (2.54 — 15.24 cm)
- Over 85% of the fibers are aligned within +5 degrees of axial direction
- Creep and fatigue behavior is comparable to continuous fiber composites
- Form complex shapes not possible with continuous fibers but retain most properties of continuous fiber composites
- Fibers are typically Kevlar, carbon or glass

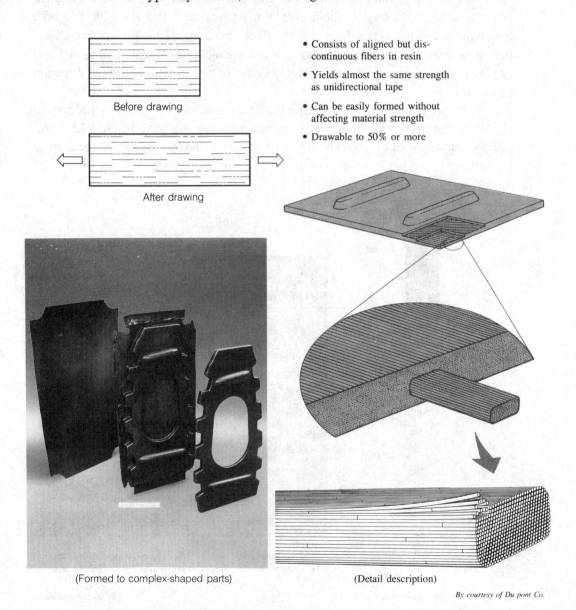

By courtesy of Du pont Co.

Fig. 2.5.3 Tapes (Discontinuous Fibers)

(c) Chopped fibers or whisker tapes — These tapes, shown in Fig. 2.5.4, are non-structural materials for secondary structural applications

Summary tables of comparison of different weaves and material properties of epoxy prepreg tapes is given in Fig. 2.5.5.

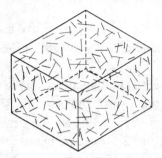

- Can be formed into complex shapes
- Lower material strength
- Not a primary load-carrying material

Fig. 2.5.4 Chopped Fibers and Whisker tapes

	Epoxy Prepreg Tapes						
Fiber Type	AS4	AS4	AS4	AS4	IM6	IM7	AS4
Resin Type	3501-6	3502	3501-5A	1919	3501-6	8551-7	8552
Use Temperature, °F, Dry	350°	350°	350°	180°	350°	225°	350°
Typical Composite Properties (62% F.V., 77°F)							
0° Tensile Strength, ksi	310	275	310	365	350	400	330
0° Tensile Modulus, Msi	20.5	20.3	21.5	20.5	23	22.4	20.4
0° Compression Strength, ksi	240	240	240	200	240	210	221
0° Flex Modulus, Msi	18.5	18.5	18.5	18.0	21.0	15.5	18.4
Short Beam Shear Strength, ksi	18.5	17.5	18.5	14.1	18.0	15.3	18.6
CAI, Compression After Impact	22	22	22	—	22	53	33

	Epoxy Prepreg Fabrics				
	Style				
Characteristics	A193-P	A280-5H	A370-5H	A370-8H	16360-5H
Weave	Plain	5-Harness Satin	5-Harness Satin	8-Harness Satin	5-Harness Satin
Fiber Types	AS4	AS4	AS4	AS4	IM6
Tow Size	3K	3K	6K	3K	12K
Areal Weight, g/m²	193	280	370	370	360
Yield, yds²/lb.	2.81	1.94	1.47	1.47	1.51
Dry Fabric Thickness, mils	10.5	16.0	21.0	21.0	21.0
Typical Composite Properties (62% F.V, 77°F)					
Tensile Strength, ksi	100	100	100	100	146
Tensile Modulus, msi	10.0	10.5	10.5	10.0	14.0
Short Beam Shear, ksi	9.5	10.0	10.0	10.0	10.2
Cured Ply Thickness, mils	7.5	10.0	13.5	13.5	13.2

(Source: Hercules Incorporated)

Fig. 2.5.5 Unidirectional Tape Preperties of Carbon Prepregs (HERCULES)

(1) WOVEN FABRICS (2-D weaving)

2-D (two-dimensional or two-directional) woven fabrics are more expensive than unidirectional tapes. However, significant cost savings are often realized in the manufacturing operation because layup labor requirements are reduced. Complex part shapes for processes requiring careful positioning of the reinforcement can benefit from the use of the more handleable woven forms of fiber.

Fabrics are generally described according to the types of weave and the number of yarns per inch, first in the warp direction (parallel to the length of the fabric), then in the fill direction (perpendicular to the warp).

(a) Unidirectional fabrics — Fabrics are essential unidirectional and in these fabrics the reinforcements are oriented in one direction (warp) and held in position by tie yarns (fill) of a non-structural nature which go over and under the structural fibers; the fibers (warp) lie straight with the (see Fig. 2.5.6) same characteristics as tape.

(b) Plain weave — In this construction, one warp yarn is repetitively woven over one fill yarn and under the next as shown in Fig. 2.5.7. It is the firmest, most stable construction, which provides minimum slippage. Plain-weave fabrics are less flexible and are suitable for flat or simply-contoured parts; a slight sacrifice in fiber property translation occurs. Strength is uniform in both directions.

(c) Satin weave — In this construction, as shown in Fig. 2.5.8, one warp yarn is woven over several successive fill yarns, then under one fill yarn. A configuration having one warp yarn passing over four and under one fill yarn is called a 5-harness satin weave. Satin weaves are less open than other weaves and strength is high in both directions. In particular the commonly used 8-harness satin retains most of the fiber characteristics of tape and can be easily draped over complex mold shapes.

In making a qualitative comparison between components made from tape (or unidirectional fabrics) and woven fabrics (2-D fabrics) the following comments can be made:

(a) A preference for a woven fabrics is usually due to one or more of the following advantages:
- Enhanced product toughness with respect to delamination — a product based on woven fabrics has fewer layers and the kinking of the fibers tends to deflect interlaminar cracking in such structures to give a higher fracture toughness.
- Woven fabric structures are more damage tolerant because of the interlocking of fibers.
- The ability to make thin, single ply laminates with balanced properties — a single ply of woven fabric may be as thick as two plies of collimated fiber but the same layup composed of tape would require four plies to produce a balanced layup.
- The close spacing of the fiber tows. The nature of the weave constrains lateral movement so that enhanced thickness uniformity is possible.
- The convenience of handling a broad-goods product is a factor which should not be underestimated. Since each ply is inspected during layup, inspection time is effectively halved since one ply of fabric equals two plies of tape.
- With woven fabric products there is no weak transverse direction.

- The drapeability of certain weaves is a special factor enhancing certain processing operations.
- Properties of special types, e.g., unidirectional fabric, can be tailored into the fabric by specifying appropriate construction.
- Opportunity is given by hybridized fabrics to tailor reinforcement use to produce high, medium and low property qualities, as required.

(b) Woven fabrics have several disadvantages:
- They are usually based on more expensive 3000 filament tows to ensure a thin uniform sheet
- They involve an additional manufacturing stage — weaving
- Because the fibers are necessarily kinked, the product cannot be expected to yield its full potential strength
- Fiber discontinuities and overlapped splices are required
- Lower fiber volume fractions than tapes
- Greater scrap rates than tapes
- Fabric distortion (bowing and skewing) causes part warping

Commonly used weaves and fabric constructions are shown in Fig. 2.5.9.

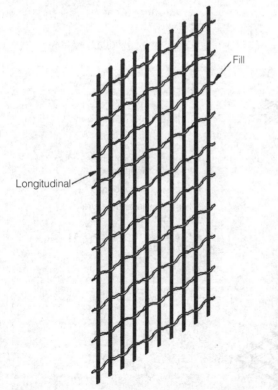
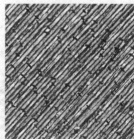

- Longitudinal strands are as straight as tape
- Improved drapability over tape
- Minimal reduction of fiber strength
- Improved fiber alignment

Fig. 2.5.6 Unidirectional Fabrics

Chapter 2.0

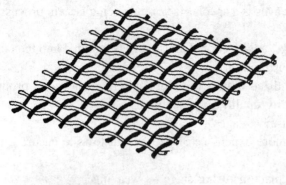

- Some loss of properties due to fiber cripming
- Provides reproducible laminate thickness
- Good drapability
- Speedier layup (reduces cost)
- Easir to handle
- Wet-out difficulties with tightly woven fabrics
- Width limitations
- Thicker than tape
- Good damage tolerance

Fig. 2.5.7 Plain Weave Fabrics

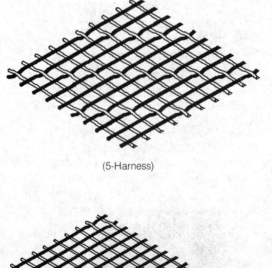

(5-Harness)

- Less crimping of fibers reduces loss of properties
- Excellent drapability
- High tensile and flexural strength
- Not as damage tolerant as plain weave

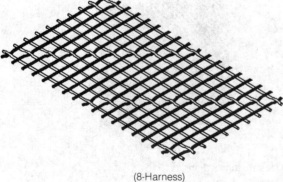

(8-Harness)

Fig. 2.5.8 Harness Satin Weave Fabrics

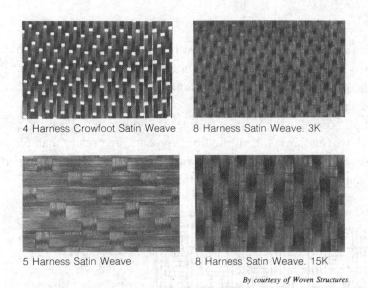

Fig. 2.5.8 Harness Satin Weave Fabrics (cont'd)

⚠ Fabric style	Weave	Thickness (in)	Cured ply thickness (in)	Weight (oz/yd²)	Width (in)	Yarn count (warp × fill)
120	Plain	0.0045		1.8	38.50	34 × 34
143	Crowfoot satin	0.0010		5.6	38	100 × 20
181	8-Harness satin	0.009		5.0	38.50	50 × 50
243	Crowfoot satin	0.013		6.7	50	38 × 18
281	Plain	0.010		5.0	38.50	17 × 17
285	Crowfoot satin	0.010		5.0	38.50	17 × 17
328	Plain	0.013		6.8	38.50	17 × 17
W-107	8-Harness satin	0.023	0.012	9.7	42	24 × 24
W-133	8-Harness satin	0.023	0.013	11.0	42	24 × 23
W-134	Plain	0.013	0.007	5.5	42	12 × 12
W-166	12-Harness satin	0.015	0.0045	7.2	42	48 × 48
W-176	5-Harness satin	0.009	0.008	3.6	42	24 × 24
W-177	Crowfoot satin	0.015	0.0078	6.4	42	24 × 12
W-184	Crowfoot	0.018	0.0098	8.2	42	24 × 12
W-185	Plain	0.038	0.025	20.5	42	4 × 6
W-186	5-Harness satin	0.006	0.003	2.7	42	24 × 12
W-190	Crowfoot satin	0.010	0.007	5.8	42	24 × 12
W-191	5-Harness satin	0.014	0.0087	7.2	42	24 × 24
W-233	8-Harness satin	0.019	0.012	10.5	42	20 × 18
W-301	Double crowfoot	0.026	0.017	14.0	36	15 × 16
W-305	8-Harness satin	0.025	0.017	14.0	36	15 × 16

⚠ Fabric style designations are assigned by the individual producers

Fig. 2.5.9 Commonly Used Fabric Weaving and Construction

(2) KNITTING

In a knit, fibers are knitted together with non-reinforcing binder fibers which go over and under the fibers (reinforcement fibers are not crimped). This means that the fibers can bear full loads, providing efficient translation of stress from matrix to fiber. Fig. 2.5.10 shows knitted fabrics.

Another purpose of using knitting is to reduce both material use and production times associated with woven fabrics. Knitting fiber contents are greater than for weaving — up to 65% compared with 50-60%. These stronger laminates with lower part weights are finding their applications in airframe structures.

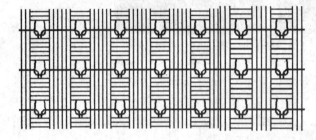

- Because fibers are staight, properties approach those of tape
- Preplied in 45/−45 or 0/90 plies
- 70% less labor intensive than tape since number of plies to be laid up is reduced
- Reduced trim scrappage

(a) Schematic of knitting

By courtesy of Composite Reinforcements Business

(b) closer look at knitted fibers

Fig. 2.5.10 Knitted Fabrics

(3) STITCHING

Stitching is a similar process to knitting except that it results in reinforcement properties in the Z direction of the finished parts. The reinforcing thread material used to hold the plies or preforms together can be virtually any fiber that will endure the sewing process. The operation involves stitching the dry or prepreg laminates together using needle and thread. Types of stitches used, either lockstitch or chainstitch, is shown in Fig. 2.5.11. The most important difference between the two stitch types is that a chainstitch is susceptible to unraveling and a lockstitch is not. This difference is very important if there is any handling of the product prior to the final molding and the possible unraveling of stitches would prove detrimental to the final product. A lockstitch requires significantly less tension during the stitch formation than a chainstitch. This factor is important to stitched laminate because too much tension could detrimental to the laminate strength capability in X or Y direction.

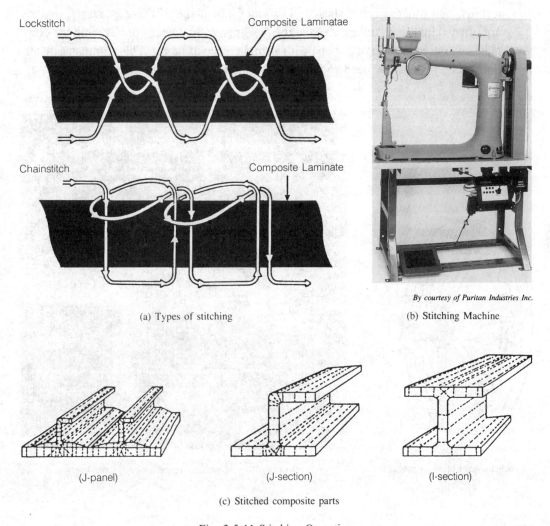

(a) Types of stitching (b) Stitching Machine

By courtesy of Puritan Industries Inc.

(J-panel)   (J-section)   (I-section)

(c) Stitched composite parts

Fig. 2.5.11 Stitching Operations

(4) BRAIDING

Braiding is a process particularly suitable for tubular parts as shown in Fig. 2.5.12. For many applications, especially complex components (see Fig. 2.5.13), braiding is more economical than filament winding and also can be automated for high speed production. A braided part is normally made by winding around a mandrel. Unlike the fibers in filament winding braid fibers are mechanically interlocked. This helps the braid to better endure twisting, shearing and impact. The structure is inherently seamless and without overlaps.

(a) Most braids are biaxial having interlocking yarns wound in two orientations. In triaxial braiding (see Fig. 2.5.14) a third fiber is added longitudinally, along the braiding axis. In biaxial braids the braiding angle can be varied between 10 and 85 degrees. Small angles are best for tensile strength, large angles for hoop strength.

(b) Contoured woven shapes, as shown in Fig. 2.5.15, are seamless fabric preforms which are available in a wide variety of contoured shapes such as hemispheres, ogives, cones, toroids, and other specific shapes. The chief advantage of these "socks", as they are often called, is that they provide the designer with a means of covering the complete contoured surface of a shape with a single piece of fabric. This eliminates distortion problems, or the need to cut and fit patterns of flat-woven fabric.

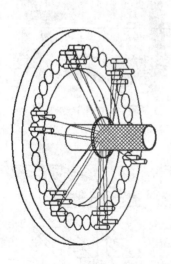

(a) Schematic of braiding process

(b) Formation of a braid around a circular mandrel

Fig. 2.5.12 Braiding

Materials

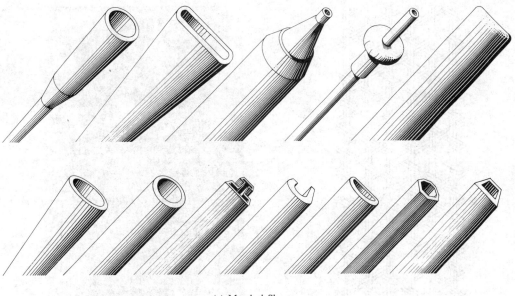

(c) Mandrel Shapes

By courtesy of Wardwell Braiding Machine Co.

Fig. 2.5.12 Braiding (cont'd)

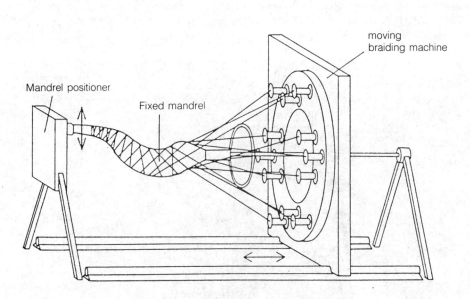

Eccentricity reduction through the vertical movement of the mandrel

By courtesy of Institute für kunststo ffverarbeitung

Fig. 2.5.13 Complex Shaped Braiding Method (Ref. 2.29)

Chapter 2.0

(a) Schematic of triaxial weaving

By courtesy of Fiber Innovations, Inc.
(b) A preform of c/c (Triaxial braiding)

Fig. 2.5.14 Triaxial Woven Fabrics

Missile Cone

Fig. 2.5.15 Contoured Woven Shapes

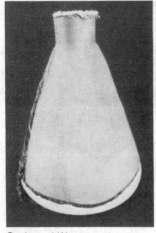

Conloured Woven carbon shapes ("sock") for large missile exit cone.

Example of bi-geometric weaving capability.

Ogive contoured shapes for small radomes.

Wedge contoured shape for tail radome on military fighter airplane.

By courtesy of Woven Structures

Fig. 2.5.15 Contoured Woven Shapes (cont'd)

(5) 3-D AND MULTI-DIRECTIONAL WEAVING PREFORMS

One of the key advantages of 3-D (three-dimensional or three directional) preform fabric composites, shown in Fig. 2.5.16 and Fig. 2.5.17, is their high impact damage tolerance, compared to laminated composites, due to their fully integrated fibrous substrates. 3-D braided fiber preforms and Z direction strength which provides superior strength to weight when impregnated with a resin or matrix. If a crack does occur, the tortuous path it has to follow, plus the need to break positive mechanical interlocks between successive fibers, tends to ensure slow propagation. These complex preforms exhibit exceptional resistance to delamination and may be suitable for advanced composite airframe structures such as fuselage barrel; spar beams, frames, ribs etc. and for other special applications; from reinforcing I or channel beams to jet engine turbine blades.

Chapter 2.0

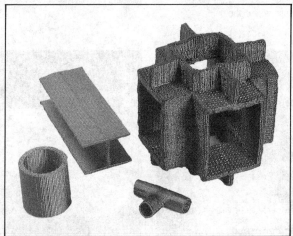

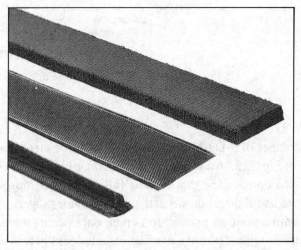

By courtesy of Shikishima Canvas Co., Ltd.

Fig. 2.5.16 Samples of 3-D Weaving Preforms

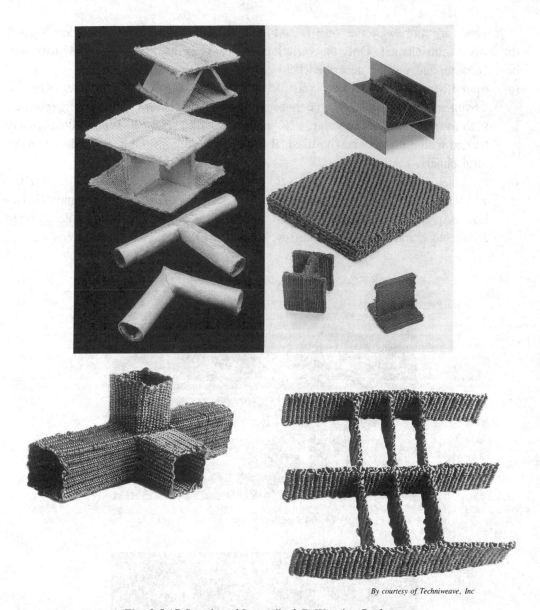

By courtesy of Techniweave, Inc

Fig. 2.5.17 Samples of Integrally 3-D Weaving Preforms

In the 1970s, it became apparent that the high cost of hand-assembled preforms would severely limit the application of 3-D or multi-directionally reinforced composites to very few airframe and aerospace applications. Therefore, a few automated and semi-automated systems have been created or are under development to design and fabricate preforms to reduce costs. The increasing quality by eliminating the human error that occurs in hand assembly is desperately needed in composite manufacturing. There are few limits on the composition of reinforcement fibers that can be woven into 3-D or multi-directional preforms. Generally the only limitation to fiber selection is the combination of brittle fibers and small yarn bend radii. High modulus fiber is particularly prone to fracture during preform construction using automated weaving machines.

Chapter 2.0

There are more varieties of 3-D and multi-directional preforms which are beyond the scope of this chapter. Only the variations of preforms that are most widely used and best characterized will be described below.

(a) Fluted-core Fabrics — These constructions shown in Fig. 2.5.18, are integrally 3-D woven and have parallel faces connected by ribs in a variety of flute configurations, such as triangular, rectangular, "X" and other cross sections. When fluted-core is treated with a matrix and rigidized, it can be used in both flat and contoured structural panels.

(b) Near-net-shape-weaving — This weaving process is defined by the ability to tailor width and multilayer cross section thickness, as well as fiber volumes and orientations. An example is the multilayered turbine blade preform featuring tailored fiber architecture and tapered thicknesses, shown in Fig. 2.5.19.

(a) Samples of products

(b) Triangular web flutted-core fabric being woven on loons

Fig. 2.5.18 Fluted-core 3-D Fabrics

Materials

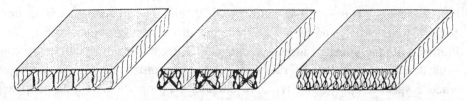

(c) Illustrates three common types of fluted-core

By courtesy of Woven Structures

Fig. 2.5.18 Fluted-core 3-D Fabrics (cont'd)

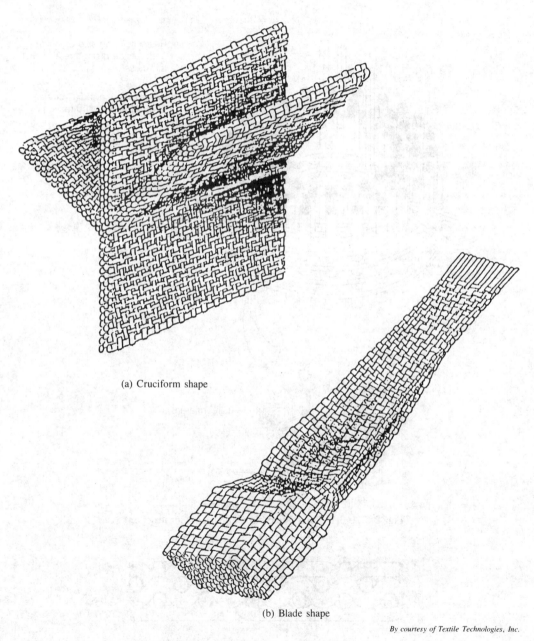

(a) Cruciform shape

(b) Blade shape

By courtesy of Textile Technologies, Inc.

Fig. 2.5.19 Artist Rendering of Actual Woven Shapes

103

(c) Polar weave — This is a 3-D preform which has reinforcement yarns in the circumferential, radial, and axial (longitudinal) directions, as shown in Fig. 2.5.20. Preforms of this geometry normally contain 50% fibers that can be introduced equally in the three directions. Some variations in relative yarn distribution can be accomplished when a specific application requires unbalanced properties. If high-hoop tensile strength, for example, is required additional fibers can be added in the circumferential direction, at some sacrifice of radial and longitudinal properties.

Figures 2.5.21 through 2.5.24 illustrate 3-D and multi-directional preforms which are especially useful in airframe or aerospace structure applications.

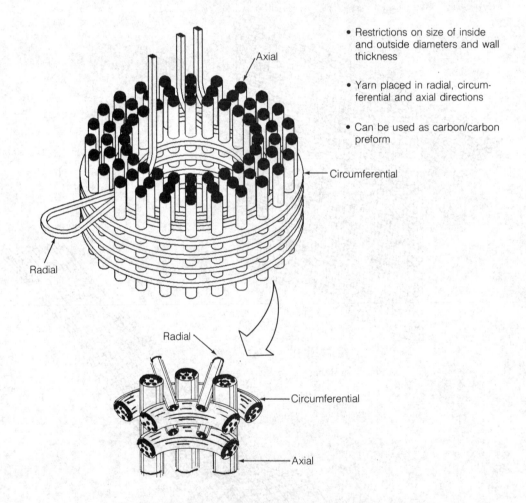

- Restrictions on size of inside and outside diameters and wall thickness
- Yarn placed in radial, circumferential and axial directions
- Can be used as carbon/carbon preform

Fig. 2.5.20 3-D Polar Weaving Cylindrical Construction)

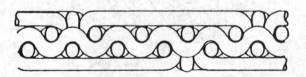

Fig. 2.5.21 5 Harness Mult-layer

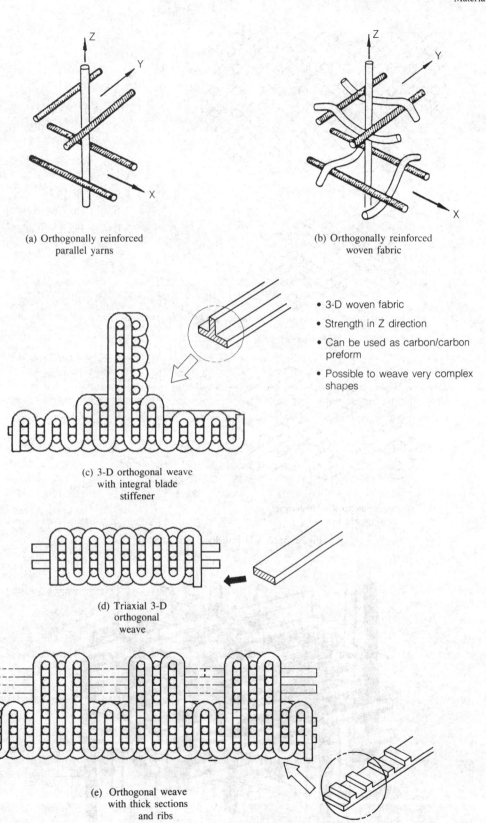

Fig. 2.5.22 3-D Orthogonally Reinforced Weaving

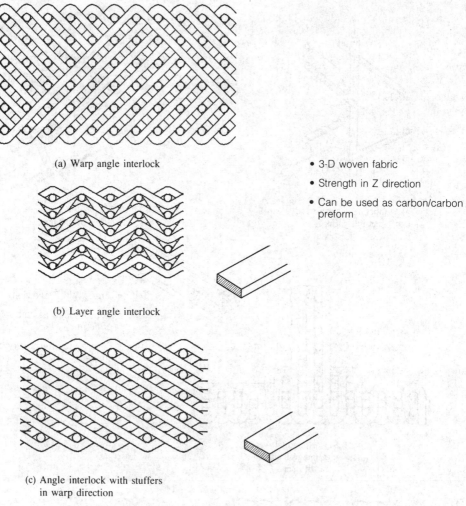

(a) Warp angle interlock

(b) Layer angle interlock

(c) Angle interlock with stuffers in warp direction

- 3-D woven fabric
- Strength in Z direction
- Can be used as carbon/carbon preform

Fig. 2.5.23 3-D interlock weaving

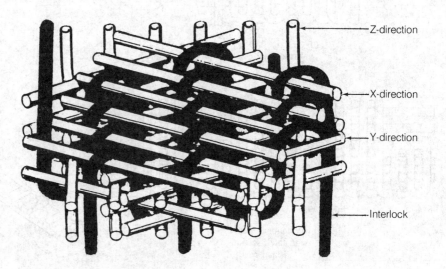

Fig. 2.5.24 4-D and Multi-directional Preforms

(6) VERY-COMPLEX-SHAPE (VCS)

It is almost impossible to construct very complex shapes, even with today's high technology textile techniques. But the ancient technique of rattan handicraft is a good example of a method which will make almost any complex preform; some examples of rattan for household use are shown in Fig. 2.5.25. The procedure of applying this technique to airframe structures is

- The first step is to manually construct a preform of the desired design concept using rattan methods
- The second step is to find or develop a semi-automated robotic system to replace hands
- The third step is to simplify the handmade preform to fit the semi-automated machine (certain requirements, e.g., strength, weight, etc.)
- The final step is to construct the material preform using fully automated system to completely cut down manufacturing cost in mass production

(7) PECULIAR FABRICS (Thermoplastic Composite Product Forms)

The key advantages of "Hybrid- fabric" (HF) forms include flexibility and drape for complex contoured shapes and deep draws, ease of processing, and quality composite part reproducibility. HF can be made into various prepreg forms such as coweaving, braiding, and knitted fabrics.

One type of HF is formed using an intimate blend of reinforcing fibers and thermoplastic-fibers (fibers made from thermoplastic matrix). HF give thermoplastics increased producibility allowing them to compete with thermosets. HF exist in the following forms:

(a) Commingled weave (most common form) — In commingled weaves, strands of thermoplastic resins and filaments reside together in the same bundle [see Fig. 2.5.26(a)], in a single yarn. This single yarn is then woven into a fabric. In general, the commingled woven fabric composites exhibit higher physical and mechanical properties than cowoven fabric composites. Commingled HF promise outstanding drape and fabric flexibility for reproducible as shown in Fig. 2.5.27.

(b) Coweaving — Thermoplastic fibers or slit film is woven side by side with reinforcing fibers to form a fabric. The thermoplastic fibers and reinforcing fibers exist in separate yarns. [see Fig. 2.5.26(b)].

(c) Plied matrix — In this method, a thermoplastic fiber is wrapped around a reinforcing yarn, resembling the spirals on a barber pole [see Fig. 2.5.26(c)].

Fig. 2.5.28 compares commingled and cowoven fabrics.

(d) Interlacing fabrics — A method of interlacing prepreg unidirectional tape (see Fig. 2.5.29) yields a special form of thermoplastic composite which offers woven-composite formability with structural properties equivalent to unidirectional tape prepregs. Because interlacing tapes are produced from prepreg unidirectional tapes, the resulting composites have very low porosity and void content.

Chapter 2.0

Fig. 2.5.25 Rattan Samples

Materials

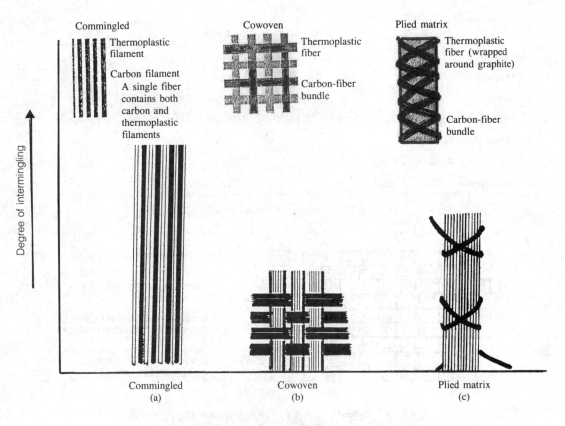

Fig. 2.5.26 Hybrid-fabrics

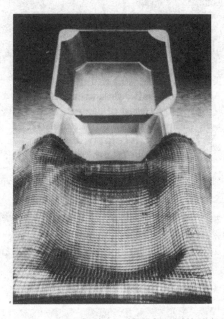

By courtesy of BASF Structural Materials

Fig. 2.5.27 Outstanding Drape and Fabric Flexibility of Commingled Hybrid Yarn Fabrics

109

	Commingled	Cowoven
Cost	More	Less
Drapable	Better	Less
Wetting	Good	Less
Mechanical properties	Good	Lower
Evenly distributed	Good	Less
Low matrix viscosity required	No	Yes
Unidirectional fabrics Available	Yes	No

Fig. 2.5.28 Commingled Versus Cowoven Fabrics

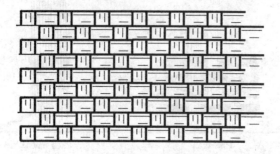

- 5 and 8 harness interlace are standard
- Strand width: 1/8; 3/16; 1/4; 3/8 and 1/2
- Quasi-isotropic laminates are achieved by balancing 0/90 and 45/−45 plies
- Individual strands are actually strands of unidirectional tape

(Closeup of APC-2 unconsolidated Quadrax Biaxial Tape, 0/90 and ±45)

By courtesy of Quadrax, Inc.

Fig. 2.5.29 Interlacing Weave Fabrics

(e) Powder prepreg — A method of dry powder processing under development impregnates fibers with powder, primarily thermoplastics, which leaves the prepreg drapable and enabling faster layup procedures. This encapsulated-powder method uses very fine powder, typically less than 20 microns, to achieve a good physical blend. Some of the important features of the powder coated prepreg are:
- Versatile for both thermoplastics and thermosets
- Operates at room temperature
- Involves no solvents
- Requires no significant refrigeration

- Can be woven, pultruded and thermoformed
- Offers a viable alternative to RTM processing of textile preform composites

An assessment of thermoplastic material forms, including of prepreg tape, commingled fabric, Quadrax interlacing fabrics and powder prepreg is given in Fig. 2.5.30.

Product Forms	Advantages	Disadvantages
Unidirectional tape	• Best strength efficiency • Most common/readily available form • No debulking (dry) • Wide-process compatibility (dry) • Autoclave process compatibility (wet)	• Limited drape (dry) • Hot irons required to seam and tack weld material (dry) • Solvent/volatile removal (wet)
Commingled Fabric	• Excellent drape/formability • Reduced layup time	• Reduced strength efficiency • Limited fiber wet out • Special handling precautions required • Processing requires high pressures for extened time periods
Quadrax	• One ply quadrax equals two plies of unitape • Reduced layup time • Improved drape over unitape	• Higher material cost • Interwoven strips limit ply slippage • Minimal strength reduction
Powder prepreg	• Good drape/formability • Good tack without solvents	• Limited matrix selection • Powder binder/moisture removal

Fig. 2.5.30 Assessment of Thermoplastic Material Forms

2.6 SANDWICH CORE MATERIALS

The challenge of making a structure as light as possible without sacrificing strength is fundamental in aircraft design. Inevitably the requirement leads to the need to stabilize thin surfaces to withstand tensile and compressive loads and combinations of the two, in shear, torsion and bending. Traditional airframe structural design has in the past, and does still to some extent, overcome this difficulty by the use of longitudinal stiffeners and stabilizing rings with stringers, and ribs or frames. But it is an inelegant solution and, in fact, the stabilization of a surface — creating a resistance to deforming forces — can be more efficiently effected by the use of twin skins with a stabilizing medium between them — what is now termed a sandwich structure, as shown in Fig. 2.6.1. It is noteworthy, however, that scientific design has been paralleled, or rather anticipated, by nature and evolution, for the structure of the human skull bears a remarkable resemblance to a sandwich structure.

Chapter 2.0

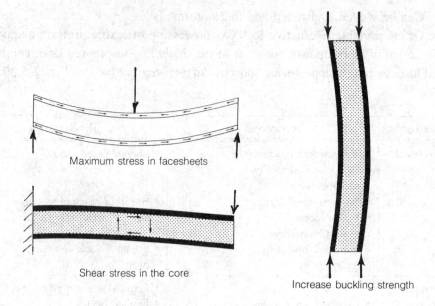

Fig. 2.6.1 Structural Effeciency of Sandwich Construction

This name is comparatively recent, but the structural sandwich has been in use over a very long period, probably before recorded history. Clearly, a sandwich offered the prospect of more precisely-controlled characteristics but, in practice, this form of sandwich had to wait until the development of suitable adhesives made possible an all-bonded structure possible.

(1) HONEYCOMB CORE

In aircraft engineering the development of honeycomb core in structural composites began in the 1940s, when there was a growing interest in sandwich construction, using thin strong skins bonded with a very low-density honeycomb core material, as shown in Fig. 2.6.2. The first honeycomb was canvas cloth impregnated with phenolic resin. In later developments, honeycomb has been made from metal, fiberglass, paper and recently, from advanced composites, aramid and carbon fibers. Honeycomb core can be made out of almost any thin sheet material and is usually made in a form in which, as the name implies, hexagonally-shaped cells (the most commonly used core in composite structures) are arranged in a geometrical pattern of horizonatal/diagonal rows, by bonding foil or sheet adhesives. See Fig. 2.6.3 for common honeycomb terminology. Other types of honeycomb core are illustrated in Figures. 2.6.4 through 2.6.9. Today, a variety of metallic foils and non-metallic materials are used for the production of honeycomb to meet a variety of requirements. The rigidity of honeycomb sandwich makes it attractive for applications where low-deflection structures of minimum weight are needed, such as aircraft nose radomes, wing leading and trailing edge panels, fuselage floor panels, etc. The beauty of using honeycomb is that because of its light weight and high strength, the end products do not require such thickness.

Characteristically, the non-metallic honeycombs are
- Vibration and fatigue
- Easily handled, formed, and machined
- Free from fatigue failures

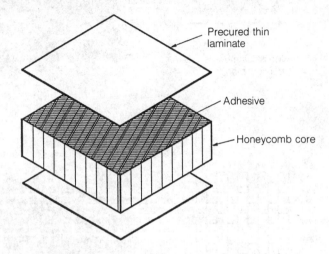

Fig. 2.6.2 Typical Honeycomb Sandwich Construction

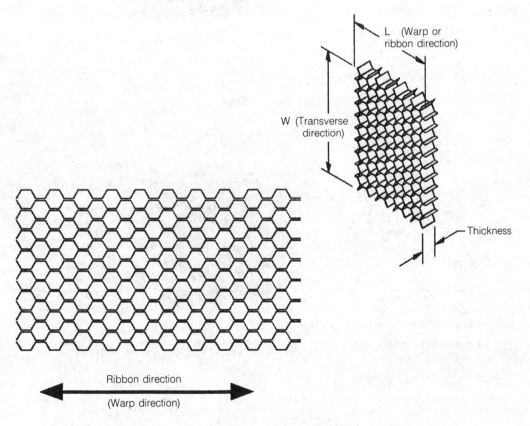

Fig. 2.6.3 Honeycomb Terminology

Chapter 2.0

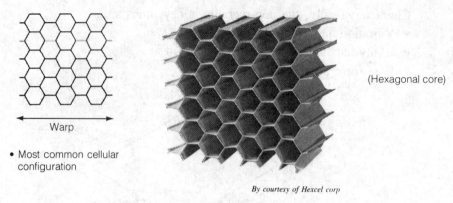

- Most common cellular configuration

(Hexagonal core)

By courtesy of Hexcel corp

Fig. 2.6.4 Standard Hexagonal core

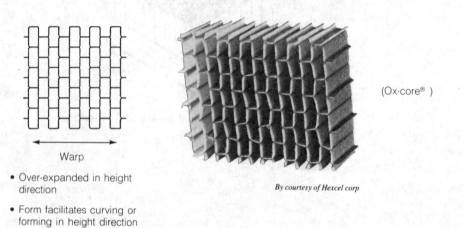

- Over-expanded in height direction
- Form facilitates curving or forming in height direction

(Ox-core®)

By courtesy of Hexcel corp

Fig. 2.6.5 Over-expended core

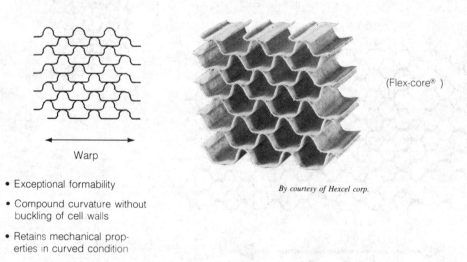

- Exceptional formability
- Compound curvature without buckling of cell walls
- Retains mechanical properties in curved condition

(Flex-core®)

By courtesy of Hexcel corp.

Fig. 2.6.6 Special Core

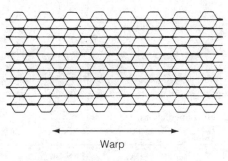

- High density core
- Very high shear strength along warp direction

Fig. 2.6.7 Reinforced Corrugated (Hexagon core)

By courtesy of Hexcel corp

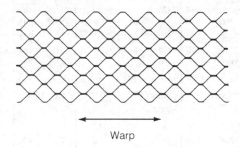

Fig. 2.6.8 Square core

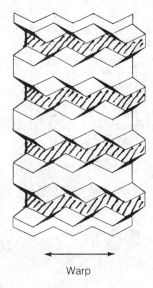

Fig. 2.6.9 Z-core (Molded from one piece sheet)

Chapter 2.0

- High strength at low density
- Small cell size at low density
- Damage resistance under normal use
- Formable
- Fire resistant (self-extinguishing)
- Water and fungus resistant
- Excellent dielectric properties
- Good bonding surfaces
- Good thermal and electrical insulator
- Available in standard, over-expanded and special cell configurations.

Honeycomb Hexcel Designation Material - Cell - Density - Gauge	Compressive						Plate shear					
	Bare		Stabilized				"L" Direction			"W" Direction		
	Strength psi		Strength psi		Modulus ksi		Strength psi		Modulus ksi	Strength psi		Modulus ksi
	typ.	min.	typ.	min.	typical		typ.	min.	typical	typ.	min.	typical
HRH 10 - 1/8 - 1.8 (1.5)	110	70	130	85	—		90	65	3.7	50	36	2.0
HRH 10 - 1/8 - 3.0 (2)	300	180	325	270	20		190	165	7.0	100	85	3.5
HRH 10 - 1/8 - 4.0 (3)	500	330	560	470	28		270	225	9.2	140	110	4.7
HRH 10 - 1/8 - 5.0 (3)	775	600	860	660	—		325	235	—	175	120	—
HRH 10 - 1/8 - 6.0 (3)	1075	800	1125	825	60		370	260	13.0	200	135	6.0
HRH 10 - 1/8 - 8.0 (3)	1840	1320	1900	1350	78		490	355	16.0	275	210	7.8
HRH 10 - 1/8 - 9.0 (3)	1870	1400	1970	1500	90		505	405	17.0	310	260	9.0
HRH 10 - 5/32 - 5.0 (4)	800p	—	900p	—	—		360p	—	11.0p	180p	—	5.0p
HRH 10 - 5/32 - 9.0 (4)	1775p	—	2050p	—	—		525p	—	18.0p	285p	—	9.5p
HRH 10 - 3/16 - 2.0 (2)	150	90	165	115	11		110	80	4.5	60	45	2.2
HRH 10 - 3/16 - 3.0 (2)	300	180	370	270	20		160	130	5.8	90	70	3.5
HRH 10 - 3/16 - 4.0 (3)	500	320	560	470	28		245	215	7.8	140	110	4.7
HRH 10 - 3/16 - 4.5 (5)	425	320	475	400	—		290	225	9.5	145	110	4.0
HRH 10 - 3/16 - 6.0 (5)	650	580	700	650	—		390	330	14.5	185	150	6.0
HRH 10 - 1/4 - 1.5 (2)	90	45	95	55	6		75	45	3.0	35	23	1.5
HRH 10 - 1/4 - 2.0 (2)	150	80	170	105	11		110	72	4.2	55	36	2.8
HRH 10 - 1/4 - 3.1 (5)	275	180	285	240	—		170	135	7.0	85	60	3.0
HRH 10 - 1/4 - 4.0 (5)	370	310	400	360	—		240	200	7.5	125	95	3.5
HRH 10 - 3/8 - 1.5 (2)	90	45	95	55	6		75	45	3.0	35	23	1.5
HRH 10 - 3/8 - 2.0 (2)	150	80	170	105	11		110	72	4.2	55	36	2.2
HRH 10 - 3/8 - 3.0 (5)	285p	—	300p	—	17p		170p	—	5.6p	95p	—	3.0p
HRH 10/OX - 3/16 - 1.8 (2)	110	70	130	—	—		60	45	2.0	60	35	3.0
HRH 10/OX - 3/16 - 3.0 (2)	365	255	400	270	17		125	95	3.0	140	105	6.0
HRH 10/OX - 3/16 - 4.0 (2)	598	420	—	—	—		132	105	4.6	158	124	8.4
HRH 10/OX - 1/4 - 3.0 (2)	350	210	385	250	17		110	90	3.0	115	90	6.0

Mechanical Properties at Room Temperture
P — Limited testing has been performed. Test data obtained at 0.50 inch thickness per MIL-STD-401.

(Source: Hexcel Corp.)

Fig. 2.6.10 Mechanical Properties of Hexcel HRH 10 Honey (Nomex/Phenolic)

- High shear properties (±45° fiber orientation)
- Small cell sizes
- Better flexibility, damage resistance and handleability
- Service temperature to 350°F (with short exposures at higher temperatures)
- Low mositure pickup
- Low smoke emission
- Available in standard and over-expanded cell configuration

The following mechanical properties are based on preliminary testing of products at 0.500 inch thickness. The tests were performed per MIL-STD-401.

Hexcel Honeycomb Designation Mat'l-Cell-Density	Compressive Stabilized			Plate shear					
				"L" Direction			"W" Direction		
	Strength psi		Modulus ksi	Strength psi		Modulus ksi	Strenght psi		Modulus ksi
	typ	min	typical	typ	min	typical	typ	min	typical
HFT - 1/8 - 3.0	350	270	23	185	150	17	95	75	7.0
HFT - 1/8 - 4.0	560	420	46	315	240	37	150	120	12
HFT - 1/8 - 5.5	890	640	69	460	360	40	230	180	14
HFT - 1/8 - 8.0	1750^p	1500^p	129^p	580^p	495^p	49^p	340^p	290^p	24^p
HFT - 3/16 - 1.8	120^p	—	14^p	105^p	—	13^p	50^p	—	4.0^p
HFT - 3/16 - 2.0	170^p	—	17^p	115^p	—	15^p	60^p	—	5.0^p
HFT - 3/16 - 3.0	365	275	34	200	155	24	100	80	9.0
HFT - 3/16 - 4.0	598	430	44	275	210	30	140	115	14
HFT - 3/16/OX - 6.0	1180^p	—	63^p	320^p	—	18^p	260^p	—	19^p
HFT - 3/8 - 4.0	500	350	—	300	210	24^p	145	100	11^p

p — Preliminary values are obtained from testing of only only one or two blocks of Honeycomb type and often only one or two specimens for each point or condition tested.

(Source: Hexcel Corp.)

Fig. 2.6.11 Mechanical Properties of Hexcel HEF Honeycomb (Fiberglass/Phenolic)

- Service temperature to 500°F
- 600°F short and intermediate term exposures
- All polyimide resin and adhesive system
- Good dielectric properties
- Good insulative properties
- Compatible with polyimide adhesives, foam and prepreg

(Mechanical Properties, see next page)

Fig. 2.6.12 Mechanical Properties of Hexcel HRH 327 Honeycomb (Fiberglass/polyimide)

Hexcel Honeycomb Designation Mat'l-Cell - Density	Compressive Stabilized		Plate shear			
			"L" Direction		"W" Direction	
	Strength psi	Modulus ksi	Strength psi	Modulus ksi	Strength psi	Modulus ksi
	typ min	typical	typ min	typical	typ min	typical
HRH - 327 - 1/8 - 3/4	310^p 220^p	—	190^p 155^p	19^p	90^p 70^p	7.5^p
HRH - 327 - 3/16 - 4.0	440^p —	50^p	280^p —	29^p	130^p —	10^p
HRH - 327 - 3/16 - 4.5	520 400	58	320 220	33	150 110	11
HRH - 327 - 3/16 - 5.0	600^p —	68^p	370^p —	37^p	180^p —	12^p
HRH - 327 - 3/16 - 6.0	780 625	87	460 345	45	230 170	15
HRH - 327 - 3/16 - 8.0	1210 1000	126	700 490	62	420 300	22
HRH - 327 - 3/8 - 4.0	440 325	50	280 195	29	150 100	12
HRH - 327 - 3/8 - 5.5	680 540	78	420 300	41	210 160	13
HRH - 327 - 3/8 - 7.0	1000^p —	106^p	550^p —	53^p	310^p —	18^p

Test data obtained at 0.500 inch thickness.
Honeycomb normally not tested for bare compressive strength.
Tolerance for nominal densities shown above ±10%.
HRH 327 — 1/4 inch cell size is available upon special request, if quantity is sufficient.
Bold print indicates readily available material.
p — preliminary values.

(Source: Hexcel Corp.)

Fig. 2.6.12 Mechanical Properties of Hexcel HRH 327 Honeycomb (Fiberglass/polyimides) (cont'd)

- Highest strength/modulus non-metallic honeycomb
- High strength retention up to 350°F continuous service
- ±45° web fiber orientation
- Low moisture pickup
- Excellent thermal stability
- Very low coefficient of thermal expansion
- No corrosion with graphite face sheets
- Available in standard cell configuration only
- For thermoplastic application

The following mechanical properties are based on preliminary testing of products at 0.500 inch thickness. The tests were per MIL-STD-401.

Hexcel Honeycomb Designation Mat'l-Cell-Density	Compressive Stabilized		Plate shear			
			"L" Direction		"W" Direction	
	Strength psi	Modulus ksi	Strength psi	Modulus ksi	Strength psi	Modulus ksi
	typ min	typical	typ min	typical	typ min	typical
HFT - G - 1/4 - 5.0	940 —	103	455 —	92	290 —	35
HFT - G - 3/8 - 4.5	852 —	100	394 —	74	271 —	26

(Source: Hexcel Corp.)

Fig. 2.6.13 Mechanical Properties of Hexcel HFT-G Honeycomb (Carbon/Phenolic)

Honeycomb core is available in a wide range of materials including advanced materials such as carbon fiber combined with thermoplastic matrix for higher temperature applications and the perforated honeycomb core for aerospace (outer-space structures) applications.

Selected honeycomb materials are described in Figures 2.6.10 through 2.6.13. The data given are typical values and should not be used for final design; designers should consult with the material manufacturer for actual values.

(2) SYNTACTIC CORE

Syntactic is a foam created by combining glass microballoons with matrix (resin) as shown in Fig. 2.6.14. It will conform to contoured surfaces and is available in moldable forms. Syntactic is used as a lightweight filler for thin sandwich constructions. Increased strength and stiffness are provided by syntactic without a marked increase in weight. Strategic use of syntactic can reduce structural weight and costs.

Syntactic core material is not usually used for thick panel sections greater than ¼ inch (6.4 mm), since its density (approximately 40 lb/ft^3 (2.51 kg/m^3)) is somewhat greater than that of conventional honeycomb core (2 — 10 lb/ft^3 (.0126 — .63 kg/m^3)). In some applications, syntactic core and honeycomb core have been used successfully in the same structure. Thus, both can be viewed as complementary products each solving different design problems and enhancing both the efficiency and cost effectiveness of structures. Syntactic core offers the following advantages over conventional honeycomb core:

- High compressive, transverse tensile and lateral strength
- No facesheet wrinkling problems
- Provides continuous support to the facesheets
- No in-service moisture ingress problems

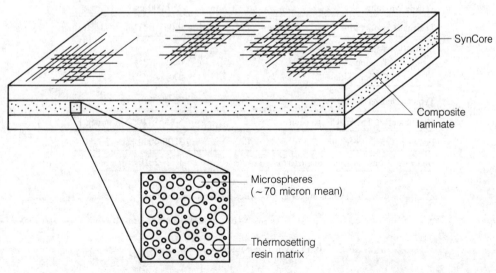

By courtesy of The Dexter Corp.

Fig. 2.6.14 Schematic Representation of A Typical Syntactic Core Sandwich Construction

Chapter 2.0

- Can be used in thin core sections while honeycomb core structures are generally limited to thicknesses in excess of ¼ inch (6.4 mm), if expensive machining is to be avoided.

Information on specific syntactic materials (thermoset systems of epoxy and BMI) are given in Fig. 2.6.15. Thermoplastic systems are also available for airframe applications.

Cure temp. (deg. F)	Matrix resin (type)	Service temp. (deg. F)	Syncore type
250	Epoxy	180-200	HC-9823
250	Epoxy (Industrial)	180-200	HC-9860
350	Epoxy	280-300	HC-9872
350	High Crush Strength Epxoy	280-300	HC 9875
350/475	BMI	400-450	HC-9802
250	Fire Retardant Epoxy	180-200	LN 89015*
350/425	Cyanate Ester	400-450	LC-8807*

*Research products.

(a) SynCore products

Shear Strength	1700-2400 psi
Compressive Strength	7400-8500 psi
Tensile Strength	3000-4000 psi
Shear Modulus	150,000 psi
Compressive Modulus	190,000 psi
Tensile Modulus	350,000 psi
Thermal Conductivity	0.043 BTU/hr.ft^2-deg F/ft
Elongation to Failure Tension Compression	0.8%-1.2% 4.0%
Typical Density Range	35-40 lb/ft^3
Coefficient of Thermal Expansion	5-15 × 10^{-6} in/in/deg F
Flatwise Tensile Strength	3000-4000 psi
Dielectric Constant (non-conductive)	1.7-2.0

(b) Typical material properties at room temperature

(Source: SynCore Design Manual, Dexter Adhesives and Structural Material Division)

Fig. 2.6.15 SynCore Material Information

(3) FOAM

Foam is a sponge-like material made of resin with either closed or connection cells, and has the same advantages as syntactic material. Foam is less dense (general applications from 2 — 20 lb/ft^3 (.0126 — .126 kg/m^3)) than syntactic and is used for the same applications that syntactic is used.

Information on foam materials made from polymethacrylimide is given in Fig. 2.6.16.

(4) POTTING COMPOUND

Potting compound is a low density material which has the following uses:
- Material inserts
- Edge filling
- To prevent crushing by local concentrated loads in honeycomb core
- Joining material in honeycomb constructions (see Fig. 2.6.17)
- To secure inserts installed in either honeycomb or foram core

Fig. 2.6.18 shows data for a selected potting compound material; other potting compound materials for particular applications, such as high temperature performance, are available.

ROHACELL' IG is a lightweight, rigid, high-quality, polymethacrylimide foam. It is especially suited for use as a core material for composite construction.

Typical mechanical properties

Properties	Dimension	Rohacell' IG				Rohacell' Pressed grades		ASTM Test Method
		311G	511G	711G	1101G	P170[1]	P190[1]	
Density	LBS./cu. ft.	1.9	3.1	4.4	6.9	10.6	11.9	D1622
Tensile strength	PSI	142	270	398	498	1,070	1,210	D638
Compressive strength	PSI	57	128	213	427	924(398)[1]	1,110(455)[1]	D1621
Flexural strength	PSI	114	228	356	640	1,490(1,420)[1]	1,780(1,710)[1]	D790
Shear strength	PSI	57	114	185	341	640(427)[1]	782(427)[1]	C273
Modulus of elasticity	PSI	5,120	9,950	13,100	22,700	45,500	54,000	D638
Shear modulus	PSI	1,990	2,990	4,270	8,250	17,000	26,300	D2236
Shear modulus	PSI	1,850	2,700	4,120	7,110	12,500	14,200	C273
Elongation at break	%	3.5	4	4.5	4.5	5	6	D638

Test conditions: 23°C (73.4°F) and 50% relative humidity

1) P170 and P190 are post-compressed, non-isotropic grades. Values in parenthesis are measured perpendicular to the plane of the sheet.

(a) Rohacell IG Grades (source from CYRO Industries)

Fig. 2.6.16 Rohacell Foam Material (Mechanical properties)

Chapter 2.0

ROHACELL' WF is a lightweight, rigid, high-quality, polymethacrylimide foam. It was developed to satisfy the demand of the aerospace industry for a high strength core material for advanced composite construction, ROHACELL WF foam will withstand autoclave processing at temperatures up to 360°F and pressures up to 100 psi over a 2-hour time period (depending upon density and heat treatment). ROHACELL WF has an isotropic structure, so pressure can be applied from all directions with excellent dimensional stability.

Typical mechanical properties

Properties	Dimension	Rohacell' WF grades					ASTM Test Method
		51 WF	71 WF	110 WF	200 WF	300 WF	
Density	LBS./cu. ft.	3.1	4.4	6.9	12.5	18.7	D1622
Tensile strength	PSI	227	312	525	852	1,491	D638
Compressive strength	PSI	128	213	511	1,190	2,272	D1621
Flexural strength	PSI	227	412	738	1,704	2,840	D790
Shear strength	PSI	114	185	341	710	1,136	C273
Modulus of elasticity	PSI	10,650	14,910	25,560	34,000	53,000	D638
Shear modulus	PSI	2,840	4,686	8,236			D2236
Shear modulus	PS1	2,698	4,118	7,100	21,000	42,600	C273
Elongation at break	%	3	3	3	3.5	3.5	D638

Test conditions: 23°C (73.4°F) and 50% relative humidity   1) Guaranteed minimum values are available upon request

(b) Rohacell WF Grades (Soruce form CYRO Industries)

Fig. 2.6.16 ROHACELL Foam Material (Mechancial properties) (cont'd)

Fig. 2.6.17 Honeycomb Core Joined by Potting Compound

Curing agent used with corfil 615 potting Compound	Specific gravity g/cc			Compressive Strength at room temperature psi (MPA)			Typical Modulus of Elasticity psi (MPA)	Typical cure cycle
	Avg.	Low	High	Avg.	Low	High		
"A" 6 pha	0.600	0.547	0.686	6,000 (42)	4,000 (28)	7,210 (51)	91,800 (642)	Two hours at room temperature plus 16 hours at 120°F (50°C)
"Z" 14.5 pha	0.668	0.655	0.684	7,625 (54)	7,500 (53)	7,750 (55)	218,000 (1,526)	16 hours at 120°F (50°C) plus one hour at 300°F (150°C)
"U" 15 to 25 pha	0.599	0.585	0.613	4,710 (33)	4,380 (31)	5,040 (35.2)	—	16 hours at room temperature
"DTA" 7 pha	0.646	0.600	0.710	4,150 (29)	3,700 (26)	4,440 (31)	115,000 (805)	24 hours at room temperature

(Source: American Cyanamid Co.)

Fig. 2.6.18 Typical Physical Properties of CORFIL 615 Potting Compound

References

2.1 Baucom, R. M. and Marchello, J.M., "Powder Towpreg Process Development", First NASA Advanced Composites Technology Conference, Seattle, WA, Oct., 1990.

2.2 Ko, F. K., "THE ATKINS & PEARCE HANDBOOK OF INDUSTRIAL BRAIDING", Atkins & Pearce, Inc., 3865 Madison Pike, Covington, KY 41017. 1988.

2.3 Weeton, J. w. and etc., "ENGINEERING GUIDE TO COMPOSITE MATERIALS", American Society for Metals, Metals Park, OH 44073. 1987.

2.4 Anon., "SYNCORE DESIGN MANUAL", Dexter Adhesives and Structural Materials Division, 2850 Willow Pass Road, Pittsbrug, CA 94565-0031.

2.5 Anon., "AEROSPACE COMPOSITES & MATERIALS", The Shephard Press Limited, 111 High Street, Burnham, Buckinghamshire SL1 7JZ, England.

2.6 Anon., "Foam and Syntactic Film give Core Support to Composites", ADVANCED COMPOSITES, Jan/Feb 1991. pp. 32-38.

2.7 Schwartz, M. M., "COMPOSITE MATERIALS HANDBOOK", McGraw Hill, New York, NY. 1984.

2.8 Anon., "Materials Keep a Low Profile", MECHANICAL ENGINEERING, June, 1988. pp. 37-41.

2.9 Anon., "ENGINEERING MATERIALS HANDBOOK, Vol. 1 — Composites", ASM International, Metals Park, Ohio 44073. pp. 43-168. 1987.

2.10 Lewis, C. F., "Materials Keep a Low Profile", MATERIALS ENGINEERING, June, 1988. pp. 37-41.

2.11 English, L. K., "BMI: Today's Resin for Tomorrow", MATERIAL ENGINEERING, April, 1989. pp. 59-62.

2.12 Hunt, M., "MMC's for Exotic Needs", MATERIAL ENGINEERING, April, 1989. pp. 53-55.

2.13 Stover, "Foam and Syntactic Film Give Core Support to Composites", ADVANCED COMPOSITES Jan/Feb, 1991. pp. 32-38.

2.14 Hunt, M., "Epic Proportions in Metal-Matrix Composites", MATERIALS ENGINEERING, Mar., 1991. pp. 24-27.

2.15 Bucci, R. J. and aMueller, L. N., "ARALL Laminates: Properties and Design Update". 33rd International SAMPE Symposium. Mar. 7-9, 1988. pp. 1237-1248.

2.16 Young, J. F. and Shane, R. S., "MATERIALS AND PROCESSES", published by Marcel Dekker, Inc. 1985.

2.17 Seymour, R. B., "POLYMERS FOR ENGINEERING APPLICATIONS", published by ASM International, Metals Park, OH 44073. 1987.

2.18 Anon., "PLASTICS — Thermoplastics and Thermosets", co-published by D.A.T.A. Inc. and The International Plastics Selector, Inc. 1988.

2.19 Roy, S. K. and Chanda, M., "PLASTICS TECHNOLOGY HANDBOOK", published by Marcel Dekker, Inc. 1987.

2.20 Anon., "COMPOSITES & LAMINATES", co-published by D.A.T.A. Inc. and The International Plastics Selector Inc. 1988.

2.21 Goodman, S. H., "HANDBOOK OF THERMOSET PLASTICS", published by Noyes Publications. 1986.

2.22 Schurmans, P. B. H. and Verhoest, J., "INORGANIC FIBERS AND COMPOSITE MATERIALS (A Survey of Recent Developments)", published by Pergamon Press, Elmsford, New York, NY. 1984.

2.23 Metcalfe, A. G., "COMPOSITE MATERIALS, Volume 1 — Interfaces in Metal Matrix Composites", published by Academic Press, New York, NY. 1974.

2.24 Kreider, K. G., "COMPOSITES MATERIALS, Volume 4 — Metallic Matrix Composites", published by Academic Press, New York, NY. 1974.

2.25 Hancox, N. L., "FIBRE COMPOSITE HYBRID MATERIALS", published by MacMillan Publishing Co., Inc., New York, NY. 1981.

2.26 Schoutens, J. E., "INTRODUCTION TO METAL MATRIX COMPOSITE MATERIAL", prepared under sponsorship of the DOD Metal Matrix Composites Information Analysis Center. 1982.

2.27 Tsai, S. W., "MATERIALS TECHNOLOGY SERIES", from Vol. 1 through 14, published by Technomic Publishing Co., Landcaster, PA. From 1974 to 1984.

2.28 Wake, W. C., "TEXTILE REINFORCEMENT OF ELASTOMERS", published by Elsevier Appied Science, England. 1982.

2.29 Michaeli, W. and Rosenbaum, U., "Structural Braiding of Complex Shaped FRP Parts", 34th International SAMPE Sympsium, May 8-11, 1989.

2.30 Stevens, T., "PMR-15 is A-OK", MATERIALS ENGINEERING, Oct., 1990. pp. 34-38.

Chapter 3.0

TOOLING

3.1 INTRODUCTION

In selection of tooling materials (see Fig. 3.1.1) one should be sensitive to the thermal expansion in the tool and try to match it to the coefficient of thermal expansion (CTE) of the composite. For elevated temperature forming and consolidation, where the tooling must go to the same temperature as the laminate, steel, carbon (graphite), or ceramic tooling material must be used. In forming operations where only the laminate is heated and then pressed into cold tooling, a variety of materials can be used, such as aluminum, wood and even rubber and silicone. Composite tooling differs from conventional (metallic) tooling as shown below:

* Tolerance build-up is much more critical
* The final machined dimensions of the tool are not necessarily the final dimensions of the composite part; the degree of disparity depends on:
  — Type of tooling
  — CTE characteristics
* Final part dimensions are those present at the ultimate gelation temperature of the matrix system

Tool Material	Caefficient Thermal Expansion	Heat Conductivity	Material Cost	Fabrication Cost	Durability
Aluminum	Poor	Good	Good	Fair	Fair
Steel	Good	Good	Good	Poor	Good
Graphite	Excellent	Good	Good	Good	Poor
Ceramics	Excellent	Poor	Good	Fair	Fair
Fiberglass Resin Composite	Poor to Good	Fair	Good	Good	Poor
Graphite Epoxy Composite	Excellent	Fair	High	Fair	Poor

Fig. 3.1.1 Tooling Material Guide

There are no hard and fast rules to ease the tooling selection decision, and while some guidance (see Fig. 3.1.2) can be offered, the most cost effective tooling choice is still evolving.

The following gives the rating of tooling properties (factor: 1 — lowest; 5 — highest):

TOOLING PROPERTIES	FACTOR
Dimensional accuracy	5
Dimensional stability	5
Durability	5
Thermal mass	4
Surface finish	3
Ease of reproducibility	3
Temperature uniformity	3
Material cost	3
Ease of tool fabrication	3
Ease of repair	3
Tool weight	3
Ease of inspection	2
Resistance to handling damage	2
Ease of thermocouple implantation	1
Release agent compatibility	1
Sealant compatibility	1

The following list shows the variety of possible tooling methods:
- Autoclave
- Out-of-autoclave
- Hard mandrel
- Washout mandrel
- Inflatable mandrel
- Pressure vessel
- Silicone rubber
- Press forming
- Diaphragm forming
- Mechanical pressure
- Integrally-heated tools
- Elastomeric tooling

Fig. 3.1.3 gives the pros and cons of each of the most commonly used tooling types.

It is obvious that tooliong for composites is a very wide field, involving many technologies. In only rare situations is there a well defined correct solution to a specific tooling problem. What is the best way to fabricate a component for one organization with considerable experience in a specific method, may not be optimum for another with a different background. Compromises are needed at almost every step of tool evolution depending on requirements, incentives, economic resources, schedules, and other such issues.

Type	Characteristics
(a) Male mold 	• Most commonly used for aircraft parts because of its low cost • Lowest layup cost • Small radius producibility ≥ .05 inch • Baseline (non-aerodynamic surfaces) • Surface control one side only • Localized control of vacuum bag surface
(b) Female mold 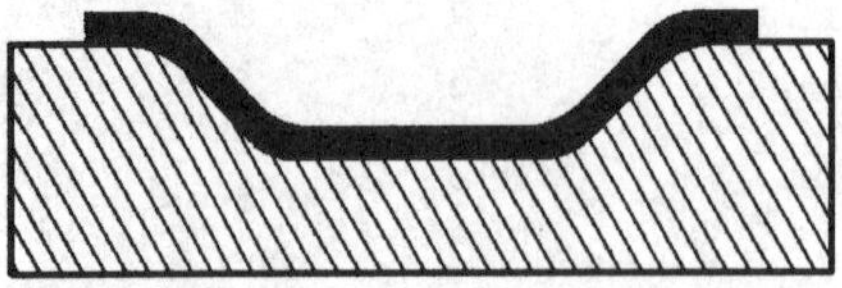	• Limited use in contour applications because of bend radius • Highest layup cost • Radius producibility ≥ .25 inch • Localized control of vacuum bag surface • Surface control one side only
(c) Matched die mold 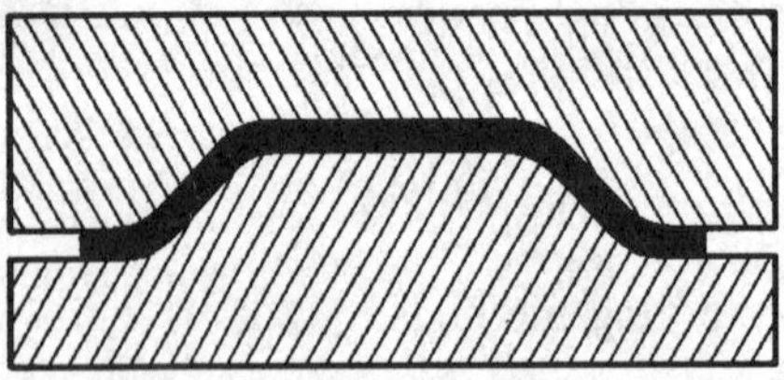	• Used male/female tooling to control laminate thickness and is very expensive • Best thickness control • Highest tooling cost • Moderate layup cost • OML/IML control (smooth surface both sides)

Fig 3.1.2 Types of Tooling

Guidelines for composites tooling are generally the same as those for sheet metal forming dies or compression molding. Tool contact with the deforming material should occur in such a way that the sheet surface pressure is uniform at all times. In geometries where this is not possible, such as those where the loading direction is perpendicular to the surface, the use of flexible tool halves is recommended to provide a sort of hydrostatic pressure. Normally, tools should be designed with a draft angle of 1 to 2 degrees to counteract the effect of "closure" or "spring-in" after cure and to facilitate ease of part removal from the tools or dies.

The demands placed upon mold tooling for curing composite parts can be very severe; the ideal tooling characteristics for composites are given below:
- CTE characteristics compatible with part to be produced
- Able to withstand severe temperature and pressure conditions without deterioration
- Dimensional stability
- Low cost
- Durable
- Reproduce pattern with high dimensional accuracy
- Retain mechanical properties at high temperatures

Type	Potential uses	Pro	Con
Autoclave	• Large components • Low volume production • Honeycomb sandwich assemblies • Cocured parts • Parts having vertical walls • Bonding	• Low cost • Internal heating possible • Undercut feasible • Vertical walls attainable • Versatility, particularly with large components • Complex, cocured parts attainable • Thermal expansion can be made to match part	• Low production rates • High labor cost due to ancillary material layup • Loose dimensional control of bag surface • Low molding pressure, relative to matched dies require more generous radii • Curing temperatures limited by ancillary materials unless internally heated tools are used with insulation installed between bag and layup • More process variables involved than with matched dies • Bag failure usually causes part to be scrapped
Matched metal dies	• Relatively small parts • Both surfaces dimensionally controlled	• High productivity • Good dimensional control • High molding temperatures • Good quality surfaces on all faces • High fabrication pressures • Durable • Internal heating feasible • Good thermal response and control • Compression molding tool technology available • Minimizing ancillary material use	• High cost due to machining, stops guides etc. • Tool thermal expansion different from composite • Limited ability to selectively reinforce • Undercuts require multi part tools • Draft angles required where vertical wall preclude part removal from tool • Large components present tool flexibility, heating uniformity and air-volatile removal difficulties • Difficult to repair or modify
Elastomeric	• Allows complex geometries • Large components feasible	• Considerable part design flexibility • More complex parts feasible than with matched metal dies due to casting of elastomeric elements • Ability to layup on numerous elastomeric mandrels and install these in the metal tools allows complex, parts to be made	• Limited life • Volatile and air removal less than ideal • 500°F processing limit • Low conductivity of elastomeric elements can cause undesirable thermal gradients in the part
Monolithic graphite	• Tight dimensional control of complex components • Rapid cure cycles (high heat up rates) • Prototype parts	• Low CTE, matched to graphite fiber composites • Temperature capability 600°F • Lower cost than metals • Easily machined in specialized facility • High thermal conductivity • Easy part release • Easy to repair or modify	• Susceptibility to impact damage • Special precautions needed when machining • Not suited for matched die molding
Ceramics	• Tight dimensional control of high temperature components	• Can be cast into complex shape • Low CTE which can be controlled • Electric and fluid heating systems easily cast into tool • Temperature capability 600°F	• Susceptible to impact damage • Difficult to repair

Fig. 3.1.3 Summary of Tooling Types (pros and cons)

Concurrent development of composite design and the tooling used to build it is the soundest foundation for product success. When the complexity of a part indicates serious problems, design modification should be sought early in the development phase. This allows mutually acceptable compromises between the design and manufacturing, to the long term advantage of both. However, after production tools have been made is almost always too late (due to both schedule and cost) to make major changes to the design. Much basic information on the design of tools for metallic structures is relevant when designing tools used to fabricate composites.

The method of heating the tools and ensuring their satisfactory thermal performance also strongly influences tool design. Inability to provide the desired temperature time history throughout the cure cycle is a basic fault of some tools. In most applications, each tool is used in conjunction with a number of ancillary materials (as discussed in Chapter 4.0) such as breathers, bleeders, bags, and sealants; their part must be considered in tool design.

Tooling cost should be kept as low as possible, particularly in prototype and research or development programs. The use of sculptured metal shapes should be limited. Composite tooling can also be expensive if tooling cast form molds is used for lay-up and cure; the final tool has to be high-temperature resistant and compatible in thermal expansion.

Life cycle costs are important to consider in tool design and tooling should be a non-recurring cost in the life of a fabrication project. Short life cycle tools or tools requiring a high degree of maintenance result in recurring costs which burden the total product cost.

Tool repair procedures must be sufficient to maintain:
- Strength
- Heat transfer
- Dimensions of the original tool
- Surface finish characteristics of the cured part

Nevertheless, tool design, construction, and operation must be consistent with the component being produced and the characteristics of the composite materials being used and their related fabrication and cure operations. It should be kept in mind that tools for composites are unique and careful consideration should be given to tool design during the fabrication of any composite component.

A tooling design should result from team effort and communication between the design, tooling, and manufacturing disciplines will minimize problems and provide a superior end product.

The following three aspects are integral components of composite structures:
- Design
- Tooling
- Manufacturing

These disciplines must work together throughout the development phase of any product, from idea inception to its culmination as hardware. The primary areas in which inter disciplinary agreement is needed concern:
- The trade-off between simplicity of tool design and number of parts produced
- The ability of composite materials to provide part count reduction

The overall reduction of part count can have an accompanying cost reduction benefit for the final structure through increased ease of assembly and minimized time flow in production. Full communication between disciplines provides optimum conditions to achieve this trade off.

All tooling systems discussed in this Chapter ralate very closely to the manufacturing methods dicussed in Chapter 4.0.

Tooling Materials

In complex, highly curved components a serious degradation in composite strength and dimensional accuracy can result when there is a serious mismatch in tool and composite CTE. Therefore, for parts that must meet close dimensional control or part mating requirements, the choice of the tooling material becomes very important and thermal change must be calculated into all tool dimensions. A prime consideration for selection of materials for fabrication of large tools used for curing composite structures is compatibility of thermal expansion between the tool and part.

Tooling materials currently used are
- Aluminum or steel — fabricated by standard metalworking techniques
- Electroformed nickel
- Composite mold tooling — Laid up and cured over a master model to produce a mold tool

Fig. 3.1.4 gives both the coefficients of thermal expansion and coefficients of thermal conductivity for most commonly used tooling materials. To compare with materials, see Fig. 2.1.3 of Chapter 2.

Material	Thermal Conductivity (BTU-in/ft^2-hr-°F)	Coefficient of Thermal Expansion (Micro-in/in/°F)
Graphite	400	1.5-2.0
Aluminum	1395	13.0
Steel	350	6.7
Nickel	500	6.6
Carbon-Fiber/Epoxy	24-42	0-1.5
Fiberglass/Epoxy	22-30	7-13
Ceramics (MgO, Al$_2$O$_3$, Gypsum)	10-80	3-6

Material	Apparent Density (g/cm^3)	Specific Heat (Cal/Gr-°C)	Thermal Mass (Cal/cm^3-°C)
Graphite	1.78	0.25	0.44
Aluminum	2.70	0.23	0.62
Steel	7.86	0.11	0.86
Nickel	8.90	0.11	0.98
Carbon-Fiber/Epoxy	1.5	0.25	0.38
Fiberglass/Epoxy	1.9	0.3	0.57
Ceramics (MgO, Al$_2$O$_3$ Gypsum)	1.6-3.9	0.84-1.50	1.2-5.3

(Source: Stackpole Carbon Co.)

Fig. 3.1.4 Typical Properties of Tooling Materials

The basic material characteristics relevant to tooling for composites depends to a great degree upon each application; factors determining choice of tooling materials are:
- Thermal Expansion compatibility
- Thermal Conductivity
- Accuracy required
- Strength
- Ease of tool fabrication
- Shop capability
- Cost per part
- Durability

Material	PRO	CON
Aluminum	• Machineability • Thermal conductivity • Pressure from expansion • Low weight and mass	• Dimensional Stability • Becomes soft when heated over 350°F • Easily distorted • Incompatible coefficient of thermal expansion
Steel	• Durability • Surface finish	• Warpage • Thermal expansion • Weight
Monolithic graphite	• Low expansion • Dimensional stability • Easily machined • Heat resistant • Thermal conductivity • Cost (for prototype parts)	• Porous • Soft surface • Bonding • Low strength • Needs back-up structure
Ceramic	• Low expansion • Dimensional stability • Heat resistant • Low shrink casting	• Porous • Soft surface • Machineability • Low strength • Thermal conductivity
Silicone rubber	• Pressure from thermal expansion • Out of autoclave cure • Can be molded to any shape • Low cost	• Hard to control, predict and measure pressure • Loses dimensional stability with repeated use

Fig. 3.1.5 Tooling Material Selection

Material	PRO	CON
Graphite/epoxy	• Excellent dimensional stability Stability • Good heat-up rate • Lightweight • Very compatible coefficient of thermal expansion • Low denisty • Ease of construction (plaster model) • Low cost tooling	• Durability • Limited strength at high temperatures • Must build master model • Not fealible for molding cocured stiffened panels
Electro-formed nickel	• Very smooth and scratch-resistant mold surface • Coefficient of thermal expansion (CTE) about 40% that of aluminum • Rapid heat-up/cool down • Light weight • Good repairability • Good release properites • Low cost to duplicate tools	• Long lead-time • High fabrication cost • No long-term durability • Substructure very heavy on large tools
Invar	• Low CTE • High thermal conductivity • Durable	• High material cost • Long lead-time • High fabrication cost
Avamid-N	• Good CTE match • Low fabrication cost	• Limited strength at high temperature • Limited high temperature capacity

Fig. 3.1.5 Tooling Material Selection (cont'd)

- Repairability
- Life Assessment
- Tool mass

High thermal conductivity is desirable:
- To minimize large thermal gradients in the tool and the part being fabricated
- When high temperature thermosets (i.e. polyimides) and thermoplastics are being cured, rapid rise in temperature may be required

Fig. 3.1.5 gives a tooling material selection comparison.

Tool Fabrication

The conventional sequence of composite tool fabrication:
- Master model (see later discussion in this chapter) is typically fabricated from plaster which computers can be used directly to machine the master model by numerical control (NC) process
- Plastic faced plaster mold is made off master model
- Graphite epoxy is laid up on plastic faced plaster mold to produce working tool

Alternative tool fabrication procedures:
- Machining of tool directly from metal
- Machining of tool directly from solid monolithic graphite
- Fabrication of tool directly from metal parts placed on a graphite base
- Casting of tool, e.g., ceramic mixture material, directly from master model

The following are general requirements for a tool design:
- Tool should extend a minimum of 2 to 3 inches (5 to 8 cm) beyond the edge of the part (actual edge designated by engineering drawing)
- Provision should be made for vacuum attachment
- Edge sealing provision should be considered to minimize the use of tape sealants (bagging problem area)

Tools always need to be adequately supported; metals used in tooling are chosen because they are inherently strong and low cost. The tool can be strengthened by a conventional "egg crating" constructed backup as shown in Fig. 3.1.6 which has two advantages:
- Lighter total weight
- Rapid heat to the back side of the tooling

For non-metallic tooling, however, the "egg-crating" or similar backup structures should be manufactured from non-metallic materials as a separate structure from the tool. It is generally more convenient to produce a return flange at right angles to the face of the tool, thereby strengthening the tool and providing support stand.

The backup structures added to provide stiffness and strength must be designed to avoid thermal distortions of the tool and localized temperature control problems.

Heating Systems of Tools

It is clear that tool thermal response must be accurately determined since temperature control is so important in the curing of composites. To accomplish this the selection of the heating systems and the tools response to it must be carefully evaluated. The three basic heating systems are:

(1) External heating by hot gases, typified by the conventional operating mode of autoclave or hydroclave (by heated water, oil, etc.) and these methods involve quite sophisticated calculations that depend on:
- The operating characteristics the autoclave heating fluid system
- Tool size
- Material and part shape

Tooling

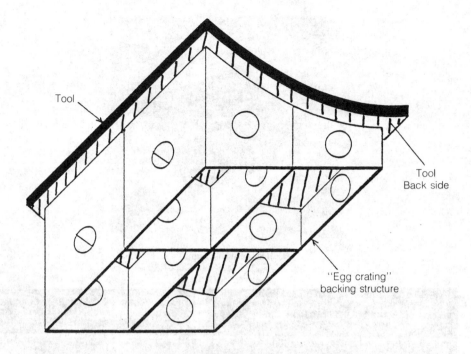

Fig. 3.1.6 Underneath View of a Tool to Illustrate "Egg Crating". Multiple Holes Allow Gas Flow to Rear of the Mold Surface.

- Heat capacity of all the tools in the autoclave during the run
- Position of the tools in the hot gas stream

A most important facet of this is developing a thorough understanding of the thermal and flow rate characteristics of each autoclave.

(2) Electrical heating of either the tool itself or the platen in intimate contact with the tool which can be accomplished by:
  - Computing tool volume, weight and surface area
  - Calculating the heat required
  - Determining the heat lost from the un-insulated side of the tool
  - Adding together the factors mentioned above provides the total number of BTU required to heat the tool to the desired temperature

This method also involves custom-designed heating elements and blankets for curing composites. The heating systems are available in single or multiple zone units, with temperature capability to 1400°F (760°C). Instead of heating the whole environment, a blanket can be used to heat only the part, saving energy:

- Silicone heat blanket, shown in Fig. 3.1.7(a), provides uniform heating up to 450 — 600°F (230 — 320°C)
- Insulating blanket, shown in Fig. 3.1.7(b), can be used with heating blanket to reduce heat lost to the environment and it can withstand up to 900 — 1400°F (480 — 760°C)
- Fig. 3.1.7(c) shows a reusable vacuum bag and heating blanket in one, which heats up to 450°F (230°C)

Chapter 3.0

(a) Heating blanket up to 450°F

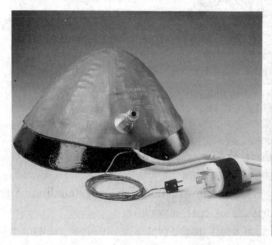
(b) Insulating blanket up to 1400°F

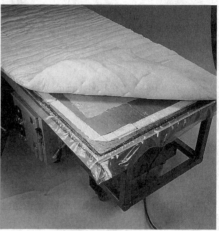
(c) Experimental compled curvature vacuum blanket up to 450°F

By courtesy of Briskheat Corp.

Fig. 3.1.7 Heating Blankets

(3) Fluid heating of either the tool or platen interior:
Steam or heated oil are the most commonly used media which are circulated within coils buried in the tool. Essentially the heating is controlled by the length of heating coil per unit volume of tool. The oil heating devices, sometimes known as Hydrotherm units, are programmable to provide heating or cooling between the temperature range of roughly 60°F (15°C) to 375°F (190°C)

(4) High-velocity jet stream heating (Moen system):
With the Moen system, shown in Fig. 3.1.8, the laminate part is subjected to high-velocity jets of air and the heat source is brought to the part, rather than the part being brought to the heat source; this heats the part faster and more efficiently and thus, requires less energy.

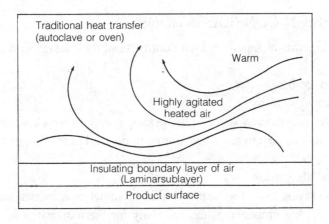

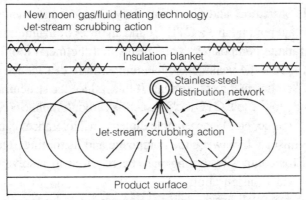

By courtesy of Heat Transfer Technologies

Fig. 3.1.8 Moen System-High-Velocity Jet-Stream Heating systems

This heating system uses a series of thin-walled tubes to distribute hot or cold jets of air through a series tiny holes along the tubes, and the tubing, serpentined over the surface of the tool, acts as a framework for the insulation barrier that surrounds the framework of tubes. This system can provide a temperature range between 300 to 2000°F (150 to 1100°C).

(5) Microwave and induction heating systems have been used successfully. These methods have been used on small parts and tend to be geometry dependent but have found little favor in the large structure field.

All of the above systems except item (5) accomplish the polymerizing step by heating the part from the outside in. This is a very inefficient method for thicker laminates and can result in laminate structures which are non-uniform. This is because the outside layers gel before the entrapped air and excess matrix at the interior of the laminate can be removed.

The heating method of item (5) cures the laminates from the inside out by activating molecules creating internal heat resulting in the polymerization of the matrix.

3.2 TOOL DESIGN CONSIDERATIONS

The three most significant factors which control final tool design concept selection are
- Cost
- Tool service life
- Dimensional stability

This Chapter will discuss some of the principal factors which should be considered during the planning and design stage for tooling to be used in the fabrication of composite structures.

(1) Thermal Expansion

The increasing use of woven and unidirectional tape, which have widely varying CTEs, has emphasized the necessity for thermal match in tool design. The problems encountered are most acute in the fabrication of long relatively slender composite structures and also complexly shaped components.

Steels and aluminum have a CTE around an order of magnitude greater than most carbon/graphite composites. This means that the metal tool contraction during cool down from the peak cure cycle temperature can induce severe residual or "built-in" strains in the component. If this reduce the structural capability it may be necessary to use low CTE tooling such as carbon/graphite composite, monolithic graphite, ceramic, etc. Unacceptable dimensional tolerances can also arise because of a CTE mismatch between the composite and its tooling. Therefore, low CTE materials for composite tooling may be required to resolve this problem.

(2) Part Size and Configuration

Occasions may well arise when the component to be fabricated is too large to fit in the available processing equipment, be it
- Autoclave
- Press
- Oven

In this case, there is little choice but to utilize a self-contained tool (the tooling and processing equipment for a specific component is built as a single unit, as discussed later in this Chapter) to fabricate the that part. The self-contained tool is facility independent. There is also a size limitation on the use of matched metal molds. Large matched metal tools are heavy, have excessive heat capacity or may be too flexible for press applications.

Tool size or complexity may require a redesign to create smaller, simpler parts which more joints and thus greater assembly costs. The compromise of design and tooling properties must be made to achieve subsystems goals varies from case to case.

Part configuration will affect the tooling selection because of the varying amounts of pressure needed to produce:
- A complexly shaped skin panel
- A thickness variation
- Taper or ply drop-offs

(3) Part Tolerance

In properly designed composite structures, the requirements for close control of part tolerances is held to a minimum. Composite aircraft components are usually tooled to the surface having an appearance or aerodynamic smoothness requirement. There are cases where components are tooled to the faying or mating surface to be subsequently mechanically fastened or adhesively bonded. For example, if only one face of the component is a controlled dimension, it is chosen as the tool face. The other surface dimension is not controlled precisely. However, there are designs where it is necessary to closely control both face dimensions. This can only be done to close tolerance by using a matched mold (see Fig. 3.1.2(c)) of some type.

Matched die or press to stop tooling will, of course, result in part thickness equivalent to the tool cavity dimensions; such tooling must be capable of compensating for the material bulk factor (thickness of preform lay-up/thickness of cured component). The bulk factor be as high as 125% even with intermediate debulking cycles during the lay-up.

On a part or assembly the tolerances play a very important role in the type of tool selected and the cost of the tool. Practically, the engineering tolerance used for a cured laminate is $\pm 10\%$ of the part thickness. Choice of tooling must anticipate assembly requirements. To summarize:
- Part tolerance is an important factor in determining cost and the type of tool selected
- Close tooling tolerances when producing thin laminates, e.g., 0.025 inch (0.63 mm), is expensive and can be impractical
- Tolerance accumulation affects composite assemblies to a greater extent than sheet metal assemblies

(4) Tool Life Expectancy

It is universally held that long-run production tooling must be made of steel. Aluminum tooling made of heavy roll formed and machined plates is the second choice. Non-metallic tooling is generally not as desirable, but is acceptable to development programs in which few units are to be fabricated.

(5) Surface Finish

Polished tool surfaces (RM63 or less) are generally not required for composite structures. Almost invariably, the preferred tool surface a parting film such as Teflon, fluorinated elastomer, or Teflon impregnated fabric. In addition, the use of peel plies, a sacrificial piece of fabric such as nylon cloth, is becoming more prevalent. This material acts as a surface bleeder in net cured resin systems and is removed just before bonding or painting.

Tooling surface requirements occur when:
- Tooling surface location is an important factor in final assembly fit and function
- Specification of surface smoothness in addition to tooling surface may be necessary to meet aerodynamic requirements on some designs such as high performance wing surfaces

(6) Repairs

All types of tooling used for composite structure curing can be repaired
- Steel and aluminum tools are repaired by welding and grinding to restore contour. Minor surface damage in steel or aluminum tools can also be repaired in using fill-in resins
- Composite material tooling can be repaired by grinding out the damaged area to ensure that a clean solid surface is reached and then rebuilding the area with the same fabrics and resin

(7) Process Effects

Tool design, construction, and operation must be consistent with the number of component to be produced and the characteristics of the structural materials being used and their related fabrication and cure operations.

The following must be taken into considerations:
- If the tooling is to be used in a blanket press or autoclave, the tool surface, pressure or vacuum bag, and bag seal must be vacuum tight. This characteristic must be designed and built into the tooling from the beginning of the tool design, since it is virtually impossible to effect a permanent leak repair.
- If internal component pressures are to be supplied by means of solid silicone elastomeric blocks, direct heating of the blocks by means of cartridge inserts may be required
- If matched molding or press operations are used, adequate means of press mounting, tool alignment, and means of heating and cooling must be incorporated

Processing considerations include:
- Temperature:
  — 350°F (180°C) — must thermoset materials
  — 350 to 700°F (180 to 370°C) — thermoplastics or polyimides
  — Above 700°F (370°C) — thermoplastics, polyimides, etc.
- High temperature tooling [i.e., above 350°F (180°C) for curing or consolidating materials of thermoplastics, polyimides, etc.]:
  — Expensive
  — Requires heat stable tooling materials
  — Thermal effects must be reduced
  — See Fig. 3.2.1 for assessment of tooling materials suitable for high temperature composites
- Pressure:
  — Vacuum only
  — Pressure chamber for higher pressure than that produced by vacuum
  — Press
  — Rubber mandrel thermal expansion
  — Aluminum block (or modules) thermal expansion

Tooling material	Process compatibility	PRO	CON
Steel	• Autoclave • Press	• Good thermal conductivity • Durable	• Warpage at high temps • High CTE • High fabrication cost • High denisty
Invar	• Autoclave • Press • Diaphragm	• Low CTE • High thermal conductivity • Durbale	• High material cost • Long lead-time material • High fabrication cost
Titanium	• Autoclave • Diaphragm	• Low CTE (closely matches composites) • Good thermal conductivity • Durable	• High material cost • High fabrication cost • Limited experience as a tooling material
Ceramic	• Autoclave • Diaphragm	• Low cofficient of thermal expansion • Low cost material • Low fabrication cost	• Low thermal conductivity • Fragile • Low fabrication cost • High density
Monolithic graphite	• Autoclave • Diaphragm	• Low CTE • Good thermal conductivity • Low density	• Fragile • Limited vacuum integrity • Special machine handling equipment required • Moderate fabrication cost
Aluminum	• Diaphragm • Press	• Low cost material • High thermal conductivity • Medium density	• High CTE • Limited strength at high temperatures
Avamid-N	• Autoclave	• Good CTE match • Low fabrication cost	• Limited strength at high temperatures • Limited high temperature capaibility

Fig. 3.2.1 Assessment of Tooling Materials Suitable for High Temperature Composite

(8) Tool Proofing

Before a tool is used for the fabrication of parts, a composite thermal survey and contour check should be performed. This can be accomplished by fabricating a representative part to the approved process specification.

(9) Tool handle provisions

If a tool weighs over 40 lbs, handling features are required which include lifting rings, fork-lift handling features and castor assembly. However, even the lightest weight tools made from composite material require special protective handling, because they are more prone to handling damage than their metal counterparts.

3.3 METALLIC TOOLING

Technological advances have made it possible for some metals to be compctitive in composite applications. Metal has always offered advantages of durability and, under good maintenance, a virtually unlimited tool life. Several metallic allows are also free of the CTE mismatch that plagues conventional steel and aluminum tooling.

High CTE tools may cause micro cracking in the composite part or fiber distortion in high temperature part. This happens during the time span right after composite gelation occurs. In many design cases thermal expansion of metal tools is used for compaction of laminates and the thermal contraction of the tool aids in the removal of the part from the tool.

Metal tooling has proven to be the most popular and heat transfer is another important characteristic which makes metals very effective. However, as composite structures grow in size and complexity, the differences in CTE make it difficult to control the final dimensions. Metal tools also have high thermal inertia which means that heat-up/cool-down takes time and energy.

However, when judged strictly on the basis of durability in mass production, metallic tooling often turn out to be the most economical choice.

Aluminum

Aluminum tooling is one of the most widely used tooling materials in composite manufacturing and it is generally used for flat laminates or small and slightly contoured laminates. It is readily cast and machined, light in weight to facilitate handling, quite cost effective, and has excellent thermal conductivity for heat transmission during forming and curing. For many applications aluminum is the material of choice for curing composites at temperatures below 400°F (200°C).

Aluminum material tools are satisfactory for simple parts without tight tolerances. Since aluminum has a relatively high CTE, this can be a problem or an advantage, depending upon the application, and this factor must be accounted for. It can be used as "modules" to create compaction pressure to fabricate skin stiffener flanges during an elevated temperature cure (see Fig. 3.3.1) a method similar to that of elastomeric rubber tooling except using aluminum. The high CTE is not a problem for the fabrication of almost planar structures such as wing cove panels.

Advantages of aluminum tooling are:
- Low cost
- Easily machined to complex shapes
- Excellent thermal conductivity for heat transmission during forming and curing
- Lower weight compared to steel
- Easily repaired by welding and grinding to restore contour

Aluminum is:
- Not applicable for highly compound curved parts due to its relative high CTE
- Not suitable for curing above 350°/400°F (177°/204°C) or consolidation of such materials as thermoplastic composites

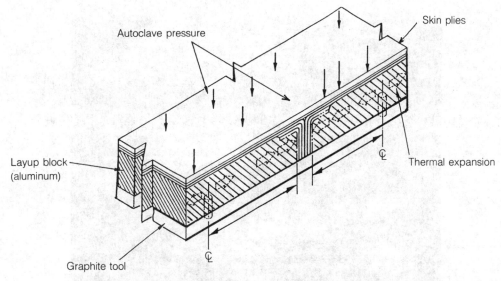

[Stiffeners and skin are compacted from thermal expansion aluminum blocks (or modules)]

Fig. 3.3.1 Aluminum Blocks Create Compaction Pressure by Their High CTE

Steel

Steel is a proven tooling composite material since large numbers of composite parts, both structural shapes and skins, have been made on steel tools. Steel has been used as the predominant tooling material because of availability, low cost, and a more compatible CTE compared to aluminum. Stainless steel is used extensively when severe radius forming is required. For a large tool, steel (or titanium) is a better choice than aluminum although still not totally compatible with the thermal expansion of composite structures.

Steel has following advantages:
- High Durability
- Weldability
- Repairability
- Vacuum integrity

Concerns are:
- High initial cost
- Difficulty of forming into complex curves and shapes
- Weight

In practice, to minimize the mass of the steel tool for relatively simple tools, sheet or plate shock can be machined to the desired configuration. If this is not feasible, rolled, or formed and/or welded and machined sheet stock as shown in Fig. 3.3.2, can be used to build up the mold surface (machined stock may be needed if close tolerances are required) tool structure. On large tools, backup structure of heavy steel tubings and angles are frequently required where these are attached to the mold surface. It is recommended to use an insulator between the mold surface and the backup structures to prevent the heat sink capacity of the backup structure from disturbing the thermal gradient over the mold surface.

Chapter 3.0

By courtesy of The Boeing Co.

Fig. 3.3.2 Rolled and Formed Steel Sheet Stock Tool for Consolidation of Thermoplastic Composites

Invar

Perhaps the best known of the metal tooling materials is Invar 36 (36% nickel) or Invar 42 (42% nickel) low-carbon austenitic steel alloy with very low CTE (from 0.5 — 6×10^{-6}in/in/°F). Invar refers to the metal's "invariable" dimensional properties. Invar is durable as steel. It has, in fact, all of steel's advantages, plus the added bonus of a CTE which matches composite materials.

(1) Invar 36 boasts a CTE of 1.5×10^{-6}in/in/°F, more suitable for thermosetting composite materials.

The material properties of Invar 36 as follows:

Density lb/in³ (g/cc)   0.29 (8.12)
Thermal conductivity   73
(32 to 212°F) BTU/ft²/hr/°F/in
CTE (10^{-6}in/in/°F):
   0.5 (room temperature to 200°F)
   1.5 (200 — 400°F)
   2.7 (400 — 600°F)
   6.0 (600 — 800°F)
Tensile strength (ksi)   70
Yield strength, (ksi)   40
Elongation   4%

(2) Invar 42, with CTE of 3.5×10^{-6} in/in/°F, can be used to build tooling for high temperature thermoplastic composite materials in the 500 to 800° (260 to 430°C) range.

Considerations in its use are:
- High initial cost
- Low thermal conductivity
- Weight
- Welding has been troublesome.
- Vacuum fittings (may have to be custom made from Invar in order to avoid leaks at the fittings)

Electro-deposited Nickel

Electro-deposited nickel tool (or electroformed or electroplated nickel tooling) is done by electro-depositing a platable metal over a mandrel or plastic model that is subsequently removed. Process steps for producing an electro-deposited nickel tool are shown in Fig. 3.3.3.

Advantages include:
- Durability
- Good part release, damage resistance, and vacuum-leak resistance
- Relative ease of repair via soldering or welding
- Tooling to be fabricated is limited in size only by the size of the electroforming tanks

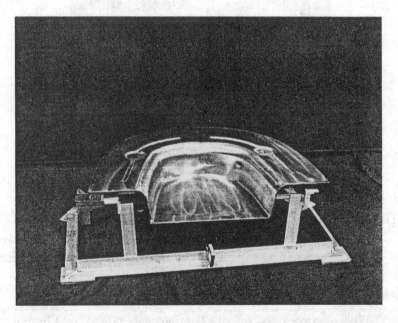

(a) Electroformed nickel mold for autoclave production of composite parts.

Fig. 3.3.3 Electro-Deposited (Electroformed) Nickel Tooling

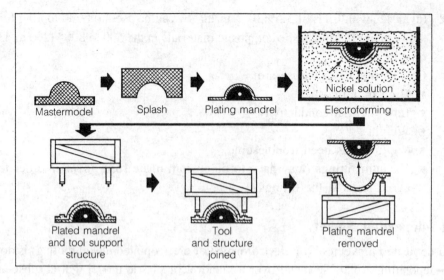

(b) Process steps for producing an electroformed nickel tool.

By courtesy of EMF Corp.

Fig. 3.3.3 Electro-Deposited (Electroformed) Nickel Tooling (cont'd)

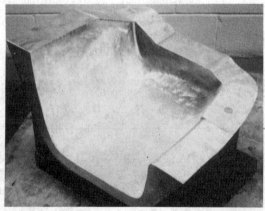

By courtesy of Internal Copper Association, Ltd

Fig. 3.3.4 Cast-to-size Copper Alloy Tooling

Considerations include:
- CTE of 7.0×10^{-6} in/in/°F (roughly the same as steel)
- Usually a correction (or shrink) factor must included in the tool design to allow the electroformed tool to expand to the correct dimension during curing

Copper Alloy

The bronzes, alloys of copper and tin, are familiar materials as used in monuments and statues. But a number of modern copper alloys, including the aluminum bronzes, beryllium coppers, and other high strength alloys, provide an excellent combination of strength and thermal conductivity for cast-to-size tools. A simple aluminum bronze tool cast directly to final size is shown in Fig. 3.3.4.

3.4 NON-METALLIC TOOLING

To date, the requirement of composite structural assembly, tolerance, economics and high-temperature applications, etc. for part size and complexity have grown considerably. Large parts, such as aircraft wing skins, or dimensional/tolerance consideration may lead to the decision to employ non-metallic tooling. Generally, non-metallic tooling fabrication is a multi-step process.

Master Model

Plaster master models, as shown in Fig. 3.4.1, have been widely used for constructing tooling masters (patterns). The main consideration with plaster master model is their inherent tendency to magnify CTE differences and inaccuracies through the reversals required to build a master model. The process is both labor intensive and time consuming.

The construction sequences are:
(1) Mounting a series of templates (also see Fig. 11.4.28 in Chapter 11)
(2) The number of templates is determined by the size of the part and amount of contour
(3) Templates are firmly fixed into position
(4) Plaster is cast to the template
(5) A finish coat of finer grain plaster applied over the surface
(6) Oven baked
(7) A seal coat is applied
(8) A resin gel coat is applied to obtain a smooth surface

Another problem centers on the temperature capability of the plastic master model. It is often difficult to make the master out of materials that will stand the high temperature needed to cure the tool. This has forced the use of intermediates:

- The first taken off the master model
- The second taken from the first
- The second intermediate, which can stand high temperatures, is then used to make the tool

On large or complexly machined master models, the contours can be checked with templates or with a coordinate-measuring machine which can also be used to verify the proper location of tooling holes, scribe lines, and reference points.

New board stock modeling materials have been developed which allow easy fabrication of tools. They are ideal for machining on three- and five-axis milling equipment directly from CAD data, as well as for hand-carving. The model may cure at room temperature and can be used to produce a prototype or short-run tool designed for use up to 350°F (170°C). One of these modeling materials is shown in Fig. 3.4.2.

Composite Mold Tooling

The complete tooling package is made up from either wet lay-up or prepreg systems using both vacuum/oven and autoclave consolidation techniques. In practice the majority of composite mold tooling methods involve impregnating glass or carbon fabric by covering them with high temperature resistant resins.

Chapter 3.0

(a) Templates are being rigged

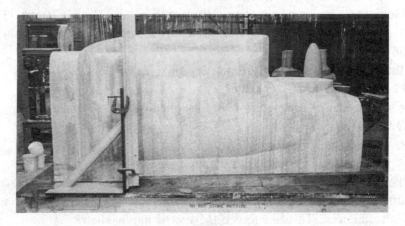

(b) Fairing complete

By courtesy of Harbor Patterns Inc

Fig. 3.4.1 Master model (plaster) for a Helicopter Lower Fuselage/Gun Turret

(a) Attach the rough mold surface to the base

Fig. 3.4.2 Ren Shape 450 (Polyurethane) Modeling Material

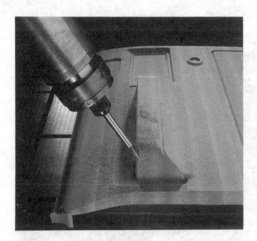

(b) Machine (or carve) contours on the model by NC   (c) Renshape 450 masters

By courtesy of CIBA-GEIGY Corp.

Fig. 3.4.2 Ren Shape 450 (Polyurethane) Modeling Material (cont'd)

Composite mold tooling has several basic advantages over counterparts in fabricating composite components:
- It is a low cost tooling method because the composite prepreg materials can be laid on plaster model and then cured
- The CTE of the tooling material is compatible
- Graphite mold tools have a relatively uniform temperature distribution, which allows the part to heat up evenly and prevents build-up of internal residual stresses
- Low density makes composite material tools easier to handle than metal counterparts.
- If damage occurs a new tool can be fabricated rapidly and economically from original master model

The weak point of composite mold tools has been the matrix, which must be made to stand multiple cycles to elevated temperatures without cracking. Basically, there are several methods of fabricating composite mold tools, as described below:

(1) Wet-layup — The wet lay-up method, shown in Fig. 3.4.3, is directly from a low temperature resistant plaster model. Dry fabric and liquid matrix are applied to the plaster model, consolidated under a vacuum bag and allowed to cure at room temperature. The tool is then removed from the plaster model and post-cured. To ensure a good tooling surface a gel coat is applied to a released model.

(2) Hot-cured Prepreg — This method of tool production, shown in Fig. 3.4.4, eliminates the problems of variability, low fiber-volume fraction and high void content found in the wet-layup method. But the cost is higher because this of extra tooling stages to produce a high temperature model (a permanent pattern) in order to process the prepreg material.

Chapter 3.0

(3) Room-temperature-curing prepreg — In this method specially-prepared prepregs are vacuum bagged on the master model until they developed sufficient strength for a free-standing postcure in an oven. This is an important feature for cheaper tooling because the master model is never subjected to heat.

By courtesy of Harbor pattern Inc

Fig. 3.4.3 Prepreg and wet layup composite Tools

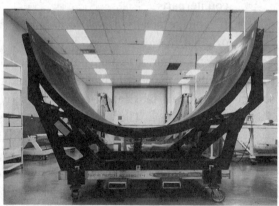

By courtesy of Airtech International Inc.

(a) Boeing B757 Fan-cowl Tool

Fig. 3.4.4 Composite Prepreg Tools

150

Tooling

By courtesy of Advanced Composites Group, Inc.

(b) Good drapability prepreg tool

Fig. 3.4.4 Composite Prepreg Tools (cont'd)

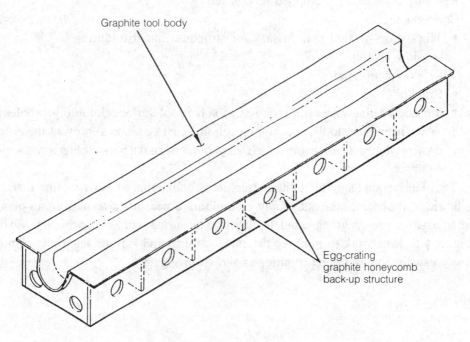

Fig. 3.4.5 Egg-crating Backing Structure for Long Aspect-ratio Composite Mold Tool

High-temperature carbon or graphite/epoxy tooling is ideal for curing laminates because it has excellent strength-to-weight characteristics as well as thermal and dimensional stability. Also because the tools eliminate heat-sink considerations during part curing and subsequent molding, autoclave time is reduced significantly and manufacturing cost is also reduced. Backup structure for tools can also be composed of composites, as described below:

(a) Composite carbon/epoxy hollow prepreg stiffeners, square tubes, and angles (see Fig. 3.4.4(a))
(b) A system of incorporating "egg-crating" backup structure made from solid laminate or honeycomb panels with series of holes on the back side of the tool face (see Fig. 3.4.4(b) and Fig. 3.4.5)

Monolithic Graphite Tooling

Monolithic graphite has a very unique blend of physical properties which make it very suitable as a composite tooling material. Among the more important of these are:
- Low CTE [1.7 to 2.2×10^{-6} in/in/°F (3.0 to 4.0×10^{-6} cm/cm/°C)]
- High thermal conductivity
- Low thermal mass
- Dimensional stability at temperatures up to 3600°F (2000°C)

Advantages:
- Easily machined
- Extremely amenable to integrally heated design
- Large tools can be made by bonding graphite segments together
- Virtually no restrictions on maximum temperature capability
- Easily repairable, if chipped or cracked

Concerns are:
- Thick cross-section is necessary for structural integrity during
  — slow heat-up
  — slow cool-down
- Fragile
- Surface coating generally required to solve tool surface durability problems
- When machining tools, special precautions must be taken to prevent the graphite dust produced from being inhaled by humans and from arcing any electrical equipment

Tool fabrication using monolithic graphite is unlike that of other tooling materials. The fabrication process consists simply of building a near-net-size machining blank of bonding together pre-cut graphite billets, followed by surface contour machining, as shown in Fig. 3.4.6. Handwork to improve the surface finish and surface sealing to minimize porosity usually follow the machining, as shown in Fig. 3.4.7.

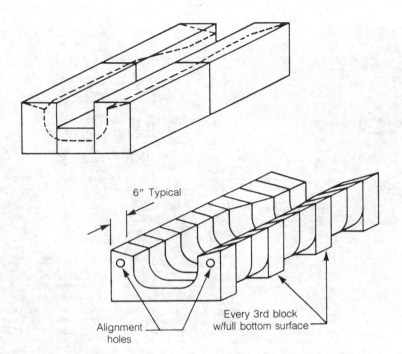

Fig. 3.4.6 Near-net-contour Monolithic Graphite Tool

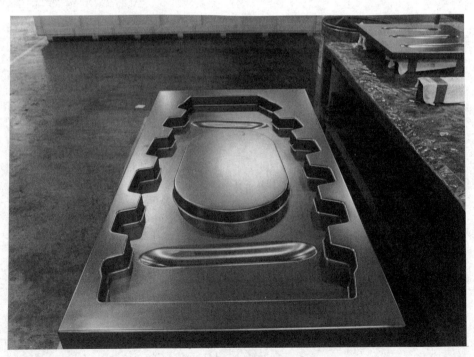

(a) Tool for formed rib

Fig. 3.4.7 Monolithic Graphite Tools

(b) Tool for formed rib with multi-beads

By courtesy of Coast Composites, Inc.

Fig. 3.4.7 Monolithic Graphite Tools (cont'd)

(Tool made by CBC product Group)

Fig. 3.4.8 Ceramic Tool (Comtek) for A-6 Rewing project

Ceramic Tooling

Ceramic materials are attractive candidates for tools because of a characteristically low CTE. Ceramics are excellent insulators, allowing heating to take place via electrical wires buried in the tool and located close to the surface which is contiguous to the composite laminate part.

Advantages of ceramic tooling are:
- Relative low CTE (0.5 to 4.5×10^6 in/in/°F)
- Critical part dimensions can be held to tighter tolerances
- Suitable for high temperature cure/consolidation such as polyimides and thermoplastic materials

Concerns are:
- High cost
- Lack of easy machinability
- Difficult repairs
- Low fracture resistance
- Low tensile strength and brittleness
- Low thermal conductivity
- Relative coarse surface
- Long heat-up and cool down times

Since ceramic material is frangible, attempts are being made to reinforce it to allow its use for large tool surfaces. Ceramics are castable and could be used as large tools for surfaces with compound contours, as shown in Fig. 3.4.8.

Fig. 3.4.9 illustrates a tool support and backup structure for a ceramic mold tool design. The tool frame (on the back side of the ceramic mold face) is mechanically and thermally isolated from the ceramic mold, by using wire tubing and fiber glass cloth insulation, to prevent stress on ceramic mold which would occur if frame and ceramic mold were mechanically attached due to the thermal expansion differential of the steel frame and the ceramic mold.

Reinforcement methods for ceramics are similar to those used for reinforced concrete:
- Reinforcements such as bars, wire, mesh, screen, fibers can be imbedded in ceramic tools
- Steel reinforcements are prohibited because of the CTE incompatibility of the two materials
- Glass and Kevlar fibers are prohibited as they do not bond well to ceramic materials

Spray-metal Tooling

Spray-metal tooling consists of a thermally sprayed metal shell [roughly 0.1 inch (2.54 mm)] which is supported by a non-laminated, isotopic composite backup structure. Pattern materials for net-shape molded parts are typically such low-cost, low-temperature materials as wood, plaster, or tooling wax. These tools are integrally heated for use in simple hydraulic or pneumatic presses to achieve increased production rates, and lower energy costs. Several materials can be used for the facing including:

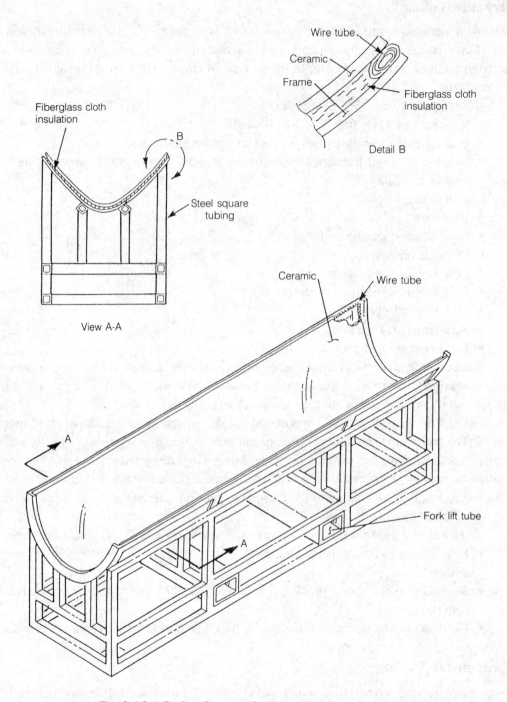

Fig. 3.4.9 A Backup Structure for Supporting Ceramic Mold Tool

- Aluminium/bronze, nickel alloys and tin/zinc alloy (CTE range from 5 to 7.5 in/in/°F) for applications up to 600°F (315°C)
- Nickel/iron alloy and titanium (CTE as low as 2.5×10^{-6} in/in/°F) for applications up to 750°F (400°C)

Wash-out Mandrel

Wash-out mandrel materials can be easily removed after processing of complexly shaped parts; see Fig. 3.4.10(a). To date, salts and plaster are the most widely developed and the most versatile mandrel materials, and removal or wash-out with water is easily accomplished. Both of these mandrel systems are described below:

(1) Mandrels from Eutectic salt or similar materials used for curing temperature under 350°F (180°C) are currently widely used to fabricate highly complex shaped ducts for various aircraft as shown in Fig. 3.4.10(b). A "hot-melt" procedure is used for form the mendrel, as described below:
   - A dry powder (i.e., Eutectic salt) is heated to 400°F (200°C), and becomes a liquid
   - The liquid is poured into a form tool (break-away-block)
   - The material solidifies from the outside in
   - When the operator feels the mandrel has a sufficient wall thickness, the molten center is emptied back into the melting pot
   - The mandrel is allowed to cool
   - The mandrel is removed from the forming tool
   - A number of composite layers (e.g., fiberglass, Kelvar, etc.) are applied over the mandrel and placed in an oven and cured
   - After cure, the mandrel is removed by placement in a hot water tank where the mandrel washes away leaving the fabricated composite part

(a) Eutectic Salt Mandrel

Fig. 3.4.10 Hot-melt Process (Under 350°F Curing)

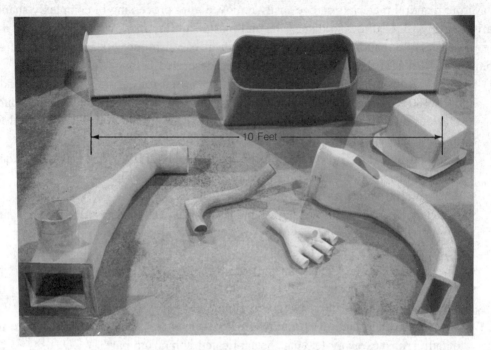

(b) Complex Shaped Ducts

By courtesy of Lockheed Aircraft Service Co.

Fig. 3.4.10 Hot-melt Process (under 350°F Curing) (cont'd)

By courtesy of Lockheed Aeronautical Systems Co.

Fig. 3.4.11 Plaster Wash-out Mandrels Used to Consolidate Hat Stiffened Curved Panel

(2) Plaster material has been used successfully to make mandrels for processing thermoplastic components which need a curing temperature above 350°F (180°C). Some difficulties are
- Low flexure strength
- Brittleness

A specially treated plaster mandrel (see Fig. 3.4.11) has been developed to improve the fabrication of high temperature composite materials. The properties of this new mandrel are described in Fig. 3.4.12.

PROPERTY	TEMPERATURE		
	(Room	600°F	800°F)
Compressive strength (psi)	290	450	490
Young's modulus (psi)	50,000	38,000	38,000
CTE (10^{-6} in/in/°F)	1.46-1.85 (up to 660°F)		
(curing temperature up to 800°F and pressure up to 400 psi)			

(Source: Composites Horizons, Inc.)

Fig. 3.4.12 High Temperature Wash-out Mandrel (CARE-MOLD)

3.5 CHAMBER SYSTEMS

Autoclave

The autoclave is a general all purpose system which can handle any configuration part which will fit within the chamber. Heating systems are available with utilize electricity, gas, fuel, oil, steam, etc. The choice depends upon which is most economical and which is available. In general, an autoclave is a heated pressure vessel, as shown in Fig. 3.5.1, and its process depends on internal heat transfer from a pressurized gas (inert gas, i.e., Nitrogen and CO_2) to cure the composite components. Most large autoclaves are built to the customer's requirements. Many smaller autoclaves are in production. Of all the heating systems the autoclave is the most popular in the composites field; its characteristics are given below:
- High acquisition cost
- Temperature up to 1,200°F (650°C)
- Pressure capability up to 500 psi
- Requires bagging
- Long cycle times

Autoclave use dictates certain criteria including:
- Low mass tools
- Promotion of efficient heat transfer (heated gas circulated by fans)
- Designing the tool to allow as much circulation of heated gas around or throughout the tool as possible (see both Fig. 3.1.6 and Fig. 3.4.4 which illustrate backup structures of "egg-crating" with a series of holes, and hollow square tubing design cases, respectively)

Chapter 3.0

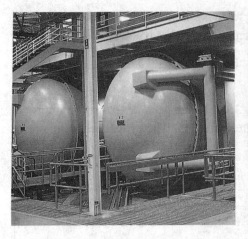

By courtesy of Thermal Equipment Corp

(a) A fleet of autcolaves. (two 15ft × 45ft-650° with 215 psi; one 9ft × 32ft-650°F with 210 psi)

By courtesy of Thermal Equipment Corp

(b) A 25ft × 60 ft Autoclave

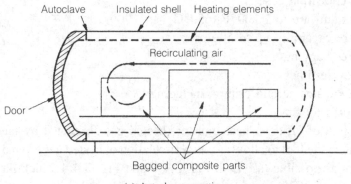

(c) Autoclave operation

Fig. 3.5.1. Autoclave Systems

The tool materials recommended for autoclave cure are:
- For a small number parts to be produced and cured at temperatures under 400°F (200°C) — use laminated tools
- For prototype or development parts which only require a very few parts be produced within a short time — use plaster or casted tools
- For long production use (durability), rapid heat transfer, and higher temperature cure systems (i.e., thermoplastic or polyimides materials) — use machined metal tools

Concerns are:
- Long process times
- Size limitation (large components needs bigger autoclaves)
- Inherently energy inefficient system
- Major difficulties caused by large thermal gradients and slow heat up times
- Bag failures

Some general knowledge about autoclave systems:
- It is mandatory to design and fabricate an autoclave to meet American Society of Mechanical Engineers (ASME) requirements, if it is to be operated in the U.S.A.
- Hydrostatic testing requirement
- The autoclave and its closures should be inspected regularly by a recognized inspector for safety considerations
- Insulation is used to greatly reduce energy costs and keep the maximum exterior autoclave shell temperature below 140°F (60°C)
- The life of an autoclave should exceed 50 years under proper maintenance
- The typical gases used in autoclave are:
  — Air: Air is inexpensive when supplied in the 100 to 150 psi (700 to 1030 kPa) range and temperature in 250°F (120°C) range. Air sustains combustion and is hazardous at temperatures above 300°F (150°C).
  — Nitrogen: Nitrogen is the most commonly used gas in autoclaves; however, nitrogen is expensive, especially for use in large autoclaves which operates at high pressure. Nitrogen can be used up to 650°F (340°C) and at more than 200 psi.
  — Cabon dioxide (CO_2): Co_2 is the second most commonly used gas. Care must be taken during operation because its high density presents a potential hazard to personnel.
  When using either nitrogen or CO_2 gas, be certain that adequate oxygen is present
- Most small autoclaves are electrically heated. Electrical heating may not be economical for large autoclave operation

Hydroclave

A hydroclave is similar to an autoclave and the same basic principles apply, except that it is hydraulically pressurized. The major difference is the higher pressures involved, up to thousands of pounds per square inch, as compared to 200 psi maximum for most autoclaves.

The use of water can be characterized as follows:
- Minimizes the pressure sealing problems
- Reduces personnel hazards to an acceptable level
- Eliminates the need for expensive, high pressure gas sources
- Any leaks that occur reduce the pressure inside the vessel to zero
- Rubber bagging is used instead of the plastic film used in autoclaves
- Due to the higher pressure involved, additional membrane elasticity and strength is necessary to prevent ruptures in the membrane in areas where bridging is likely to occur
- If used as a pressure cooler by using water as a coolant and tools are integrally-heated to provide the temperature specified by the cure cycle and common rubber bags can be used and re-used, for a great number of cycles

The standard water heating methods are slow, and both the temperature rise rates and maximum temperature requirements necessary for most composites are unattainable when using a hydroclave. The heating capabilities of water over 450°F (230°C) is limited and this forces the use of integrally-heated tooling systems. Obviously this system is used in only special applications. The common knowledge and safety considerations which apply to autoclaves are also applicable to this system.

The differences between an autoclave and hydroclave are:

FUNCTION	AUTOCLAVE	HYDROCLAVE
Fluid medium	Inert gas	Water
Pressure	up to 500 psi	up to 12,000 psi
Temperature	750°F (400°C)	450°F (230°C) (under pressure)

Concerns are same as autoclave.

Pressure Vessel

While autoclaves have a high initial acquisition cost, the pressure vessel (see Fig. 3.5.2) provides pressure only a low acquisition cost and the needed heat can be supplied by an integrally-heated tool system or in a similar manner. The common knowledge and safety considerations listed for autoclaves also apply to this system.

Advantages are:
- Low acquisition cost (simple tube construction without complicated Heating systems — costly items)
- Can provide much higher pressure than autoclaves
- Use of heating blankets or integrally-heated tooling provides an economical heat source
- Good dimensional control (mold line control) possible
- Short cycle time

Concerns are:
- need separate heating system

Tooling

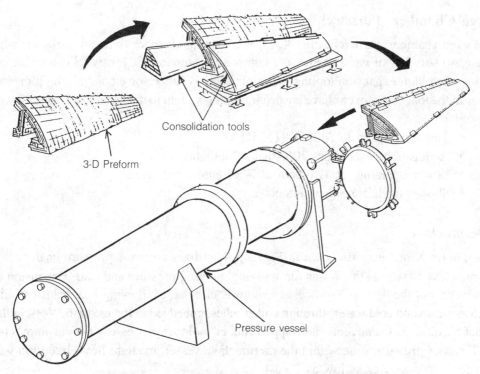

Fig. 3.5.2 Pressure Vessel (Heating provided by Consolidation Tooling itself)

By courtesy of AOV Industries

Fig. 3.5.3 An Oven Heats and Cures Composites (Without pressure except vacuum)

163

Oven Chamber (Furnace)

The oven chamber, shown in Fig. 3.5.3, is a heating chamber (without pressure) which is used to cure low or vacuum pressure composite components. Pressures can be increased by using high expansion tooling materials such as rubber or employing high pressure gases in the tool. Most ovens have an air-circulating system to distribute the heat uniformly. Advantages are:
- Temperature up to 2400°F (1315°C)
- Acquisition costs about 20% that of autoclaves
- Lower operating costs (about 50% of autoclave costs)
- Shorter cycle time than autoclave

Thermoclave

The Therm-X process, as shown in Fig. 3.5.4, utilizes a unique pressure chamber, the thermoclave vessel, as the means for applying molding pressure and heat. Expansion and contraction of the Therm-X process medium is effected by flowing heat transfer fluids, such as steam and cold water, through coiled tubes embedded in the medium. Vessel ullage is not required as in conventional trapped rubber molding processes. The medium's high CTE causes pressurization within the thermoclave vessel, medium flows like fluid when subjected to pressures as low as 10 psi.

The Therm-X process has following advantages:
- Vacuum bag leaking problems are eliminated
- Employs particles of silicone elastomer which can be used at temperatures up to 1500°F (820°C) and can provide pressures up to 3000 psi
- It stores less energy than pressurized gases, so the vessel is less likely to catastrophically burst
- When pressure is released, the silicone returns to its original powder form

Concerns are:
- Equipment size
- Cost

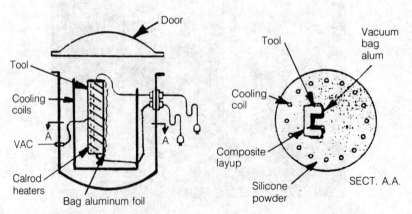

(a) Thermoclave vessel pressure is a function of the average temperature of all the medium (particute silicone elastomeric rubber-silicone powder) within the vessel.

Fig. 3.5.4 Thermoclave and It's Therm-x Process

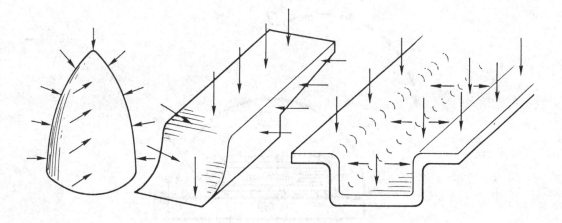

(b) Medium exerts uniform pressure on all surfaces of even the most complex shapes

By courtesy of United Technologies Chemical systems

Fig. 3.5.4 Thermoclave and Its Therm-x Process (cont'd)

3.6 SELF-CONTAINED TOOLING SYSTEM

As mentioned earlier, the autoclave utilizes heat and pressure to cure composite parts. Several autoclave limitations can be eliminated by use of self-contained tooling systems as illustrated in Fig. 3.6.1. In this system the heating and pressurizing systems are part of the tool.

Self-contained tooling is a system in which the pressurization and heating are integral to the tool itself or a dedicated part of the tooling package.
   (a) Heating can be accomplished electrically by incorporating wires or blankets in the tool.
   (b) Pressurization can be accomplished with the aid of expanding rubber or hydraulic systems.

Advantages are:
* Allow processing of parts too large to fit in an existing processing facility, i.e. an autoclave
* Allows alternate processes on same production floor without disrupting existing facilites
* A small increase in manufacturing capability is possible without a commitment to purchase new fixed asset equipment
* The heat source can be electricity, heated oil or steam
* Pressurization can be accomplished hydraulically or with compressed gas

Benefits gained from self-contained tooling compared to autoclave use are:
* Lower cost
* The small volume of area to be pressurized within the tool means far more rapid pressurization than can be attained with an autoclave
* Accurate temperature control is much more readily achieved than with autoclave use

Chapter 3.0

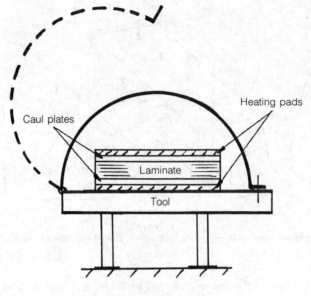

(a) simple plate laminate

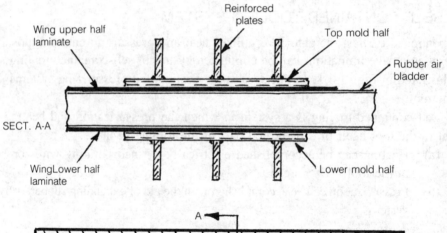

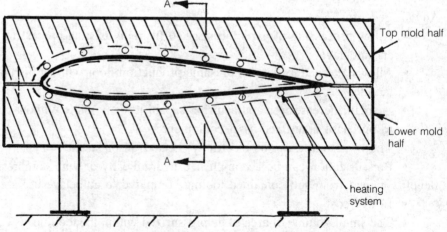

(b) Mold halves of fabricating wing skin laminates

Fig. 3.6.1 Self-contained Tools

Concerns are:
- Little experience in the design and fabrication
- Tool mass
- Design details must to be frozen at the time tool fabrication is started, otherwise the cost incurred in tool revision is impractical

Fig. 3.6.1(b) shows mold halves which are made of carbon prepregs. The prepreg assembly is placed in the lower mold half, a rubber air bladder which will be used for pressurization during curing process is inserted, and the top mold half is lowered into position and clamped. The mold halves are independently heated by integral-heating systems. Cycle times need not pace any other equipment in the plant, and the mold may be at any place along the assembly line. Parts can be finished on line and immediately placed into the airframe assembly.

Integrally-heated Tooling

Integral heating systems, shown in Fig. 3.6.2, eliminate the need for an autoclave or oven and they can be used in a pressure vessel (see Fig. 3.5.2) to cure high quality composite structures. This system offers the following advantages:
- The ability to handle large or complex shapes
- Reduced energy consumption
- Improved heat distribution and control
- Eliminates the need for costly autoclave time
- Can be incorporated into prepreg, ceramic, or wet-layup resin tools
- Permits part fabrication to be done along side of the airframe assembly

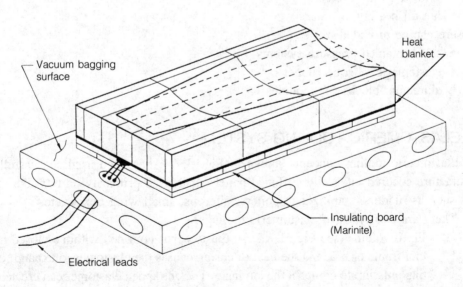

(a) Tool used to cure a skin component

Fig. 3.6.2 Schematic of Integrally-Heated Monolitic Graphite Tools

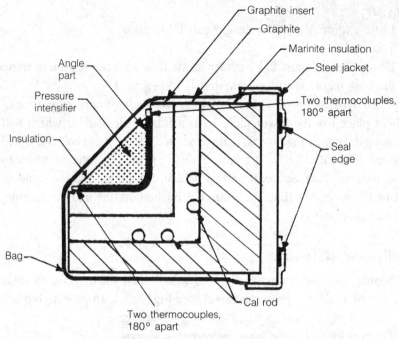

(b) Cross-section of the tool curing Angle part

Fig. 3.6.2 Schematic of Integrally-heated Monolitic Graphite Tools (cont'd)

Heating systems can be provided by:
- electrical resistance heaters
- heating blankets
- printed circuit systems
- hot oil heating

Pressure can be provided by:
- self-contained pressure systems
- elastomeric silicone rubber
- aluminum blocks (or modules)

3.7 ELASTOMERIC TOOLING SYSTEM

The Elastomeric tooling concept, shown in Fig. 3.7.1, has been used successfully to manufacture cocured integrally stiffened panels. It is used primarily on box-like structures such as rudders, stabilizers, spoilers, ailerons, small wing boxes, etc.

There are two types of elastomeric toolings:
(a) Fixed volume (see Fig. 3.7.2) — Elastomer is confined within a closed metal tool frame or box and the heat of curing causes the elastomer to expand, exerting pressure to compact the laminate. The entrapped elastomer can create over 1000 psi pressure on itself and surrounding metal tooling at high curing or consolidation temperature.

Tooling

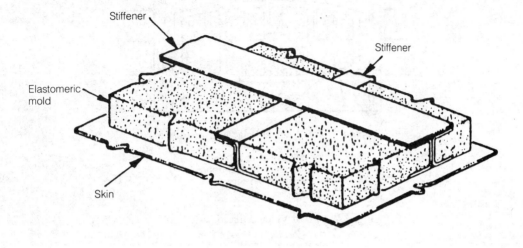

(a) J-stiffened panel

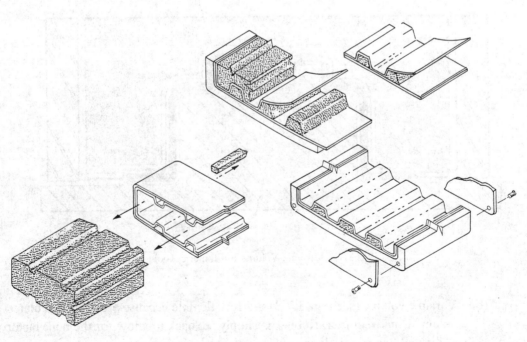

(b) Corrugation panel

Fig. 3.7.1 Elastomeric Tooling

Chapter 3.0

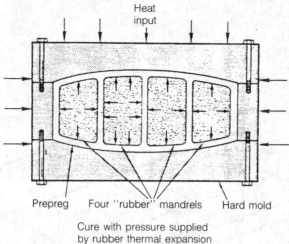

Fig. 3.7.2 Fixed-Volume Elastomeric Tool

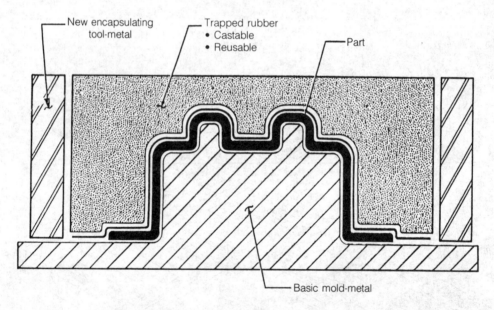

Fig. 3.7.3 Variable-Volume Elastomeric Tool

(b) Variable volume (see Fig. 3.7.3) — More flexible because precisely calculated volume is not required. Rubber is simply set back to allow for the bulk factor of the molding material; any excess pressure is vented against the vacuum bag.

The simplest form of an elastomeric molding tool:
- Consists of one or more silicone rubber mandrels
- Fits into a closed box or frame work
- Leaves ample room for the composite prepreg plies lay-up
- Is heated to cause the silicone rubber mandrels to expand
- Generates pressure against the walls of the box or framework and also the composite laminates
- Compacts and forms the part to its final configuration

Elastomeric tools can be designed for an heated-oven (no pressure) or integrally-heated tooling system cure processes. Other methods may also be included, such as heated-platen press or autoclave molding.

In some design cases, this is the most cost effective process for fabricating components. See Fig. 3.7.4.

The advantages of elastomeric tooling:
- No rework — low cost
- Shorter cure — improved turn around time
- No bag, no sealant, no tool cleanup
- Castable
- Can be reinforced by mixed with chopped fibers

Concerns are:
- Requires fabrication of encapsulating tool — additional cost
- Durability — rubber does have a limited life when subjected to repeated thermal cycling. (30 cycles or less for some cases)
- In compatibility with curing materials (in some cases)

When used in conjunction with matched die molding to produce complex components that cannot be made by match dies alone. Fig. 3.7.5 shows a design which is not feasible with metal molding along. The consolidation pressure for the horizontal flanges is provided by matched metal pressure while that for the vertical flanges is provided by the elastomeric mold. It should also be recognized that there are temperature limitations for this type of tooling.

The thermal conductivity of rubber is much lower than that of metal tools and the heat up lag of the elastomers can cause composite curing problems. To solve this, the elastomer volume must be kept to a minimum and/or the cure cycle of the composites must also be modified. However, if elastomeric tooling is the only viable process, there are some design situations, such as shown in Fig. 3.7.5, in which inserting metal plugs into the rubber mass to reduce rubber volume and improve heat-up rate will reduce cost and improve performance.

Inflatable Mandrels

A variation of the elastomeric tooling system is the inflatable mandrel (or bladder). They are inflated from an outside source and transfer pressure to the laminate panel surface. Fig. 3.7.6 shows inflatable mandrel tooling for the fabrication of wing type structures.

Chapter 3.0

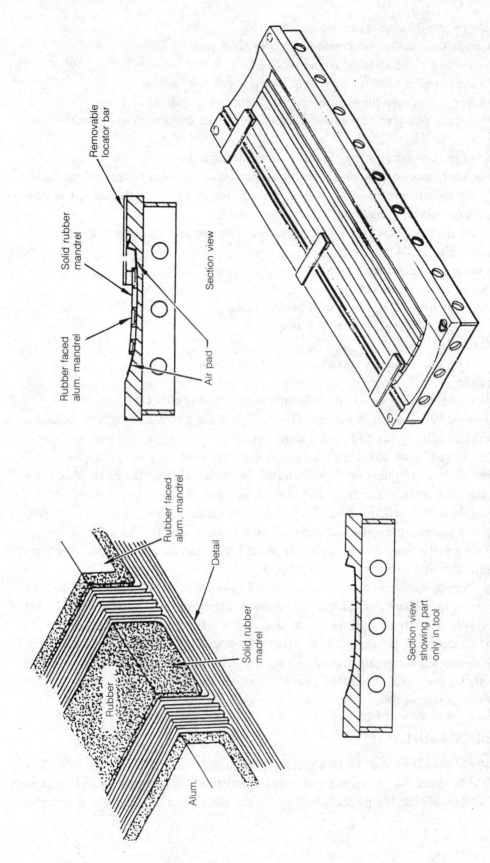

Fig. 3.7.4 Cocure Blade Stiffened Panel by Combination of Elastomeric Rubber and Aluminum Mandrels

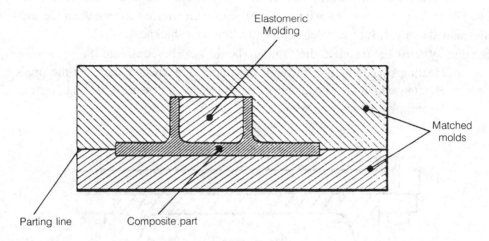

Fig. 3.7.5 Cross Section of a Bracket to Illustrate Local Use of Elastomeric Molding

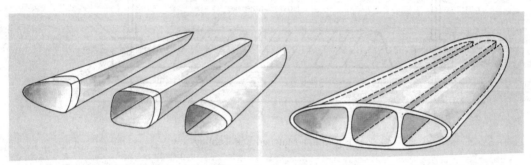

Tooling innovation for the fabrication of wing type structures. Inflatable elastomeric bladders provide smooth and precise interior sections during single cure cycle

By courtesy of Witco Corp.

Fig. 3.7.6. Inflatable Elastomeric Mandrels (Bladders)

3.8 MATCHED DIE TOOLING

A matched die tool, as shown in Fig. 3.8.1, is usually fabricated from metal (most are steel) and is also called matched metal mold tool. They are used to fabricate composite components by the following manufacturing methods (see Chapter 4.0 for more detail discussion):

- Press forming
- Compression molding
- Resin Transfer Molding (RTM)
- Injection molding

Matched die tools are composed of upper and lower halves:

(a) The lower half usually is made of a base plate to which is affixed a male die (plug) machined from a aluminum or steel material
(b) The upper half has a matching female die (cavity) which is machined from the same material

Accurate alignment of the two mold halves is essential and is accomplished by two pins; rest buttons are used to maintain the designed clearance between the male and female dies and thereby ensure correct composite laminate thickness.

Heating systems for matched die tooling should also be considered:
- Heating by conduction through the upper and lower platens of the press
- Heating which is transferred to the press from a nearby heating source
- Heating by an internal tool heating system (costly)

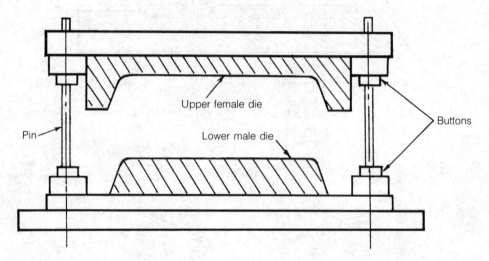

Fig. 3.8.1 Matched Die Tooling

References

3.1 Cremen W. S., "Thermal Expansion Molding Process for Aircraft Composite Structures", SAE Paper No. 800612. 1980.

3.2 Anon., "Engineering Materials Handbook, Vol. 1 — Composites", ASM International, Metals Park, Ohio 44073. pp. 575-601. 1987.

3.3 Burden, J. H., "An Overview of Monolithic Graphite Tooling", 34th International SAMPE Symposium, May 8-11, 1989.

3.4 Anon., "Tooling for Composites: New Materials May Solve Some Old Problems", ADVANCED COMPOSITES, July/Aug 1990.

3.5 Anon., "Goodbye Intermediates", AEROSPACE COMPOSITES & MATERIALS, 1990.

3.6 Anon., "Tooling for Composites with Monolithic Graphite", ADVANCED COMPOSITES, Jan/Feb 1990.

3.7 Anon., "Tooling for Composites: The Blue-collar Trade Moves Up", ADVANCED COMPOSITES, Nov/Dec 1989.

3.8 Anon., "Improving Composite Tooling", AEROSPACE COMPOSITES & MATERIALS, 1988.

3.9 Brain, N., "Epoxy Resin in Composite Tooling", AIRCRAFT ENGINEERING, June, 1987.

3.10 Klein. A. J., "Curing Techniques for Composites", ADVANCED COMPOSITES, Mar/Apr 1988.

3.11 Brown, A. S., "Materials Pace ATF Design", AEROSPACE AMERICA, April, 1987.

3.12 Anon., "Tooling in Graphite, for Graphite", ADVANCED COMPOSITES, Nov/Dec 1986.

Chapter 4.0

MANUFACTURING

4.1 INTRODUCTION

Manufacturing processes and tooling are the elements which control the success and cost of a composite component and it is therefore mandatory that they should be considered as an integral part of the design process. This Chapter, which concerns composite manufacturing processes, will concentrate on organic matrix materials because they have been widely used as airframe primary structures for several decades. The cure (thermoset)/consolidation (thermoplastic) phase of the composite manufacturing process, as shown in Fig. 4.1.1, involves application of heat and pressure to a layup laminate in controlled cycle. Manufacturing of other composites, such as metal matrix, carbon/carbon, and ceramic matrix composites, which have limited applications and were briefly discussed in Chapter 2.0 will not be addressed in this Chapter.

In this chapter, only those manufacturing methods which are applicable to airframe structures will be discussed.

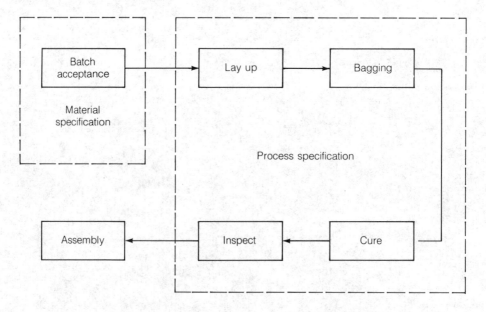

Fig. 4.1.1 A Flow Chart of Composite Laminate Part Fabrication

In composite manufacturing, it is difficult to draw a clear line between manufacturing and tooling systems because they are so closely related each other. Therefore, the tooling information in Chapter 3 should be considered simultaneously with composite manufacturing. The basic problem facing manufacturing engineers is producing composite hardware at an acceptable cost while ensuring its reliability. As matter of fact, almost any composite hardware can be fabricated if economic constrains are removed.

Preparation for a composite manufacturing program should include the early involvement of manufacturing personnel including:

- Active participation in the development of early design concepts — review and evaluation of preliminary engineering drawings
- Support of engineering in selection of materials:
  — Fabrication of test panels
  — Fabrication of basic part configuration specimens
- Coordination of manufacturing and quality assurance organizations;
  — Formulation of tooling concepts
  — Development of inspection, property tests, and non-destructive inspection (NDI) criteria
  — Establishment of the level of planning documentation required
  — Issuring cost and schedule quotes
- Development of a manufacturing plan
- Fabrication of developmental and engineering test configurations
  — Process development specifications
  — Engineering test requirements
- Proof of production tooling concepts
  — Production readiness verification tests
  — Full scale tool prove acceptance
- Support of production during fabrication of hardware

Usually, composite components contain fewer parts than their metallic counterparts. This feature, plus the reduced number of mechanical fasteners required in most composite designs, is a basis for cost reduction. Of course, the part size, geometry, complexity, and required quantity are all considerations in the selection of a fabrication process.

Generally, the most labor-intensive step in composite fabrication, as shown in Fig. 4.1.2, has been the lay up of uncured tape plies, which is normally done by hand. On parts amenable to the tape form, advantage can be taken of the automated methods of dispensing and laying tape with pre-determined orientation and ply sequencing. Some companies have directed a large portion of their research efforts into the manufacturing process so that the implementation of automated systems could be cost-effective. Some of these advances are beginning to be incorporated into the industry's building blocks for the so-called "Factories of the future". One of the processes which has received a great deal of attention in manufacturing research is automated composite systems. The automated system is a must in composite manufacturing to reduce cost.

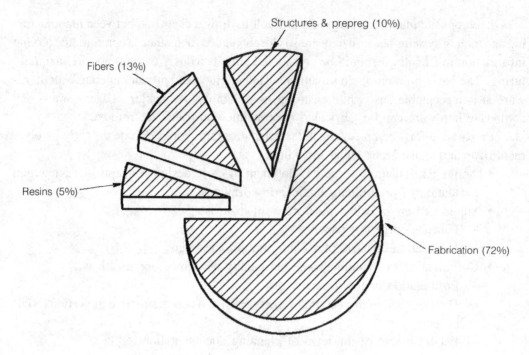

Fig. 4.1.2 Today's composite Fabrication cost Distribution

Material Selection

When selecting composite materials for manufacturing, the following items should be considered:
- Cost
- Ease of processing, fabrication, and handling
- Immediate or near term commercial availability (except for developmental programs)
- Multiple material sources
- Potential to be used in an automated manufacturing process

Material specifications generally include:
- Qualification or source approval requirements
- Incoming batch acceptance requirements

The selection of material forms is quite varied:
  (a) Material forms:
  - Matrix types
  - Reinforcements (fibers):
    — Individual fibers
    — Chopped fibers
    — Woven fabric
    — Tape
    — Mat
    — Preforms: 2-D, 3-D, multi-directional preforms
    — Others: knits, braids, hybrids

(b) Wet layups
(c) Prepregs:
- Solvent coated
- Hot melt coated
- Powder impregnated (thermoplastics)

(d) Commingled (thermoplastics)
(e) Fiber bundles or shapes

Fabrication methods for thermoset and thermoplastic composites may appear to be similar. Heat and pressure are applied to both materials to transform basic prepregs and other material forms into cured or consolidated laminates. However, there are some major processing differences as summarized below:

PROCESS	THERMOSETS	THERMOPLASTICS
Chemical reaction	Yes	No
Temperature range	250-400°F (121-204°C)	500-800°F (260-427°C)
Cycle time	3-7 hours	Can be less than 30 minutes
Viscosity	Low	High
Pressure required	50-100 psi	200 psi and higher
Fabrication	Batch	Batch or continuous process
Scrap rate	High	Potentially low (can be recycled)

Unlike thermosets, thermoplastics need only be heated to the melting point, consolidated, and then immediately cooled, a forming cycle which can result in high productivity and makes low unit cost possible. Therefore, automated equipment and continuous processing may be utilized during forming of thermoplastics.

Tool Selection

The most important key to a successful fabrication program for composite structures rests on the tooling. When selecting tools for composite manufacturing, there are several factors which should be considered and usually one is of major importance, although others may also be crucial. Tool design and requirements were discussed in Chapter 3; and consideration will now be given to selecting tooling to facilitate fabrication of composite components. Several important factors are:

(1) Part configuration, as shown in Fig. 4.1.3, and cure or consolidation are the primary factors in tool selection.
(2) To determine the most important functions of the tool, rate the relative importance:
- Cost
- Repeatability
- Durability
- Potential for further development

Whether the tool is intended to make a prototype part or production parts and whether it is intended to make demonstration parts or be used for flight test purposes are the most important factors in tool selection.

(3) Many factors must be considered when choosing a tooling system:
- Laminate material to be used

Chapter 4.0

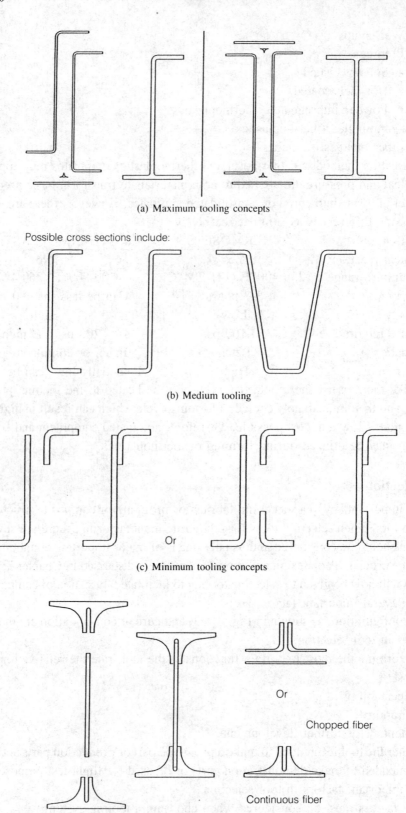

(a) Maximum tooling concepts

Possible cross sections include:

(b) Medium tooling

(c) Minimum tooling concepts

Chopped fiber

Continuous fiber

(d) Minimum tooling concepts

Fig. 4.1.3 Part Configuration Affects Tool Selection

- Vacuum system requirements
- Tolerance requirements
- Effect of thermal expansion
- Bonding of mating surfaces
- Strength and stiffness requirements of the tool while minimizing its bulk and heat capacity
- Size of the composite component being cured or consolidated

(4) For high temperature curing/consolidation of composite (e.g., thermoplastics), the prime concerns in tool selection are:
- Heat-up rate
- Durability
- Thermal expansion

(5) Selection of male vs. female tooling systems — This is a very important consideration in selecting a tooling system for fabricating composite structures, especially for large components. Fig. 3.1.2 in Chapter 3 shows the comparisons between the two. The male tool provides assembly advantages that may be hard to comprehend or envision unless work has been performed on both systems.

If tools are designed without considering the processing requirements of the tool/prepreg system as a whole during cure, the following may result:
- Prepreg problems conforming to the tool, resulting in bridging, or unequal buildups may form and cause pressure gradients, causing the part to warp as it cures
- Porosity and wrinkles in the finished part

(6) The tool and the work instructions should be "tried out" and any needed changes be accomplished before release to production.

Thermal Expansion Effects

Thermal expansion in composite manufacturing becomes a significant factor even at the conceptual design stage because:
- The CTE of composites are directional and vary with fiber orientation (see Fig. 4.1.4)
- Many composites have a coefficient of thermal expansion (CTE) near zero in the fiber direction and stresses may be induced where composites are joined to metals

Thermal expansion consideration include:
- Particular attention must be paid to structures which are bonded, cocured or co-consolidated and which will experience high or low service temperature
- Stresses or deflections can be induced in structures consisting of materials which have very dissimilar CTEs
- Hybrid laminates, consisting of two or more types of material in the composite layup, may experience detrimental thermal expansion effects

Methods of reducing thermal expansion effects:
- Symmetrical laminates (see Fig. 4.2.7) should always be used to minimize warpage and deflections within a laminate

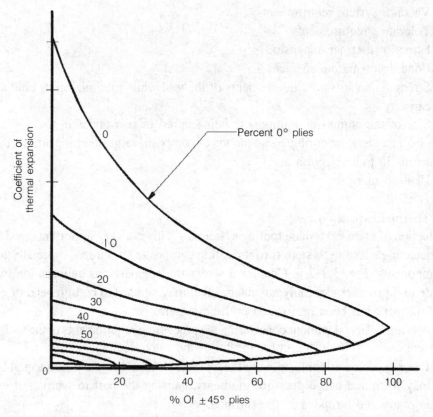

Fig. 4.1.4 CTE Varies with Percentage of 0° plies

- CTE may be increased in the 0° direction by increasing the percentage of plies in the 90° direction of a laminate. Conversely, this reduces the CTE in the 90° direction
- Tailor the laminates give the required CTE
- Metal matrix composites and aluminum (due to its greater CTE) must be given special consideration; the CTE of boron is closer to that of aluminum

Processing

Process selection is dependent on part configuration, design requirements and manufacturing capability. Before final judgments can be made as to the preferred process method, the whole system, including pros and cons should be clearly understood.
Consideration is given to:
- Process specifications — General or testing reference documents
- Process bulletins — project or individual
- Cure or consolidation cycle control:
  — Vacuum debulking
  — Definition of bleeder system (if required)
  — Cure or consolidation definition: heat-up, cool-down rates, dwell cycles, cure or consolidation temperature and time, pressure, vacuum, etc.

- Material control
- Tooling
- Manufacturing procedure and controls in sufficient detail to ensure part conformance to engineering requirements
- Pilot part — initial part fabrication prior to production to verify process
- Workmanship
- Cure or consolidation cycle control
- Records:
  - Vacuum
  - Autoclave pressure
  - Temperature records based on use of multiple thermocouples on the part as defined in the specification
- Quality assurance (Q.A.) requirements and procedures
- Destructive testing requirements

The process of cure/consolidation generally involves
- Viscosity of resin decreases as heat is applied
- Tool expansion
- Vacuum is applied to remove volatiles
- Pressure is applied and laminate cure/consolidation begins
- Vacuum is vented (released)
- Resin gels and part geometry is set
- Heat is continued until resin is fully cured
- Part and tool cool down
- Tool shrinks — it could damage or trapped part

There are numbers of fabrication processes for composite manufacturing and they are identified by the facility use to provide the cure cycle or the type of tool used. It must be kept in mind that some processes will result in increased production costs and some less frequently used processes have merit for unique designs. It is fundamental to realize that there is no unique cure/consolidation cycle even for any one particular resin formulation.

Solvents and water, as absorbed moisture (usually only about 1% for thermoset and .2% for the thermoplastic of prepreg weight), are vaporized into volatiles and along with the air trapped in the layup, must be removed under high temperature and pressure processing. High pressure is usually not applied till well into the cure cycle, after most of the volatiles have escaped. If the pressure is applied prior to the removal of volatiles and air, it is entirely possible for the pressure to effectively seal off local highly curved areas and prevent the trapped vapors from escaping.

Porosity in composite parts:
- Porosity is microscopic interfacial voids dispersed throughout the thickness of a laminate
- Porosity is caused by:
  - Low resin content
  - Failure to remove volatile matter
  - Failure to compact/consolidate the laminate

- Effects of porosity: (> 2%):
  — Slight reduction in static compression strength
  — Significant reduction in static interlaminar shear strength
  — Significant reduction in interlaminar shear fatigue strength
  — No reduction in compression fatigue strength

The following should be considered when selecting a manufacturing process:

(1) Composite material and form selections will greatly influence the processing choice; prepregs are usually used because:
  - Elimination of formulation problems and the mess of wet-layup
  - Consistent quality and resin content
  - Adaptability to assembly line technique for increased production at lower unit costs
  - Fewer rejects
  - Less variance in mechanical properties
  - Best quality materials
  - Design freedom because of the ease with which prepregs can be formed into irregular shapes
  - Lower inventory levels since no resins or catalysts need be stocked

(2) Tool selection should be one of the prime factors in process selection.

(3) The tightness of tolerances and which surfaces of the laminate part are controlled often exert a strong influence on the selection process. Tight tolerances almost invariably lead to additional expense and the control surfaces have a strong influence on whether male or female tools are used.

(4) Always consider the option of making a single relatively complicated part to perform the same function as a number of simpler components which must be joined. In addition to increasing structural efficiency, this results in a reduction in planning and tool costs, because one complex tool will usually not cost as much as a greater number of simpler tools. This in turns reduces production control, tool inventory and therefore total production costs.

(5) Cost reduction is possible when secondary bonding and/or mechanical fastening is eliminated. However, this approach must be balanced by operational requirements and need for future repair or replacement of damaged parts.

(6) More automatic computer control autoclaves would increase consistency and reduce operator induced variances. Statistical process control data should be collected for traceability, analysis and historical record as well. Cure cycle data could be optimized and downloaded to the autoclave system.

(7) Processability trials
  - Many composite manufacturing specifications require fabrication of a trial part representing the production design
  - Fabrication of a trial part will determine if the design, material and process can be used for production parts

Possible manufacturing processes include:

(1) Autoclave
  - Costly — US$3000 to 10,000 per run depending on part size and complexity

- Requires compatible cure or consolidation process systems (temperature and pressure)
- Requires compatible tool thermodynamics
- Long cycle times

(2) Oven
- Much lower cost than autoclave
- Usually requires trapped rubber tooling to provide pressure
- Shorter cycle time than autoclave
- Low pressure capability

(3) Press forming
- Very short cycle time
- Good dimensional control since inner and outer mold lines are controlled
- Common net resin system
- Net shape parts

Manufacturing Cost

Besides the structural weight savings, manufacturing cost has become as important as performance in determining the success or failure of a particular airframe design. The factors involved in controlling cost are:
- The use of accurate, realistic cost estimating tehcniques
- Identification and resolution of potential cost problems early in the design phase, avoiding large cost overruns or schedule slippage
- Manufacturing process selection
- Material choices
- Geometry complexity
- Assembly process
- Layup process

Several guidelines are useful when trying to estimate actual costs:
- Know what drives the cost of a particular design
- Optimize the configuration for minimum cost while still meeting performance goals
- Determine the actual cost using available estimation techniques and historical data (eliminate conservatism in cost estimation)
- Compare cost target
- Use of the comparison of material placement costs shown in Fig. 4.1.5

Safety and Health

Many of the chemicals used in the manufacture of composites may be harmful and/or toxic (see Ref. 1.2 and Ref. 4.60); therefore care should be execised when using these materials. Some basic points to consider are:
- The means to better monitor and control the clean room environment must be obtained to meet specifications and improve continuity of composites fabrication processes

- Skin contact with liquid resin should be avoided by using gloves, eye protection glasses, and protective clothing
- Oxygen masks
- Material handling safety sheets should be published

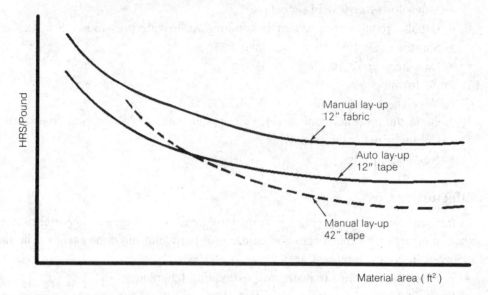

Fig. 4.1.5 Comparison of Material Placement Cost

4.2 MANUFACTURING GUIDELINES AND PRACTICES

Current issures with composites include difficulties in processing, lack of past experience and the high cost of raw materials. All of these factors must be balanced in the development of a composite manufacturing program, as well as the cost effect on the overall airframe structure as a whole.

Guidelines

(1) Size limitations — The available facilities impose limits on the size of advanced composite assemblies; autoclave size (diameter and length) is one such factor. Fig. 4.2.1 shows a one-piece wing skin panel tool which fills into a given autoclave.

Size limitation for composite structures is related to equipment capabilities and the ability to provide the required time-temperature-pressure cure or consolidation cycle established for the particular matrix of the composite system. For laminated parts, the planform size may be limited by the layup area capability or the limits of the tape-layup machine. The size of filament wound structures is limited by the size of the winder and the autoclave or oven.

- Hand layup process size unlimited but a special facility may be required (see Fig. 4.2.2)
- Out of autoclave methods should be considered for processing large parts

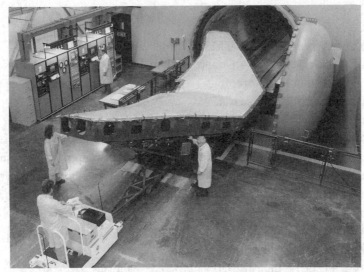

By courtesy of McDonnell-Douglas Corp.

Fig. 4.2.1 One-piece Wing Skin Panel of AV-8B11 Prior to Entering Autoclave

By courtesy of Heath Tecan Aerospace Co.

Fig. 4.2.2 MD-11 Center-engine Inlet Duct Halve

- Automated equipment may be required for fabricating large components since hand layup is slow and costly

(2) Component shape — The shape limitations during construction result partially from the drape quality of the tape and fabric. Tight radii and abrupt changes in surface features should be avoided since they usually cause bridging between plies. Although composite materials can be easily fabricated into a complexly shaped parts, certain limitations still exist and attention should be paid to the following:
- Fabrics are generally easier to form
- Shape limitations are the result of the material's drape capability
- Avoid shape changes in surface contours
- Avoid tight radii (bridging will result)
- Shape restrictions are also related to access requirements for layup and tooling

Graphite/epoxy prepreg material has a drape and tack and can be used to fabricate parts with greater changes in contour than is possible with thermoplastics, see Fig. 4.2.3. Thermoplastic prepregs require different methods such as thermoforming, to fabricate contoured parts.

(3) Tolerance:

Within practical limits, tolerances should always be as large as the use function of the part will allow. Length and width tolerances for composite parts should be kept to the same standard production tolerances used for metal counterparts. However, each material and method of fabrication will produce parts with some thickness variation.

- Normal cured or consolidated ply thickness
    0.00xx±0.0003 inch (0.0076 mm), e.g., tape
    0.0xxx±0.0012 inch (0.0305 mm), e.g., fabrics.
- Nominal cured thickness should be reference
- Factors affecting thickness are:
    — Amount of bleed during cure cycle

(a) Closeout aluminum tool

(b) GR/EP closeout parts

Fig. 4.2.3 Hat-Section Closeout

- Cure pressure
- Tooling (matched die or bagged curing)
- Resin content

- Controlled thickness is very difficult to control except by expensive matched die tooling
- Male or female tool selection affects tolerance as shown in Fig. 4.2.4
- Fastener grips should be specified for max laminate thickness
- Structural joint pull-up or nesting parts should be analyzed for min/max tolerance conditions and shims provided as required

Fig. 4.2.5 gives tolerance requirements for composite airframe structure applications.

(4) Surface smoothness and flatness (see Fig. 4.2.6) — A smooth surface may be required on a composite part for aerodynamic considerations or to ensure an adequate roughness for adhesive bond or painting Required surface smoothness and/or flatness is general achieved through:
- Specifying the tool surface side of the composite laminates
- Specifying requirements for a defined area
- Laminate surface varies with the material, assembly method and choice of tool side for layup and cure
- Aerodynamically smooth surfaces are more easily produced when outer plies are tape rather than fabric
- Fabrication process does not insure mating surface flatness on the bag side
- Specify requirements for a defined area (manufacturing can use option of caul plates)

(5) Laminates thickness — For most parts, control of thickness on the drawing to tolerances closer than that provided by specification is not necessary

(6) Drilling/countersink — Drilling and countersinking of carbon or graphite thermoset laminates requires use of special carbide tools. Kevlar and thermoplastic composite laminates may need a special drilling procedure and tool (see Fig. 4.2.7). Generally, an outer ply fabric is used to control breakout.

(7) Engineering drawing practices — Basic requirements for standard composite drawings (Refer to Appendix B) are:
- Ply orientation symbol
- Ply orientation callout
- Tooling surface designation, if required
- Ply stacking "ply view"
- Do not call out laminate thickness on drawing except when tighter tolerance is required
- Ply stacking sequence is usually started from tooling surface for convenience

(8) Material form selection — The most desirable material form is one which has the required strength and allows net shape forming in a matter of minutes.

(9) Symmetrical balanced laminate — Laminate should be symmetrical about the midplane of the laminate, as shown in Fig. 4.2.8, because unsymmetrical laminates experience:
- Warping during care or consolidation
- Extension-bending coupling

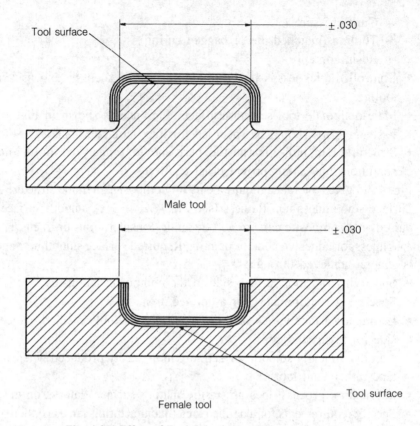

Fig. 4.2.4 Effect of Tool Selection on Tolerance

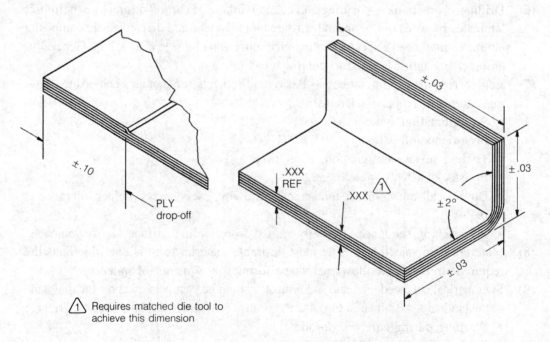

⚠ 1 Requires matched die tool to achieve this dimension

Fig. 4.2.5 General Tolerance Requirements of airframe Laminate

Manufacturing

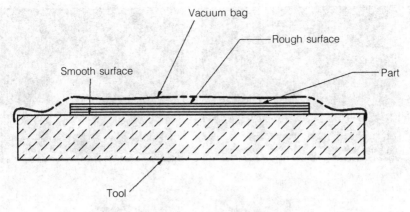

(a) Smooth surface on tool side only

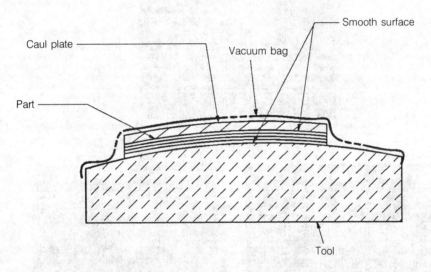

(b) Smooth surfaces on both Sides because of caul plate

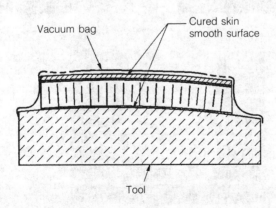

(c) Smooth surface on both sides because of pre-cured skins were used

Fig. 4.2.6 Methods of Obtaining Surface Smoothness and Flatness

Chapter 4.0

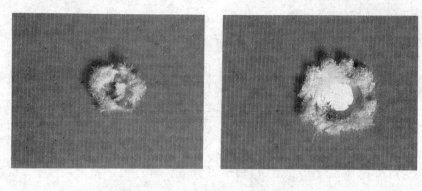

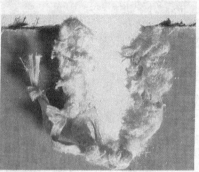

(a) Drilled by conventional tools

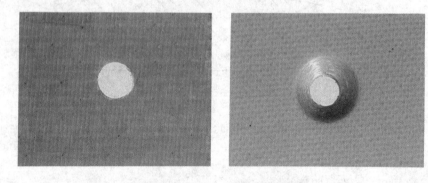

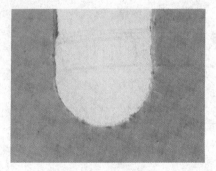

(b) Drilled by carbide tools

By courtesy of Rolf Klenk GmbH & Co. KG

Fig. 4.2.7 Special Carbide Drill Tools needed for Kevlar (Aramid fibers)

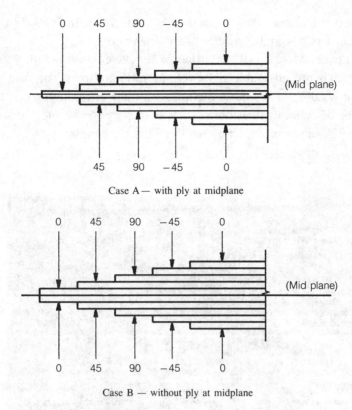

Fig. 4.2.8 Symmetical and Balanced Laminate

A symmetrical balanced laminate has the following characteristics:
- It is symmetrical about the midplane
- Ply drop-offs are also symmetric about the midplane
- Each 45° ply should be balanced by a −45° ply
- Use 45° or −45° plies on outer surfaces of the laminate
- Avoid grouping more than six plies of the same orientation
- Avoid grouping of 90° plies
- For thicker laminates [e.g. greater than 0.030 inch (0.762 mm) based on 0.005 inch (0.127 mm)/ply thickness], adjacent ply angles should not differ by more than 60° i.e., do not combine 0° and 90° or 45° and −45° plies
- Minimum of 10% plies in each direction
- Minimum number of plies for a typical laminate is seven plies

Since current methods for fabricating composite components are time consuming and subject to the skill of the worker, numerous alternative manufacturing technologies have been developed to replace or assist manual production. Methods, such as pultrusion and filament winding, which use neat resin fibers benefit from a substantial materials cost reduction over those which utilize prepreg materials. In addition, since these methods generally have lower equipment costs and cycle times, they are able to compete with manual production even without a savings in material costs. However, trade-offs between cost effectiveness, geometric constraints and quality characteristics of the raw material should be conducted for each method considered. For example:

- Pultrusion is the cost effective method but current technology is limited to constant cross sections and restricted fiber orientations
- Resin Transfer Molding (RTM) offers lower cycle times, but there are limitations on the maximum attainable fiber volume ratio and part planform size
- Filament winding offers more flexibility in fiber orientation but is limited to applications of cylindrical sections, such as rocket motors, etc. New development of fiber-placement (see Section 4.3) in filament winding system offers promise for airplane structures

(a) Lear fan 2100

By courtesy of P.T. Cipta Restu Sarana Svaha

(b) AA200

Fig. 4.2.9 Aircrafts with High Level Integration and Less-part-count Low cost Assembly

Since a high level of integration and low assembly costs, as shown in Fig. 4.2.9, enable composite structures to successfully compete with other counterparts, composite component complexity is an important issue. Because the highly complex parts require more manual labor automation is an attractive option. Technologies which have been developed for other industries can often be directly used without modification for composite manufacturing (see Fig. 1.1.11).

Practices

(1) Bend radius:
- Fig. 4.2.10 gives minimum bend radii for composite fabrication
- Fig. 4.2.11 shows bend radius problem areas
- Fig. 4.2.12 shows the junction void is usually filled with roving or rolled unidirectional tape. The size of the void should be kept to a minimum

(2) Bend angle springback (see Fig. 4.2.13):
- Degree of variation between part and tool depends on:
  — Resin
  — Debulking
  — Fiber orientation
  — Thickness
- Closure occurs due to differential expansion between the inner and other surfaces
- This should be taken into consideration when designing tool configuration (tool should be designed with slightly opened angle)

(3) Favor automated manufacturing whenever possible to reduce labor cost

(4) Use prepreg materials unless clear advantages favor "wet" layups

(5) If sharp contour changes cannot be avoided, use woven fabrics (see Fig. 4.2.14)

(6) Use cocured or co-consolidated assemblies when possible. Elimination of fasteners:
- Reduces labor
- Reduces part count
- Reduces cost
- Improves supportability

(7) Avoid joining materials with highly different CTEs

(8) For laminated parts thickness dimensions shall be referenced, unless specifically required for fitup or mating on assembly to avoid excessive clamp-up force to break the parts due to the unforgiving nature of composites. Traditional plastic deformation which redistributes loads in metallic structures does not occur in composite structures. Determination of final thickness of a laminate includes a cumulative tolerance based on the number of plies used in the laminate.

(9) Do not splice normal to fiber direction of unidirectional tape/fabric

(10) Splice requirements (see Fig. 4.2.15):
- Use butt splice for unidirectional tape with a gap of 0.00-0.03 inches (0-0.762 mm)
- Use an overlap splice for fabric, with an overlap of 1.00 to 1.5 inch (2.54 to 3.81 cm) between adjacent plies

Chapter 4.0

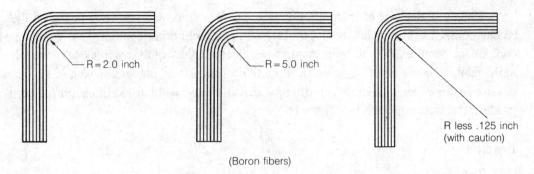

(Boron fibers)

Fig. 4.2.10 Minimum Bend Radii

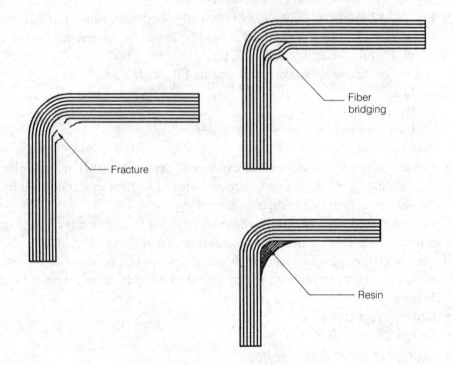

Fig. 4.2.11 Bend Radius Problem Areas

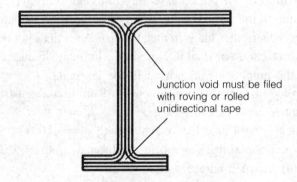

Fig. 4.2.12 Junction Void Fillet

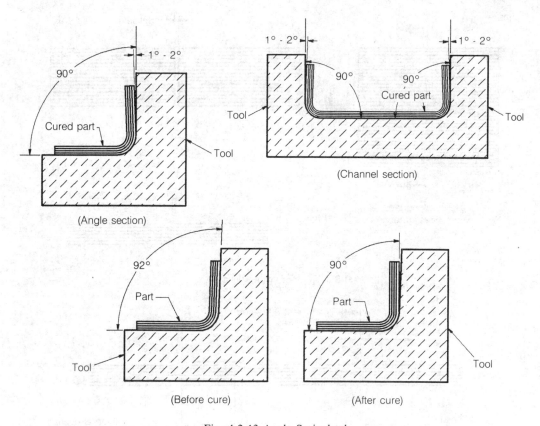

Fig. 4.2.13 Angle Springback

By courtesy of Beech Aircraft Corp.

Fig. 4.2.14 Drapcable Woven Fabrics

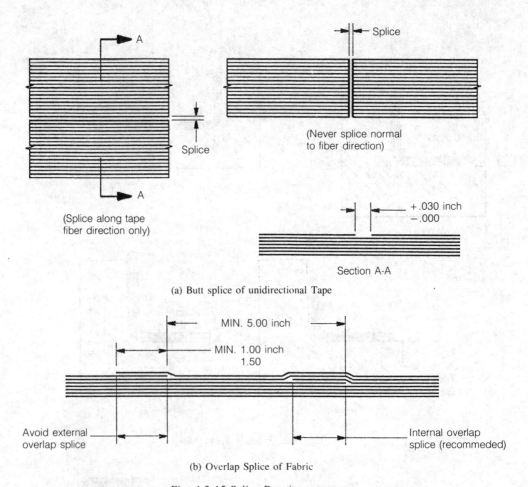

Fig. 4.2.15 Splice Requirements

(11) Interlaminar slip concern — On complex contour laminates allowance should be given to slippage between plies when forming radiused parts as shown in Fig. 4.2.16
(12) In most cases, the changes of laminate thickness are by add-on or drop-off plies (ADP) as shown in Fig. 4.2.17.
  • ADP should also be as symmetrical and balanced as possible
  • Distance between ADP steps should be at least .15-.25 inch (3.81-6.35 mm) as shown in Fig. 4.2.17(a) with a large tolerance if possible
  • ADP should always be tapered and the slope angle should not exceed 10 degrees as shown in Fig. 4.2.17(b)
  • ADP should not involve over 6 to 8 plies based on ply thickness of 0.005 inch (0.127 mm)/ply or over 2 to 3 plies for thicker ply thicknesses
  • ADP should not occur on outer surface plies to avoid peeling
  • Pre-cured or consolidated inserts [see Fig. 4.2.17(c)] or machined cured inserts [see Fig. 4.2.17(d)] should be compatible with ADP guidelines
  • Cover all ply steps with at least one continuous outer ply [see Fig. 4.2.17(c)] to:
    — Aid in load redistribution
    — Prevent edge delaminations

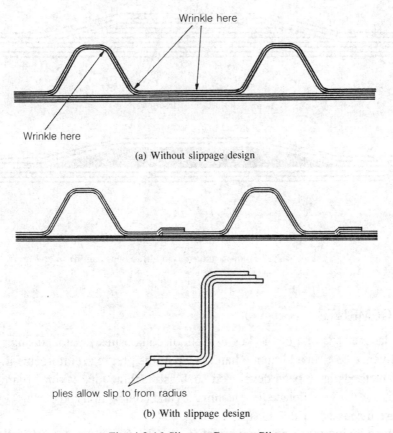

(a) Without slippage design

(b) With slippage design

Fig. 4.2.16 Slippage Between Plies

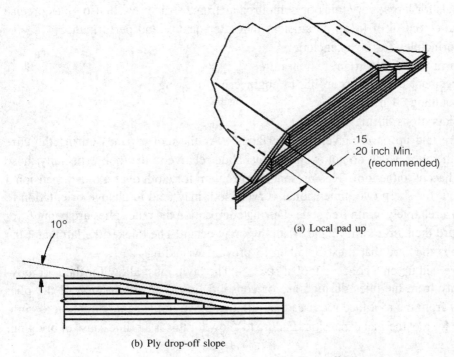

(a) Local pad up

(b) Ply drop-off slope

Fig. 4.2.17 Add-on and/or Drop-off Plies

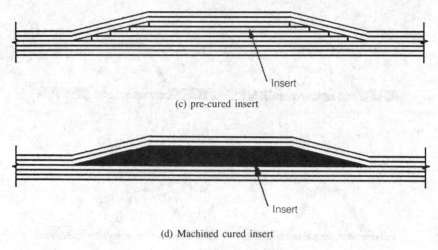

(c) pre-cured insert

(d) Machined cured insert

Fig. 4.2.17 Add-on and/or Drop-off Plies (cont'd)

4.3 PLACEMENTS

Material placement is one of the key aspects of composites manufacturing since it can greatly influence cost. In addition to hand layup of the prepreg reinforcement, automated placement methods have been developed such as machine tape laying, filament or tape winding, etc. A variety of placement techniques and compaction and curing/consolidation methods are discussed in this Chapter.

The tape-laying machine, shown in Fig. 4.3.1, uses computer-controlled automation techniques to replace manual layup methods and greatly reduces the cost of using tape materials for large, very thick or contoured airframe structures. It also gives precise and automatic control of layup parameters that alter quality and performance.
The basic principles for layup include:
- Warm prepreg to room temperature
- Placement of plies according to engineering drawing
- Debulking, if necessary

Some considerations during layup are:
(1) Plies are laid up on a convex tool surface — As the tool expands during the cure cycle, the plies which do not expand, must slide relative to the tool, especially those plies adjacent to the tool. This presents no problem for cloth over a one dimensional curve but for sharp two dimensional curves cloth may tend to change orientation to take up a relatively strain free state. During compaction the outer plies are being forced toward the hard tool face and wrinkling may occur. The thicker the laminate, the greater is the care that must be taken to prevent wrinkling.
(2) Plies are laid up on a concave tool surface — The layip must allow for the tool moving away from the plies during cure or consolidation. The compaction of the plies furthest from the tool face requires that all plies move in the same direction to compensate for tool thermal expansion. Failure to allow for this is a prime cause of bridging.

(3) Plies are laid up over a sharply curved tool surface — In the layup of unidirectional tape care must be taken due to its relative weakness normal to the fibers. Movement of the plies laterally in that weak direction can easily result in tape failure during cure or consolidation with resulting degradation of strength.

Hand Layup

For most cloth configurations using templates to cut out the developed shape, prior to laying it up on the tool, is the simplest method (see Fig. 4.3.1). Each ply is applied separately, with darts being cut only where needed to assure smooth layup at sharp corners and sometimes a rubbing tool is used to move air and wrinkles. Sometimes a heated tool is needed to soften the prepreg and ease its working into tight radii and doubly curved surfaces. A heat tool is often required to work with thermoplastic prepreg because of its boardly characteristics.

The different material forms and their layup characteristics are listed below:
- Tape [see Fig. 4.3.1(a)]:
  — Contours difficult
  — High trim scrappage
  — Many butt splices required (0.000/+0.030 inch)
- Fabric [see Fig. 4.3.1(b)(c)]:
  — Contours easy
  — Less labor intensive
  — Fewer splices required
  — Lower properties than tape

By courtesy of Advanced Composite Products, Inc.

(a) Prepreg tapes

Fig. 4.3.1 Methods of Hand Layup

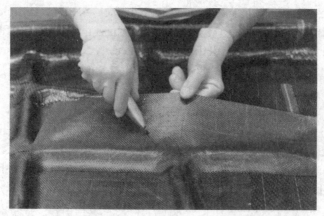

By courtesy of Ren Plastics

(b) Prepreg fabrics

By courtesy of Lockheed Aeronautical Systems Co.

(c) Commingled fabrics

By courtesy of The Boeing Co.

(d) Honeycomb core and prepreg fabrics are laid up by hand on 767's elevator

Fig. 4.3.1 Methods of Hand Layup (cont'd)

- Knits
  — Straight fibers — same properties as tape
  — Pre-plied in ±45° or 0°/90° plies
  — Less labor intensive than tape
  — Reduced quality assurance inspection
  — Reduced trim scrappage

Tape-laying Machine

The development of the tape-laying machine offers the potential for significantly lowering the costs of fabrication composite parts for flat (see Fig. 4.3.2), single-curvature (see Fig. 4.3.3), and limited compound contour parts. The laying machine is economically viable for multi-unit runs of large aspect ratio, relatively planar configurations. In practice, the machine dispenses tape from a supply roll, places it directly on the underlying mold or tape-covered surface, applies shoe pressure to seat and debulk the tape, takes up and stores the release paper (thermosets), and at the end of each movement across or along the mold, cuts the tape to the required length and angle. It also changes the angle from layer to layer so as to provide the advantages of cross-ply construction. In addition, it adds extra tape lengths at selected locations/angles to give local reinforcement or attachment provisions.

Fig. 4.3.2 shows a thermoplastic tape laying process in which prepreg tape is heated by a shoe-type tape layer as it is ironed or rolled down onto a mold surface, and consolidation pressure is maintained. The temperature must be above the melting point of the matrix to maintain an effective consolidation state.

By courtesy of Cincinnati Milacron

Fig. 4.3.2 Thermoplastic Tape Laying Machine

Chapter 4.0

Because tape-laying machines use preimpregnated unidirectional tape, this is probably the most attractive material form in terms of cost and properties for the preduction of airframe primary structures such as planar wing, fuselage, control surfaces skin, etc. Advantages of the tape laying machine are:
- High production
- Flat to simple contours with constant of variable thickness
- Low layup cost
- Up to 3 to 6 inches (7.6 to 15.2 cm) wide tape is laid down at the rate of 1000 inches (2540 cm) per minute
- No debulking required
- Automated trimming and layup trim scrap reduced

Factors which should considered include:
- Initial cost
- Limited compound contour capabilities
- Setup labor required

Fiber-placement Laying Machine

Unlike the tape-laying method, the fiber-placement method can be used to meet the need for adding and dropping fibers, winding concave surfaces, changing dimensions, winding cutouts, etc. Fig. 4.3.4(a) shows a seven-axis fiber-placement machine which can individually control up to 24 tows with 1/8 inch (3.2 mm) wide tow laid down in one path. Fig. 4.3.4(b) shows a robotic wrist, individual feed rollers for starting each tow, and a segmented compaction roller which permits the prepreg tows to be laid into or around complex contours, and compacted directly onto the tool surface.

By courtesy of Cincinnati Milacron

(a) Stepped contoured surface

Fig. 4.3.3 A Contoured Tape Laying Machines

Manufacturing

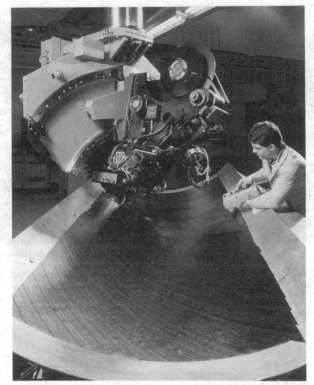

By courtesy of The LTV Corp.

(b) Curved contoured surface

Fig. 4.3.3 A Contoured Tape Laying Machines (cont'd)

By courtesy of Cincinnati Milacron

(a) FPX Fiber-placement Machine

Fig. 4.3.4 Typical Fiber-placement Machine

205

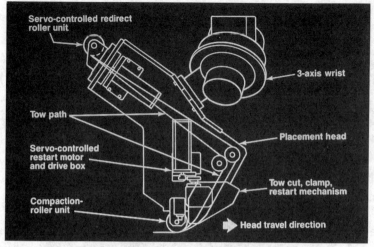

By courtesy of Cincinnati Milacron

(b) Schematic of a fiber-placement unit shows several of the elements that lend the process its unique capability to wind complex shapes and structures

Fig. 4.3.4 Typical Fiber-placement Machine (cont'd)

Benefits of fiber-placement include:
- Filament winding
- Contour tape laying
- Precise ply thickness control
- In-process compaction
- Low void content (less than 1%)
- Unlimited fiber angles
- Yields much less material scrap than the filament winding process

Fig. 4.3.5 shows a comparison of filament winding (wet), hand layup (prepreg tape), and fiber placement (prepreg tow).

Hot-roll Consolidation

Fig. 4.3.6 shows the hot-roll consolidation method which has the following characteristics:
- pre-consolidation of laminates
- High volume, low cost consolidation of flat laminates
- Size limitations

Filament (Tape) Winding

Providing the highest strength-to-weight ratio, filament winding consists of feeding reinforcement filament, or roving through a matrix or resin bath or using preimpregnated roving and winding it on a mandrel as shown in Fig. 4.3.7. Special winding machinery lays down the impregnated roving in predetermined patterns, giving maximum strength where required. After the appropriate layers are applied, the wound mandrel is cured, and the molded part removed from the mandrel. Filament winding provides the greatest control over orientation and uniformity, but it usually restricted to surfaces of revolution.

Manufacturing

Item	Filament winding (wet)	Hand layup (prepreg tape)	Fiber placement (prepreg tow)
Void content	4 to 8%	<1%	<1%
Thickness flexibility	0.010 to 0.025 inch/ply not constant tapered parts	0.005 to 0.010 inch/ply controlled by prepreg spec	0.005 to 0.015 inch/ply varibale/ programmable
Winding angle flexibility	>15° 15 to 90°	0 to 90° flat pattern	0 to 90° variable/ programmable
Tow cut and add	No	Hand splice/cut only	Programmable
Lap and gaps	0.125 inch	0.030 inch	<0.030 inch
Geometry	Best for bodies of revolution	Compound flat pattern limited hand located/placed	Compound machine placed/ cut/add
Material scrap rate	20 to 40%	50 to 200%	5 to 20%

(Source: Hercules Inc.)

Fig. 4.3.5 Filament Winding/Hand layup/Fiber-placement Comparison

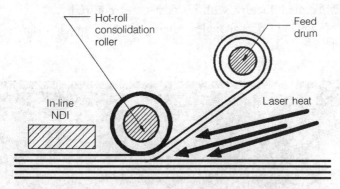

Fig. 4.3.6 Hot-roll Consoldation (Thermoplastic)

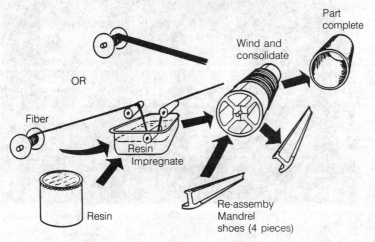

Fig. 4.3.7 Filament Winding Technology

Chapter 4.0

Most filament winding technology has been developed for the rocket-motor industry; the cylindrical shape (see Fig. 4.3.8) and large size rocket-motor cases make filament winding particularly attractive for this application. Filament or tape winding is an important method of manufacturing composite structures and the advances in computer techniques have improved its flexibility and brought down costs. The basic technique is similar to that used by a silk worm winding its cocoon.

The advantages are as follows:
- More rapid surface coverage
- Hoop tension loading uniformity of fibers
- No prepregging cost
- Can lay 30-250 lbs/hour
- No cold storage required
- Can over wind bulkheads, ribs, stringers, etc., and cocure
- Uses lowest cost material form available
- Automated
- High production rates
- High fiber volume
- Moderate to high elongation resin
- Low viscosity resin
- High quality manufacture with no post-cure required
- Capable of winding fibers at angles ranging from 11° to 89° from the axis of rotation

By courtesy of Hercules Inc.

Fig. 4.3.8 The 12 ft. Dia. and 12 ft Long Filament-Wound Section For The Space Shuttle's Solid Rock Motors

Typical types of filament winding patterns are shown in Fig. 4.3.9; the four basic winding patterns are described below:

(1) Helical winding [see Fig. 4.3.9(a)] — Mandrel is rotated about a horizontal axis and winding is carried out at any angle from 25°-85° to the axis of rotation. The mandrel rotates continuously while the fiber feed carriage moves back and forth.

(2) Orbital winding [see Fig. 4.3.9(b)] — The feed carriage moves in a racetrack pattern-around the mandrel. It is suitable for surfaces with slopes over 20° to the axis of rotation when using wet fibers, or 30° when using dry fibers.

(3) Polar winding [see Fig. 4.3.9(c)] — Mandrel is rotated perpendicular to its longitudinal axis and the mandrel remains stationary while the fiber feed arm rotates about the longitudinal axis inclined at a slight angle, and the mandrel is indexed to advance one fiber band width for each rotation of the feed arm.

(4) Whirling winding [see Fig. 4.3.9(d)] — Mandrel is rotated about a vertical axis.

In filament or tape winding each tool or mandrel takes on slightly different designs, mainly due to the necessity of winding mandrels and end domes for fiber turnaround. These have to be designed with regard to winding machine capabilities, part design requirements, and processing characteristics. For these reasons, initial tooling expenses are generally higher for filament or tape winding than for hand layup.

The first step of filament winding fabrication is to prepare a suitable mandrel:

(1) Filament or tape winding must be performed on a "male" tool (winding mandrel)
(2) The mandrel may form part of the finished product, or may simply be an aid to production which is removed to be destroyed or reused
(3) Washout (disposable) mandrels (see Tooling in Chapter 3.0), used in closed shapes, include soluble plaster or salt, and foamed plastics, inflatable bags, and metal mandrels which can be reassembled, as shown in Fig. 4.3.7.
(4) The vacuum system method uses the fact that materials like sand lock into a hard mass under compression, but revert to a free-flowing state when the pressure is removed.

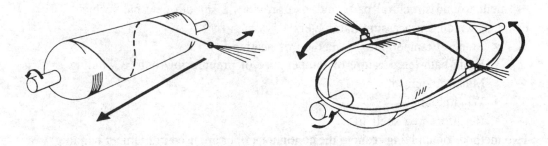

(a) Lathe or helical winder   (b) Racetrack or orbital winder

Fig. 4.3.9 Types of Filament Winding Process

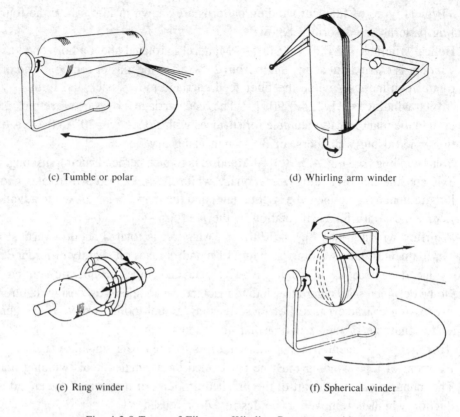

(c) Tumble or polar (d) Whirling arm winder

(e) Ring winder (f) Spherical winder

Fig. 4.3.9 Types of Filament Winding Process (cont'd)

(5) Fibers can be wrapped over a thin-walled seamless aluminum liner as shown in Fig. 4.3.10; the liner serves as the mandrel for the filament over warp. The liner is pressurized during winding, gel, and cure.
(6) An external clamshell molding technique produces a cocured external surface on filament wound structural parts and also provides a smooth external surface

Filament winding processing parameters:
- Mandrel temperature — method of heating
- Resin bath temperature or temperature of prepreg tow before winding
- Mandrel spin rate
- Winding angles
- Resin/prepreg shelf life

Two methods of applying resin to the continuous fiber are used in filament winding:
(1) In the wet system, the fiber picks up resin as it travel through a trough. A variation of this system is the "controlled wet" procedure in which the resin is metered onto the fiber. Although the wet system is the lower-cost method, it is messy, slow, wastes resin, and requires protection of personnel against odor and fumes.
(2) The prepreg system overcomes the limitations of wet system, but at a high material cost. Also, cold storage is required for thermoset prepreg material to prevent resin advancement.

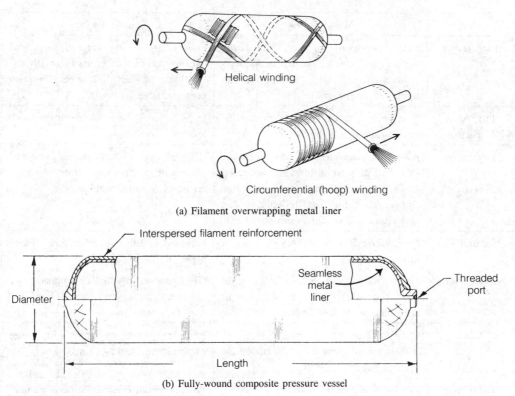

Fig. 4.3.10 Overwrapping a Thin-Walled Seamless Alumium Liner

For large parts, filament or tape winding on a cylindrical tool (usually more than 10 feet in diameter and 10 feet in length) for example is a much better procedure than attempting to hand lay up the same part on the same tools because the accessability problem over the large working area. For a comparison between these two methods is shown in Fig. 4.3.11.

Thermoplastic matrix composites offer exciting possibilities for filament or tape winding of complex shapes. Unlike thermoset matrices, which need to be post-cured in an oven or autoclave, thermoplastics can be consolidated during the winding process. The process still under development uses a heated mandrel with preheating of the tape just before it touches the mandrel surface. In an other alternate method, a hot gas torch heats the composite materials as it is laid down, compacting and consolidating it, then it is cooled immediately. This operation can eliminate oven or autoclave curing, has indefinite material out life, is compatible with other thermplastic components, has low residual stresses, provides continuous winding of thick-walled parts, and has minimal ply waviness. Because thermoplastic filament or tape winding is in effect a continuous welding operation, each element of the tow or tape adheres to the composite beneath giving a very high friction coefficient. This gives the designer greater freedom to choose highly non-geodesic trajectories. Furthermore, with filament or tape wound thermoplastic composites, material can be pressed into concave surfaces during winding, whereas with most placement methods fiber bridging would occur.

Item	Male Tooling	Female Tooling
Workability	Easy access everthing is in front of the workers and can be rotated at workers convenience. Entire component is fabricated as one unit.	Poor access — workers must be suspended over work area, worker must move to new locations, product is fabricated in two separate halves.
In-Process Inspection	Easy access, product can be rotated for visual inspection and accessibility.	Inspector must be suspended over work area.
Processing	Allows for automated material placement, i.e. filament-winding of skin, automated application of film adhesives. Tool acts as additional helper as it rotates for honeycomb and reinforcement placement.	Difficult to use automated material placement techniques. Material often separates from vertical surfaces.
Structural Risk	Circumferential splice is lower in risk due to lower stresses.	Longitudinal splice is higher in risk due to higher stresses.
Surface Finish	Gives good smooth surface finish inside and outside when used in conjunction with clamshells formed to OML.	Gives good smooth surface in outside only unless an internal caul sheet is used.
Cure	Option of autoclave or non-autoclave cure with pressure. Possible elimination of vacuum bagging operation.	Autoclave cure required with large cumbersome vacuum bags.
Debulking	Winding operation performs debulking operation in process.	Probable requirement of various vacuum debulking operations. First ply vacuum debulk to maintain contact of ply to tool.

Fig. 4.3.11 Filament Winding Male Tool Vs. Hand Layup Female Tool (Fuselage Shell)

Thermoplastics as well as thermosets can be wound to complex non-axisymmetric shapes with an enhanced robotic system which can manipulate an end effector for winding around a stationary mandrel. The seven to ten-axis system (five or six axes on the robot and two or four on the end effector), depending on whether thermoplastics or thermosets are being wound, can make T and elbow shapes, as shown in Fig. 4.3.12, that could not be made on conventional filament winders. This system can wind tape instead of tow (filament) and the end effector is also equipped with a mechanism for dispensing tape and rewinding its backup paper on a roll. Because tape (thermoplastic tape or tow must be heated by the dispensing head just prior to winding) is stiffer, however, tape winding, like tape laying, is much more limited in the complexity of shapes that can be fabricated and tape conformance decreases with increased tape width.

General concerns about filament or tape winding method are:
- High initial cost
- Limited shapes
- Fibers in the 0° direction to the axes of rotation cannot be wound efficiently
- Moderate to high void content
- Resin content is difficult to control on large diameter parts
- Restricted to surface of revolution

- Difficult to maintain filament angle for variable contoured parts
- Resin content in thermosets is difficult to control with large diameter parts

The winding machine itself constitutes probably less than 50% of the overall winding process. The other critical elements of the winding system are the creels, the tensioning system, the impregnation system, and the winding eyes. All these systems must be coordinated to assure the fiber and resin and applied on the mandrel with a desired and repeatable tension, fiber volume, and without damaging the fiber. Lack of anticipating the time, technology, and money involved in solving these problems, along with the problems of a complicated winding machine, are the major reasons that most winding machines remain idle after initial trials.

Controlling tension of deploying fibers in filament or tape winding during the winding process is one of the most critical requirements for successful implementation of this method. Controlling tension involves the following:

- Ability to hold set tension at the mandrel at varying mandrel speeds
- Ability to hold set tension at the mandrel when the winding process tends to cause strands to go slack in the system
- Ability to handle filaments or tapes with minimal degradation
- Ability to accurately and rapidly adjust a tension setting

The most stable way to wind a fiber over a surface is to follow the geodesic line between two points at which the line traces the shortest path between these two points. This limits the choice of angles at which the fibers may lie on the surface, dictating the structural properties of the shape and limiting the range of shapes possible.

By courtesy of Automated Dynamics Corp

Fig. 4.3.12 Tubular T-shapes Can Be Wound on Robotic Winding Systems (ROWS)

Chapter 4.0

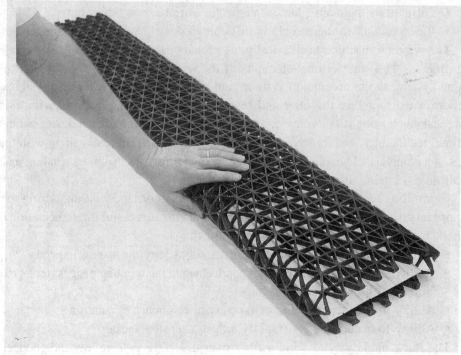

By courtesy of Lockheed Aeronautical Systems Co.

(a) Isogrid helicopter blade trail

By courtesy of Alcoa/Goldsworthy Engineering

(b) A geodesic orthogrid fuselage structure

fig. 4.3.13 Geodesic Structure Formed by Filament Winding

The filament winding method is an ideal way to make geodesic winding patterns for geodesic composites structures (orthogrid and isogrid structural panels) to achieve ultralight weight with both very high strength and stiffness. Geodesic structures commonly use orbital, helical and planar winding, either alone or in combination. Fig. 4.3.13 shows an isogrid helicopter blade, and an orthogrid geodesic fuselage barrel which allows windows to be cut out of the skin structure.

Multi-axis Fiber Winding

Fiber tow placement is an automated fabrication process which can be used to create complex shaped, high performance composite structures. Fiber tow placement consists of the automated placement and compaction of tows of preimpregnated fibers to a predetermined thickness on predetermined geometry. Multiple tows are laid down as a band, as shown in Fig. 4.3.14, with band location and angle precisely controlled. Fig. 4.3.15 shows a 7-axis fiber placement machine fabricating the aft fuselage of the V-22 tilt rotor airplane. Compared to hand layup, it achieves a 50% reduction in direct labor costs.

Fiber tow placement involves four key elements:
- Multi-axis computer controlled placement machine
- Off-line modeling and programming
- Prepreg tow material form
- Mandrel tooling

By courtesy McClean-Anderson Inc

Fig. 4.3.14 Multiple Tows Are Laid Down As A Band

Chapter 4.0

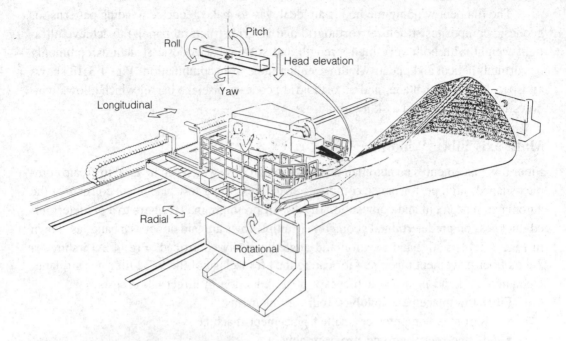

By courtesy of Hercules Inc.

Fig. 4.3.15. 7-axis Multi-axis Fiber Placement Machine Fabricating V-22 Tilt Roto Airplane Aft Fuselage

4.4 BAGGING SYSTEM

Bagging systems for autoclave forming (molding) techniques have been developed for fabricating composite airframe structural components over the past decades. The standard autoclave bagging systems, shown in Fig. 4.4.1, are well developed for the production of thermoset composite parts, and have been adapted for thermoplastics as well. The main differences is in the use of high temperature bagging and sealing materials capable of operating at up to 800°F (427°C). The primary purpose of the bagging system is to hold the laminate parts in position and to position the vacuum bag during cure. At the beginning of the cure cycle, a vacuum is drawn from the inside of the vacuum bag. Autoclave pressure is then applied to the outside of the bag. The pressure consolidates and holds the mating surfaces of the assembly in close contact while the composite material melts, flows, and cures.

Bagging materials are generally non-reusable materials which are consumed in the processing stage and they are:
- Bagging film
- Bag sealant
- Breather plies
- Bleeder plies
- Dam
- Separator film
- Mold release ply

(a) Vacuum bagging with vacuum port and vacuum vent line mounted

Fig. 4.4.1 Samples of Autoclave Bagging Systems

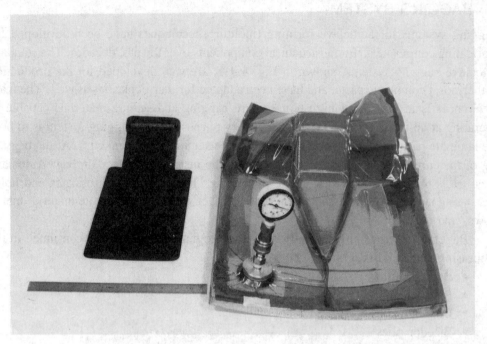

By courtesy of Composites Horizons, Inc.

(b) Sample I of vacuum bagging with vacuum port and pressure/vacuum gauge mounted

By courtesy of Airtech International Inc.

(c) Sample II of vacuum bagging with vacuum port and pressure/vacuum gauge mounted

Fig. 4.4.1 Samples of Autoclave Bagging Systems (cont'd)

Bagging systems have the following characteristics:
- High expendable material cost
- Labor intensive
- Forces fibers against tool
- Applies pressure to all surfaces
- Debulks
- Cures at high temperature
- Prepreg material generally used
- Size limitations
- Controls time-temperature-pressure

The tool serves as a bed and causes a back pressure to be imposed against the bottom surfaces of the assembly. If the assembly were not bagged and sealed, vacuum cold not be drawn. In this case, the autoclave pressure would be the same at all surfaces in the assembly. The pressure pushing outward on the surfaces would equal that pushing inward. Therefore, no difference in pressure would exist and the pressure would serve no purpose at all. Fig. 4.4.2 shows a typical vacuum bag system. The components of a bagging system are described in detail below:

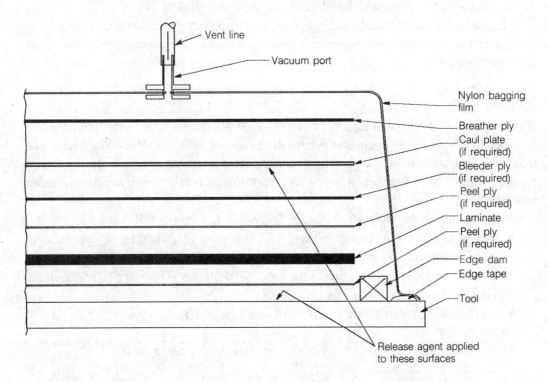

Fig. 4.4.2 Schematic Showing Sequence of A Typical Bagging System

Release Film

These films are used to prevent the composite part from adhering to tool surfaces.

Release film is also placed between the bleeder and breather plies as a separator film. It is used to prevent resin flow into the breather plies, to which the vacuum system vents. Resin in the vacuum system can clog it and render it inoperative.

- For resins that produce volatiles during cure or consolidation, a release film with small preforations which are widely spaced is used to prevent the breather from becoming clogged with resin and unable to perform its function
- For resins that produce no volatiles during cure/consolidation, an unperforated release film is frequently used so that resin removal can be controlled

Peel Ply

Peel ply film or Teflon coated fabric materials are placed on top of the layup serve to restrict but not prevent resin flow from the laminate. Most peel plies are porous or perforated and the size and spacing of perforations or the porosity of the material determines the amount of resin flow from the surface of the laid up laminate. If no resin removal during cure is desired, unperforated peel ply film should be used. The peel ply also serves to control the surface roughness and finish of the composite.

Surfaces that are to be adhesively bonded as a secondary operation after curing often use a peel ply to obtain a rough surface finish and also to keep them clean until ready for joining operation.

Bleeder Plies

Bleeder plies (used for bleed bagging system), often fiberglass cloth fabric, are normally required to absorb excess resin and permit the escape of volatiles. Various solvents and other volatile chemicals in the prepreg that take part in chemical reactions during layups must be vented, or an unacceptably porous structure will result. There are basically two options:

(1) Bleedout may occur through the edges (edge bleed) with the following advantages:
- It is used on smaller laminate parts i.e., less than 12 inches (30.5 cm) because there is less chance of bridging
- Bagging costs are reduced due to the absence of bleeder, separator and vent cloths
- There is less chance of surface wrinkles
- It is much easier to debag when there is no bleeder to be removed

Considerations are:
- A greater chance exists of volatiles being left in the cured part
- Variations in resin flow are very likely to result in the laminate being resin rich in the center and resin starved near the edges

(2) Bleedout may occur vertically through the top surface, or in a combination of both edge and vertical bleedout, which means the volatiles and excess resin flow from the mold face out towards the bag. The vertical bleedout method is applicable only to bagged laminates, and is definitely the preferred method. Vertical bleedout bleedout is widely used throughout the composites industry for autoclaved parts and the reasons for its general preference are:
- Resin flow is more easily controlled since it is flowing only a distance roughly equal to the laminate thickness
- Better removal of trapped air and volatiles is provided

It is recommended that the vertical bleedout system be used for any large laminates [i.e., 12 inches (30.5 cm)] cured in an autoclave.

Concerns are:
- Materials cost is increased because it involves use of a bleeder cloth, a barrier layer and the vent cloth
- Labor costs increase
- There is a chance of wrinkling as the bag length is reduced over any convex surface. Bag wrinkles may be prevented by using a caul plate which is a thin metal formed to the shape of finished part
- As the increased pressure causes the various laminate plies under the bag to decrease in thickness, bridging occurs across the radius
- There may be local volatiles and resin trapped at the radius in sharp corners
- Dams or boundary supports must be provides to prevent edge bleeding

The resin flow into the bleeder is controllable to an extent by autoclave pressure time history. It is desirable to have most of the volatiles removed prior to applying pressure because their flow into the vent cloth is restricted once the pressure is applied. If excessive resin is forced into the bleeder and resin does not flow back into the laminate when the pressure is applied, the resin-starved laminate may result in a deficient part.

Today, the non-bleed bagging system has become widely used because prepregs have a more controlled resin content than that of the older prepreg materials and there is no need to bleed out extra resin. Fig. 4.4.3 schematically compares bleed and non-bleed systems, which are described blow:

(1) Bleed system [see Fig. 4.4.3(a)]:
- 2 to 10% lighter because of resin loss
- Layup/cure costs 15 to 20% higher (may be justified if weight is critical)
- Usually prepreg contains approximately 40% resin by weight
- Excess resin absorbed by bleeder plies (cloth) during cure
- Good controls required to assure reproducible parts

(2) Non-bleed system [see Fig. 4.4.3(b)]:
- Recommended for general use
- Prepreg contains 32 to 35% resin by weight and no resin is removed during cure
- Lower cost system because no bleed plies are required
- Use when cocuring skins to core for honeycomb panels since surface bleed is not possible

Chapter 4.0

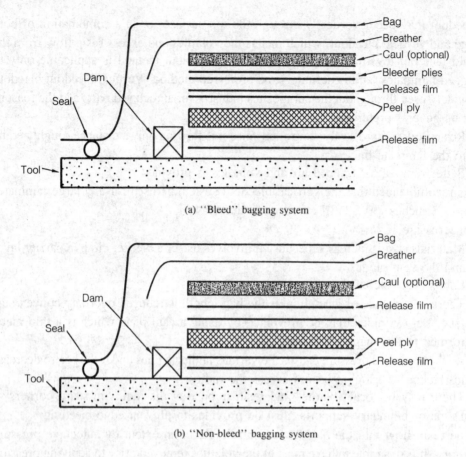

Fig. 4.4.3 Comparison of Bleed and Non-bleed Bagging Systems

Caul Plate (optional)

A caul plate is a rigid smooth sheet or plate (metal or hard rubber materials) with the same size and shape as the finished part and in direct contact with the composite layup in order to:
- Provide a smooth bag-side surface
- Prevent excessive resin washout
- Smooth the angle and prevent wrinkles at tapered areas
- Regulate configuration of joggles
- Prevent wrinkles and surface irregularities from forming when an elastomeric caul plate is used. Mosites rubber cauls are used for high temperature thermoplatics
- Help distribute pressure evenly over the part

Breather Plies

Breather plies are usually fiberglass or synthetic fabric which is placed on top of the release film to allow dispersion of vacuum pressure over the layup and removal of entrapped air or volatiles during cure/consolidation. Coarse, open weave fabrics are used, otherwise bridging and bag failure may occur.

Dam

Dam height should be slightly thicker than the layup laminate after cure or approximately the same as the layup thickness including release and bleeder plies, to prevent rounding off the part edge by the action of the vacuum bag. The dam is located peripherally to minimize edge bleeding and it must maintain its position throughout the cure (it may be integrally machined as part of the tool).

Vacuum Bags

The vacuum bag provides the means of removing vapors, and encouraging the required resin flow, so its design and implementation are important.
- The vacuum is used to assist removal of trapped air or other volatiles
- The vacuum and pressure-temperature cycles are adjusted to permit maximum removal of air with maximum resin flow

The vacuum bag is the final item in a composite layup for autoclave or oven cure. The bag must be vacuum tight so no autoclave gas can enter the layup. Should bag or edge sealant failure occur, the autoclave gas enters the layup, neutralizing air and volatile removal. The choice of bag material and its compatibility with the tool shape and sealing arrangements are major decisions in the composites manufacturing process.

Bags that are stretched too tight tend to lead, break or, on convex radii, span the layup, reducing effective autoclave pressure. In any event, leaks in the film or in the sealing of the bag are definitely unacceptable especially when curing thermoset parts because it results in the scrap of the part.

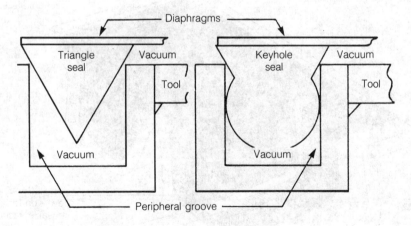

(a) Two types of reusable vacuum bags contain a groove around the tool, eliminating the need for a bulky frame or seal rib. Both designs allow the part to be removed easily, while the keyhole design provides a little better seal.

Fig. 4.4.4 Examples of Reusable Bags

Chapter 4.0

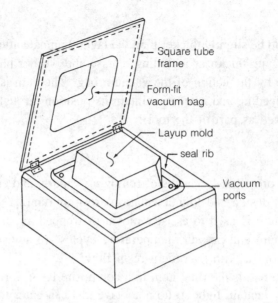

(b) A resuable vacuum bag contains a frame to which a silicone rubber diaphragm is bonded. The system is easy to add to existing tooling, and does not require close tolerance between the frame and the tool.

By courtesy of Bondline products

Fig. 4.4.4 Example of Resuable Bags (cont'd)

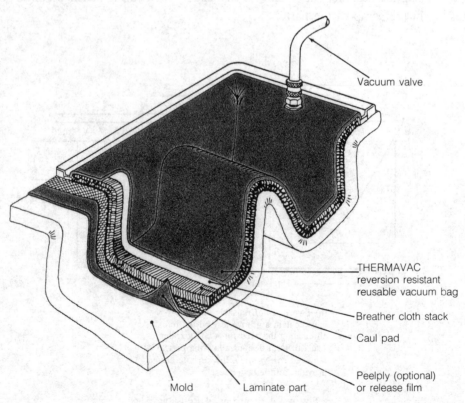

By courtesy of Arlon, Silicone Technologies Division

Fig. 4.4.5 Reuable Bag Tailored to A Specific Shape

Types of vacuum bags are described below:
(1) Nylon film bags — This is by far the most commonly used bagging material (normally 1 to 3 mils thick). Other films such as silicone, Mylar, Kapton, Upilex-R, etc., can all be used, depending on the temperature resistance of the film and the ability to seal to the molding tool surface. Extreme care must be used when handling bagging film as pin holes are easily formed.

Factors to consider when using Nylon film bags:
- Elongation under heat and pressure may occur during cure cycle
- Wrinkles in the vacuum bag generally transfer to the layup, so care must be taken in positioning the film and arranging the bag.
- Film bags are limited to one cure and they undergo destruction when removal from the cured or consolidated part
- Requires the use of tacky sealant tape, usually ½ inch (1.27 cm) thick, which goes around the part like caulking and held down the vacuum bag

(2) Reusable bags — Reusable bags such as elastomeric bags, thin metal bags, etc., require a different approach to sealing than film bags as shown in Fig. 4.4.4. These bags usually become a part of the tooling approach. If the part is a flat or non-complex shape, use a low-modulus rubber. Uncured B-stage silicone, which is sticky and flexible, can be formed over the tool to customize the bag and then heat is applied to cure it. Another method is to use an already-cured sheet of an ultra-low modulus silicone to form the bag. It is tailored to a specific shape as shown in Fig. 4.4.5 to ensure intimate contact and it provides even pressure on parts that have complex shapes or undercuts.

Factors to consider when using reusable bags:
- Requires some type of clamping mechanism to provide a vacuum seal to the tool mold surface
- Mechanical seal of bag to tool mold surface is preferred
- Seal must work at both room and curing temperature
- Sealing reusable bags requires considerable ingenuity
- The CTEs of the various components used in the bags sealing should be accounted for at the tool design stage
- Reusable bags are generally molded to the shape of the part being cured
- Bag must be replaced when it eventually ages after a number of heat cycles

Monitoring Systems

During curing, a monitoring system is required to monitor all necessary data within the entire process to ensure a good quality composite part which meets specifications.
(1) Temperature and pressure measurement — Thermocouples or sensors are used to measure composite processing temperature. One thermocouple should be used for each approximate 3 to 5 square feet of the layup and thermocouples (minimum two in general practice) shall be placed in contact with the edge of the layup and spaced to provide representative monitoring of the entire layup laminate. Fig. 4.4.6 shows a pressure sensor (disposable item) which can be placed directly at points within or on a laminate, wherever processing pressure data is required.

Chapter 4.0

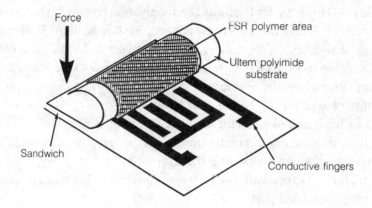

By courtesy of American Composite Technology

Fig. 4.4.6 Disposable Pressure Sensor

(2) Composite curing — Dielectric sensing is a technique most often used to monitor and control composite curing. Usually the sensor is placed at the location where the cure is to be monitored. During cure the capacitance and conductance of the material are recorded.

Vacuum/vent lines

Vacuum ports as shown in Fig. 4.4.7(b) shall be placed beneath the vacuum bag or diaphragm at intervals which will ensure that any portion inside of the layup liminate can be vented out through the vacuum/vent lines [see Fig. 4.4.1(a)]. Some pressure/vacuum gauge [see Fig. 4.4.1(b)] are required to monitor the vacuum pressures.

A vacuum of a minimum of 20 inches (50.8 cm) of mercury (Hg) shall be used to check for leaks prior to autoclave processing. During the check, a loss of greater than 2 inches (5.08 cm) of Hg in 5 minutes is unacceptable.

Debulking

When initially laid up, any number of plies are bound to trap a certain amount of air between them, and on contours their elasticity may prevent their laying smoothly on top of each other. The more plies that are laid up, the more likely is it that wrinkles will result and be cured into the laminate, resulting in an unacceptable laminate. A debulking procedure used in conjunction with squeezing the plies onto the tool surface is repeated as often as necessary during the layup of thick laminates to reduce wrinkles and to fit the prepreg smoothly into the tool. However, applying debulking on thin laminates or slightly contoured laminates results in little improvement in structural performance.

Prior to cure or consolidation, the layup laminate may need to be debulked under 20 inches Hg vacuum after every 5 to 10 plies are laid down. The debulking bagging arrangement usually conforms to the general configuration shown in Fig. 4.4.1 except under special conditions.

Manufacturing

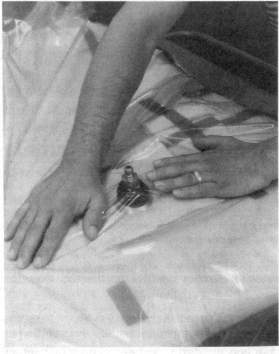
By courtesy of Ren plastics

(a) Breather layer with one vacuum port for each square yard of tool surface before bagging

© 1991 Swagelok Co., all rights reserved
By courtesy of Swagelok Co.

(b) Vacuum port, vacuum connector, and vacuum vent line

Fig. 4.4.7 Vacuum Port and Vent Line

But knowing when and where not to debulk is as important as any other aspect of composite fabrication. Therefore, the decision to debulk should not be taken lightly, and should be done only when understanding of both cost and technical value is clear. Sometimes the structural value obtained from debulking is nearly nil, particularly for thin, large radius laminates.

However, if the debulked laminate is not continuously being processed and is allowed to sit idle, air will again penetrate the laminate and much of the effect of the debulking is lost. This is especially true for laminates of complex shapes where the fiber wants to return to its original state.

Both hot and cold debulking can be used:
- Cold debulking — This is the most commonly used since it is faster and less costly. It removes most of the air, but very little resin displacement occurs.
- Hot debulking — This is done at various temperatures. However, it must be noted that each heat debulk reduces the pot life of thermoset resins and causes the gel point to occur sooner than normal — the cure cycle may therefore have to be modified. Hot debulking does not cause any degradation of thermoplastic resins.

Rigorously established rules for debulking, applicable to all situations, do not exist and judgment must be used.

4.5 MOLD FORMING

The forming method used, in reference to the quantity of parts to be made, generally determines the cost of production and the quality of the part. The choice of a particular method depends on the level of capital investment and the effect of the deformation path on the fiber distribution in the resultant part.

Again, one of the most important decisions to be made in developing a system for producing a composite part is the material form selection. For structural applications where controlled ply layup and fiber orientations are required the following material forms may be used:
- Tape (or unidirectional tape)
- Woven fabrics
- Discontinuous fiber sheets (blanks)

The choice of either tape or fabric laminates depends on the component's:
- Configuration and complexity of the component
- Mechanical properties requirements

(1) Tape is usually reserved for simple curvatures requiring high mechanical strength. The diaphragm forming process retains the high strength properties of tape even with complex shapes, such as deep drawn or multi-cavity components. For thermoplastics, handling is an issue since a layup heated at high temperatures must be transferred to the die for forming. A matter of major concern during forming is fiber interaction and displacement. Tape and slit tape (i.e., Quadrax, see Fig. 2.5.31) allow adequate intraply slippage and do not buckle in thermoplastic composites.

(2) Where stringent property translation is not required, especially in quasi-isotropic layups, woven fabrics are preferred because of ease of handling. Fabrics tend to prevent gross fiber motion due to restricted lateral fiber displacement. This can result in fiber wrinkling where intraply slippage is required for the molding of complex shapes.

(3) Commingled fabrics [thermoplastics, see Fig. 4.3.1(c)] offer several fabrication benefits in that they are dry, drapeable, easily drawn to complex shapes and multiple cavities, and can be formed on the tooling at room temperature (see Materials in Chapter 2).

Autoclave Forming

Autoclave forming is the most commonly used method of applying heat and pressure simultaneously. A layup assembly on the mold in the autoclave is shown in Fig. 4.5.1. The autoclave is a heated, pressure chamber. Parts of significant complexity can be cured in an autoclave.

The essential steps in the process are:
- Layup
- Debulking, if necessary
- Preparation of a bagging system (see Fig. 4.4.2)
- Cure in an autoclave, hydroclave, etc. by controlled heat and pressure (see Section 3.5 of Chapter 3)

Manufacturing

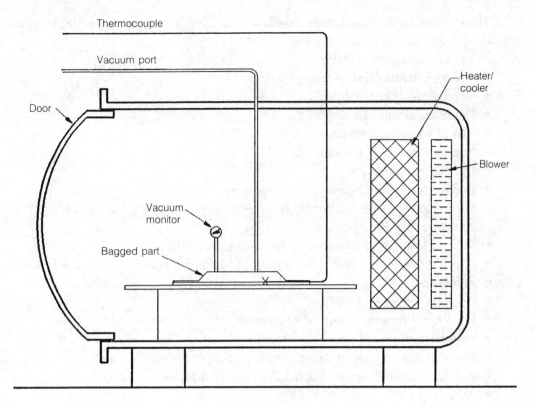

(a) Autoclave and curing part arrangement

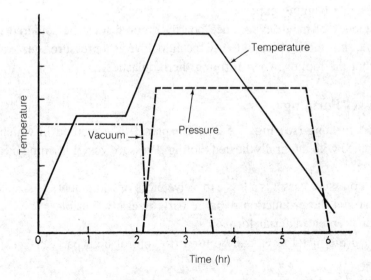

(b) Typical autoclave cure cycle — thermoset composite.

Fig. 4.5.1 Autoclave Forming Processing

The vast experience gained in the autoclave processing of thermosets has been transferred into the autoclaving of thermoplastic parts.

Advantages of using an autoclave:
- Uniform deformation pressure
- Good dimensional stability
- Processes all materials
- Flat to complex contours
- Variable thickness laminates
- Cure or consolidation
- Bonded or co-cured honeycomb structures
- Large parts can be accommodated (large autoclaves), e.g., a 25 ft. × 60 ft. autoclave, see Fig. 3.5.1(b)
- Method lends itself to cocuring or co-consolidation
- Existing technology (there is a wealth hands-on experience and lesson learned)

Considerations are:
- Complex, expensive tooling
- Long cycle times/low rate of production
- Expensive disposable bagging expensive
- Bag and sealant process failures add to cost
- Potential material movement in cure or consolidation
- Autoclave size

Considerations when autoclaving thermoplastics:
- High thermal inertia of the tooling and the autoclave
- Expendable bagging expense
- Inefficiency of hand layup since boardy prepreg must be soldered in place

If possible, it is more efficient to use the autoclave as a pressure source and integrally heat and cool the tooling when forming thermoplastics.

Pressure-vessel Forming

Pressure-vessel forming (see Fig. 3.5.2 of Chapter 3) is a method in which a pressure vessel is combined with integrally-heated tooling. Pressure vessel forming has the following advantages:
- Since a pressure vessel is a less expensive piece of equipment, its size is not limited by cost and the production of large parts is easily feasible
- Dry or prepreg material forms
- Matched die metal tools can be used (to control both part surfaces)

Vacuum/oven Cure

Vacuum/oven cure, shown in Fig. 4.5.2, is a less expensive fabrication method than autoclave molding since the vacuum pressure is directed only on the part and heat is provided by an unpressurized heated oven (see Fig. 3.5.3 of Chapter 3).

Manufacturing

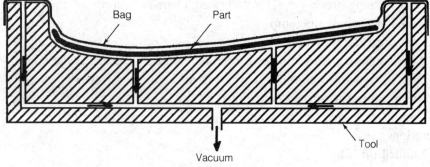

Fig. 4.5.2 Vacuum Forming

Concerns about this method include:
- Low pressure (vacuum)
- High temperature bagging materials
- A non-adhering film is placed over the layup and sealed at the edge
- Uses a vacuum to eliminate entrapped air and excess resin
- Resin rich/resin starved conditions
- Loss of parts due to bag failure

Recommended alternate methods:
- Use trapped rubber — press cure
- Pre-molded elastomeric overlay method — vacuum/oven cure
- Elastomeric vacuum bag method — vacuum/oven cure

Therm-X Forming

The Therm-X forming process, shown in Fig. 3.5.4 of Chapter 3, can be described as follows:
- Tool and part are surrounded by silicone powder
- Pressure is produced by heating the powder
- Cooling coils are also installed if curing thermoplastic material

Advantages are:
- No bag required
- High temperature [1500°F (816°C)]
- High pressure [3000 psi (20.7 Mpa)]

Considerations are:
- Developmental process
- Equipment size limited
- Integrally-heated tools required

Vacuum Forming

Vacuum forming, shown in Fig. 4.5.3, only uses a single male or female tool plus an elastic diaphragm that is forced to conform to the tool via vacuum pressure. Relatively low pressure can be used if applied in a constant, compliant manner over the entire surface area.

231

The vacuum forming process has the following characteristics:
- Cured at room temperature
- Low pressure (vacuum)
- A non-adhering film is placed over the layup and sealed at the edge
- Uses a vacuum to eliminate entrapped air and excess resin
- Resin rich/resin starved conditions
- Loss of parts due to bag failure

Considerations are:
- Limited pressure
- Temperature limitation
- Unable to form complex parts

Pressure-bag Molding

Pressure-bag molding, shown in Fig. 4.5.3, is similar to vacuum-bag molding except that air pressure, usually more than 30 psi, is applied to a rubber bag.

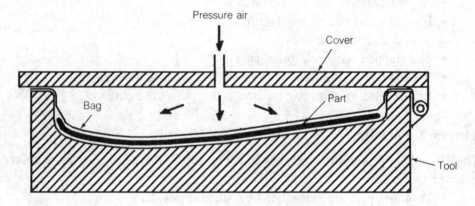

Fig. 4.5.3 Pressure — Bag Molding

Elastomeric Forming

Elastomeric rubber forming is shown in Figures 4.5.4 and 4.5.7 and the process is as follows:
- Temperature only process
- Cure or consolidation pressure is created by thermal expansion of the rubber (silicone rubber)
- May use the combination of both temperature and pressure (or vacuum and applied additional pressure)

Creep Forming

Creep forming is shown in Fig. 4.5.8 and the process is described below:
- The die is an integrally heated and cooled system
- The panel is loaded into the die and covered with a vacuum bag to apply atmospheric pressure

Manufacturing

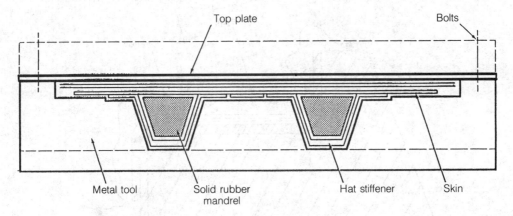

(a) Matched die/solid rubber combination

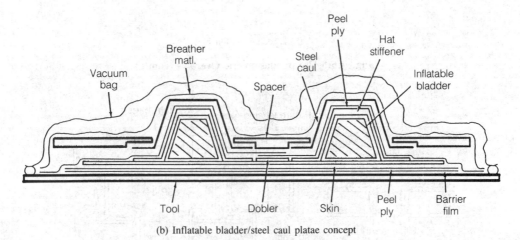

(b) Inflatable bladder/steel caul platae concept

Fig. 4.5.4 Fabrication of Hat-Section Panels

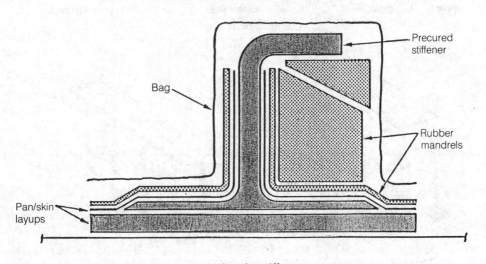

(a) J-section stiffener

Fig. 4.5.5 Use of Elastomeric Overlays

233

Chapter 4.0

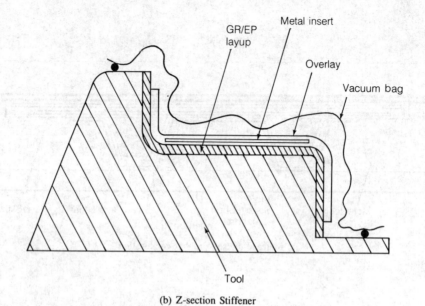

(b) Z-section Stiffener

Fig. 4.5.5 Use of Elastomeric Overlays (cont'd)

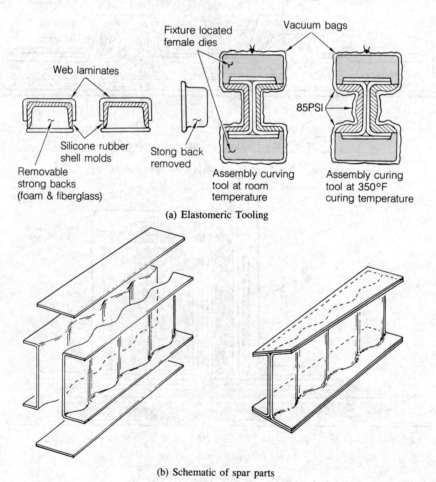

(b) Schematic of spar parts

Fig. 4.5.6 Sine-wave Spar Fabricated by Elastomeric Tooling

Manufacturing

(c) Final spar

By courtesy of Rockwell International

Fig. 4.5.6 Sine-Wave Spar Fabricated by Elastomeric Tooling (cont'd)

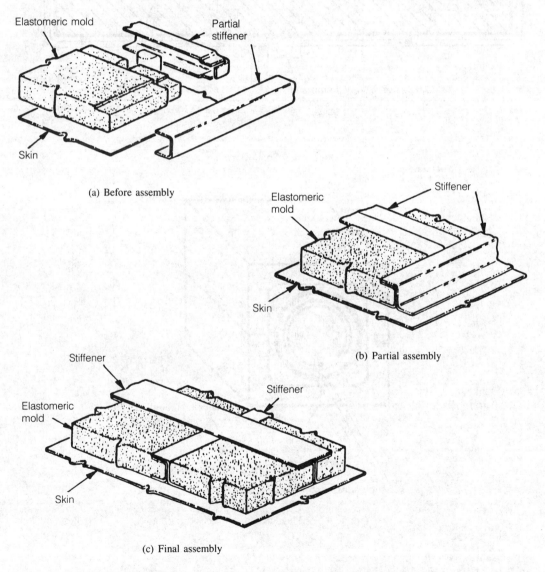

(a) Before assembly

(b) Partial assembly

(c) Final assembly

Fig. 4.5.7 Elastomeric Tooling of An Integrally Stiffened Pancel

235

Chapter 4.0

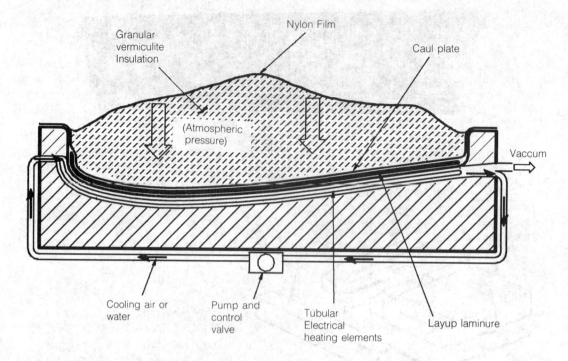

Fig. 4.5.8 Creep Forming

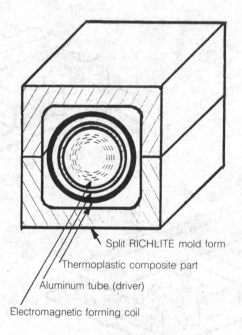

Fig. 4.5.9 Magneforming Process

- The combination of heat and pressure allows the panel to creep formed into shape
- Ideally suited for thick thermoplastic skins
- Low cost forming of simple contoured skin panels
- Shape limitations

Magneforming

Magneforming is shown in Fig. 4.5.9. This method has been patened by Grumman Corp. and the process works as follows:
- Pressure pulse is produced in the workpiece by the magneformer
- Pulse is propelled against a dielectric mating surface and the permanent deformation of the workpiece occurs
- Since thermoplastic material has low electrical conductivity, it should be formed by using a metal driver
- Thermoplastic material must be heated to melt temperature before forming
- Final shape depends on the mating tool surface
- Low cost
- Rapid cycles
- Limited size

Deepdrawing Forming

- Male die required
- Heated thermoplastic fabric sheets are clamped by a pair of aluminum pressure plates
- The male die pushes through the heated fabric sheet and forms the part as shown in Fig. 4.5.10
- Low cost forming

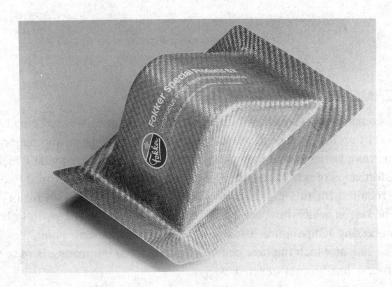

(a) Product manufactured by Fokker Special Products B.V.

Fig. 4.5.10 Deepdrawing Forming

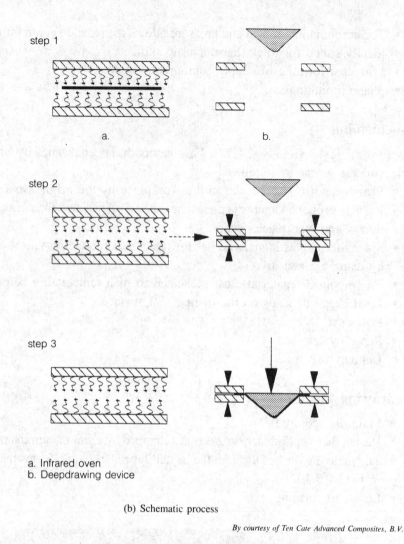

a. Infrared oven
b. Deepdrawing device

(b) Schematic process

By courtesy of Ten Cate Advanced Composites, B.V.

Fig. 4.5.10 Deepdrawing Forming (cont'd)

4.6 THERMOFORMING (PLATEN PRESS)

Advanced composite systems based on thermoplastic composites are being considered as replacements for sheet metal in new airframe structures. One of the major benefits offered by thermoplastics is that they have characteristics, as does metal which allows the rapid transformation of raw materials into finished parts when press formed (see Fig. 4.6.1). The time required for this rapid conversion is limited only the two factors:
- The rate at which heat can be added to the thermoplastic resin to bring it up to processing temperature (usually at or above its melting point)
- The rate at which the heat can be removed from the materials once the forming process has been completed

The actual forming step represents a small portion of the total time for heat-up and cool-down. Processing time is not governed by the time required to complete a chemical reaction as it is with thermosets.

Manufacturing

By courtesy of Wabash Metal products, Inc.

Fig. 4.6.1 400 Ton Compression Press Machine

When matched die tools (see Fig. 3.8.1) are used, heat up rates are faster than in an autoclave and this reduces the processing time, post-curing occurs in an oven after removal from the press. There is no vacuum to remove air and volatiles when using matched die tooling. The upper half die is not put in contact until temperatures are reached which allow free passage of air and volatiles. Much of the air and volatiles remaining after die closure are squeezed out with the resin into the flashing which is later trimmed off the part. Matched die tools are processed in this manner simply to increase the production rate.

Thermoforming (a general term of fabrication for thermoplatic materials) consists of laminates in flat-sheet form, which are assembled from individual plies of woven prepreg fabrics or unidirectional prepreg tape to the desired thickness and reinforcement (fibers) orientation, and then consolidated with heat and pressure. This allows considerable flexibility in arranging the plies to achieve isotropic or anisotropic mechanical properties, and the sheet can also have non-uniform thickness if desired. Thermoforming has three key elements.

(1) A laminate support frame, which carries the laminate into the heat source, supports the laminate during and after softening of the matrix, rapidly transfers the softened part from the heat source to the forming tool, and then releases the laminate onto the lower tool
(2) A heat source capable of evenly heating the laminate to its processing temperature in a short period of time
(3) A thermoformer capable of rapid closing speeds with sufficient clamp pressure form the laminate

Chapter 4.0

Since the laminate must slip against the tool during processing, there can be difficulties in maintaining an adequate gas seal to hold vacuum below the laminate and positive air pressure above it. Therefore, an elastomeric or thin aluminum bladder or diaphragm is clamped over the laminate. Pressure can then be applied directly to the diaphragm (hydrostatsic forming or hydroforming process).

Thermoforming has the following advantages:
- Complex shapes are possible
- High dimensional control (all surfaces)
- Minimum machining requirement
- Good process control
- Good heat-up rate
- Good cool-down rate
- Fibers can be oriented as required
- High production potential

Factors which should be considered:
- Press size
- Expensive tooling
- Preform material required
- Precise die required
- Lower mechanical properties obtained

Matched Die Forming

Matched-die forming, as shown in Fig. 4.6.2, is a most widely used forming system for thermoplastics because it is available in a vary from small simply operated hand processes to large computer controlled presses (see Fig. 4.6.3). The dies used are generally made of metal which can be internally heated and/or cooled. The dies are designed to fixed gap of close tolerance.

Advantages of matched die forming:
- Tolerance can be as close as design requires

Factors which should be considered:
- High forming pressure required for good solidation
- High fabrication costs
- Friction at die interface
- Long heating and cooling times
- Non-uniform deformation/pressure, if thickness mismatch exists

Stamping

Stampable thermoplastic composite sheets are reinforced with continuous swirl mat or chopped-fiber mat, rather than woven or unidirectional fibrics. The stamping process, as shown in Fig. 4.6.3, is basically the same as thermoforming except that it requires higher pressure forming equipment and more expensive matched-steel molds, similar in design to those used for compression molding thermosets. Unlike thermoforming, stampable composites are capable of a good deal of flow and the process is recommended for parts requiring good surface finish.

- Sheets are reinforced with continuous strands
- Precut blanks are heated to above the resin melt temperature (thermoplastic materials) and then moved to a matched die metal press tool where they are compressed in a mold shape
- Lower forming pressures permit the use of lower cost equipment and tooling
- Owing to incorporation of random continuous fibers, stamping comes closer to the versatility of injection molding, but offers much higher strength

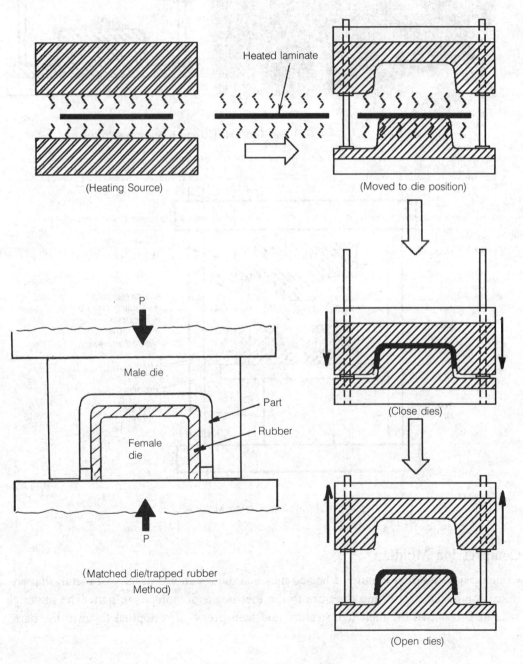

Fig. 4.6.2 Match Die Forming

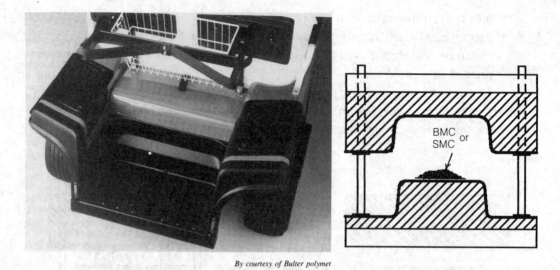

By courtesy of Bulter polymet

Fig. 4.6.3 Bay Well Of A Golf Cart Fabricated Using A Stamping Process

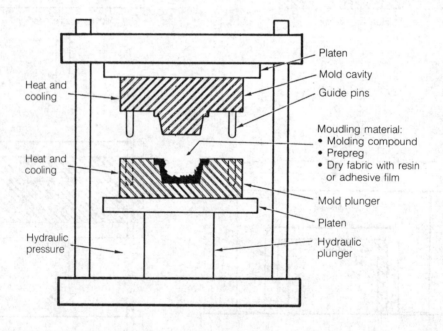

Fig. 4.6.4 Compression Molding

Compression Molding

Compression molding features a heated matched die (or mold, usually of metal) that applies simultaneous heat and pressure to the thermoplastic composite part. The material is heated to above the melt temperature and then pressure is applied to form the part.

This is high volume, high pressure method suitable for molding complex, high strength composite components. Bulk-molding compound (BMC), sheet-molding compound (SMC) pre-forms, or mat molding may all be processed by compression molding as shown in Fig. 4.6.4. Part configurations and cross-sections are extremely flexible, and inserts and attachments may be accommodated. The process is capable of holding very high tolerance. The process is summarized below:

- Sheet-molding compound (SMC)
- Bulk-molding compound (BMC)
- Preform mat
- Heat and pressure applied to cure resin
- Typical resin is thermoset
- High-volume, high-pressure
- Complex shapes
- Reduced material properties (since using discontinuous fibers)
- Matched die mold tools can be cast or forged steel, cast iron, or cast aluminum.

Diaphragm Forming

Diaphragm forming, shown in Fig. 4.6.5, has been used to fabricate complexly shaped parts from prepreg thermoplastics. Forming with diaphragms (materials with excellent elongation characteristics at deformation temperature) of aluminum [600-840°F (360-500°C)], Upilex-R film [390-840°F (200-450°C)], Kapton [390-840°F (200-450°C)], or polyimide [570-750°F (300-400°C)], or rubber press forms, has successfully produced large complexly shaped parts. The geometric complexity of parts made is limited by the deformation of the diaphragm and since the diaphragm material deforms in a finite temperature range, only materials which can withstand these temperatures can be used. The diaphragm forming process is as follows:

- Layup is held between two disposable plastically deformable diaphragms
- Diaphragms and layup are heated to the processing temperature
- The deformed material is placed on a tool half using a combination of air pressure and the movement of the tool to contact the layup

The current "Superform" process uses aluminum diaphragms to form complexly shaped parts out of thermoplastics as shown in Fig. 4.6.6 through 4.6.8. Its only limitation is deformation temperature. Formed aluminum diaphragms [see Fig. 4.6.6(b)] can be reused during co-consolidation processing or for other bagging operations. Superforming is described below:

- Part fully formed in a single step
- Good fiber placement control
- Complex shapes with reducing wrinkling
- Unconsolidated layup is simultaneously consolidated and formed into the desired shape
- Similar to superplastic forming of metals
- Minimize splits, wrinkles and thinning
- Relatively low pressure operation

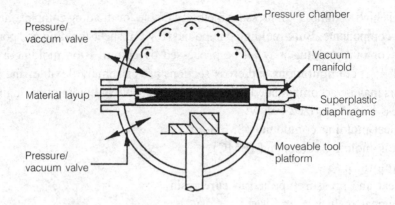
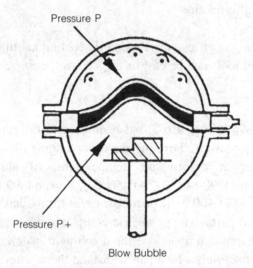

Blow Bubble

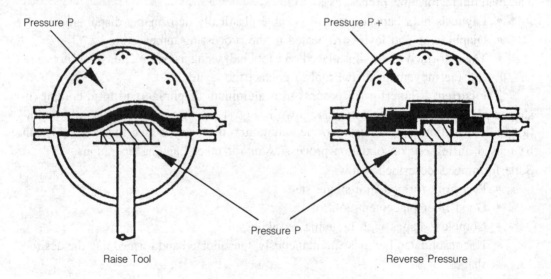

Raise Tool · Reverse Pressure

Fig. 4.6.5 The Diaphragm Forming Process

Manufacturing

(a) Diaphragm formed panel skin

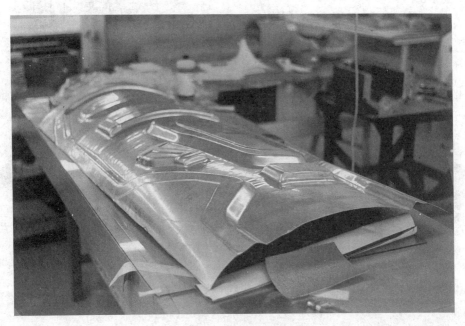

(b) Aluminum diaphragm sheet

By courtesy of Lockheed Aeronautical Systems Co.

Fig. 4.6.6. Panel Fabricated by Diaphragm Forming

Chapter 4.0

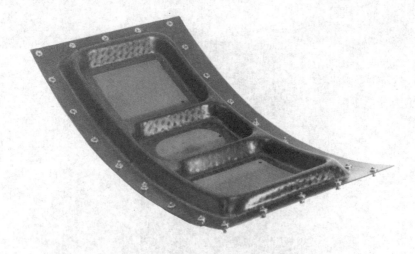

(Note: Inner formed skin and outer flat skin were welded together by ultrasonic welding

By courtesy of Lockheed Aeronautical Systems Co.

Fig. 4.6.7 Fuselage Service Door Fabricated by Diaphragm Forming

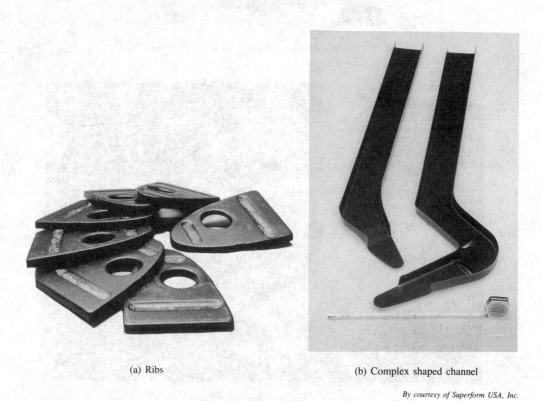

(a) Ribs (b) Complex shaped channel

By courtesy of Superform USA, Inc.

Fig. 4.6.8 Rib and Channel Fabricated by Disphragm Forming

Advantages are:
- Complex shapes possible
- Good fiber placement control
- Uniform deformation pressure
- Co-consolidation

Considerations are:
- Long cycle time
- Temperature limitation
- Small radii limitations
- Size limit
- Expandible caul materials required
- Thickness variations restricted

Hydroforming

This is a well-known metal forming process that has been successfully used to shape thermoplastic composites using preheated composite blanks as shown in Fig. 4.6.9. This is the simplest and most efficient method for forming parts of simple geometry. It is essentially a two-step process:
- Thermoplastic material is elevated to the working temperature.
- Layup is deformed in a cooler die to shape utilizing the hydrostatic pressure of a trapped rubber punch

This method takes full advantage of the benefits of thermoplastics and only requires one solid die, usually female, and a silicone rubber male punch. It is possible to form more complex details including contours, joggles, and a variety of stiffener shapes, etc. in a single operation.

- Only one solid die required
- High pressures [10,000 psi (68.9 Mpa)] possible
- Undercuts can be formed
- Relatively even pressure over the surface of the molding, minimizing fiber damage in high spots and sharp corners
- Complex shape thermoplastic parts
- The mold should have a well-polished surface to allow ply slippage during forming
- Use adequate vent holes to ensure evacuation of all the air trapped between the laminate and the mold surface
- Rapid cycles
- Moderate tooling costs
- No peripheral equipment needed such as heating and cooling lines

Factors which should be considered:
- Machine size limited
- Low temperature

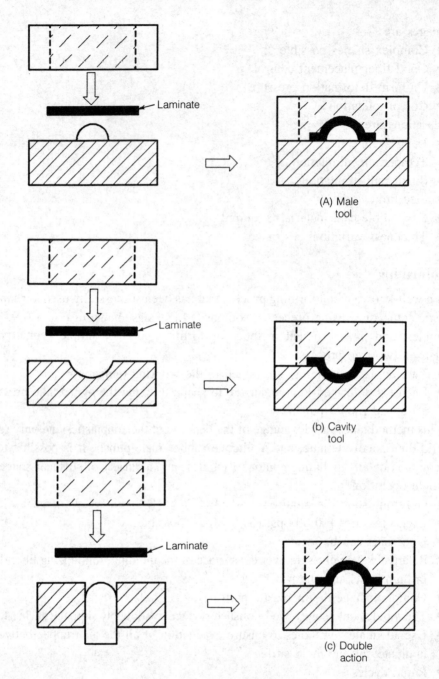

Fig. 4.6.9 Hydroforming

Rubber Forming

Fig. 4.6.10 shows a process in which one tool half is replaced by a thick pad of rubber that conforms to the solid tool half under pressure in a forming process. The rubber pad can be profiled to the tool geometry and is usually the upper tool half, which is generally much larger and is permanently attached to the press platen. The pressure is determined by the local extent of deformation and approaches uniformity for parts with very shallow draw.

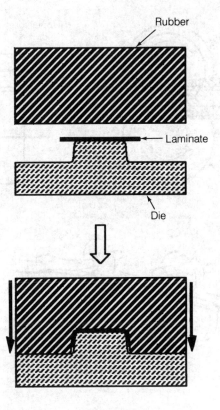

Fig. 4.6.10 Rubber Forming

- Only one solid die required
- Very complex shapes possible
- High pressure possible

Advantages are:
- One solid die
- Rapid process
- Heat limitation

Factors which should be considered:
- Forming pressure not uniform over the laminate
- Very shallow draw
- Good level of draw can be made of more compliant foam materials which imposes temperature limitations
- May require post-cure
- May have equipment size limitations

Roll Forming

Roll forming, shown in Fig. 4.6.11, is an automated, continuous process whereby continuous lengths of metal strips are progressively formed through a set of matched roller dies, producing a constant cross section of continuous shapes, e.g., T, Z, L, channel, hat section, etc. Roll forming can also produce long thermoplastic structural sections as well.

Chapter 4.0

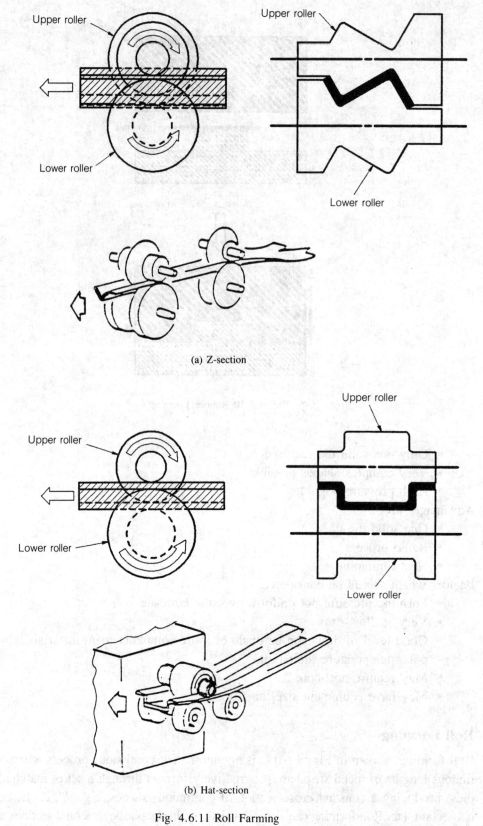

(a) Z-section

(b) Hat-section

Fig. 4.6.11 Roll Farming

Roll forming involves the following:
- Pre-consolidated sheet stock is heated to the molding temperature by an infra-red preheating oven
- A series of matched rollers that are driven by the roll forming machinery forms and consolidates
- Stiffeners, stringers, longerons, spars, etc. can be made using this process
- Low manufacturing/tooling costs
- No roll lubrication or surface release agent is required during roll forming because a smooth polished surface does not encourage adhesion
- Tapered thicknesses are possible

Advantages are:
- Dimensional accuracy
- High pressure possible

Considerations are:
- Limited shapes
- High setup costs
- Constant cross-section required

4.7 MOLDING

Resin Injection Molding (RIM)

Resin injection molding (RIM), shown in Fig. 4.7.1, is the highest-volume method using single or multi-cavity molds to produce complex parts at very high production rates. Thermoset molding compounds may be injection molded with the injection screw or plunger and chamber of the molding machine held at low temperatures but when the mold temperature is elevated to 300-350°F (150-163°C) the process is improved.

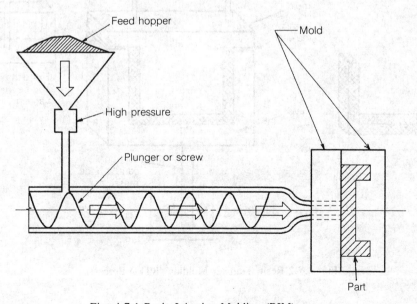

Fig. 4.7.1 Resin Injection Molding (RIM)

The RIM process works as follows:
- Pellets of resin containing fiber reinforcement (about .5 inch long) and fed into a hopper and then into a heated barrel
- Heated resin is then forced at high pressure into a matched die metal tool mold
- Parts can be very precise and complex
- Limited size
- Substructured components
- High tooling costs
- Reduced material properties (low strength part)

Resin Transfer Molding (RTM)

Resin transfer molding (RTM), shown in Fig. 4.7.2, has been in use for nearly half a century and recently it has been used as a lower-cost alternative to compete with the prepreg layup and autoclave cure process. With RTM, the fiber preform is placed in a closed mold and resin is injected at low pressure (less than 100 psi for thermosets), although for some applications a vacuum assist is appropriate. The resulting part is mold controlled on all sides, without bagging labor or material waste. In addition, inserts, ribs, frames, stiffeners and core materials can be molded in place (see Fig. 4.7.3). Since fiber content can range from 0 to 50% by weight, both primary and secondary structures can be molded to specific requirements. Although these advantages are profound, there are several processing concerns which must be overcome in order to successfully mold a high quality airframe structural composite parts with RTM.

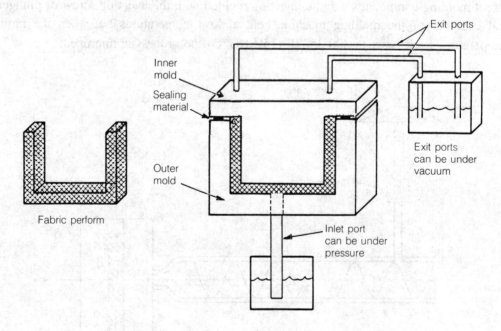

Fig. 4.7.2 Resin Transfer Molding (RTM) Process

By courtesy of TERTM Technology Corp.

Fig. 4.7.3 Fuselage Door of Cessna 206 Aircraft Has A Molded-in Hinge Reinforcement And Armrest

Resin transfer molding process variables are:
- Resin viscosity
  — Temperature
  — Pot life
  — Integral heating or oven
- Dry fiber preform weave must provide ease of fiber wetting (see example from Fig. 4.7.4 and Section 2.5 of Chapter 2)
- Use and amount of:
  — Pressure
  — Vacuum
- Placement and number of inlet and outlet ports
- Degassing initial resin
- Evacuation of air from preform
- Low pressure process for moderate-volume production quantities
- Mold is closed and clamped
- A low-viscosity and catalyzed resin is pumped in and to replace the air cavities

RTM requires a matching mold set (both male and female tools) and RTM tooling materials selection depends on application usage. The following materials are used for RTM tooling:
- Mass cast epoxy
- Kirksite casting
- Polymer/fiber composites

Chapter 4.0

- Electroformed nickel (see Fig. 3.3.3)
- Machined aluminum or steel materials

There are many variables, as mentioned previously, which influence the success of the RTM process:

(1) Pressure — Determines mold material requirements; relatively high pressure, e.g., greater than 50 psi may require aluminum or steel molds and high clamping forces. Too high a pressure can also cause fiber wash and incomplete fiber wet out.

(2) Temperature — Depends upon the resin system cure cycle and minimum viscosity temperature.

(3) Viscosity — The viscosity of the resin during infusion is a major factor in determining both pressure and temperature requirements.

(4) Fiber content — The maximum viscosity allowable with RTM depends on fiber content, which is generally 50%, 70% can be achieved with difficulty using special molds and high pressure resin injection equipment.

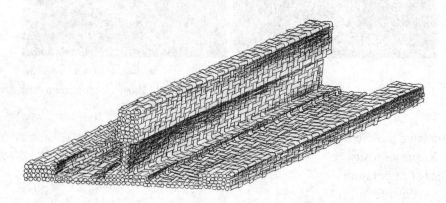

(a) Multi-layer/Multi-Thickness T-section (Variation in thickness in base as well as in blade stiffener)

(b) Radome Shape (Tapered thickness from base to tip and through the thickness reinforcement)

By courtesy of Testile Technologies, Inc.

Fig. 4.7.4 Fiber Architectures — Preforms

(5) Fiber orientation — The fiber angle of the preforms relative to the flow path of injection affects the resin flow velocity and the wet-out the fibers.

(6) Part thickness — This may also have an effect on fiber wetout and the void content of the molded part.

Advantages of resin transfer molding include:
- Using preforms can provide increased damage tolerance for finished parts
- Produces complexly shaped parts with good surface detail and accuracy (see Fig. 4.7.5)
- Complex structures can be formed, e.g., enclosed wing or fuselage box with molded-in ribs or frames
- Make near-net-shape parts requiring a minimum of trimming
- Mat and woven reinforcement is laid up dry
- Obtain smooth surfaces on both side of part
- Removes the need for shipping and storing refrigerated prepreg (thermosets)
- By monitoring the process itself very closely, inspection at the end can be eliminated
- Exposuring of shop personnel to chemicals is greatly reduced because hazardous materials are enclosed inside the process equipment

Factors which should be considered:
- Process still very much in its infacny for airframe applications
- Greater initial expense of two-side mold tooling
- Presently only thermoset (low resin viscosity) resin is used and it must be preheated before injection
- Since a catalyzed thermoset resin would not have a long pot life, the resin and catalyst are usually heated individually and mixed before injection
- It is difficult to place and hold the fiber preform in the proper position as the mold closes
- Potential problems in the tool's ability to withstand clamping-up forces (sealant)

The net-shape molding presents difficulties in mold construction and tolerances must be tight. In this molding process one cannot be assured that fiber content (or distribution) will be uniform and too often a resin rich edge is produced. The pros and cons of net-shape molding are given below:
- No trimming required
- Mold construction more complex
- Resin rich edges occur
- Difficult to hold preform in proper position
- Difficult to trim preform with precision

The RTM method appears at first glance to be fairly simple in nature, but as previously explained this is not often the case. The challenge in the future of RTM is to resolve the existing problems in the process which mainly involve:
- Fiber reinforcement preforms
- Tooling complexity
- Resin flow and wet-out of preform
- Full automatization of the process

By courtesy of Bentley — Harris Mfg., Co.

Fig. 4.7.5 Helicopter Drive-shaft Coupling Fabricated By RTM

Wash-out (disposable) Mandrel Molding

This is a very interesting process which will play a very important role in future innovative composite design because conventional mandrel must be left inside the enclosed structure after cure or consolidation, but the wash-out mandrel can be removed. Wash-out mandrel molding was discussed in Section 3.4 of Chapter 3 and the process is briefly summarized below:

(1) For curing temperatures under 350°F (180°C), use the Hot-melt process which usually uses a Sodium-nitrate mixture, e.g., Eutectic salt, to form complex shaped parts as shown in Fig. 4.7.6 and Fig. 3.4.10 of Chapter 3.0.

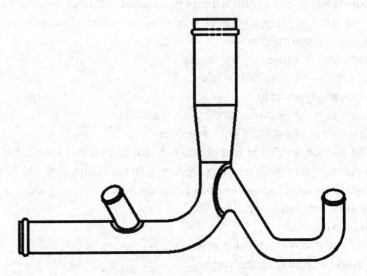

Fig. 4.7.6 Complex Shaped Duct Can Be Fabricated By The Hot Melt System

(2) For curing temperatures above 350°F (180°C), use plaster. This has been successfully used to produce the unique panel shown in Fig. 4.7.7 which is a co-cured PMR-15 panel. The CARE-MOLD process was utilized to form closed-end hollow ribs.

By courtesy of Composites Horizons, Inc.

Fig. 4.7.7 The CARE-MOLD (U.S. patent No. 4,552,329) Process Was Used To Form Closed-end Hollow Rib Panels

4.8 PULTRUSION

In the pultrusion process, as shown in Fig. 4.8.1 and Fig. 4.8.2, resin-impregnated filaments are fed into a heated die (heated by microwaves or resistance heaters). Pultrusion is a very efficient and economical means of producing unlimited lengths of constant cross-section parts (see samples shown in Fig. 4.8.3). The process has developed around the rapid-addition reaction chemistry exhibited by thermosets, which initiates an exothermic reaction when heat is introduced which results in solidification of the resin. In addition to the resin, continuous fibers are integral to the process. The cured section emerging from the die is grasped, pulling the remaining filaments through at a constant rate. The composite profile emerges from the die as a hot, constant cross-section profile that cools sufficiently to be clamped and pulled by the action of a pulling mechanism. At the end of the process the pultrusion is automatically cut to specified lengths.

Pultrusion process parameters:
- Wet resin system
- No prepregging cost
- No cold storage
- Integral cure
- Debulk of incoming material
- Resin bath size and method of fiber wet out

- Resin:
  — Temperature
  — Viscosity
  — Pot life
  — Fillers
  — Release agents
  — Anti-foaming agents

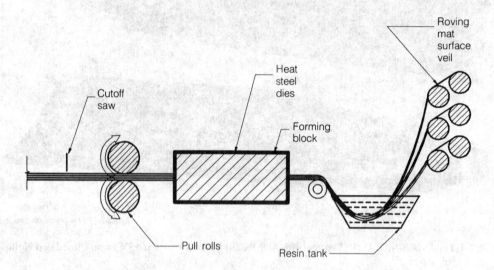

Fig. 4.8.1 Typical Pultrusion Process

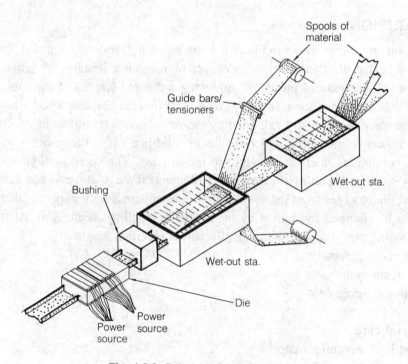

Fig. 4.8.2. Pultrusion Method Utilizing Tapes

Manufacturing

By courtesy of Lockheed Aeronautical Systems Co.

(a) I-section

By courtesy of Lockheed Aeronautical Systems Co.

(b) Wing airfoil section

Fig. 4.8.3 Samples of Pultrusion Products

Chapter 4.0

By courtesy of Lockheed Aeronautical Systems Co.

(c) J-section panels

By courtesy of Alcoa/Goldsworthy Engineering

(d) Complex sections

Fig. 4.8.3 Samples of Pultrusion Products (cont'd)

By courtesy of Alcoa/Goldsworthy Engineering

(d) Complex sections (cont'd)

Fig. 4.8.3 Samples of Pultrusion Products (cont'd)

- Die and mandrels
  - Length
  - Shape
  - Type and zone of heaters
- Pultrusion rate (5-8 ft/min.)
- Load on pullers

In general, basic elements of the pultrusion process are as follows:

(1) Begins with reinforcement dispensing creel system
(2) Reinforcement forming and resin impregnation utilizing one of the following procedures:
   - Resin dip tank — Fibers dip down into the resin tank and are thoroughly impregnated and then moved into a forming guide(s) that preforms preshaping and strips off excess resin before the wet fibers enter the die.
   - Straight-through resin tank — Fibers enter the tank through holes and slots and then move through the forming guide(s) to the die.
   - Resin injection system — This system injects the resin directly into the die to wet the fibers. This system controls resin content better than the other two methods to achieve the desired properties of the pultrusion. The resin is mixed just before it enters the die and there are therefore few environmental problems because the die can be covered and capture vapor emissions.
(3) Heated die station — With thermoset resins, it is important to control how much and how fast heat is directed on the resin. Also, it is important to control the rate at which heat is removed from the part. Otherwise, an incomplete cure or cracking may result as the part cools.
(4) Process controls — monitor process data to ensure quality
(5) Cooling station
(6) Clamping/pulling device
(7) Ends with automatic cut saw to specified lengths

In pultrusion processing, axial reinforcing fibers may be dispensed from a frame structures holding numerous creels. The process is widely used for making complex shapes, but is currently limited to items with constant cross-section. This restriction may soon be eliminated by further development of possible variable thickness/cross-section pultrusion. Advantages of pultrusion:
- continuous method for producing constant-section, reinforced shapes
- Low cost
- Excellent dimensional control
- Excellent fiber alignment
- High volume
- Automated
- Low scrap

Factors to consider:
- Limited to constant cross section
- Require pultrudable prepreg form
- May require post cure
- Most ply orientations difficult (other than 0° direction)

For thermoplastic resin pultrusion, the die must be heated above the resin melting point. Because thermoplastics shrink much less than thermoset resins, expect higher drag forces in the die. One of the key advantage of thermoplastic pultrusion is the ability to post-form the part after consolidation.

Tapered-thickness Pultrusion

A new concept envisions pultrusion with tapered or variable thickness, as shown in Fig. 4.8.4, which is achieved by add-on or drop-off of plies (ADP) (see Fig. 4.2.16). The pultrusion is enclosed in a chamber and surrounded by a series of sliding formers which are used to pressurize and heat the pultrusion surfaces. The constant pressure and temperature can be controlled throughout the entire operation by a computer-controlled automation system.

Pulforming

The pulforming process, shown in Fig. 4.8.5, produces profiles that have a constant cross-section area instead of the constant cross-section shape anywhere along the length of the part (pultrusion). The process utilizes a series of female molds with built-in heating systems mounted onto a continuous ring or belt it operates, the ring rotates and pulls the wetted fibers into the female mold. At the same time, a flexible steel belt covers the female mold and compacts the wetted fibers inside the mold and applies heat. The wetted fibers within the closed mold are cured to the contoured mold profile (shape) as shown in Fig. 4.8.5(c). The pulforming process is summarized below:
- A process related to pultrusion
- Able to produce other than straight, constant-sectional reinforced products (allows constant volume/changing shape or changing volume/changing shape)

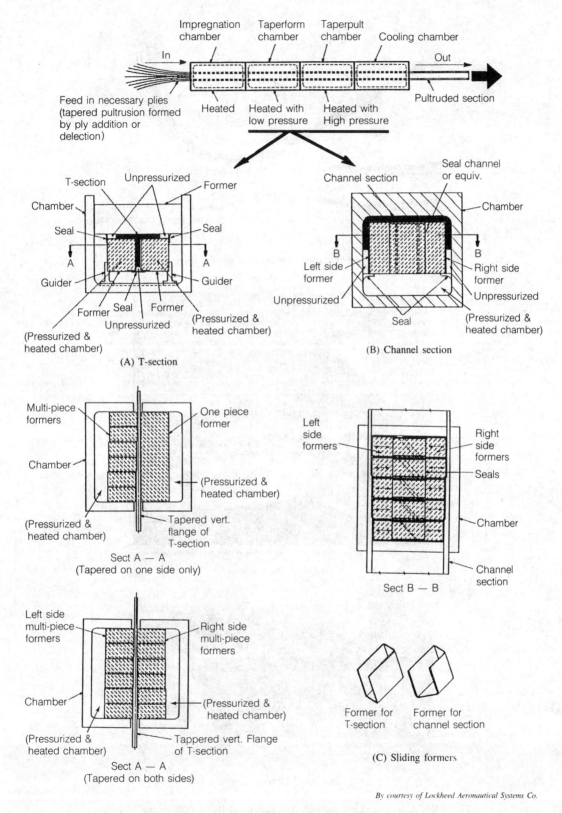

Fig. 4.8.4 Tapered Pultrusion Concept

Chapter 4.0

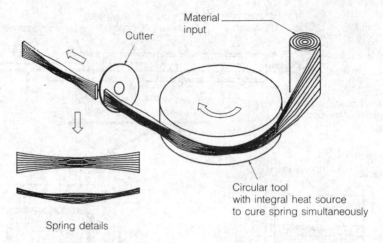

(a) Schematic of pulforming process

By courtesy of Alcoa/Goldworthy Engineering

(b) Pulforming machine

(c) Pulforming parts (Leaf spring)

Fig. 4.8.5 Automatic Leaf Springs Are Pultruded By A Pulforming Machine

Braided Pultrusion

The braided pultrusion process is shown in Fig. 4.8.6 and is described below:
- ±45 degree cross-plies possible
- Constant cross-section
- Tubular cross-section is pulled through forming dies which change the circular section into the desired shape

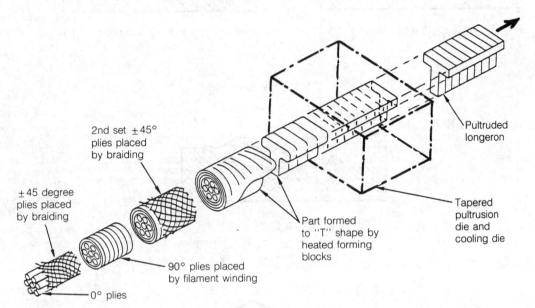

Fig. 4.8.6 Braided Pultrusion

4.9 SANDWICH STRUCTURES

One of the most effective structures for increasing the stiffness-to-weight ratio is the sandwich design which has found widespread application in airframe structures. Sandwich cores can be made in a variety of types and materials (see Section 2.6 of Chapter 2), namely:
- Honeycomb core
- Foam or syntactic core

In space applications, perforated honeycomb core materials should be used with one face sheet having widely spaced small perforations to release trapped air. If air trapped prior to cure cannot escape, unacceptable pressures will develop inside the core areas when the external atmospheric pressure is reduced to zero in outer-space environments. Fabrication of honeycomb sandwich structures as shown in Fig. 4.9.1, may be accomplished by cocuring the components together (with or without adhesive, see Fig. 4.9.2) or utilizing secondary adhesive bonding, as shown in Fig. 4.9.3.
- Thermoset laminates may be cocured together with the honeycomb core
- Thermoplastic laminates cannot be cocured with thermoset core, since the honeycomb core cannot be exposed to the high temperature needed to consolidate thermoplastic parts

Chapter 4.0

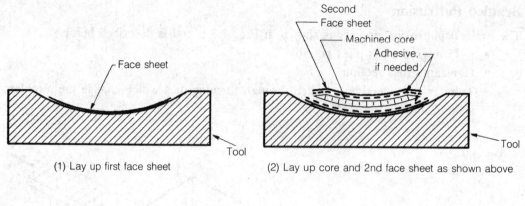

(1) Lay up first face sheet

(2) Lay up core and 2nd face sheet as shown above

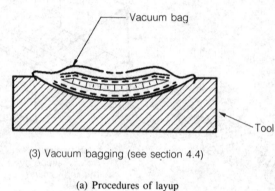

(3) Vacuum bagging (see section 4.4)

(a) Procedures of layup

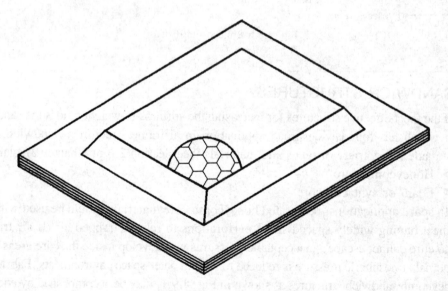

(b) Typical honeycomb panel with edge closeouts

Fig. 4.9.1 Prepreg Face Sheets (Or Add Adhesive As Required) And Core Laid Up Directly on Tool And Cured Together

Manufacturing

By courtesy of Beech Aircraft Corp.

Fig. 4.9.2 Aramid Honeycomb Cord And Carbon Face Sheets Held Together With Structural Adhesive Films (3M Scotch-weld AF163-2) on Starship Aircraft

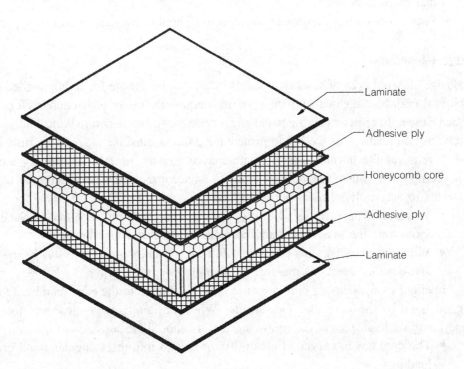

(Detail precured and subsequently bonded together)

Fig. 4.9.3 Secondary Curing

- If cocuring is not possible, an adhesive layer is applied between the pre-cured laminates and the core, and this assembly is then heated to cure the adhesive layer
- Cure temperature and pressure must be limited to avoid damage to the honeycomb core

(1) Cocuring:
- Least expensive method (fewer cure cycles)
- Avoids shimming and thick bond line requirements (eliminates fitup problems)
- Lightest (no additional adhesive required)
- Aerodynamically smooth surface on tool side only
- Size and shape limited by equipment
- Material properties are reduced because dimpling occurs on the bagged face sheets that are cocured with the honeycomb core
- Use of a caul plate is usually required for a smooth surface on the bag side as it minimizes dimpling

(2) Secondary curing:
- Allows inspection of parts prior to assembly
- Smooth surface on both sides is possible (on both bagged and tool surfaces)
- Size and shape limited by equipment
- If cocuring is not possible, an adhesive layer is applied between the pre-cured laminates and the core, and this assembly is then heated to cure the adhesive layer
- Hot bonding or curing may result in warpage due to different coefficients of thermal expansion
- Face sheet quality improved over cocured honeycomb face sheets

Edge Closeouts

The edge closeout area of sandwich panels may be quite simple for lightly loaded panels such as aircraft fuselage interior panels, or may require the use of edge doublers for heavily loaded panels. In either case, the panel edges must be strengthened to withstand fastener, screw, or bolt loading, as well as to protect the panel against damage during installation, and/or removal for normal aircraft maintenance, repair, etc. Most primary structure honeycomb must transmit significant loads to adjacent members and designing the edge closeouts to accomplish this is not a easy task:

- When cocuring (see Fig. 4.9.4) the panel edge-closeouts are cured in place at the same time the face-sheets and core are being cured together
- With secondary bonding (see Fig. 4.9.5), it is usually possible to attach face-sheets and edge-closeouts to the core in a single bonding cycle

In either case, adequate cure pressure must be applied to the edge member to assure a good bond to the core and face-sheets. When designing edge-closeouts for either honeycomb or foam panels the following factors should be considered:

- The core has to carved at about 30° or less to minimize angular bond pressure loading
- Z or channel shaped edge closeouts, as shown in Fig. 4.9.5(a) and (b), are made as separate members prior to bonding and the inside core face must be bonded to the Z's vertical face to transfer shear load without bending the skins

- Doublers are almost always required at the edges to help transfer loads (see Fig. 4.9.4 for example)
- Any design requiring the machining of both core faces is much more expensive than those machined on only one face

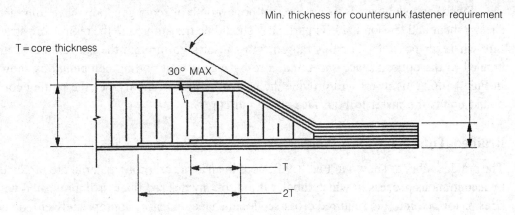

Fig. 4.9.4 Typical Honeycomb Panel Edge Closeout (Cocuring)

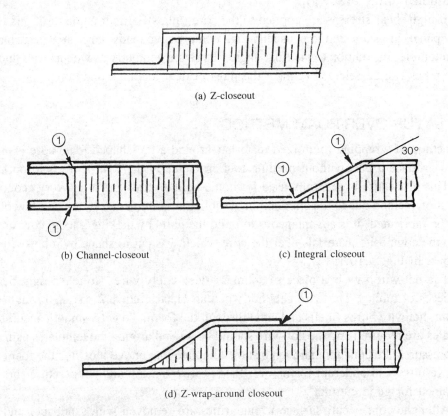

(a) Z-closeout

(b) Channel-closeout

(c) Integral closeout

(d) Z-wrap-around closeout

Note (1) Locations with fitup problems

Fig. 4.9.5 Typical Honeycomb Panel Edge Closeout (Secondary Curing)

Inserts

The purpose of using inserts is to pick up secondary moderately concentrated loads on sandwich panel where the load must be applied away from the edges. It is necessary to stiffen the core locally to aid in distribution of the load from the insert. This is normally accomplished by filling the area around the insert with potting compound (see Chapter 2).

Since sandwich panels using cores of honeycomb or foam generally have thin face-sheets, shear and tension loads in particular should be transmitted to the entire core structure wherever possible. For this reason, most inserts go through the entire core or are bonded to the opposite face-sheet and core by means of a potting compound as shown in Fig. 4.9.6. This insert will provide the best structural strength since the potting compound bonds the insert to both face-sheet and core.

Bradied Tube Sandwich Panel

The braided tube, as shown in Fig. 4.9.7, used in an array as a core material are produced by a continuous process in which tubular braid, of any desired fiber, is impregnated with a resin, formed into the required cross-sectional shape and size, appropriately cured and then cut to length. Typical cross-sectional shapes include squares, triangles and circles as illustrated in Fig. 4.9.7(a).

If high shear stresses are applied to the core (tubes), a fiber angle of 45° is preferred, if panels are subjected to unusually high compressive loads, large angles such as 60° are preferable. In addition to the fiber angle, the shape geometry as well as wall thickness of the tubes can be selected to meet the most critical loading conditions.

4.10 LAYUP-OVER-FOAM METHOD

The method of wrapping reinforced materials around a pre-shaped foam core is an easy way to produce a part without needing tooling (actually the foam core is a tool).

This procedure is generally used for home-built aircraft and it is a very economical fabrication method due to the elimination of tooling requirements. If one or a few of parts are to be fabricated, it is advantageous to build the parts by making a foam core and laying up the composite materials over the core which was cut to shape by a hot-wire saw, as shown in Fig. 4.10.1.

The hot-wire saw is a piece of stainless steel safety wire, stretched tight between two pieces of tubing. The wire gets hot when an electric current passes through it and this thin, hot wire burns (melts and cuts) through the foam. To get a smooth accurate cut, templates are required and the hot-wire should be guided around the templates using only light pressures. Pushing too hard against the template may move it or flex the foam block which results in an undercut foam core. Proper wire tension and temperature should be maintained for good cutting.

After the core is cut, surface irregularities are removed with sandpaper and prior to using wet layup laminates the surface is sealed with micro/epoxy resin and then the laminates are cured at room temperature (Refer to Ref. 4.6 and Ref. 4.10 for further details).

Manufacturing

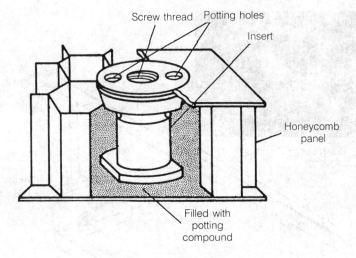

Fig. 4.9.6 Typical Honeycomb Potted Insert

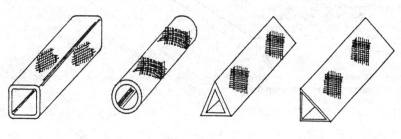

(a) Shapes of braided tube

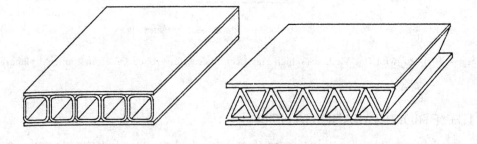

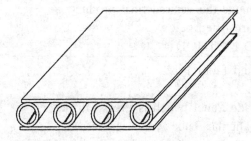

(b) Panel assemblies

Fig. 4.9.7 Braided Tube-Sandwich Panel

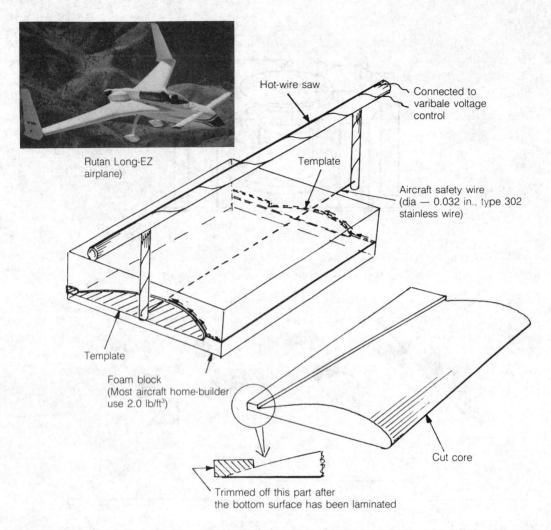

Fig. 4.10.1 Hot-wire Cut Method (Which Has Been Successfully Used On Home-built Airplanes)

4.11 HYBRIDIZED FABRICATION

Hybridized fabrication is nothing more than a finished product fabricated, under cost-effective circumstances, by more than one operation which combines any two or more fabrication methods to achieve the desired final product

Hybrid fabrication cases are listed below:

HYBRID METHOD	EXAMPLES
(1) Double Filament Wound Floor Post	Fig. 4.11.1
(2) Multi-spar Fabrication	Fig. 4.11.2
(3) Integrally-stiffened Blade Panel which is Press Formed	Fig. 4.11.3
(4) Filament Wound Rectangular Tube which is Pressed	Fig. 4.11.4
(5) A Filament Wound Tube Pressed into a J-shaphted Part	Fig. 4.11.5
(6) Press Forming Four Components and Co-consolidating them into a J-section stiffener	Fig. 4.11.6

Manufacturing

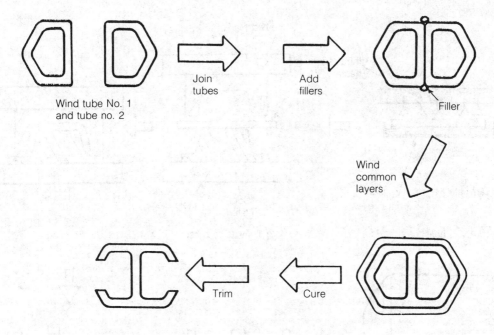

Fig. 4.11.1 Double Filament Wond Floor Post Part

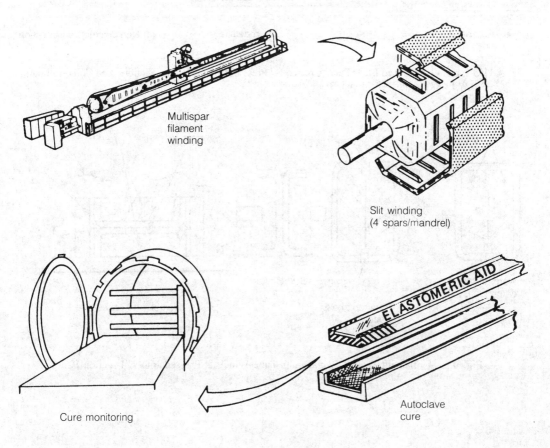

Fig. 4.11.2 Multi-spar Fabrication

Chapter 4.0

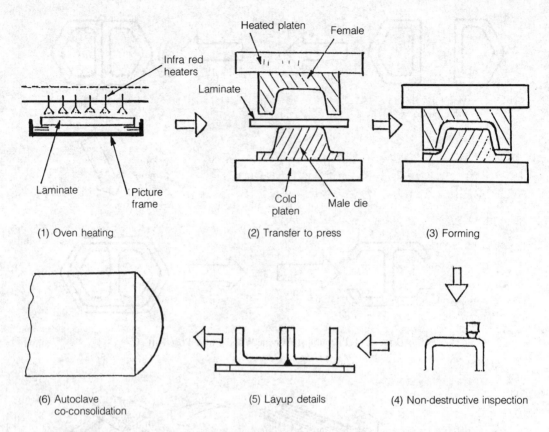

Fig. 4.11.3 Integrally-stiffened Blade Panel Created By Press Forming Channel Sections And Co-consolidating With Skins Into The Final Blade Panel

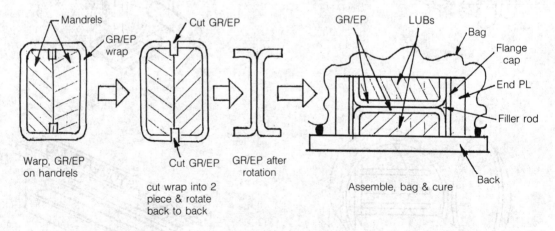

Fig. 4.11.4 Filament Wound Rectangular Tube Which is Pressed Into A I-section Part

Manufacturing

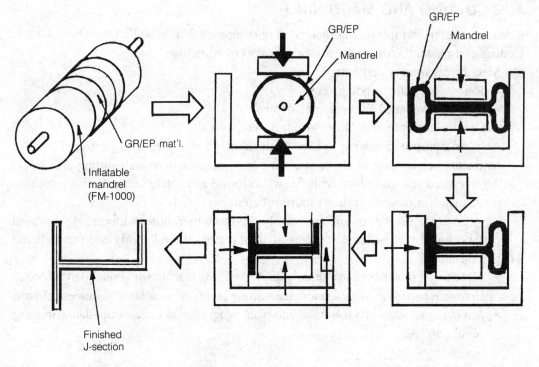

Fig. 4.11.5 Filament Wound Tube Which is Pressed Into A J-shaped Part

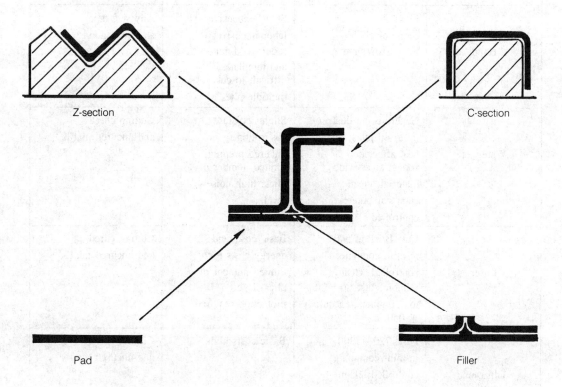

Fig. 4.11.6 Press Forming Four Components (C-form, Z-form, Pad and Filler) And Co-consoliding Them Into A J-section Part

275

4.12 CUTTING AND MACHINING

Common cutting and machining methods are compared in Fig. 4.12.1 and Fig. 4.12.2. Cutting and machining methods are briefly described below:

(1) Mechanical routers and cutters:
- Rout, saw, mill, sanding, etc.
- Can be automated by computer control
- Creates contaminates (dust, grinding, etc.)

An automated sawing system with diamond grit blades is preferred for flat laminates, which typically have straight edges, because of overall cutting efficiency. An overhead saw combined with fixture to locate and rotate laminate parts can perform such trimming operations more efficiency

Complex edge trims and are usually produced by routing with carbide-diamond cutters. For trimming large volumes of similar parts the following two methods can be considered:
- Mounting a router head in a three-axis, NC milling machine or similar drive system. Fixture is used to provide a reference starting point and to control placement of parts
- Mounting a router in a robot that has been programmed to accommodate trimming requirements.

Method	Pros	Cons	Application
Hand	Flexibility; economical for narrow tape	Slow; expensive; labor-intensive; requires Mylar and templates; difficult to cut multiple plies	Any size tape and broadgoods
Water jet	Generates no dust; up to 40 piles; no heat-affected zones; takes wide material; clean cuts; computer controlled	Slight moisture absorption in uncured prepreg; limited number of plies; high noise level	Cutting cured and uncured mat'ls
Laser	Cuts B/EP at 540 in/min. computer controlled; clean reliable cuts; up to 20 plies; accuracy ±.030 in.	Basic costs and energy costs high; limited number of plies; Eye protection required	Cutting cured and uncured matl's
Ultrasonic cutting	Fast; 720 - 1200 in/min. computer controlled; clean up to 20 plies; accuracy ±.030 in	Basic costs high	Cutting cured and uncured matl's

Fig. 4.12.1 Cutting Methods

Sawing	• Conventional cutting methods • Low cost equipment • Finishing cuts required
Bandsawing	• Rough cutting • Finishing cuts required • Cutting tolerances is a function of operator skill
Radial sawing	• Fast, accurate cuts • Limited to straight line cuts • High quality finish on cuts

Fig. 4.12.2 Sawing Methods

(2) Water-jet cutter:
Water-jet cutting along pre-programmed paths allows very close-tolerance machining of cured parts or precision cutting of prepreg plies for robotic or manual layup of composite parts:
- Good accuracy
- Can accommodate very thick parts
- No dust or grinding created
- Good finish, no secondary operations required
- Can be automated by computer control or numerical control (NC)

There are two types of waterjet cutting:
(a) Nonabrasive waterjet cutting — Material thickness is a limiting factor in the ability of this system to cut some types of material effectively.
(b) Abrasive waterjet cutting — This system has the ability to cut nearly any type of material, but is best applied to denser and thicker materials of both composites and metals as shown in Fig. 4.12.3.

Considerations are:
- The jet stream tends to angle away from the direction of the cutting and this effect become more pronounced as the thickness the panel increases
- Abrasive grit size is an important factor in determining the cutting surface finish and also dependent on the type of material being processed
- Waterjet cutting can result in a kerf width that is about 0.001 inch (0.025 mm) and the kerf wider at the waterjet entrance than it is at the exit of the panel
- It is recommended that each individual testing be conducted to establish cutting parameters and acceptance results.

(3) Laser cutting — Laser cutting is considered to be a thermal process because when laser light impinges on a surface and rises the temperature of the material. The resulting temperature increase causes melting, vaporization, and decomposition.

Chapter 4.0

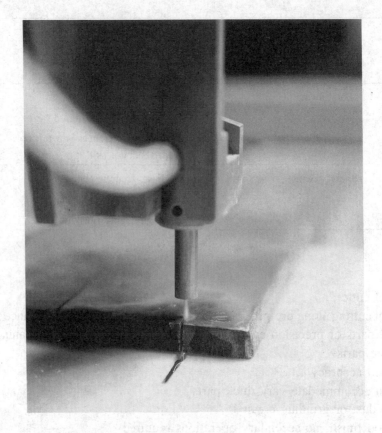

(a) ½ inch titanium being cut

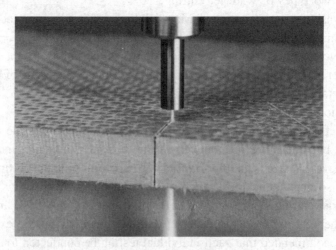

(b) ¾ inch Kevlar® being cut

By courtesy of Ingersoll-Rand Co.

Fig. 4.12.3 Abrasive Waterjet Cutter

Laser cutting is a non-contact thermal process and the absence of contact allows creation of intricate, fragile parts, (and simplifies fixture). This method is independent of the strength or hardness of the composite constituents but, due to its thermal nature, laser cutting has limited use where charring or thermal degradation are not acceptable.

- Power limitations restrict usage to only thin laminates
- Cutting heat cures edge of material causing problems (charring); however, it is good for cutting materials such as commingled fabric, Quadrax, etc.
- In a graphite/epoxy laminate, graphite fibers are good thermal conductors, and cause the matrix to cure near the edge of the cut (undesirable)
- The cutting motion of the laser is a non-contact action, and since there is no surface friction, the fiber alignment is not disturbed.
- Since the laser is a programmable gantry robot, excellent repeatability and precise tolerance control is achieved.

(4) Ultrasonic cutting:

Ultrasonic energy is a simple mechanic vibratory energy operating in frequencies which is imparted to a part without large mechanical displacements or forces. These vibratory energy is in turn amplified by an amplitude transformer (booster) and/or horn before being applied to a workpiece. Ultrasonic vibratory energy can be employed to activate the cutting medium blade which allows faster, easier, and more precise cutting of composite materials, including thin cured laminates, thick cured Aramid laminates, boron prepregs, and honeycomb core materials.

(a) Considerations are:
- Require carbide/coated blades for adequate blade life
  — In general, honeycomb, fiberglass/epoxy, and Aramid/epoxy prepregs cut well with sharp, uncoated blades
  — Graphite thermoplastic prepregs require carbide-coated blades to achieve adequate blade life
- The higher the power, the better the cut
- High power availability, however, does not necessarily increase the effectiveness of a system
- The more brittle (high-modulus) the material is, the lower the required amplitude to cut it
- Amplitude remain unchanged during cutting to maintain speed and control

(b) The factors that can affect the cutting task are:
- Material type
- Material thickness
- Available amplitude
- Sharpness or coating on the blade
- Operator skill

Chapter 4.0

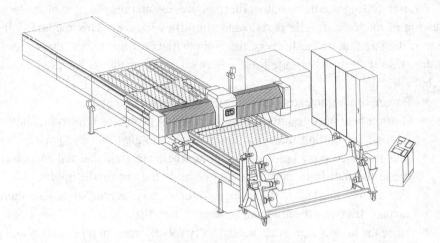

(a) Schematic of cutter process

Positioning accuracy	± .003"	±0.075mm
Repeatability	± .001"	±0.025mm
Contouring accuracy	± .005"	±0.125mm
Speed	2400"/min	up to 60M/min

By courtesy of American GFM

(b) Model US40 ultrasonice cutting machine
Fig. 4.12.4 Composite Prepreg Cutting Machine

(c) Composite prepreg cutting:

Composite prepreg cutting, shown in Fig. 4.12.4, is a high-speed, fully automated cutting system used for cutting composite prepreg plies. The process is capable of cutting two dimensional shapes on a production basis, while also serving as a system for marking and tracking the parts cut. The process first involves creating the geometry of the patterns on a CAD/CAM computer, and the information is then downloaded to the cutter machine automatically. This cutter saves labor, time and cost.
- Automated — CNC programable
- Accommodates tape, fabric, knits
- Moderate capital cost

(6) Gantry robot (see Fig. 4.12.5):

The gantry robot is outfitted with a computerized numerically controlled processor, which helps to lower processing costs by reducing setup time and improving the accuracy and efficiency of the work it performs, such as drilling, deburring, routing, cutting, fastening, etc.

(7) Trim scrap shredder:

The shredder (a modified conventional paper shredder) or pelletizing equipment can turn scrap thermoplastic prepreg into short fibers or recycle processed laminates into useful reinforced molding compound.

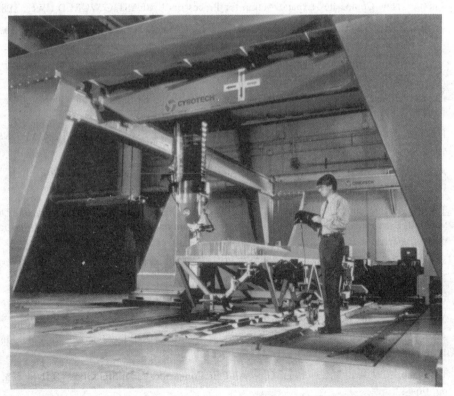

By courtesy of Lockheed Aeronautical Systems Co

Fig. 4.12.5 Multi-functional Gantry Robot

References

4.1 Krolewski, S.M., "Study of the Application of Automation to Composite Manufacture". MTL-TR89-47. 1989.

4.2 Anon., "Structural Fabrication Guide for Advanced Composites", Manufacturing Technology Division, Air Force Materials Laboratory, Wright-Patterson Air Force Base, OH. 1974.

4.3 Anon., "DoD/NASA Structural Composites Fabrication Guide", Manufacturing Technology Division, Air Force Material Laboratory, Wright-Patterson Air Force Base, Ohio. 1982.

4.4 Anon., "Engineering Materials Handbook, Vol. 1 — Composites", ASM International, Metals Park, Ohio 44073. pp. 497-680. 1987

4.5 Brahney, J.H., "A half-century in Composites", AEROSPACE ENGINEERING, Oct., 1990. pp. 15-18.

4.6 Anon., "Long-EZ Plans", Rutan Aircraft Factory Inc., Bldg 13, Majave Airport, CA 93501. 1980.

4.7 Sheppard, L.M., "The Revolution of Filament Winding", ADVANCED MATERIALS & PROCESSING, (July, 1987). pp. 31-41.

4.8 Anon., "Filament Winding: Beyong the Symmetrical", ADVANCED COMPOSITES, (Jan/Feb, 1987).

4.9 Meyer, R.W., "Handbook of Pultrusion Technology", Chapman and Hall, 29 W. 35th., New York, NY 10001. 1985.

4.10 Hollmann, M., "Composite Aircraft Design", 11082 Bel Aire Court, Cupertino, CA 95014. 1982.

4.11 Anon., "New Composites Expand Action for Processors", PLASTIC WORLD (Dec., 1985).

4.12 English, L.K. "Fabricating the Future with Composite Materials", MECHANIC ENGINEERING:
Part I : The Basic (Nov., 1986)
Part II : Reinforcements (Jan., 1987)
Part III: Matrix Resins (Feb., 1987)

4.13 Anon., "Filament Winding and Fiber Placement: Stretching the Bonds of an Automated Process", ADVANCED COMPOSITES, Nov/Dec, 1990. pp. 20-35.

4.14 Darchuk, J. and Migliore, "Guidelines for Laser Cutting", LASERS AND APPLICATIONS, Sept 1985. pp. 91-97.

4.15 Anon., "Engineering Materials Handbook, Volume 1 — Composites", ASM International, Metals Park, OH 44073. 1987.

4.16 Talbott, Margaret and Miller, A.K., "A Model of Laminate Bending under Arbitrary Curvature Distribution with Extensive Interply Sliding", ASME Symposium on Mechanics of Plastics and Plastic Composites, Dec., 1989.

4.17 Miller, A.K., and etc., "Die-Less Forming of Thermoplastic-Matrix, Continuous-Fiber Composites", J. COMPOSITE MATERIALS, 1989.

4.18 Weeton, J.W. & etc., "Engineers' Guide to Composite Materials", American Society for Metals, Metals Park, Ohio 44073. 1987. pp. 6-9 to 6-24.

4.19 Anon., "The Challenge of Manufacturing Composites", MACHINE DESIGN, Penton Publishing Inc., 1100 Superior Ave., Cleveland, OH 44114. Oct. 22, 1987.

4.20 Tortolano, F.W. and Chamberlain, G., "All-Star Composites", DESIGN NEWS, Nov. 23, 1987. pp. 96-112.

4.21 DeYoung, H.G., "Plastic Composites Fight for Status", HIGH TECHNOLOGY, Oct. 1983. pp. 63-68.

4.22 Hearons, J.S., "Process Refinements Present New FRP Opportunities", DESIGN NEWS, July 5, 1982. pp. 45-47.

4.23 Krolewski, S.M., "Study of the Application of Automation to Composites Manufacture", MTL-TR 89-47. May, 1989.

4.24 Witzler, S., "Laying Thermoplastic Tape Prepregs: A Progress Report", ADVANCED COMPOSITES, Jan/Feb, 1987. pp. 53-54.

4.25 Klein, A.J., "Automated Tape Laying", ADVANCED COMPOSITES, Jan/Feb, 1989. pp. 44-52.

4.26 McDermott, J., "The Structure of the Advanced Composites Industry", ADVANCED COMPOSITES, 1990 Bluebook.

4.27 Prairie, M., "Fabricating Large Structures", AEROSPACE COMPOSITES & MATERIALS, 1988.

4.28 Klein, A.J., "Composites: Moving Toward Automation", ADVANCED COMPOSITES, Mar/Apr 1987. pp. 93-97.

4.29 Kulkarni, S.B., "New Applications for Filament Winding", MACHINE DESIGN, Nov. 11, 1982. pp. 66-71.

4.30 Stover, D., "Filament Winding and Fiber Placement: Stretching the Bonding of an Automated Process", ADVANCED COMPOSITES, Nov/Dec 1990. pp. 20-35.

4.31 Leonard, L., "Microdilelectrometry: Ting Sensors Close the Cure Control Loop", ADVANCED COMPOSITES, Jan/Feb 1987. pp. 49-52.

4.32 Leonard, L., "Continuous Processing: Pultrusion Sets the Pace", ADVANCED COMPOSITES, July/Aug 1988. pp. 28-35.

4.33 Klein, A.H., "Curing Techniques for Composites", ADVANCED COMPOSITES, Mar/Apr, 1988. pp. 32-44.

4.34 Sikes, S., "Stamplable Plastic Composites Gaining Ground", MACHINE DESIGN, May 22, 1986. pp. 68-72.

4.35 Okine, R.K., "Analysis of Forming Parts from Advanced Thermoplastic Composite Sheet Materials", SAMPE Journal, Vol. 25, No. 3. May/June 1989. pp. 9-19.

4.36 Maass, D. and Bertolet, J., "Forming Thermoplastic Composites for Next-generation Applications", COMPOSITES IN MANUFACTURING, Dec, 1986. pp. 12-13.

4.37 Krone, J.R. and Walker, J.H., "Processing Thermoplastic Advanced Composites", PLASTIC TECHNOLOGY, Nov. 1986. pp. 61-66.

4.38 Witzler, S., "Pressure-molding Methods for Thermoplastic Composites", ADVANCED COMPOSITES, Sept/Oct 1988. pp. 49-55.

4.39 Stover, D., "Resin-transfer Molding for Advanced Composites", ADVANCED COMPOSITES, Mar/Apr 1990. pp. 60-80.

4.40 Becker, W., "Developing RTM", AEROSPACE COMPOSITES & MATERIALS, 1989.

4.41 Becker, W., "Resin Transfer Molding: Principles of Success", Conference of "Resin Transfer Molding for the Aerospace Industry", Mar. 6-7, 1990, Radisson Plaza Hotel, Manhattan Beach, CA.

4.42 Hansen, R.S., "RTM Processing and Applications", Conference of "Resin Transfer Molding for the Aerospace Industry", Mar, 6-7, 1990, Radisson Plaza Hotel, Manhattan Beach, CA.

4.43 Wadsworth, M., "Resin Transfer Molding of Composite Structures", AEROSPACE ENGINEERING, Dec. 1989. pp. 23-26.

4.44 Price, A.L., "Composites Design and Manufacture: One and the Same", MATERIALS ENGINEERING, Jan. 1991. pp. 25-28.

4.45 Anon., "New Carbon Composite Material Developed", FLIGHT INTERNATIONAL, Apr. 17, 1982. pp. 988-989.

4.46 Anon., "The Starship Enterprise", AEROSPACE COMPOSITES & MATERIALS, 1988.

4.47 Galli, E., "RIM Applications Increase with New Resin Systems", PLASTICS DESIGN FORUM, Jan/Feb 1991. pp. 55-60.

4.48 Barth, J.R., "Fabrication of Complex Composite Structure Using Advanced Fiber Placement Technology", SAMPE, Vol. 35, 1990. pp. 710-720.

4.49 English, L.K., "Automated Composites Fabrication: The Challenge and the Promise", MATERIALS ENGINEERING, June, 1988. pp. 47-49.

4.50 Lewis, C.F., "Consistent Performance from Pultrusions". MATERIALS ENGINEERING. Mar., 1991. pp. 19-22.

4.51 Williams, D.J. "High Performance Sandwich Panels Made From Braided Tubes". 33rd International SAMPE Symposium. Mar. 7-9, 1988. pp. 324-334.

4.52 Sweency, F.M., "REACTION INJECTION MOLDING MACHINERY AND PROCESSES", published by Marcel Dekker, Inc. 1987.

4.53 Florian, J., "PRACTICAL THERMOFORMING — Principles and Applications", published by Marcel Dekker, Inc. 1987.

4.54 Schwartz, M.M., "FABRICATION OF COMPOSITE MATERIALS, SOURCE BOOK", an ASM source book — a collection of outstanding articles from the technical literature. 1985.

4.55 Clauser, H.R., "ENCYCLOPEDIA/HANDBOOK OF MATERIALS, PARTS, AND FINISHES", published by Technomic Publishing Co., Landcaster, PA. 1976.

4.56 Grayson, M., "ENCYCLOPEDIA OF COMPOSITE MATERIALS AND COMPONENTS", published by John Wiley & Sons, New York, NY. 1983.

4.57 English, L.K., "Fabricating the Future with Composite Materials: A Primer", MATERIALS ENGINEERING, Oct, 1990. pp. 41-45.

4.58 Dreger, D., "The Challenge of Manufacturing Composites", MACHINE DESIGN, Oct. 22, 1987.

4.59 Peters, S.T., Humphrey, W.D. and Foral, R.F., "FILAMENT WINDING COMPOSITE STRUCTURE FABRICATION", published by SAMPE International Business Office, P O Box 2459, Coveina, CA 91722. 1991.

4.60 Leonard, L., "Health and Safety in Composites Plants, Part I" ADVANCED COMPOSITES, May/June, 1991. pp. 37-46.

Chapter 5.0

JOINING

5.1 INTRODUCTION

A complete airplane structure is manufactured from many parts such as skins, stiffeners, frames, spars, etc. These parts must be joined together by fastening, bonding, welding, etc., to form subassemblies are joined together to form larger assemblies and then finally the completed airplane. Many parts of the completed airplane must be arranged so that they can be disassembled for shipping, inspection, repair, or replacement. Fasteners are usually used to join such parts. In order to facilitate the assembly and disassembly of the airplane, it is desirable for such fastened joints to contain as few fasteners as possible. Fig. 5.1.1 gives a comparison of joining methods.

Joints are perhaps the most common source of failure in aircraft structure and therefore it is most important that all aspects of joint design are given consideration during the structural design. Failures may occur for various reasons, such as secondary stresses due to eccentricities, stress concentrations (especially for fastened joints as shown in Fig. 5.1.2) excessive deflections, etc., or some combination of conditions, all of which are difficult to evaluate to an exact degree. These factors directly affect the strength of joints, especially the fastened joints which are greatly weakened by notch effects.

Assembly joints, which occur when any two components are assembled, are a major source of stress concentrations. In the case of bonded joints, stress concentrations occur to maintain strain compatibility between bonding components. In the case of mechanical joints, they are a result of the decreased area at the hole and the loaded hole itself. The primary purpose of this section is to acquaint the engineer with some of the problem areas encountered, introduce some of the joint design allowables generated on the subject, and show a few examples of how typical problems have been solved.

To fully realize the potential of advanced composites in lightweight aircraft structure, it is particularly important to ensure that the joints, either bonded or fastened, do not impose a reduced efficiency on the structure. This problem is far more severe with composite materials than with conventional metals because the high-specific-strength composite filaments are relatively brittle. Composites have very little capacity to redistribute loads as shown in Fig. 5.1.3 and practically none of the forgiveness of a yielding metal to mask a multitude of design approximations. This is the reason why greater efforts are devoted to understanding joints in composite materials and to providing reliable design techniques, particularly for thicker sections and for multiple fastener pattern design cases.

Chapter 5.0

Method		Anticipated •Benefits	Limitations
	Mechanical fastening	• Mature Technology • Baseline for cost data • Could supplement weld/bond assembly methods	• Low risk • Increased weight • Labor Intensive • Requires secondary seal • Shimming fit-up stress
	Adhesive bonding	• Reduced fastener count/weight	• Moderate risk • Cure cycle required • Tooling
Thermoplastic Welding	• Resistance	• Can be automated process • Continuous weld • Reduced fastener count/weight	• Moderate risk • Requires 2 side access
	• Ultrasonic	• Can be automated process • Possible continuous weld • Reduced fastener count/weight	• Moderate risk • Requires 2 side access
	• Induction	• Requires 1 side access • Can be automated process • Continuous weld • Reduced fastener count/weight	• Moderate — high risk • Requires magnetic susceptor mat'l
	Cocuring	• Total homogeneous weld joint • Probable elimination of seal	• Low risk • Part size/shape limited

Fig. 5.1.1 Comparison of Joining Methods

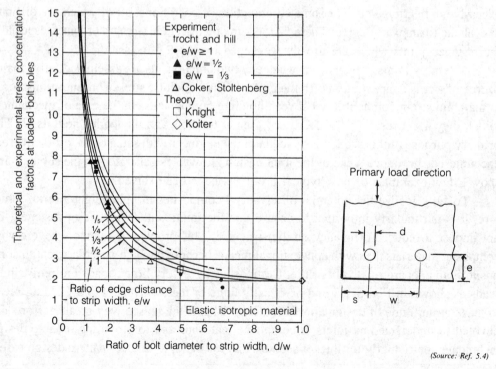

(Source: Ref. 5.4)

Fig. 5.1.2 Stress Concentration Levels Rise Rapidly for Fastener Holes Smaller Than 0.2 Times The Strip Width

Joining

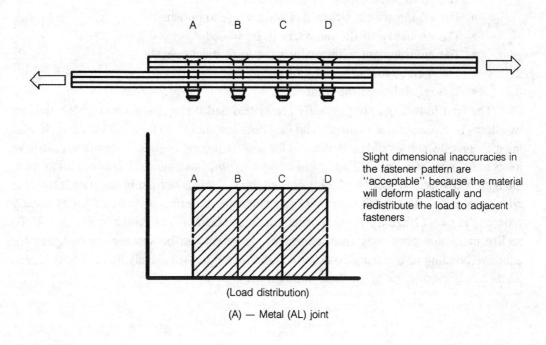

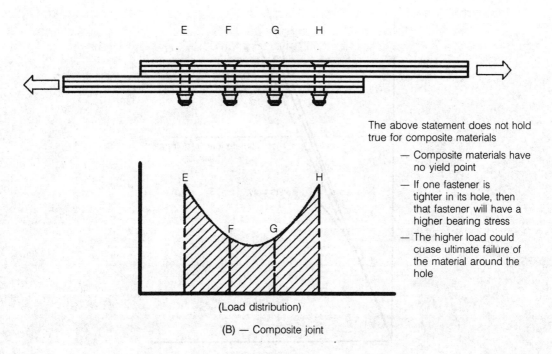

Fig. 5.1.3 Comparison Between Metal and Composite Joints

Chapter 5.0

There are six basic factors to be considered in the design of a composite joint:
- The loads which must be transferred
- The region within which this must be accomplished
- The geometry of the members to be joined
- The environment within which the joint must operate
- The weight/cost efficiency of the joint
- The reliability of the joint

The first four items are generally prescribed and it remains then to satisfy the last two items in some optimal manner. The first decision should be to select the class of joining techniques which should be studied. In the past, structural engineers considered adhesive joints to be more efficient for lightly-loaded joints, while mechanically fastened joints were thought to be more efficient for highly-loaded joints. This is definitely not always the case with composites. The notch sensitivity of composite materials at fastened joints greatly reduces the joint efficiency in composite structural assembly, as shown in Fig. 5.1.4. To realize maximum efficiency from adhesives, joints should be specifically designed for adhesive bonding or cocuring/co-consolidation methods. Adhesively bonded or co-cured joints can overcome many of these limitations.

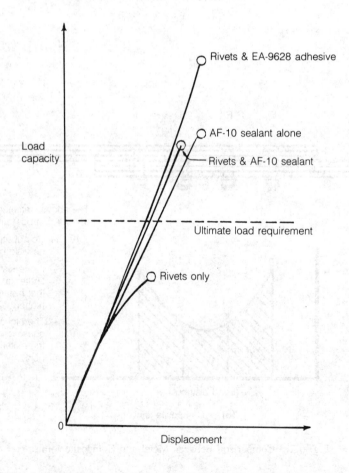

Fig. 5.1.4 Inadequacy of Rivetted Splices (Used on Lear Fan Fuselage)

It is a general rule that the most efficient composiste joints are scarf and stepped lap joints in which there is relatively little change in the load path. Double-lap and single-lap joints are quite a bit less efficient, in that order.

To achieve the goal of both structural weight reduction and cost savings, as many mechanically fastened joints and/or bonded joints as possible must be eliminated. Fig. 5.1.5 shows an example of an one piece composite wing skin panel for a fighter aircraft on which wing spanwise splice and skin-spar joints were eliminated.

By courtesy of The Boeing Co.

(a) One piece thermoplastic wing skin.

By courtesy of British Aerospace (Military Aircraft) Ltd.

(b) EFA wing spars co-bonded to the lower skin surface

Fig. 5.1.5 Use One Piece and Co-bonded Method to Eliminate Fasteners

In the end there will be some requirements for mechanical fasteners to hold access doors, removable covers (e.g., a fighter wing upper cover which must be removed for maintenance purposes), detachable parts, etc., or to join components in final assembly. Since composite materials have some unique properties and characteristics, most fasteners selected for joining composites are tailored to the these materials to avoid problems.

Because of the many factors and unknowns involved in designing composite laminate joints, tests should be conducted which simulate the operational environment and loading in order to insure joint reliability.

5.2 MECHANICAL FASTENING

The use of mechanical fasteners to assemble airframe structures is a mature technology. Composites are not an exception. Failure modes for advanced composite mechanical joints are similar to those for conventional metallic mechanically fastened joints. But the behavior of composite joints differs significantly from those of metallic joints and deserves special attention of a number of reasons:

(a) Relative brittleness of material, which results in high stress concentrations at hole edges
(b) Laminate failure is a function of stacking sequence, fiber volume, porosity, etc.

When fasteners are required, composites present special design considerations. Composite materials derive their properties from both the fibers and the matrix, and are not homogeneous. They do not respond to fasteners in the same way as metals. Therefore, it is not possible to design fasteners that are universally applicable to all composites. Composites possess different characteristics than their aluminum counterparts: even though they are very strong, they can be very delicate if not treated properly. Therefore, the selection of the correct fastened joint is critical.

Fig. 5.2.1 presents typical simplified representations of the following failure modes: shear out, net tension, bearing, and combined tension and shear out. The shear out failure mode can also be sometimes characterized by a single-plane "cleavage" failure, where the apparent laminate transverse tensile strength is less than the corresponding in-plane shear strength. In addition, bearing or shear failure of the fastener, and fastener pulling through (especially with countersunk head fasteners) are other possible failure modes.

The following equations should be used to determine allowable joint strengths:

- Bearing: $P^{br} = (d)(t)(F^{br})$
- shear out: $P^s = 2[(e/d) - 0.5](d)(t)(F^s)$
- Net tension: $P^t = 2[(s/d) - 0.5](d)(t)(F^t)$

where 0.5 — inch

F^{br} — design bearing allowable
F^s — design shear out allowable
F^t — design net tension allowable
d — fastener diameter
t — Laminate thickness
e — edge distance
s — fastener side distance (see Fig. 5.1.2)

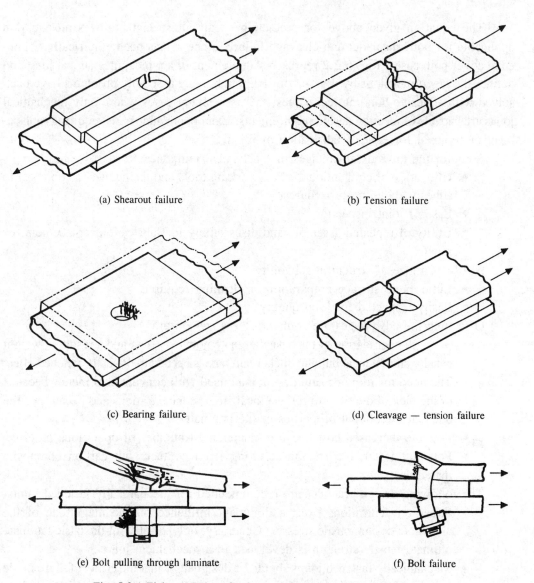

(a) Shearout failure
(b) Tension failure
(c) Bearing failure
(d) Cleavage — tension failure
(e) Bolt pulling through laminate
(f) Bolt failure

Fig. 5.2.1 Fialure Modes of Advanced Composite Mechanical Joints.

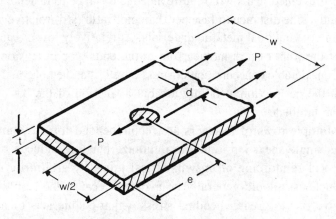

Fig. 5.2.2 Typical Mechanical Joint Element.

The equation given above for predicting shear out strength utilizes an equation applicable for both shear out and cleavage failure, since F^s has been empirically obtained to cover both cases. Fig. 5.2.2 represents an element of a typical mechanical joint and defines the key dimensions by illustration. Although many assembly problems have been solved with adhesive-bonding techniques, there are many cases where only mechanical joints are capable of meeting design requirements. Example include parts requiring replacement or removal for ease of fabrication or repair.

Some of the obvious advantages of mechanical joints are:
- Utilization of conventional metal-working tools and techniques, as opposed to adhesive-bonding procedures
- Ease of joint inspection
- Utility of repeated assembly and disassembly for fabrication replacement, or repair
- Assurance of structural reliability
- Little or no surface preparation or cleaning required
- Easily inspected for joint quality

Offsetting the advantages are some concerns, which are:
- Strength degradation of the basic laminate notched effect and a resultant weight penalty since a local buildup thicker laminate is needed to offset the notched effect
- The need for more careful design than used with conventional metals because of the lack of ductility to relieve local stress concentrations and because of the unequal directional properties of the laminate
- Possible increased cost because of increased number of operations required
- Potential fastener corrosion resulting from contact with carbon composite materials

The following design practices are recommended for mechanically fastened joints:
- Stress concentrations exert a dominant influence on the magnitude of the allowable design tensile stresses. Generally, only 20-50% of the basic laminate ultimate tensile strength is developed in a mechanical joint
- Mechanically fastened joints should be designed so that the critical failure mode is in bearing, rather than shear out or net tension, so that catastrophic failure is prevented. This will require an edge distance to fastener diameter ratio (e/d) and a side distance to fastener diameter ratio (s/d) relatively greater than those for conventional metallic materials. At relatively low e/d and s/d ratios, failure of the joint occurs in shear out at the ends , or in tension at the net section. Considerable concentration of stress develops at the hole, and the average stresses at the net section at failure are but a fraction of the basic tensile strength of the laminate
- Multiple rows of fasteners are recommended for unsymmetrical joints, such as single shear lap joints, to minimize bending induced by eccentric loading
- Local reinforcing of unsymmetrical joints by arbitrarily increasing laminate thickness should generally be avoided because the resulting eccentricity can give rise to greater bending stress which counteracts or negates the increase in material area

- Since stress concentrations and eccentricity effects cannot be calculated with a consistent degree of accuracy, it is advisable to verify all critical joint designs by testing a representative sample joint

Mechanical Joint Design Guidelines

In general, the best fastened joints in fibrous composites still impose a loss in strength of about half the basic material allowable (although there is test evidence to indicate that interference-fit fasteners can alleviate this reduction toward a net area loss):

(1) If a laminate is dominated by 0° fibers with few 90° fibers it is most likely to fail by shear out. Unlike metals, in which shear out resistance can be increased by placing the hole further from the edge, laminates are weakened by fastener holes regardless of distance from the edge. Reinforcing plies at 90° to the load helps prevent both shear out [see Fig. 5.2.1(a)] and cleavage [see Fig. 5.2.1(b)] failures:
   - Use larger fastener edge distances than with aluminum design, such as e/d > 3
   - Use a minimum of 40% of ±45° plies; see Fig. 5.2.3 for the effect of layup on the bearing stress at failure
   - Use a minimum of 10% of 90° plies

(2) Net tension failure [see Fig. 5.2.1(b)] is influenced by the tensile strength of the fibers at fastened joints, which is maximized when the fastener spacing is approximately four times the fastener diameter (see Fig. 5.2.4). Smaller spacings result in the cutting of too many fibers, while larger spacings result in bearing failures, in which the material is compressed by excessive pressure caused by a small bearing area:
   - Use minimum fastener spacing as shown in Fig. 5.2.5
   - Pad up to reduce net section stress

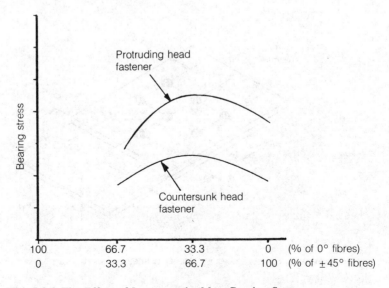

Fig. 5.2.3 The Effect of Layup on the Max. Bearing Stress.

Chapter 5.0

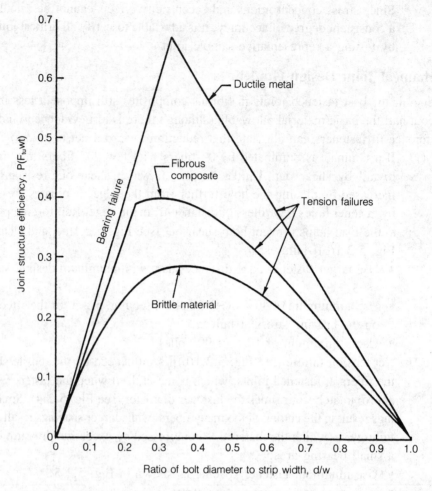

Fig. 5.2.4 Relation Between Strengths of Fastened Joints in Ductile, Brittle, and Composite Materials

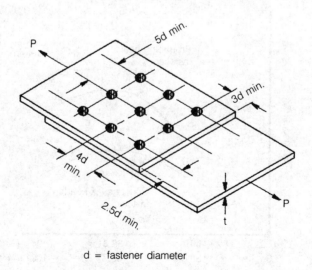

d = fastener diameter

Fig. 5.2.5 Minimum Fastener Spacing and Edge Distance

(3) Fastener pull-through from progressive crushing/bearing failure [see Fig. 5.2.1(c)]:
- Design joint as critical in bearing
- Use padup
- Use a minimum of 40% of $\pm 45°$ plies
- Use washer under collar or wide bearing head fasteners
- Use tension protruding heads when possible

(4) Fastener shear failure:
- Use large diameter fastener
- Use higher shear strength fastener
- Never use a design in which failure will occur in shear

(5) Use two row joints when possible. The low ductility of advanced composite material confines most of the load transfer to the outer rows of fasteners (see Fig. 5.2.6)

(6) The use of carbon composites in conjunction with aircraft metals is a critical design factor. Improper coupling can cause serious corrosion problems for metals. Materials such as titanium, corrosion-resistant steels, nickel and cobalt alloys can be coupled to carbon composites without such corrosive effects. Aluminum, magnesium, cadmium plate and steel will be most adversely affected because of the difference of electrical potential between these materials and carbon. Fig. 5.2.7 shows the galvanic compatibility of fastener materials with carbon composites.

(7) The choice of optimum layup pattern for maximized fastener strength is simplified by the experimentally established fact that quasi-isotropic patterns $(0/\pm 45/90)_s$ or $(0/45/90/-45)_s$ are close to optimum. This reduces experimental costs and simplifies the analysis and design of most fastened joints.

(8) One of the key factors governing fastened joint behavior in advanced composite structure is the vast difference between double-lap and single-lap joint efficiencies. The eccentricity in the load path for single-lap joints leads to non-uniform bearing stresses across the thickness of the laminate (see Fig. 5.2.8). This, in turn, leads to the development of the critical bearing stress and bypass stress around hole at the laminate interface at an even lower than average bearing stress because of the brittle nature of composite materials. Fig. 5.2.9 shows bearing stress distribution at the fastener hole, and the use of the bearing reduction factor to account for this effect. It is difficult to define the reduction factor because it is a function of fastener material characteristics, composite material and layup sequence, fastener fit, etc. Currently, an arbitrary value of 1.5 to 2.0 is used for the reduction factor until results from testing are established for each particular design case.

(9) Develop a bearing/bypass stress interaction envelope curve (function of laminate material, laminate thickness and ply layup sequence or tacking, fastener diameter, etc.) to size mechanical joints as shown in Fig. 5.2.10.

Chapter 5.0

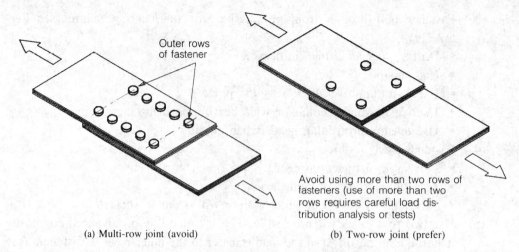

(a) Multi-row joint (avoid) (b) Two-row joint (prefer)

Fig. 5.2.6 Outer Rows of Fastener Carry Most of Load Due to The Low Ductility of The Composite Materials

Fastener Material	Compatibility with Graphite/Epoxy and Application Guidelines
Titanium, Ti Alloys, Ti-CP	Fasteners of these materials are compatible with graphite/epoxy composites. Permanent fasteners should be sealed to prevent water intrusion but removable fastener may be used with no supplement protection.
MP-35N (AMS 5758) Inco 600 (AMS 5687)	These materials are compatible with graphite/epoxy components.
A286 (AMS 5731, AMS 5737) PH13-8Mo (AMS 5629)	These CRES alloys and some other austenetic and semi-austenetic alloys are marginally acceptable in contact with graphite composites. In a severe marine/industrial corrosion environment, superficial rusting and stains develop on the fastener. Although loss of fastener integrity has not been established, this staining is usually objectionable. Permanent fasteners that can be installed with sealant and overcoated with sealant are usually satisfactory. Removable fasteners are not acceptable to some design activities.
Monel	Marginally acceptable in contact with graphite/epoxy composites. Significant current flow and material loss.
Low Alloy Steel, Martensitic Stainless Steels	Not compatible with graphite/epoxy materials. Severe rusting.
Silver Plate, Chromium Plate, Nickel Plate	These plating materials are compatible with graphite but are not adequate to protect steel in contact with graphite/epoxy composites. Silver plated A286 or PH13-8Mo would be compatible with graphite and suitable if there is no aluminum or titanium in the joint.
Cadmium Plate, Zinc Plate, Aluminum Coatings	Not compatible with graphite/epoxy composite materials. Rapid deterioration of plating or coating.
Aluminum, Aluminum Alloys, Magnesium Alloys	Not compatible with graphite/epoxy composite materials. Generally, it is feasible to adequately protect fasteners of these materials from severe corrosion if in contact or close proximity to graphite.

Fig. 5.2.7 Galvanic Compatibility of Fastener Materials with Graphite Composite.

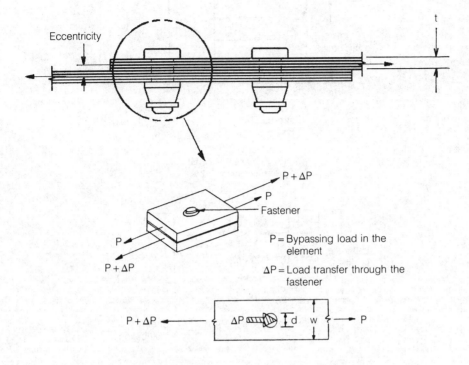

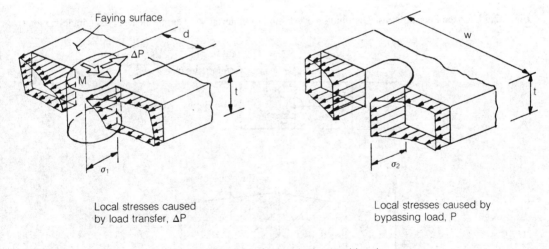

The maximum local stress in the considered element,

$$\sigma_{max} = \sigma_1 + \sigma_2$$

(Note: For the σ_{max} calculation refer to chapter 7.0 of Ref. 5.30)

Fig. 5.2.8 Local Peak Stresses Caused by Load Transfer and Bypass Load

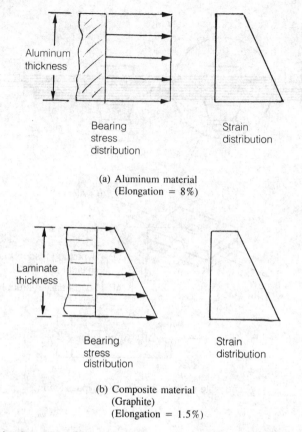

Fig. 5.2.9 Ultimate Bearing Stress Distribution of Aluminum Material Vs. Composite Material (single-lap joint).

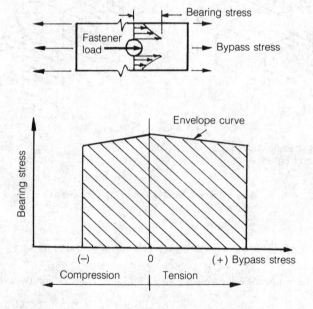

Fig. 5.2.10 Bearing/bypass Stress Interaction Envelope Curve (notched and wetted test data).

(10) Eccentricities and their effect on the joint and the surrounding structures:
  - If eccentricities exist in a joint, the moment produced must be resisted by the adjacent structures
  - Eccentrically loaded fastener patterns may produce excessive stresses if eccentricity is not considered
(11) Mixed fastener types — It is not good practice to employ both permanent fasteners and screws in combination in a joint. Due to the better fit of the permanent fasteners, the screws will not pick up their proportionate share of the load until the permanent fasteners have deflected enough to take up the clearance of the screws in their holes
(12) Do not use a long string of fasteners in a splice. In such cases, the end fasteners will load up first and yield early. Three, or at most four, fasteners per side is the upper limit unless a carefully tapered, thoroughly analyzed splice is used (wherever possible use a double shear splice). Study cases are shown in Fig. 5.2.11.
(13) Use tension head fasteners (potentially high bearing stress under the fastener head cause failure) for all applications. Shear head fasteners may be used in special applications.
(14) Driven fasteners, e.g., MS20470, MS20426 and DD rivets, should never be used to assemble composites
(15) If local buildup is needed for fastener bearing strength, total layup should be at least 40% $\pm 45°$ plies
(16) Install fasteners wet with corrosion inhibitor may be required
(17) Use large diameter fastener in thicker composite assemblies (e.g., to transfer critical joint loads, fastener diameter should typically be about the same size as the laminate thickness) to avoid peak bearing stress due to fastener bending. Fastener bending is much more significant for composites than for metals, because composite laminates are thicker (for a given load) and more sensitive to non-uniform bearing stresses (due to brittle failure modes)
(18) Don't buck rivets and conventional enlarged end blind fasteners
(19) The best fastened joints can barely exceed half the strength of unnotched laminate
(20) Peak hoop tension stress around fastener holes is roughly equal to the average bearing stress
(21) Fastener bearing strength is sensitive to through-the-thickness clamping force of laminates (see Fig. 5.2.12)
(22) For blind fasteners, use big-foot fasteners wherever possible
(23) Production tolerance build ups:
  - Proper tolerance should be given to minimize the need for shimming
  - Shim allowance should be called out on engineering drawing
  - Since production tolerances can easily be exceeded in the thickness tolerance, fastener grid length is affected
(24) In fuel tank, to prevent fuel tank leakage of a groove seal design, the max. seal gap is 3 mils

Chapter 5.0

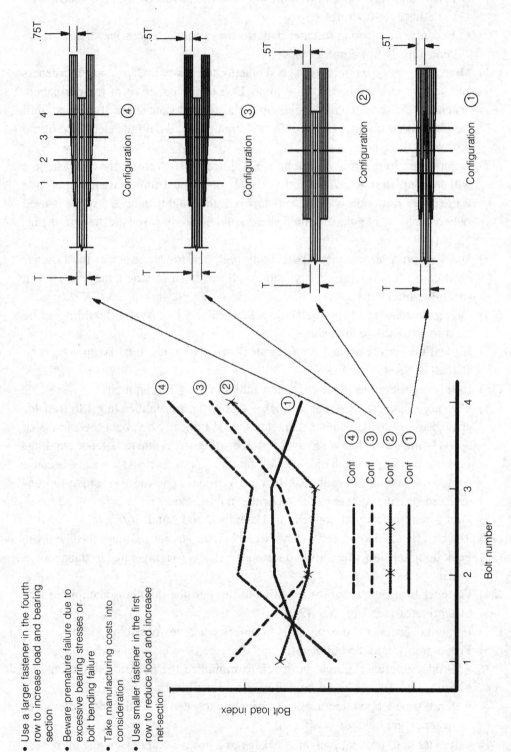

Fig. 5.2.11 Effect of Joint Configuration on Fastener Load Distribution

- Use a larger fastener in the fourth row to increase load and bearing section
- Beware premature failure due to excessive bearing stresses or bolt bending failure
- Take manufacturing costs into consideration
- Use smaller fastener in the first row to reduce load and increase net-section

300

(25) Sparking and arcing between metal fasteners in fuel tank environments (due to lightning strike) requires special design attection (see Chapter 6)

Fasteners (Metallic)

The fasteners used with composite assemblies are generally titanium alloy, (to prevent galvanic corrosion) tension heads (to avoid fastener pull-through).

(1) Fastener materials — Fastener materials have to be environmentally compatible with the laminate material to avoid galvanic corrosion; see Fig.5.2.7.
   - Materials that have been found to be compatible are: titanium, inconel, and A286 steel
   - Alloys which can be used with caution are: Stainless steels, monel, and PH steels
   - Non-compatible alloys such as aluminums and low alloy steels may be considered if reliable coatings or environmental considerations permit their use.

(2) Fastener configuration — Composite joint strength is sensitive to fastener configuration. For primary structures, a shallow head and large diameter shear fasteners will be more efficient in thin laminates where thickness is less than one diameter. However, the standard tension head fastener is more efficient for thick laminate joints where t/d is greater than 1.0. An interference fit (e.g., sleeve fastener system) should be considered in critical joint applications.

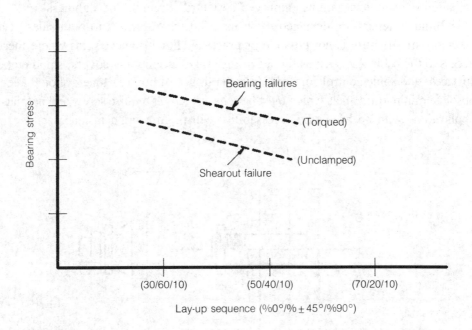

Fig. 5.2.12 Effects of Fiber Layup/Sequence Variation on Bearing Strength

Chapter 5.0

Fig. 5.2.13 through 5.2.15 shows a metallic fastener system selected off-the-shelf to be the primary fastener for composites in the early 1970's. They are still used in many applications. These fasteners are used in clearance holes and clamp-up forces are not modified for composites. A concerned issue encountered when conventional fasteners are not modified specifically for use in composites involves clamping forces. These forces, if not controlled, can cause the fastener to crush or deform the matrix that binds the fibers together, allowing the fastener to pull through after repeated loads. Therefore, do not over-torque the fastener during installation. Fig. 5.2.16 shows a fastener systems which is similar to that of the HI-LOK fastener system except that the collar torque system is slightly different. HI-LOK uses a torqued-off collar wrenching device while Eddie bolt 2 uses the Eddie lobes for torquing and when the predetermined clamp-up force is reached the lobes deform into and acorss the pin thread flutes to form a positive mechanical lock.

Fig. 5.2.17 through 5.2.20 show the HUCK-COMP fastener which is a specially designed version of the HUCK lightweight titanium lockbolt system. It is designed specifically for composite structure applications. It is an all titanium system with a Ti-6Al-4V pin and a flanged commercially pure titanium collar. The fastener comes in both a pull system version for installation with common pull tools or an automatic drill machine version for use with DRIVMATIC machines. Fig. 5.2.21 and 5.2.22 show HUCK-TITE fastener with a sleeve which is designed to be installed with interference fits (an interference-fit range of 0.001 to 0.006 inch would be possible) in composite structure without causing any installation damage and gaining improved structural strength and tightness.

Blind fasteners are designed to be installed where access to both sides of a sheet assembly or structure is not possible or practical. For instance, skins where there is no access to the back-side such as fighter or general aviation wing covers, small empennage surfaces, and some control surfaces. In general, use of blind fasteners should keep to the absolute minimum. Panels which must be removable will require screw and nutplats; other applications with no back side accessability will require blind fasteners.

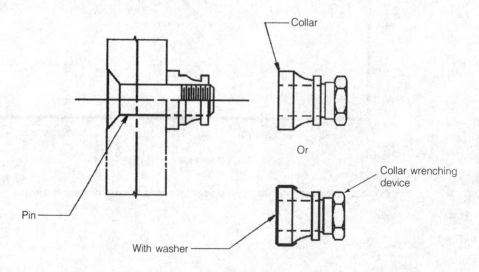

Fig. 5.2.13 HI-LOK Fastener

Joining

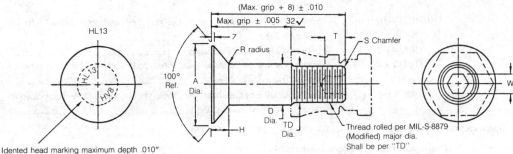

Idented head marking maximum depth .010″
"hs" or "H" No letter indicates hi-shear trademark.
"VS" Indicates voi-shan trademark.
"SPS" Indicates standard pressed steel trademark.
"V" After trademark indicates 6AL-4V titanium alloy material.
The number or numbers following the "V" indicates first dash number.

First dash No.	Nom. Dia.	A Dia.	B Ref.	D Dia. [8] Without solid film lube	D Dia. [8] With solid film lube	TD Dia.	F	H	R Rad.	Z	S Chamfer	Thread	SOCKET W Hex	SOCKET T Depth
−5	5/32	.3304 .3256	.312	.1635 .1630	.1635 .1625	.1595 .1570	.004	.0700 .0680	.025 .015	.010 .005	1/32″×45°	8-32UNJC-3A Modified	.0645 .0635	.135 .115
−6	3/16	.3813 .3765	.325	.1895 .1890	.1895 .1885	.1840 .1810	.005	.0805 .0785	.030 .020	.015 .005	1/32″×45°	10-32UNJF-3A Modified	.0806 .0791	.135 .115
−8	1/4	.5066 .5018	.395	.2495 .2490	.2495 .2485	.2440 .2410	.006	.1080 .1060	.030 .020	.015 .005	1/32″×45°	1/4-28UNJF-3A Modified	.0967 .0947	.150 .130
−10	5/16	.6335 .6287	.500	.3120 .3115	.3120 .3110	.3060 .3020	.007	.1350 .1330	.040 .030	.015 .005	3/64″×45°	5/16-24UNJE-3A Modified	.1295 .1270	.170 .150
−12	3/8	.7604 .7556	.545	.3745 .3740	.3745 .3735	.3680 .3640	.008	.1620 .1600	.040 .030	.015 .005	3/64″×45°	3/8-24UNJF-3A Modified	.1617 .1582	.200 .180
−14	7/16	.8884 .8812	.635	.4370 .4365	.4370 .4360	.4310 .4260	.009	.1895 .1865	.050 .040	.022 .005	3/64″×45°	7/16-20UNJF-3A Modified	.1930 .1895	.230 .210
−16	1/2	1.0139 1.0068	.685	.4995 .4990	.4995 .4985	.4830 .4880	.010	.2160 .2130	.050 .040	.022 .005	3/64″×45°	1/2-2UNJF-3A Modified	.2242 .2207	.260 .240
−18	9/16	1.1408 1.1337	.770	.5615 .5610	.5615 .5605	.5550 .5500	.010	.2430 .2400	.050 .040	.025 .005	1/18″×45°	9/16-18UNJF.3A Modified	.2555 .2520	.290 .270
−20	5/8	1.2723 1.2651	.825	.6240 .6230	.6240 .6230	.6180 .6120	.010	.2720 .2690	.050 .040	.025 .005	1/16″×45°	5/8-18UNJF-3A Modified	.2555 .2520	.330 .305

General notes:
1. Head edge out of roundnes shall not exceed "F".
2. Concentricity: conical surface of head to "D" diameter within .005 tir
3. "H" is dimensioned from maximum "D" diameter.
4. Dimensions of solid film lubed parts to be met after lube.
5. Surface texture per USASI B46.1.
6. Hole preparation per NAS618.
7. Use HL113 for oversize replacement.
8. Maximum "D" diameter may be increased by .0002 to allow for solid film application.

Material: 6AL-4V Titanium alloy per spec. AMS4928 or AMS4967.
Heat treat: 160,000 PSI tensile minimum (95,000 PSI shear minimum).
Finish: HL13V-( )-( ) — Cetyl alcohol lube per hi-shear spec. 305.
HL13VT-( )-( ) — Surface coating per hi-shear spec. 306, type I, color pink, and cetyl alchohol luber per hi-shear spec. 305.

(All diamensions are in inches)

Fig. 5.2.14 HI-LOK Fastener Data (HL13)

Chapter 5.0

HL13VUE-( )-( ) — Surface coating per hi-shear spec. 306, typeII, and cetyl alcohol lube per hi-shear. 305.
HL13VV-( )-( ) — Lubeco #2123 solid film lubricant.
HL13VR-( )-( ) — Surface coating per hi-shear spec. 306, type II, and solid film lube per electrofilm, inc., spec. 4396
HL13VF-( )-( ) — Surface coating per hi-shear spec. 306, type I, color blue, and cetyl alcohol lube per hi-shear spec. 305.

Specification: Hi-lok product spec. 340, section 4.

−5 Size must be installed using a torque controlled hex key.
See reference collar standards page for detail dimensions.

Code: First dash number indicates nominal diameter in 1/32NDS.
Second dash number indicates maximum in 1/16THS.
See "finish" note for explanation of code letters.

(Source: HI-Shear Corp.)

Fig. 5.2.14 HI-LOK Fastener Data (HL13) (cont'd)

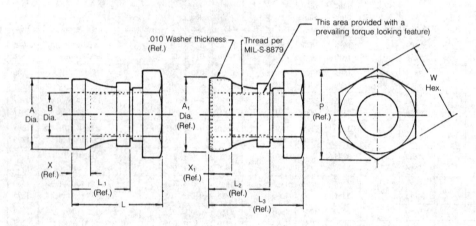

Dash No.	Pin nom. Dia.	Thread	A Dia.	A_1 Dia. (Ref.)	B Dia.	L.	L_1 (Ref.)	L_2 (Ref.)	L_3 (Ref.)	P (Ref.)	W Hex.	X (Ref.)	X_1 (Ref.)
−5	5/32"	8-32UNJC-3B	note: for −5 diameter pin, use HL86-5.										
−6	13/64"	10-32UNJF-3B	.307 .303	.340	.212 .218	.457 .437	.275	.290	.470	3.44	.314 .302	.107	.120
−8	17/64"	1/4-28UNJF-3B	.412 .408	.442	.272 .268	.552 .532	.340	.355	.565	.380	.346 .332	.112	.125
−10	21/64"	5/16-24UNJF-3B	.518 .512	.552	.336 .330	.672 .652	.430	.445	.685	.484	.440 .425	.122	.135
−12	25/64"	3/8-24UNJF-3B	.628 .622	.665	.398 .392	.744 .724	.475	.490	.755	.557	.503 .488	.122	.135
−14	25/64"	7/16-20UNJF-3B	.733 .727	.775	.463 .457	.862 .842	.560	.575	.875	.840	.753 .736	.137	.150
−16	33/64"	1/2-20UNJF-3B	.848 .842	.895	.528 .522	.942 .922	.610	.625	.955	.970	.878 .861	.137	.150
−18	37/64"	9/16-18UNJF-3B	.878 .872	.910	.598 .592	1.029 1.009	.670	.685	1.042	.970	.878 .861	.145	.158
−20	41/64"	5/8-18UNJF-3B	1.005 .995	1.040	.661 .665	1.123 1.103	.735	.750	1.136	1.120	1.003 .986	.145	.158
−24	49/64"	3/4-16UNJF-3B	1.105 1.095	1.040	.786 .780	1.371 1.351	.940	.955	1.384	1.240	1.128 1.110	.156	.169
−28	57/64"	7/8-14UNJF-3B	1.295 1.285	1.330	.911 .905	1.571 1.551	1.060	1.075	1.584	1.450	1.315 1.292	.169	.182
−32	1-1/64"	1-12UNJF-3B	1.475 1.465	1.510	1.036 1.030	1.836 1.816	1.225	1.240	1.849	1.650	1.504 1.480	.186	.199

(All dimensions are in inches)

Fig. 5.2.15 HI-LOK Collar Data (HL87)

Voi-shan
1 raised bead indicates
voi-shan identification

Standard pressed steel
2 raised beads indicate standard
pressed steel identification

Notes: 1. Go thread gage penetration shall be 3/4 of one revolution minimum.
2. Dimensions apply after plating or solid film lube.

Material: Collar — 302 se stainless steel per QQ-S-763.
Washer — 302 stainless steel per MIL-S-5059.

Finish: HL87-( )   — Collar only with cadmium plate per QQ-P-416, type II, class 3, and lauric acid or cetyl alcohol lube per hi-shear spec. 305.
HL87D-( )   — Collar only with cadmium plate per QQ-P-416, type II, class 3, and solid film lubricant per MIL-L-8937.
HL87DU-( )   — Collar only with solid film lube per MIL-L-8937.
HL87W-( )   — Collar with cadmium plate per QQ-P-416, type II, class 3, and lauric acid or cetyl alchohol lube per hi-shear spec. 305, and washer with cadmium plate per QQ-P-416, type I, class 3, no lube.

Specification: HI-LOK product spec. 340, section 101.
   Code: Dash number indicates nominal thread size in 1/32NDS.
     See "finish" note for explanation of code letters.
Example: HL87W-8 — 1/4-28 HI-LOK collar and washer.

(Source: HI SHEAR CORP.)

Fig. 5.2.15 HI-LOK Collar Data (HL87) (cont'd)

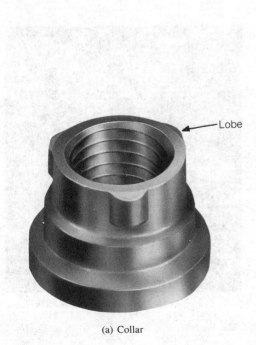

(a) Collar   (b) Pin

By courtesy of Fairchild Aerospace Fastener Division, Fairchild Corp.

Fig. 5.2.16 Eddie-Bolt 2

Chapter 5.0

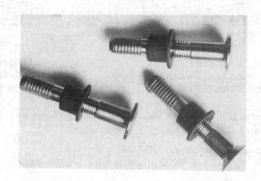

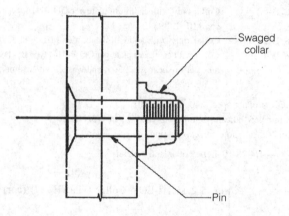

LGPL9SC (Clearance fit 0.001-0.003 inch)

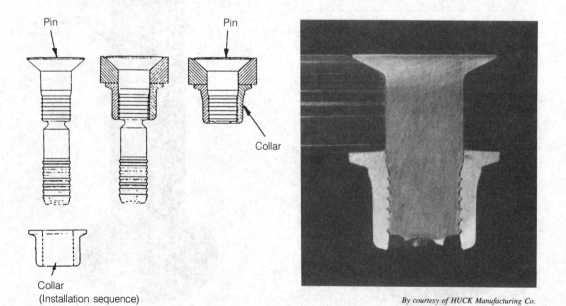

By courtesy of HUCK Manufacturing Co.

Fig. 5.2.17 HUCK-COMP Fastener

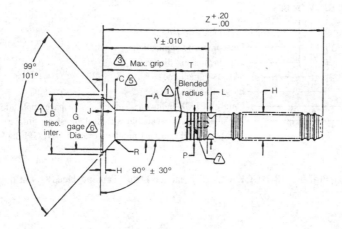

LGP pin Family number	Nom. size	A Shank Dia. Max.	A Shank Dia. Min.	8 head Dia. Max.	C head height Max.	G Gage Dia. Max.	G Gage Dia. Min.	H Gage height Max.	H Gage height Min.	J Max.	L Ref.	M Dia. Max.	P Max.	R Radius Max.	R Radius Min.	S	T Ref.
LGPL8SC-V05B( )	.164	.1635	.1630	.332	.073	.2832	.2830	.0202	.0175	.010	.126	.156	.156	.025	.015	.0045	.175
LGPL8SC-V06V( )	.190	.1895	.1890	.383	.084	.3272	.3270	.0230	.0200	.013	.150	.184	.184	.030	.020	.0045	.176
LGPL8SC-V08B( )	.250	.2495	.2490	.510	.110	.4320	.4318	.0322	.0288	.017	.187	.244	.244	.030	.020	.0045	.244
LGPL8SC-V108( )	.312	.3120	.3115	.637	.137	.5451	.5449	.0378	.0342	.020	.244	.306	.306	.040	.030	.0045	.313
LGPL8SC-V12B( )	.375	.3745	.3740	.765	.165	.6582	.6580	.0439	.0401	.023	.298	.368	.370	.040	.030	.0060	.374

△1 Concentricity: Conical surface of head to "A" diameter to be within .005 TIR.

△2 Shank straightness: Within "S" values TIR per inch of shank length.

△3 Grip length number: To determine the grip length number divide the dimensional thickness of parts being joined by .0625 inch. Grip length: To determine the grip length. Multiply the grip length number by .0625 inch. The grip length is measured from the top of the head to end the end of the full cylindrical portion of the shank.

△4 See coding under usage and application for complete standard number.

△5 Dimensions B & C for engineering reference only. Not for inspection purposes.

△6 Drill center dimple permitted in head.

Material: 6AL-4V titanium alloy per AMS4967.
Minimum shear strength: 95 KSI.
Surface texture: Ra max. per ANSI B46. 1: Bearing surface, head to shank radius, and shank & lead-in radius −32; other surfaces −125.
Cetyl alcohol lube (chlorine free).
Head marking: Pin head shall be marked with HUCK'S basic symbol, basic number, and material designator, depressed .010 max. sealant escape groove: pin head shall be marked with HUCK'S basic symbol, basic number, material designator & H, depressed .010 max. arrangement optional. Example:

Example: (LGP8 VB) △7 (LGP8 VBH)

△4 Usage and application information

Coding: The first set of letters designate the family of lightweight GP pins for shear or tension applications (LGPL).
The second set of letters & number 8SC designate head size, style, and load application.
The next letter, "V", is the material designator for 6AL-4V titanium alloy, min. Shear strength: 95 KSI.
The numbers following the material designator designates the nominal pin shank diameter in .0312 inch increments.
The final number(s) designates the grip length number or the nominal pin shank length in .0625 inch increments.

(All dimensions are in inches)

Fig. 5.2.18 HUCK-COMP Fastener Data (Tension Head)

Chapter 5.0

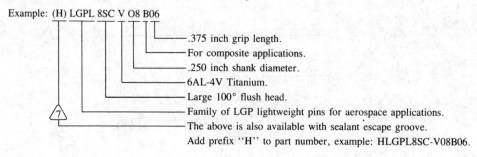

Fig. 5.2.18 HUCK-COMP Fastener Data (Tension Head) (cont'd)

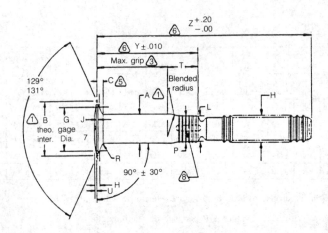

LGP pin Family number	Nom. size	A Shank Dia. Max.	A Shank Dia. Min.	B head Dia. Max.	C head height Max.	G Gage Dia. Max.	G Gage Dia. Min.	H Gage height Max.	H Gage height Min.	J Max.	L Ref.	M Max.	P Dia. Max.	R Radius Max.	R Radius Min.	S	T Ref.	U Ref.
LGPL9SC-V05B( )	.164	.1635	.1630	.332	.040	.2560	.2550	.0263	.0235	.010	.126	.156	.156	.025	.015	.0045	.175	.006
LGPL9SC-V06V( )	.190	.1895	.1890	.383	.045	.2982	.2980	.0257	.2980	.013	.150	.184	.184	.030	.020	.0045	.176	.004
LGPL9SC-V08B( )	.250	.2495	.2490	.510	.110	.4320	.4318	.0322	.0288	.017	.187	.244	.244	.030	.020	.0045	.244	.005
LGPL9SC-V108( )	.312	.3120	.3115	.595	.066	.4791	.4789	.0350	.0308	.020	.244	.306	.306	.040	.030	.0045	.313	006
LGPL9SC-V12B( )	.375	.3745	.3740	.701	.076	.5942	.5940	.0343	.0293	.023	.298	.368	.370	.040	.030	.0060	.374	.007

/1\ Concentricity: Conical surface of head to "A" diameter to be within .005 TIR.

/2\ Shank straightness: Within "S" values TIR per inch of shank length.

/3\ Grip length number: To determine the grip length number divide the dimensional thickness of parts being joined by .0625 inch. Grip length: To determine the grip length. Multiply the grip length number by .0625 inch. The grip length is measured from the theoretical intersection of crown & head angle to the end the of the full cylindrical portion of the shank.

/4\ See coding under usage and application for complete standard number.

/5\ Dimensions B & C for engineering reference only. Not for inspection purposes.

/6\ "Y" & "Z" are located from theoretical intersection of crown & head angle.

/7\ Drill center dimple permitted in head.

(All dimensions are in inches)

Fig. 5.2.19 HUCK-COMP Fastener Data

Material: 6AL-4V titanium alloy per AMS4967.
Minimum shear strength: 95 KSI.
Surface texture: Ra max. per ANSI B46. 1: Bearing surface, head to shank radius, and shank & lead-in radius −32; other surfaces −125.
Cetyl alcohol lube (chlorine free).
Head marking: Pin head shall be marked with HUCK's basic symbol, basic number, and material designator, depressed .010 max. sealant escape groove: pin head shall be marked with HUCK'S basic symbol, basic number, material designator & H, depressed .010 max. arrangement optional. Example:

Example:

/4\ Usage and application information

Coding: The first set of letters designate the family of lightweight GP pins for shear or tension applications (LGPL).
The second set of letters & number 9SC designate head size, style, and load application.
The next letter, "V", is the material designator for 6AL-4V titanium alloy, min. Shear strength: 95 KSI.
The numbers following the material designator designates the nominal pin shank diameter in .0312 inch increments.
The final number(s) designates the grip length number or the nominal pin shank length in .0625 inch increments.

Example: (H) LGPL 9SC V O8 B06

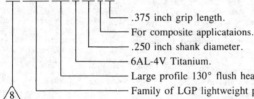

- .375 inch grip length.
- For composite applicataions.
- .250 inch shank diameter.
- 6AL-4V Titanium.
- Large profile 130° flush head, for shear applications.
- Family of LGP lightweight pins for aerospace applications.
- The above is also available with sealant escape groove.
  Add prefix "H" to part number, example: HLGPL9SC-V08B06.

(Source: HUCK Manufacturing Co.)

Fig. 5.2.19 HUCK-COMP Fastener Data (cont'd)

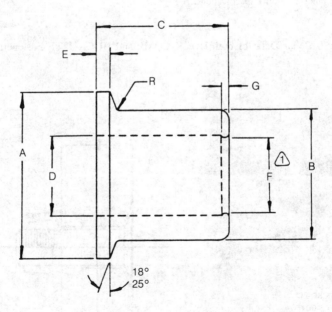

Fig. 5.2.20 Collar Data for Both HUCK-COMP and HUCK-TITE Fasteners

Chapter 5.0

Identification: Manufacturing option to identify collar with X.
⚠ Permissible punchout burr permitted within limits of "F" & "G".

Part number	Nominal size	A	B	C	D	E	F	G	R	Install with
SLFC-MV05	.164	.334 .314	.248 .244	.235 .215	.1625 .1655	.020 .030	.159	.016	.008 .018	LGP( )8S( )-V05B LGP( )9S( )-V05B
SLFC-MV06	.190	.388 .368	.274 .270	.240 .220	.1880 .1910	.020 .030	.186	.016	.008 .018	LGP( )8S( )-V06B LGP( )9S( )-V06B
SLFC-MV08	.250	.477 .457	.357 .353	.300 .280	.2485 .2515	.025 .035	.246	.016	.008 .018	LGP( )8S( )-V08B LGP( )9S( )-V08B
SLFC-MV10	.312	.604 .584	.448 .444	.370 .350	.3100 .3130	.025 .035	.308	.031	.013 .023	LGP( )8S( )-V10B LGP( )9S( )-V10B
SLFC-MV12	.375	.701 .681	.533 .529	.420 .400	.3725 .3755	.030 .040	.370	.031	.013 .023	LGP( )8S( )-V12B LGP( )9S( )-V12B

Material: Cp titanium
Finish: None
Lubrication: Dry film per MIL-L-8937 plus chlorine free cetyl alcholo.
Surface texture: AA max. per ANSI B46.1-125,
Suggested temperature limitations: Room temperature to 250°F.

Shear flanged Collar   SLFC-MV06
part no. code   └─── Nominal pin shank in 32nd inches.
(and example)   └────── Material: cp titanium
                         └───────── Shear flanged collar.

Quality control requirements as applicable per HUCK quality control manual, established by the requirements of specification MIL-Q-9858 and other governing specifications.

(Source: HUCK Manufacturing Co.)

Fig. 5.2.20 Collar Data for both HUCK-COMP and HUCK-TITE Fasteners (cont'd)

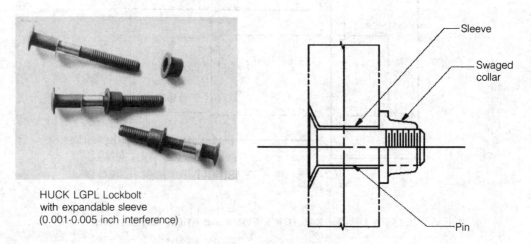

HUCK LGPL Lockbolt
with expandable sleeve
(0.001-0.005 inch interference)

Fig. 5.2.21 HUCK-TITE Interference Fit Fastener

Joining

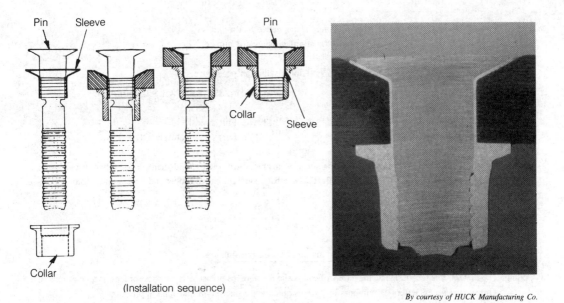

(Installation sequence)

By courtesy of HUCK Manufacturing Co.

Fig. 5.2.21 HUCK-TITE Interference Fit Fastener (cont'd)

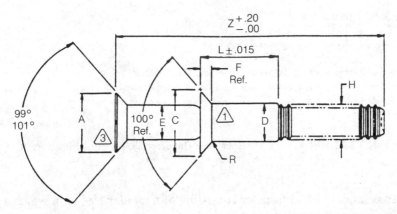

CIL fastener part number	A	C Theo. int. Ref.	D	E Min.	F Ref.	M Max.	A Radus	
							Max.	Min
CIL8SC-VC05-( )	.317 .307	.353	.1755 .1745	.1650	.075	.156	.022	.012
CIL8SC-VC06-( )	.367 .357	.403	.2015 .2005	.1910	.085	.184	.027	.017
CIL8SC-VC08-( )	.486 .476	.529	.2625 .2615	.2518	.112	.244	.027	.017
CIL8SC-VC10-( )	.605 .595	.656	.3235 .3225	.3128	.140	.306	.037	.027
CIL8SC-VC12-( )	.725 .715	.783	.3875 .3865	.3767	.166	.368	.037	.027

Fig. 5.2.22 HUCK-TITE Fastener Data (Interference Fit Flush Head)

311

⚠️ Sleeve O.D. may be distorted due to crimping the sleeve to the pin.
⚠️ See coding under usage and application for complete standard number.
⚠️ Drill center dimple permitted in head.
⚠️ Use with CIFC-MV( ) collar. see SK12296-4 for installation procedure.

Material pin: 6AL-4V titanium alloy per AMS4967.
Minimum shear strength pin: 95 KSI.
Finish pin: drifilm per MIL-L-8937 plus cetyl alcohol.

Sleeve: A286 CRES (A.I.S.I. 660) per AMS5525.
Sleeve: Full anneal.
Sleeve: Bare. passivate per QQ-P-35.

Surface finish: Ra max. per ANSI B46.1: bearing surface, head to shank radius, and shank -32; other surfaces -125.
Head marking: Fastener head shall be marked with HUCK'S basic symbol, basic number and material designator, depressed .010 max.

Example: (CIL8 V)

⚠️ Usage and application information

Coding: The first set of letters designate the family of composite interference fit fasteners for shear or tension applications (CIL)
   The second set of letters & numbers (8SC) designate head size, style and load application.
   The next letters "VC" are the material designators for 6AL-4V titanium alloy pin and A-286 CRES sleeve.
   The numbers following the material designator designate the shank diameter.
   The final number(s) designates the grip length number in .0625 inch increments.

Example: CIL 8SC-VC 08-06
   — .375 inch grip length.
   — .262 inch shank diameter.
   — 6AL-4V titanium pinia-286 CRES sleeve.
   — Large 100° flush head for shear applications.
   — Family of composite interference fit fasteners.

(Source: HUCK Manufacturing Co.)

Fig. 5.2.22 HUCK-TITE Fastener Data (Interference Fit Flush Head) (cont'd)

The conventional blind fastener consisting of a tubular sleeve in which a stem having an enlarged end cannot be used for composite structural assembly due to insufficient blind head expansion. In addition, they may damage the composite laminates during installation. This damage is caused by either excessive clamp load for the available bearing area or by radial fastener expansion within the fastener hole, which casues the composite plies to delaminated on the blind-side surface.

Fig. 5.2.23 and 5.2.24 show a COMPOSI-LOK II blind fastener specifically designed to have a large, blind side footprint for application in carbon composite structures. It has a stainless stem and the back-side clamping sleeve folds under when it hits the laminate surface to provide an axial-flow bulbing action which is enhanced semi-hydraulically by a trapped internal plastic sleelve.

Fig. 5.2.25 shows a screw-stem COMP-TITE blind fastener system designed specifically for composites with a titanium threaded sleeve and stainless stem, back side sleeve and washer. The blind side footprint is formed by expansion of a flat-faced helical washer backed up by an expansion sleeve. This washer forms a large bearing surface that will not crush or delaminate composites during installation.

Assembly

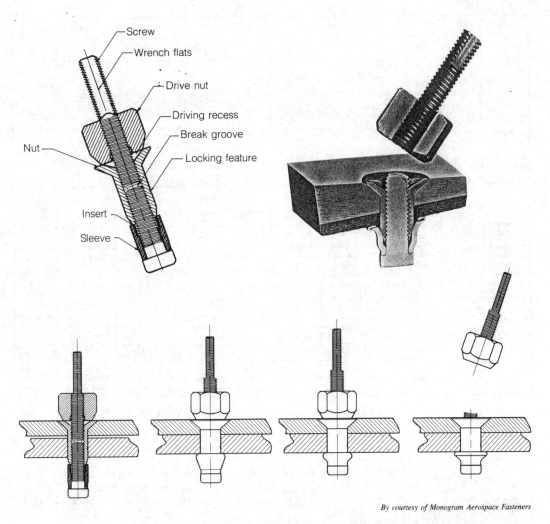

Fig. 5.2.23 COMPOSI-LOK II Big Foot Blind Fastener

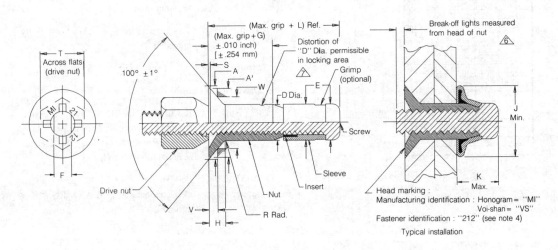

Fig. 5.2.24 COMPOSI-LOK II Blind Fastener Data

313

Chapter 5.0

TABLE I

PART NUMBER	A DIA. THEO.		A' DIA. MIN.		D DIA.		E DIA. MAX.		F WRENCH FLATS		G		H REF.		L		R RAD. MAX.		S MAX.		T ACROSS HEX.	
	INCH	mm	INCH	mm	INCH	mm	INCH	mm	INCH	mm	INCH	mm	INCH	mm	INCH	mm	INCH	mm	INCH	mm	INCH	mm
MBF2112-5-( )	.332/.325	8,43/8,25	.296	7,52	.1645/.1625	4,170/4,128	.1640	4,166	.085/.080	2,16/2,03	.017	0,43	.070	1,78	.512	13,00	.030	0,76	.015	0,38	.375	9,52
MBF2112-6-( )	.385/.378	9,78/9,60	.342	8,69	.1985/.1965	5,042/4,991	.1985	5,042	.113/.108	2,87/2,74	.027	0,68	.077	1,96	.575	14,61	.030	0,76	.019	0,48	.375	9,52
MBF2112-7-( )	.416/.409	10,57/10,39	.373	9,47	.2275/.2255	5,778/5,728	.2275	5,778	.121/.116	3,07/2,95	.035	0,89	.077	1,96	.635	16,13	.030	0,76	.020	0,51	.375	9,52
MBF2112-8-( )	.507/.499	12,88/12,67	.463	11,76	.2595/.2575	6,591/6,541	.2595	6,591	.135/.130	3,43/3,30	.055	1,40	.104	2,64	.700	17,78	.030	0,76	.020	0,51	.375	9,52
MBF2112-9-( )	.538/.530	13,66/13,46	.494	12,55	.2895/.2875	7,353/7,303	.2895	7,353	.152/.147	3,86/3,73	.065	1,65	.104	2,64	.815	20,70	.030	0,76	.020	0,51	.500	12,70
MBF2112-10-( )	.635/.626	16,13/15,90	.577	14,66	.3115/.3095	7,912/7,861	.3110	7,899	.152/.147	3,86/3,73	.070	1,78	.136	3,45	.892	22,66	.040	1,02	.026	0,66	.500	12,70
MBF2112-11-( )	.666/.657	16,92/16,69	.608	15,44	.3435/.3415	8,725/8,674	.3433	8,720	.185/.180	4,70/4,57	.075	1,90	.136	3,45	.941	23,90	.040	1,02	.026	0,66	.500	12,70
MBF2112-12-( )	.762/.752	19,35/19,10	.696	17,68	.3745/.3725	9,512/9,462	.3740	9,500	.185/.180	4,70/4,57	.080	2,03	.162	4,11	1.090	27,69	.040	1,02	.029	0,74	.500	12,70

TABLE I (CONT)

PART NUMBER	MINIMUM AVAILABLE GRIP DASH NO.	INSTALLED DIMENSIONS								MECHANICAL PROPERTIES										
		RECOMMENDED HOLE SIZE		J DIA. MIN.		K MAX.		BREAK-OFF LIMITS /6\		TENSILE STRUCTURAL FAILURE (MIN.)		DOUBLE SHEAR MIN.		LOCKING TORQUE MIN.		V GAGE PROT.		W GAGE DIA.	
		INCH	mm	INCH	mm	INCH	mm	INCH	mm	LBS.	N	LBS.	N	IN-LBS	Nm	INCH	mm	INCH	mm
MBF2112-5-( )	-150	.168/.165	4,27/4,19	.250	6,35	.300	7,62	+.103/-.000	+2,62/-0,00	900	4000	3150	14010	1.0	0,113	.0207/.0174	0,526/0,442	.2832/.2830	7,193/7,188
MBF2112-6-( )	-150	.202/.199	5,13/5,05	.300	7,62	.350	8,89	+.103/-.000	+2,62/-0,00	1400	6230	4600	20460	1.5	0,170	.0245/.0212	0,622/0,538	.3272/.3270	8,311/8,306
MBF2112-7-( )	-150	.231/.228	5,88/5,79	.350	8,89	.400	10,16	+.103/-.000	+2,62/-0,00	1600	7120	6050	26910	2.0	0,226	.0358/.0324	0,909/0,823	.3315/.3313	8,420/8,415
MBF2112-8-( )	-200	.263/.260	6,68/6,60	.400	10,16	.450	11,43	+.103/-.000	+2,62/-0,00	2100	9340	7900	35140	2.5	0,282	.0318/.0279	0,808/0,709	.4320/.4318	10,973/10,968
MBF2112-9-( )	-200	.293/.290	7,44/7,37	.450	11,43	.500	12,70	+.103/-.000	+2,62/-0,00	2600	11565	9800	43590	3.0	0,339	.0446/.0407	1,133/1,034	.4320/.4318	10,973/10,968
MBF2112-10-( )	-250	.315/.312	8,00/7,92	.475	12,06	.550	13,97	+.103/-.000	+2,62/-0,00	3600	16010	11350	50480	3.5	0,400	.0405/.0365	1,029/0,927	.5389/.5385	13,688/13,678
MBF2112-11-( )	-250	.347/.344	8,81/8,74	.525	13,33	.575	14,60	+.103/-.000	+2,62/-0,00	4400	19570	13850	61600	4.0	0,452	.0539/.0500	1,369/1,270	.5389/.5385	13,688/13,678
MBF2112-12-( )	-250	.378/.375	9,60/9,52	.575	14,60	.625	15,87	+.103/-.000	+2,62/-0,00	5000	22240	16450	73170	4.0	0,452	.0458/.0415	1,163/1,054	.6532/.6528	16,591/16,581

TABLE II

2ND DASH NO. (GRIP)	GRIP RANGE			
	MIN. GRIP		MAX. GRIP	
	INCH	mm	INCH	mm
100	.050	1,27	.100	2,54
150	.100	2,54	.150	3,81
200	.150	3,81	.200	5,08
250	.200	5,08	.250	6,35
300	.250	6,35	.300	7,62
350	.300	7,62	.350	8,89
400	.350	8,89	.400	10,16
450	.400	10,16	.450	11,43
500	.450	11,43	.500	12,70
550	.500	12,70	.550	13,97
600	.550	13,97	.600	15,24
650	.600	15,24	.650	16,51
700	.650	16,51	.700	17,78
750	.700	17,78	.750	19,05
800	.750	19,05	.800	20,32
850	.800	20,32	.850	21,59
900	.850	21,59	.900	22,86
950	.900	22,86	.950	24,13
1000	.950	24,13	1.000	25,40
1050	1.000	25,40	1.050	26,67
1100	1.050	26,67	1.100	27,94
1150	1.100	27,94	1.150	29,21

Procurement specification: MBF 2000

Installation & inspection specification: MBF 2001

Material and heat treat: Nut: 6AL-4V titanium per MIL-T-9047, STA, or AMS 4928 or AMS 4957. Heat treated per MIL-H-81299. Maximum hydrogen 125 PPM.

Screw: A-286 per AMS 5732, AMS 5731 or AMS 5737 heat treated to 175 KSI tensile minimum.

Sleeve: 304 stainless steel per AMS 5639. Fully annealed.

Insert: Acetal per ASTMD 4181.

Drive Nut: Mild steel (colorgray)

Finish: (—) Nut: Kal-gard ANN-RQ #1012 conversion coating or phosphate fluoride per boeing specification PS 741 may be used at manufacturer's option.

Sleeve & Screw: Passive per QQ-P-35, Kal-gard conversion coating Kao-gard ANN-RO #1013 optional.

Insert: None.

Drive Nut: Colorgray

Fig. 5.2.24 COMPOSI-LOK II Blind Fastener Data (cont'd)

Lubricants: Tio-lube 460, dry film lube per Mil-L-8937 paraffin wax, cetyl alcohol, used as required for performance.

General Notes:
1.) Example of part number:

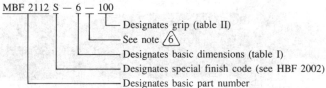

2.) Looking feature consists of three (3) indentations located 120° apart on the periphery of the nut component and approximately .040 above the intersection of the nut nose angle and O.D.

3.) See MBF 2003 for installation and removal information.

4.) Alternate head marking: "MI" (or "VS") and "2012".

5.) COMPOSI-LOK fasteners with selected combinations of the above lubricants and finishes are specially coded and may be substituted for equivalent non-coded parts at manufacturer's option. see interchangeability specification MBF 2007.

6.) An "L" in place of the dash (—) between the diameter dash number and the grip dash number designates modified break-off limits of +.053/−.050. "e.g. MBF 2112 ( )-6L200".

7.) Distortion shall not prevent insertion of the fastener into a ring gauge of length equal to max. grip and diameter equal to minimum recommended hole.
Force for insertion shall not exceed 5.0 pounds.

Fig. 5.2.24 COMPOSI-LOK II Blind Fastener Data (cont'd)

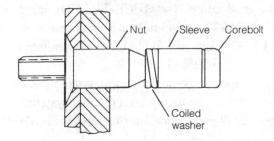

(1) Using NAS 1675-type installation tooling, the nuts is restrained from turning while the corebolt is driven.

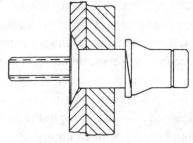

(2) The advance of the croebolt forces the washer and sleeve over the taper, expanding and uncoiling the washer to its maximum diameter.

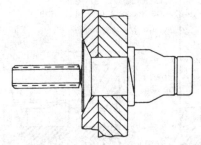

(3) Continued advanced of the corebolt draws the washer and sleeve against the joint surface, preloading the structure.

At a torque level controlled by the break groove, the slabbed portion of the corebolt separates, and installation is complete.

(blind fastener installation sequence.)

Fig. 5.2.25 COMP-TITE Blind Fastener

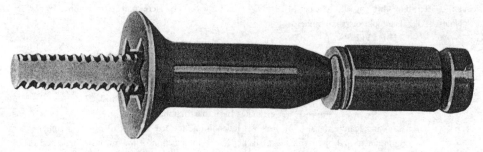

By courtesy of SPS Technologies Aerospace products Division

Fig. 5.2.25 COMP-TITE Blind Fastener (cont'd)

Fig. 5.2.26 shows a VISU-LOK blind fastener which is a version of the screw-stem bolt with a separate back side clamping sleeve. It consists of a titanium basic sleeve with a titanium stem and stainless back-side clamping sleeve.

Fig. 5.2.27 shows a COMPOSI-BOLT blind bolt which is a blind fastener with a pull-stem, single-action and single sleeve bulbing action on the blind side. It is designed specifically for composite structures. Large bearing area available in the 100° and 120° shear and tension head are suitable for composite applications. The HUCK-TIMATIC blind bolt is a titanium version of the single action UNIMATIC blind bolt. It is available with a CP titanium sleeve and an A286 pin or a 15-3-3-3 titanium pin which was designed specifically for use in composite structures.

Only a few fasteners are available to be used for composite structural assembly and this Section provides basic design data for several of the commonly used fastener systems. However, the data given in Fig. 5.2.13 through 5.2.27 are vendor information and should only be used with caution.

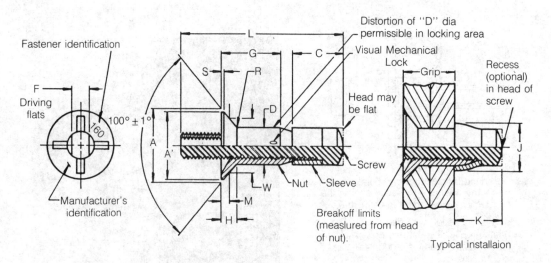

By courtesy of Fairchild Aerospace Fastener Division, Fairchild Corp.

Fig. 5.2.26 VISU-LOK Blind Fastener

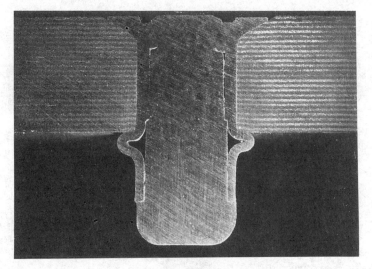
(Installed in Graphite Composite)
(a) UNIMATIC Blind Bolt

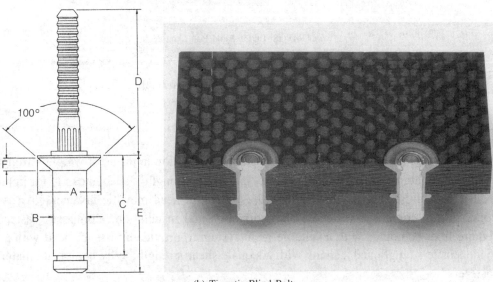
(b) Ti-matic Blind Bolt

(c) COMPOSI-BOLT Blind Bolt

Fig. 5.2.27 HUCK Blind Fastener

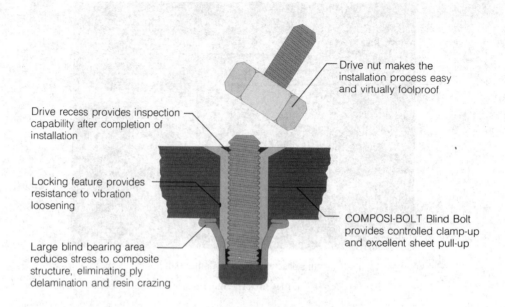

(c) COMPOSI-BOLT Blind Bolt (cont'd)

By courtesy of HUCK Manufacturing Co.

Fig. 5.2.27 HUCK Blind Fastener (cont'd)

Composite Fasteners (Non-metallic)

Fasteners made of composite materials are an ideal solution in terms of weight, strength, corrosion, lightning strike, radar signature, etc. Their strength is adequate only for lightly-loaded structures. Heavily-loaded structures may still need metallic fasteners. Areas of concern with composite fasteners are fabrication and installation. Composite fasteners require fiber combined with a suitable resin system to provide a fastener head with adequate tensile strength and a shank with adequate shear strength. Why composite material fasteners?

- Use would avoid fuel tank arcing concerns during lightning strike
- Reduce weight
- Electromagnetic transparency
- Eliminates dissimilar material corrosion
- Minimize radar signature (stealth)

In 1970's, U. S. Man Tech Program studied two types of composite material fasteners:

(A) Thermoset sleeve fasteners (see Fig. 5.2.28):
- Cut to length
- Feed and charge mold
- Form and cure in place
- Cool down and eject adhesive

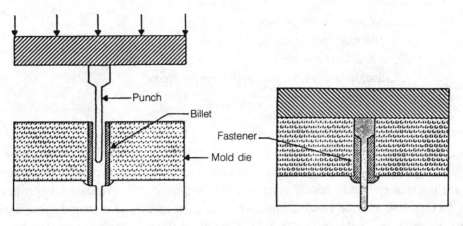

(a) Fastener sleeve molding sequence

- Reinforced epoxy, cured sleeve and pin
- Pin-bonded in place

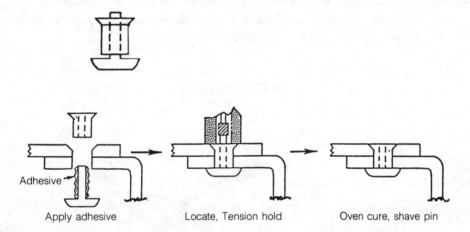

(b) Installation sequence

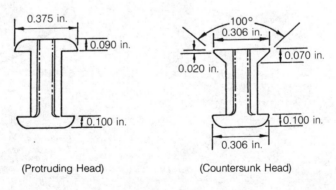

(Protruding Head)   (Countersunk Head)

(c) Fastener detail

Fig. 5.2.30 Thermoset Sleeve Fastener

Chapter 5.0

    (B)   Thermoplastic fasteners (see Fig. 5.2.29):
- Cut thermoplastic pultrusion bar to length
- Feed and Charge mold
- Ultrasonic die set to heat and form to shape
- Fuse resin
- Cool down and eject

Fig. 5.2.30 shows several composite material fasteners which are either made from pultrusion bar stock and then formed into rivets, or nuts and bolts machined from woven graphite/polyimide composite rod stock. They have shear strength comparable to aluminum and are not susceptible to galvanic corrosion when used with graphite composites.

A composite material blind fastener which consists of a metal stem and a thermoplastic sleeve (or shell) as shown in Fig. 5.2.31. This blind fastener has been used on the wing leading edge of the Beech Starship aircraft.

Fig. 5.2.32 shows an adhesive-bonded studs which is to attach to the surface of composite panel for removable bolted joints. The disposable polypropylene fixture with adhesive-faced foam base provides temporary support while the stud adhesive cures.

Fig. 5.2.33 shows a stud installation which utilizes a ratcheting device in its receptacle that protects against composite delamination.

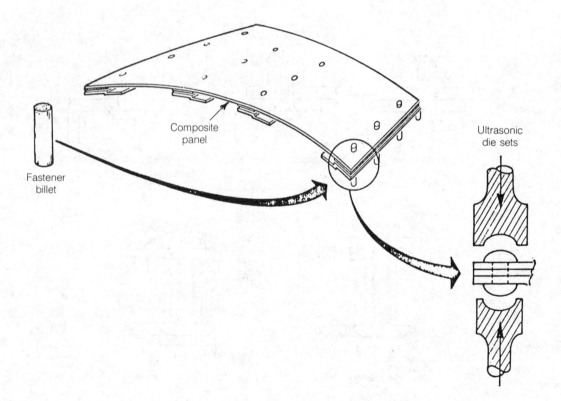

Fig. 5.2.29 Ultrasonic Thermoplastic Fastener

Joining

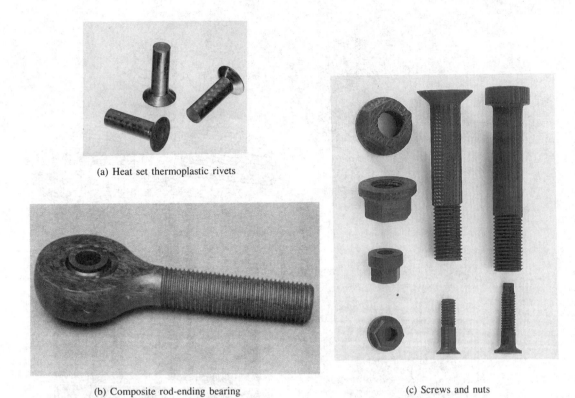

(a) Heat set thermoplastic rivets

(b) Composite rod-ending bearing

(c) Screws and nuts

By courtesy of TIODIZE Corp.

Fig. 5.2.30 Composite Material Fasteners

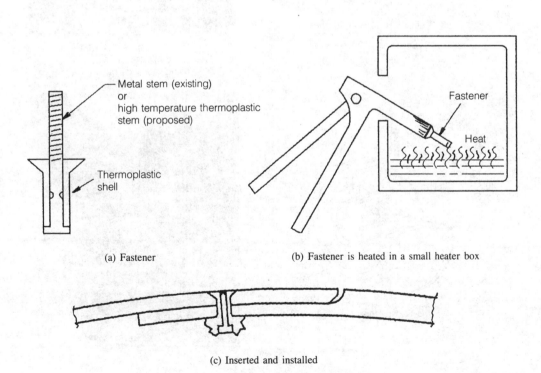

(a) Fastener

(b) Fastener is heated in a small heater box

(c) Inserted and installed

Fig. 5.2.31 Composite Material Blind Fastener

Chapter 5.0

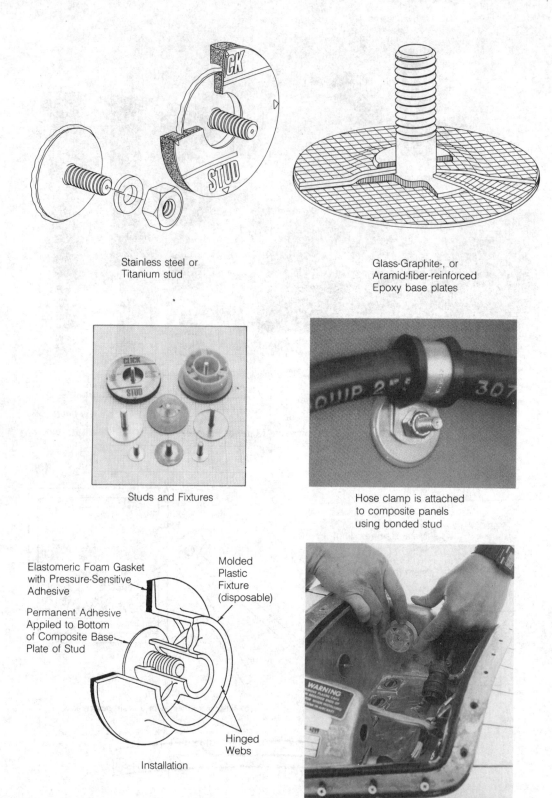

(a) Click studs

Fig. 5.2 32 Adhesive-Bonded Click Studs and Nuts

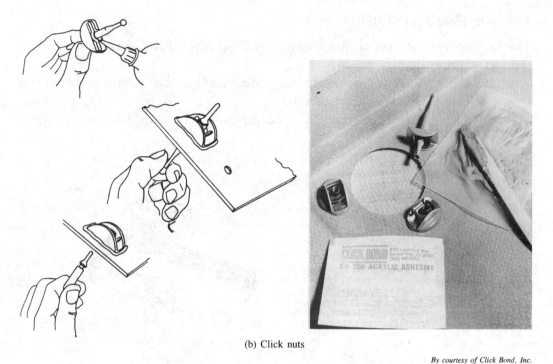

(b) Click nuts

By courtesy of Click Bond, Inc.

Fig. 5.2.32 Adhesive-Bonded Click Studs and Nuts (cont'd)

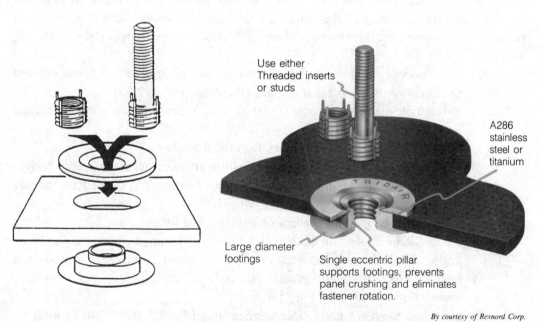

By courtesy of Rexnord Corp.

Fig. 5.2.33 CFC System (Insert and stud give Composite Panel a Solid Attachment Point)

Fastener Hole Preparation

Tape laminates require special attention during drilling to avoid problems at hole entrance and exit as shown in Fig. 5.2.34.
  (a) Carbon/Epoxy fabric outer plies — Acceptable surface quality is obtained on both the entrance and exit sides
  (b) Carbon/Epoxy tape outer plies — Backup tooling is required for acceptable quality on the exit side

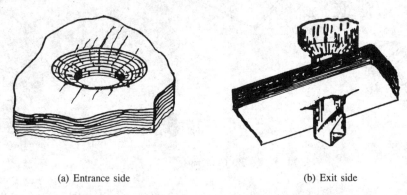

(a) Entrance side (b) Exit side

Fig. 5.2.34 Care Required to Avoid Problems at Hole Entrance and Exit

There is experimental evidence that some fastener/hole interference actually improves the load-carrying capability of the composite laminate: However, fiber damage which occurs when fasteners are installed can drastically affect success unless sleeve fasteners are used
  (a) Conventional push through interference ft installation of fasteners will damage the laminate unless precautions are taken (see Fig. 5.2.35):
   • Use hole size clearance of 0.000 to +0.003 inch (0.076 mm) for normal structural joints
   • Use standard close-ream holes for crtical joints
   • Be careful to use interference or transition fit holes and use specially designed fasteners to obtain interference fit such as expanded sleeve type fastener (see Fig. 5.2.21)
   • Do not use tapered-fasteners or cold-worked holes
  (b) Avoid "knife edges" on countersunk fasteners. Laminate thickness should be greater than 1.5 times countersunk depth, or laminate thickness should be at least 0.03 inch (0.76 mm) greater than fastener head height — including deep countersink as shown in Fig. 5.2.36
  (c) Use close tolerance holes. Interference fits of 0.0005 inch (0.0127 mm) interference are acceptable. Loose fits result in poor fatigue life (see Fig. 5.2.37)

(d) Automated drilling and fastening
   - Highly accurate and reproducible
   - Labor reduced
   - Requres use long life drill e.g., 8-facet drill
(e) When drilling aluminum or titanium/composite stackups, call out pilot holes, then separately drill full size holes in each part.Full size holes may be called out if the metal part is on the near side of the carbon/metal combination; otherwise use a computer-controlled drill, as shown in Fig. 5.2.38, which automatically adjusts spindle speed and penetration rate as the bit passes through stacked materials which vary in hardness (e.g., multi-material stack of carbon, titanium, aluminum, etc.).

There are particular challenges with mechanical fasteners that must be installed and removed frequently. This type of fastener must be insulated from the composite material to prevent abrasive wear on the fastener, as well as delamination of the composite, from the rubbing and rotating of the fastener. In addition, the laminate panel must be protected from crushing due to excessive clamping forces. Fig. 5.2.39 shows a type of fastener system which incorporates grommets for the fastener or stud to travel through. These fastener systems are used mainly for access panels which require frequent removal.

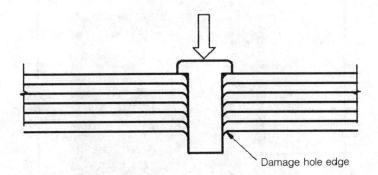

Fig. 5.2.35 Fiber Damage Due to Push Through Interference-fit Fastener

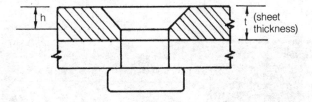

Fig. 5.3.36 Feather-edge in Countersunk Sheet

Chapter 5.0

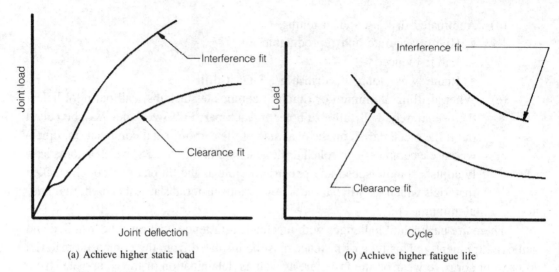

(a) Achieve higher static load

(b) Achieve higher fatigue life

Fig. 5.2.37 Improvement Load Carrying Capability with Interference-fit Fasteners

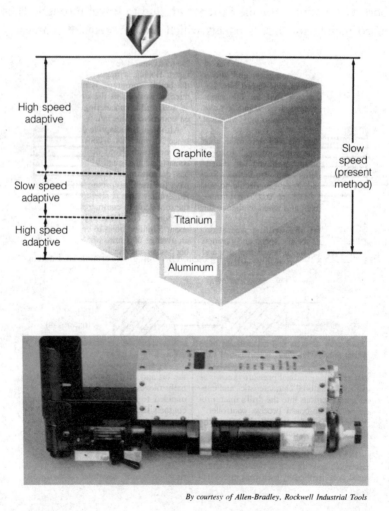

By courtesy of Allen-Bradley, Rockwell Industrial Tools

Fig. 5.2.38 Adaptive Control Drill

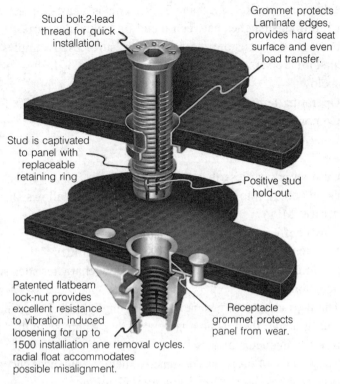

Fig. 5.3.39 Grommet Protects Laminate Fastener Hole Edges

By courtesy of Rexnord Corp.

Shim Requirements

Shims are used in airframe production to control structural fit-up and to maintain contour or alignment. With composite joints, unshimmed gaps that are only 1/4 as large as these allowed for similar aluminum counterparts can be tolerated. Therefore, the assembly of composite components will generally require more extensive use of shims than comparable metal components.

Engineering can reduce both cost and waste by controlling shim usage through design and specifications. Where to shim, what the shim taper and thickness should be, what gap to allow, and whether the gap should be shimmed or pulled up with fasteners must all be decided.

Shim materials:
(1) Solid shims: titanium, stainless steel, precured composite laminates, etc.
(2) Laminated (or peelable) shims [with a laminate thickness of about 0.003 inches (76.2 microns) ± 0.0003 inches (7.62 microns)]:
- Laminated titanium shims
- Laminated stainless steel shims
- Laminated Kapton shims

(3) Moldable shim (see Ref. 2.28 and 2.29): This is a cast-in-place plastic designed for use in filling mismatches between metal or composite parts. It can be used at any location to produce custom molded mating surfaces. Examples of applications for molded shim are shown in Fig. 5.3.40. Ground rules for use of moldable shim are given below:
- Operating temperature region is generally -65 to $400°F$
- Compression or bearing stress $> 10,000$ psi
- Long time oxidation resistance
- Resistant to aircraft fluids
- Cure shrinkage less than 1%
- Good adhesion to aluminum, titanium and steel allows shear strength > 1000 psi (69 MPa)
- Low coefficient of thermal expansion
- Low temperature, low pressure cure capabilities
- Minimum pot life of four hours with flow characteristics suitable for use with injection guns
- The degradation of fastener shear strength with increasing shim thickness is very gradual; there is no appreciable loss where shim thickness is one half the fastener diameter or less

Fig. 5.2.41 shows a list of properties designed to help the engineer choose the shim material and catalyst system best suited for a given application.

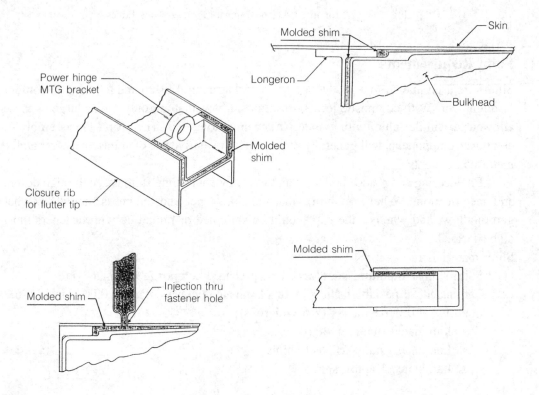

Fig. 5.2.40 Moldable Shim Applications

Joining

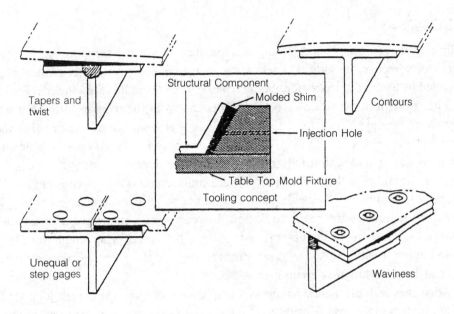

Fig. 5.2.40 Moldable Shim Applications (cont'd)

Product designation	Application time at 75°F ±5°	Assembly time at 75°F ±5°	Drill time	Cure time	Viscoslty cps (Bass & Catalyst) at 75°F
DMS-4-828	1 HR.	2 HR.	4-5 HR. at 75°F ±5°	8-9 HR. at 75°F ±5°	55,000
DMS-4-828B	2 HR.	4 HR.	9-10 HR. at 75°F ±5° or 7 HR. at 100°F ±5°	36 HR. at 75°F ±5° or 2-3 HR. at 120°F ±5°	55,000
DMS-4-828L	3-4 HR.	7-8 HR.	36 HR. at 75°F ±5° or 14 HR. 100°F ±5°	72 HR. at 75°F ±5° or 1 HR. at 180°F	55,000
MS-26A	40 Min.	2 HR.	7 HR. at 75°F ±5° or 5 HR. at 100°F ±5°	9 HR. at 75°F ±5°	220,000
MS-26B	1.5 HR.	4 HR.	20 HR at 75°F 5° or 9 HR. at 100°F ±5°	65 HR. at 75°F ±5° or 1 HR. at 180°F	220,000
MS-26	5-6 HR.	7-8 HR.	80 HR at 75°F ±5° or 20 HR. at 100°F ±5°	100 HR. at 75° ±5° or 2 HR. at 180°F ±5°	220,000

By courtesy of Dynamold, Inc.

Fig. 5.2.41 Moldable Shim Material

329

5.3 BONDING

Bonding of discrete panels and joining of separately fabricated components are standard practice for composite structures. There are numerous applications where laminated panels are bonded by a cocuring operation. Some typical examples of this are shown in Fig. 5.3.1.

The strength of the adhesive bond is independent of the overlap for all but impractically short overlaps. Therefore, the overlap is set at a sufficiently high value that the adhesive in the middle of the overlap will not creep under the worst of circumstances. Shorter overlaps could permit failure by creep-rupture; longer overlaps do not add to the strength or durability of the bonded joint. These considerations are explained in Fig. 5.3.2. The lightly loaded trough in the diagram should not be looked upon as just a joint inefficiency to be eliminated by improved design. Its presence is vital to ensuring an adequate resistance to failure of the joint by creep-rupture. The total overlap length must be sufficient to ensure that the adhesive shear stress in the middle of the overlap is essentially zero or at least so low that creep cannot occur.

Adhesives with the highest strengths do not always produce the highest joint strengths. Ductility improves the load distribution by reducing stress peaking at the end of the joint. The relative comparison between a brittle and ductile adhesive is shown in Fig. 5.3.3.

Adhesive bonding is often the best way to permanently join composites to each other. Adhesives have many advantages over mechanical fasteners for composite assembly. They can often be of the same family of materials as the matrix, assuring compatibility. They resist corrosion, seal as well as join, and weigh less than fastened joining. They can create a strong joint, with stress distributed over large areas, and require no drilling.

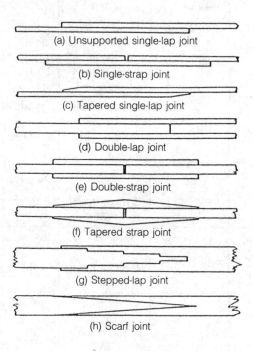

Fig. 5.3.1 Types of Bonded Joints

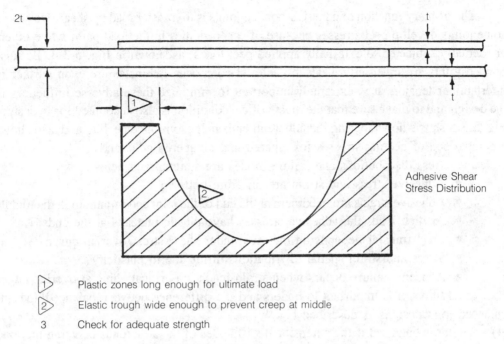

1. Plastic zones long enough for ultimate load
2. Elastic trough wide enough to prevent creep at middle
3. Check for adequate strength

Fig. 5.3.2 Elastic Trough of Double-lap Joint

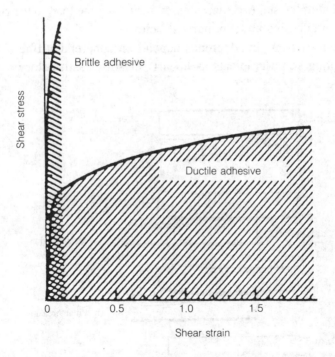

Fig. 5.3.3 Adhesive Shear Behavior Comparison

The primary function of adhesive-bonded joints is to transfer load by shear. However, some joints develop peel stresses because of eccentricities in the load path, while others are actually subjected to externally applied peel loads, as shown in Fig. 5.3.4. Because bonded joints are inherently weak in peel and composite laminates are even weaker in interlaminar tension, it is extremely important to minimize these adverse influences in the design and to make sure that the most critical condition is also accounted for in design. Fig. 5.3.5 shows how tapering the adherend ends helps to reduce peeling and also shows a comparison of peeling stresses for tapered and untapered adherends.

Adhesive stress distributions and failure modes are summarized below:
- Adhesive stress and strain are far from uniform
- Adhesive strains are maximum at the end of the joint and minimum in the middle
- At high loads the adhesive stresses have plastic plateaus at the ends
- Load transfer depends on the area under shear stress/overlap curve
- Stress and strain spatial distributions differ due to plasticity
- Adhesive failure occurs when strain reaches a critical value at overlap edges

In addition, it is important to understand the difference between balanced and unbalanced joint designs as described below:

(1) A joint is balanced if the extensional stiffnesses of the adherends carrying the load in the two directions are equal. Fig. 5.3.6 shows a double-lap joint in which the outer two adherends transmitting the load to the left have a combined membrane stiffness equal to that of the single center adherend carrying the load to the right. Shear stress behavior in adhesive are summarized below:
- Strain gradients in adherends depend on adherend stiffness
- Treating strain gradients as linearly elastic is satisfactory

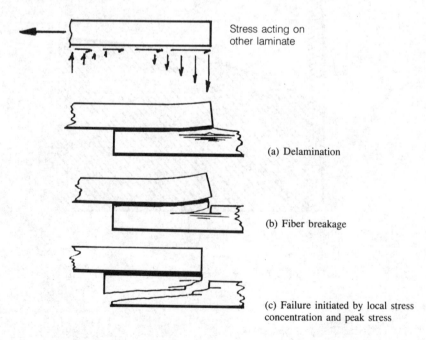

Fig. 5.3.4 Peel Stress Failure of Thick Composite Joints

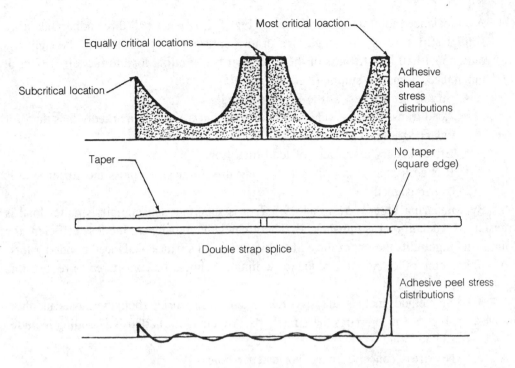

Fig. 5.3.5 Adhesive Stress Distribution in Tapered Vs. Untapered Bonded Joints

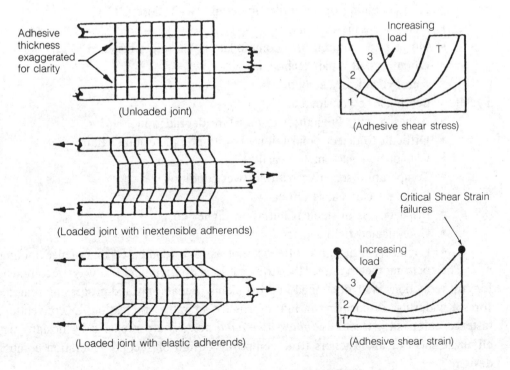

Fig. 5.3.6 Shear Stress/Strain Behavior in Balanced Bonded Joints

(2) As unbalanced joint design is shown in Fig. 5.3.7, where all three adherends have similar stiffnesses so the single one in the center carrying the load to the right has only about half the stiffness of the outer pair transmitting load to the left. Effect of unbalanced joint are summarized below:
- Strains are highest at the least stiff end
- Load transfer (area under stress/overlap curve) can be markedly less than for balanced joints
- Effect ocuurs regardless of load direction
- It is advisable to consider thickening the softer side unless the stiffer side is over-designed

By comparing Figures 5.3.6 and 5.3.7, it is seen that substantially smaller load is transferred when a joint is unbalanced. The greater the unbalance, the less effficient the joint. This highlights the importance of trying to design stiffness-balanced bonded joints. Local thickening of thinner adherends (or laminates) should be considered to relieve this problem.

Bonding methods include adhesive, dual resin, and cocuring (both thermoset and thermoplastic); Fig. 5.3.8 compares these methods. Advantages of adhesive bonding include:
- Lighter joint
- No stress concentrations (no fastener holes)
- Optimizes composite joint strength
- Can join thin sheets
- Allows joining of dissimilar materials (with close CTE)
- Acts as a joint sealant
- No pad-up required to accommodate holes (e.g., fastened joint)
- Often a good repair technique
- Aerodynamically smooth

Factors which must be considered:
- Adhesive may be subject to moisture degradation
- Difficult to detect bond failure before complete joint failure
- Cannot be disassembled easily
- Temperature sensitive with hot/wet conditions
- Surface treatment is critical
- Inspection is difficult resulting in higher costs
- Complicated field repairs
- Engineering design confidence not as high (compared to fastened joining)

The cocuring method gives the strongest joint strength and is always recommended for cost reduction. With continued improvements in materials and processing, engineers foresee wider use of cocuring in future primary airframe applications as conventionally fastened composite structures (*black aluminum design*) are replaced by designs which eliminate most or all fasteners (true composite design or second generation composite design).

One of the main shortcomings when using either adhesive bonding or cocuring for aircraft manufacturing is that to date there is no means of testing nondestructively to verify the strength of a bond.

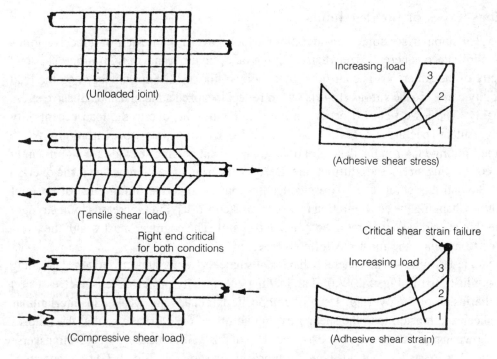

Fig. 5.3.7 Shear Stress/Strain Behavior in Unbalanced Bonded Joints

Bonding methods	Pros	Cons
Adhesive	• low temp processing • full faying surface bond	• poor hot-wet properties* • requires special surface preparation • properties less than welding* • requires bond fixtures
Dual resin bonding	• high properties • lower temp than co-consolidation • no size limitations • less surface preparation than adhesive bonding • full surface joint • minimal tooling	• requires special surfaces • requires fully consolidated preforms • must incorporate resin interlayer at initial part forming
Cocuring or Co-consolidation	• high properties • no size limitations • full surface joint	• requires extensive tooling and entire structure remelted • expensive

* There are not general conclusions

Fig. 5.3.8 Bonding Methods Comparisons

Various Types of Bonded Joints

Fig. 5.3.9 summarizes some considerations which govern the design of adhesive joints and the influence adherend size when selecting on optimum joint configuration. The load capacity of the adherends is controlled by adherend thickness. Superimposed on the load capability curve are the varous cut-offs which reflect localized adherend and adhesive stress criticality. The lowest cut-off occurs in the single-lap joint, due to the load eccentricity which results in peeling stresses as well as adhesive shearing stress concentrations. While still the limiting factor in the design of double-lap joints, peeling stresses are significantly reduced. Peeling stresses are further alleviated by tapering the adherends near the overlap ends. Beyond this point, shear-stress concentrations become the limiting factor on joint strength. Changing the configuration to a scarf or stepped-lap joint increases joint strength further but such designs are more complex and costly. The engineer must weigh the alternatives to evaluate the most cost-effective design.

(1) Single-lap joint — The general behavior of single-lap joints subjected to tensile loading is illustrated in Fig. 5.3.9(a). The loading eccentricity produces tensile stresses on the inside of the overlap. Usually composite laminates are weaker in interlaminar shear and tension than the shear carrying ability of the adhesive so that the failures shown in Fig. 5.3.10 are to be expected. Use of this joint at locations where a transverse member exists to react this bending moment, as shown in Fig. 5.3.11, is one solution to this problem. When the support structure is sufficiently stiff in bending to react much of the moment due to joint eccentricity, the shear stress distribution is simplified.

Single-lap joints usually require greater overlap lengths to alleviate adherend bending stresses associated with the eccentricities in the load path.

(2) Double-lap joint (or double-strap joint) — This joint [see Fig. 5.3.9(b)] is suitable for thin to moderately thick laminates. Actually, the shear load transfer is accomplished through a narrow effective zone at the ends of the overlap. The shear stress distribution in the middle of the adhesive carries the same small load regardless of the length of the total overlap. This joint is also highly tolerant of manufacturing imperfections throughout its structurally efficient range. As shown in both Fig. 5.3.6 and Fig. 5.3.7, the strength of double-lap joint is affected adversely by the presence of a laminate stiffness unbalance in the joint.

(3) Scarf joint — The shear stress in the adhesive of a scarf joint, shown in Fig. 5.3.9(e), which is subject to membrane applied loads, is virtually constant. This means that a sufficiently small scarf angle can be selected to provide 100% joint efficiency (the joint adhesive strength equals the laminate failure load). To insure that laminate failure occurs outside the joint (for optimum strength) the scarf angle should be between 6 to 10 degrees. At large angles, failure will occur prematurely. When extremely small scarf angles are used there is the danger of breaking off the tip of the stiffer laminate. Actual joint efficiencies are less than 100% because the manufacturing constraints often require angles greater than optimal.

Joining

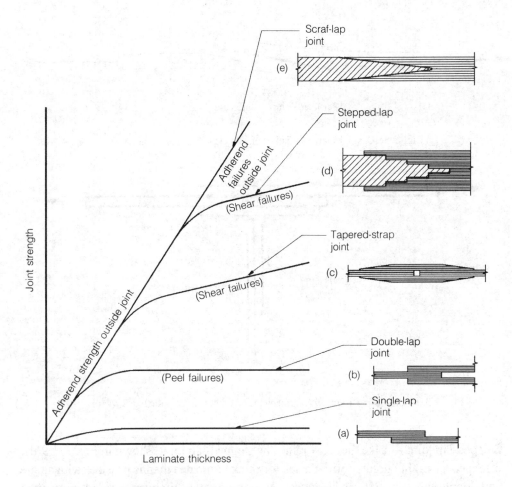

Fig. 5.3.9 Bonding Joint Configuration Vs. Strength

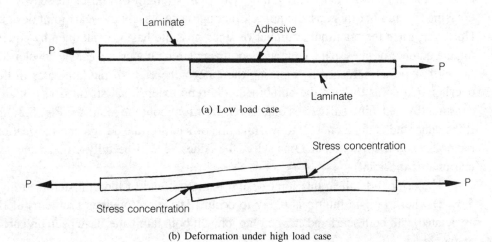

Fig. 5.3.10 Single-lap Bonded Joint Vs. Load Cases

(c) Interlaminar failure case

Fig. 5.3.10 Single-lap Bonded Joint Vs. Load Cases (cont'd)

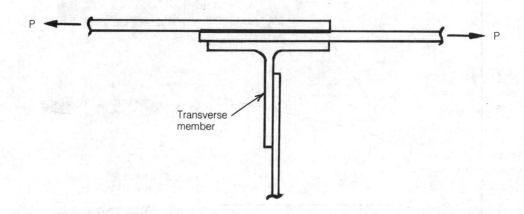

Fig. 5.3.11 Single-lap Joint with Transverse Member as Supporting Structure

(4) Stepped-lap joint — Like the scarf joint, the stepped-lap joint, shown in Fig. 5.3.9(d), offers considerably greater efficiencies for thick laminates than is possible with single and double-lap joints. It has been used for composite-to-titanium bonded joints with both carbon and boron epoxy laminates. This joint shares some of the characteristics of both the double-lap and scarf joints. The joint strength can be increased by increasing the number of steps and approximating more closely the scarf joint design. However, once the maximum number of steps possible has been attained by equaling the number of plies in the laminate, no further strength increase can be developed. Strength may be increased by tapering the outer adherends of the laminates in the overlap area so as to keep the combined adherend extensional stiffness essentially constant (by matching stiffnesses of adherends) throughout the joint; see Fig. 5.3.12. If the adherend is thick enough to warrant this procedure, the design and fabrication methods should allow for it. Otherwise, the adhesive will be fully utilized only at one end of the joint.

The critical detail in this joint design is the tail end of the thin step (see Fig. 5.3.13) where fatigue failure is likely to occur. There are obvious problems of fit associated with both scarf and stepped-lap joints if both laminates have been precured or machined.

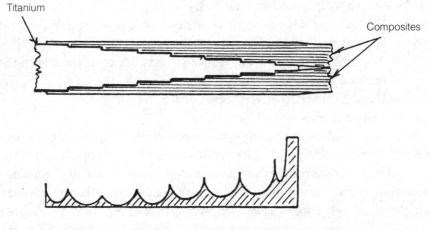

(a) Constant thickness joint (stiffness-unbalanced)

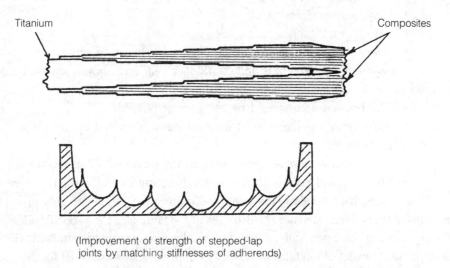

(Improvement of strength of stepped-lap joints by matching stiffnesses of adherends)

(b) Constant stiffness joint (stiffness-balanced)

Fig. 5.3.12 Adhesive Shear Stress Distribution of a Double Stepped-lap Joint

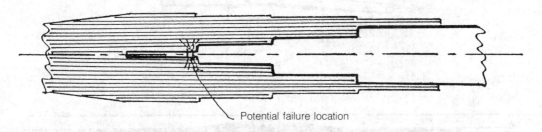

Fig. 5.3.13 Potential Failure Location with Too long an End step over a stepped-lap Joint

(5) Flush single strap joint (see Fig. 5.3.14):
- Variations of this configuration are often found where a flush surface is needed
- Highest peel stress of common joint types
- Advisable to make $t_s > t$ to reduce stresses due to moment even though it does increase eccentricity
- Increased overlap is very beneficial in reducing peel
- Laminate ends should be tapered

In summary, adhesive bonding is most appropriate for thin laminates while mechanic fasteners are more suitable for thick laminates (unless a stepped-lap joint is used).

Bonding of structures provides an alternative to mechanically fastening assemblies. Some of the factors affecting joint strength are assembly types, layup and type of adhesive. Generally, adhesively bonded assembled structures are lighter and have reduced manufacturing and maintenance costs over mechanically fastened structures. Some characteristics of secondary bonding are:
- Requires excellent fit-up
- Requires additional cure cycles
- Allows pre-cured parts to be inspected prior to assembly

For maximum effectiveness and confidence, adhesive bonds should be designed in accordance with following general principles:
- The bonded area should be as large as possible
- A maximum percentage of the bonded area should contribute to the strength of the joint
- The adhesive should be stressed in the direction of its maximum strength
- Stress should be minimized in the direction in which the adhesive is weakest

The strength of bonded single-lap and double-lap joints depends primarily on the overlap length and the extensional stiffnesses of the laminates for a specific adhesive system. Theory and test have generally shown that the highest strength is attained when the value of Et [laminate material modulus of elasticity (E) times its thickness (t)] for the two laminates are equal to one another.

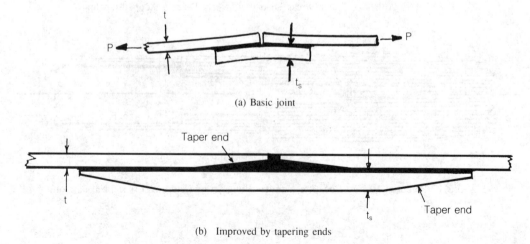

(a) Basic joint

(b) Improved by tapering ends

Fig. 5.3.14 Flush Single Strap Banded Joints

Surface Treatment

In the fabrication of composite laminates, there are many potential sources of contamination which could have a deleterious effect on the adhesion, including dust, dirt, fingerprints, lubricants, release agents, etc. The commonly used release agents, which prevent the composite part from adhering to the tool surface, may contain small amounts of organic contaminants which can transfer to the part surface. Solvent wiped on and left to evaporate is not generally satisfactory due to solid residues which are left on the surface. Good shop practice employs a "two hands" method: one hand to apply the solvent and the other to remove soon after application. Two methods of surface treatment are described below:

(1) Mechanical abrasion — Use sanding or grit blasting to prepare small areas for secondary bonding or for field repairs. The choice is generally determined by available production facilities and cost. However, careful control must be maintained to prevent damage to the laminate. Sanding is generally the most suitable surface treatment for field repairs. It is used both to remove any organic finishes that are present and to prepare the surface. Keep in mind that paint removers should not be used on composites because these attack and weaken the laminates.

(2) Peel ply — A heavy fabric (peel ply) can be applied during the original layup of the laminate. The peel ply layer is removed just prior to bonding. Peel ply cloth that has been coated with release agents may cause poor adhesion and should not be used. Pre-impregnated peel ply has proven to be the most satisfactory surface treatment for bonding large areas during manufacture. In addition, the texture of the peel ply surface gives the bond a mechanical grip and provides surface consistency (superior to any other method). Two types of Kevlar peel ply are available, regular and calendered, and the calendered is the preferred material as it is less porous. Other peel plies include nylon fabrics and Teflon coated glass fabrics.

Bonded Joint Design Guidelines

(a) Joints with short overlap lengths are more efficient in avoiding adhesive shear failure from peak shear stress (see Fig. 5.3.15)

(b) To minimize peel stress consider the following:
- In double-lap joints peel stress is a function of outer laminate bending stiffness
- In unsupported single-lap joints, peel stress is an inverse function of overlap
- In eccentric joints, much larger overlaps are needed for shear strength are often required to allow alleviation of eccentricities by gentle deformations
- Taper ends [see Fig. 5.3.16(a)] to a thickness of about 0.03 inch (0.76 mm) and a slope of 1/10
- Thicken adhesives at adherend ends as shown in Fig. 5.3.16(b)
- If peel stresses are critical in a joint, redesign it to alleviate this problem

(c) Reduce joint eccentricity:
- Thick laminates are affected more than thin laminates
- Double shear joints reduce peeling effect
- Chamfer or taper laminate thickness to reduce peeling

(d) Do not use 90° plies on outer surfaces of the laminate (use ±45° or 0° plies)

Chapter 5.0

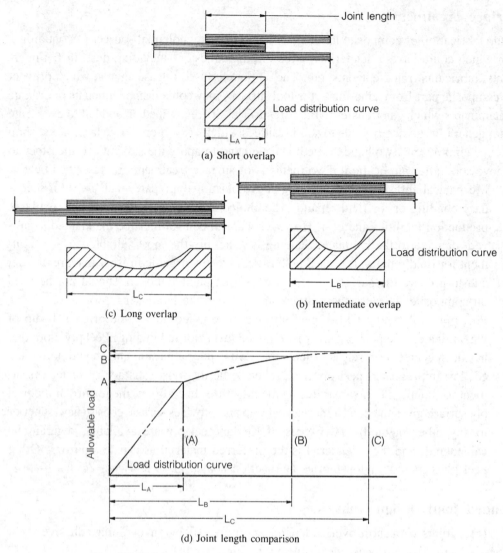

Fig. 5.3.15 Efficient Bonded Joint with Short Overlap Length

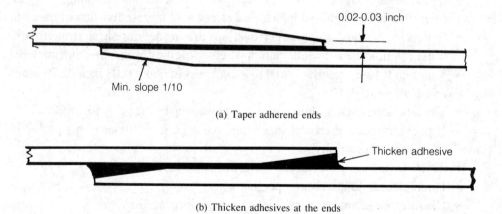

Fig. 5.3.16 Taper Ends to Reduce Stress Concentrations

(e) Thermal stresses are induced between bonded the laminate materials. Such stresses are characteristically negligible at the cure temperature but become progressively worse in proportion to the square of the temperature differential between curing and operating temperatures. However, at the bonded joint between laminates of composite materials and metals such as titanium attention should be paid to the thermal expansion difference, especially in high temperature applications. In most cases, steel or titanium may be acceptably adhesive-bonded with either carbon or boron composites. Minimize metal-to-composite thermal mismatch by judicious selection of the laminate ply orientation and of the bond joint configuration.

(f) Because adhesive bonding is a surface phenomenon, surface preparation is one of the keys to successful bonding. The durability of an adhesive bond is very much dependent on the surface treatment of the laminates.

(g) Laminate stiffness is most important and should be maximized to reduce adhesive stresses

(h) Environmentally induced degradation may be particularly significant at high temperatures and must be evaluated experimentally

(i) All tooling used in bonding operations must be carefully designed to minimize thermal distortions and residual stresses in the joint

(j) Film adhesives are recommended to ensure repeatability and fabrication control of the bonding operation

(k) Thermosetting resins are commonly used for adhesive bonding of composite structures, e.g., epoxies, polyesters, phenolics, polyimides, silicones, etc. The most commonly used is epoxy, which is often modified with additives to provide high-strength bonds, and good resistance to the environment.

(l) Paste adhesives are often used where pressure cannot be applied.

(m) Environmental effects
- Corrosion protection needed for metal part surfaces
- If bonding a carbon laminate to aluminum, use an intermediate fiberglass ply to prevent galvanic corrosion.
- Hot-wet or cold-dry conditions may be the most critical for composite bonded joints
- Dry out damaged areas prior to bonded repairs

(n) Very little strength increase is added to joints with bonded overlap lengths greater than 1.0 inch (2.54 cm).

(o) Attention should be given to the local tolerance build-ups. Bondline thickness should be controlled by the tooling and uniform bonding pressure must be precise

(p) Bonded joint testing is vital and should consider environmental sensitivity

(q) Bonded joints must always be designed to be stronger than the adjacent structure — otherwise the bond can act as a weak-link fuse

(r) When there is no need for repair (e.g., missiles and unmanned aircraft) bonded joints can be designed which permit extremely high structural efficiencies to be obtained.

Chapter 5.0

(s) Design of simple, uniformly thick, nearly quasi-isotropic carbon laminate, bonded splices should use the following guidelines:
   • 80t overlap in single-lap shear
   • 30t overlap in double shear
   • 1/50 slope for scarf joint
(t) Select the most ductile adhesive which meets the design requirements
(u) Give consideration to repair techniques
(v) Beware of problem when bonding an angle-stiffener to a skin, see Fig. 5.3.17
(w) Tension loading of T-stiffener bonded to a skin, see Fig. 5.3.18:
   • The failure mode is often in peel
   • Taper the ends of the T-flanges
   • The problem is much more complex when the skin (thin) buckles and/or the T-stiffener rotates when subjected to loading

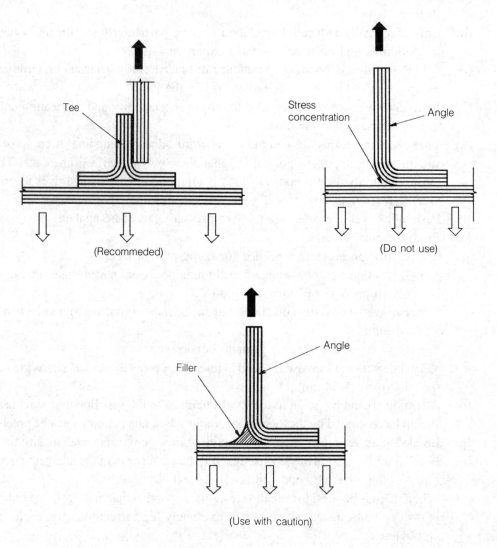

Fig. 5.3.17 Angle Stiffener Bonded to a Skin

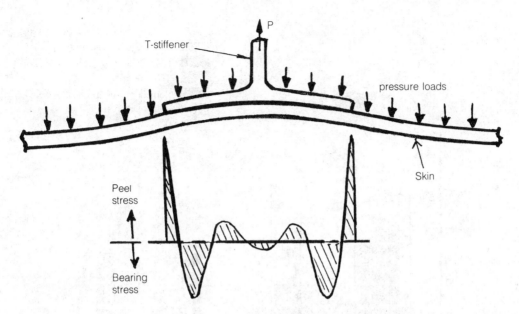

Fig. 5.3.18 Typical Stress Distribution under th Flanges of T-Stiffener

Dual-resin Bonding

Bonded joints requires tedious surface treatment and most adhesives tend to lose a high percentage of bond strength after being subjected to environmental exposure. However, with the increasing use of thermoplastics in composite structures comes some new and improved methods of bonding. Since thermoplastic composite materials soften or melt at high temperatures, parts can be co-consolidated and fused together with a sound bonded joint.

An alternative method to co-consolidation is dual resin bonding (or amorphous bonding), a technique which has acquired a wide interest in the composites industry. In order to maximize the blending/fusion of the bonding surfaces, a layer of amorphous bonding resin agent is incorporated into these surfaces during the laminate consolidation process. This bonding resin agent has to be chemically compatible to the bonded laminate's resin and capable of diffusing into the surface layer to achieve optimum bonding strength. It is also necessary that the glass transition temperature (T_g) of the bonding resin agent be below the melting point of the laminate resin to minimize distortion of the fibers. Results indicate that the dual resin bonded joint has strength comparable to co-consolidated joint and in addition exhibits superior hot/wet properties, as shown in Fig. 5.3.19. The results also indicate that most bonded joints with a bondline thickness less than 0.01 inch (0.254 mm) exhibit higher bond strength.

Advantages of dual-resin bonding include:
- Precise tooling is not required to maintain the size and shape of the bonded laminates
- Less surface treatment is required than is needed for conventional adhesive bonding
- processing at lower temperatures than co-consolidation still provides compatible joint strength

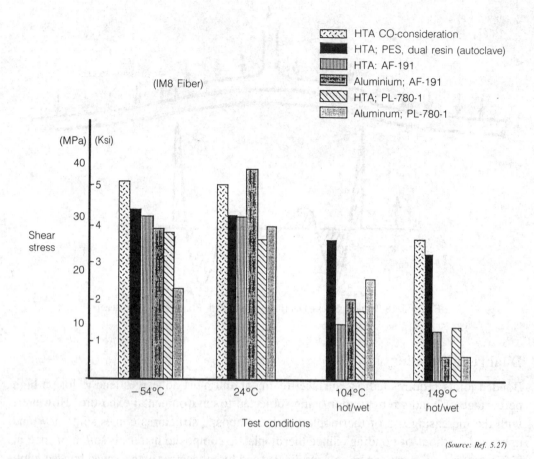

Fig. 5.3.19 Dual-resin Bonding Allowable Comparison

Cocuring

One of the most attractive features of composite materials is that sections can be molded into complex shapes. Thus, panels that normally would have to be fabricated individually can often be cocured into single parts to eliminate the need for fasteners entirely. Cocuring (cocuring is used for thermosets and co-consolidation for thermoplastics; but as used herein, cocuring will refer to both) can be defined as the curing of a laminate with stiffeners and attachments all in one process operation. Structures which integrally stiffen laminates, such as stringers of Z-sections, J-sections, hat-sections, and even skin covers, ribs and spars are prime candidates for cocuring. There are many advantages to cocuring parts in airframe structures; in addition to the integrity of a one-piece construction, there are weight savings and most important is the savings in the cost of assembly labor. A typical commercial transport for example, may have as many as two million different types of fasteners. Thus, the elimination of fasteners, which includes fastener cost and installation cost, can result in substantial savings.

Various combinations of resins can be cocured, but it is important to match the cure cycles, especially the temperature and speed of reaction. How the two resins react with each other must be examined, and they should be compatible, because the prepreg resin serves as the adhesive. Design considerations for cocuring include:

(a) Cocuring is inherently more complex, so there is a greater probability of a bad layup, with significantly more at risk.
(b) Misalignment of parts is possible, so tooling to hold the part properly becomes very important. As the component gets more complex, the parts are more difficult to keep in their proper place, therefore tooling needs to be more innovative.
(c) Part shrikage during cooling must be accounted for. The tooling should have the same coefficient of thermal expansion (CTE) as the part. Hence, composite tooling is preferred, especially for large parts.
(d) In cocuring, some areas are difficult to inspect because they are hidden.
(e) Controlling dimensions in certain areas may be more difficult when cocuring
(f) If the part is too thick, applied heat combined with the heat generated from the chemical reaction (for themosets) may damage the part. Therefore, the cure cycle should be adjusted to meet this phenomenon.

Rivet-bonded Joint

A rivet-bond, as shown in Fig. 5.3.20, clamp together components by means of small, permanently installed, lightweight mechanical fasteners before the components are bonded together. Once the structure is bonded, the fasteners are no longer needed but are usually kept in place, both to furnish added assurance against disbonds and in order to preserve smooth contours, which would otherwise be disrupted by holes.

Composite to Metal splice Joining

(1) Bonding composites to metals (consider thermal expansion):
- Titanium (preferred)
- Steel (acceptable)
- Aluminum (not recommended) — composite to aluminum (corrosion resistant aluminum) splice joints may be used under special circumstances
(2) Bonded step joints are preferred to scarf joints — more consistent results, design flexibility and lower cost for thicker laminate joints.
(3) Where possible, 0° plies (primary load direction) should be placed adjacent to the bondline; ±45° plies are also acceptable. 90° plies should never be placed adjacent to the bondline unless it is also the primary load direction.
(4) For a stepped joint, the metal thickness at the end step should be thicken enough to prevent metal failure (see Fig. 5.3.13).
(5) If possible, have ±45° plies end on first and last step of bonded step joints to reduce peak interlaminar shear stresses at the end steps.
(6) If possible, do not end too many of 0° plies on any step surface to avoid peak stress.

(7) Examples of stepped-lap joints which have been designed to hold the composite laminate wings and tail on titanium root fittings for modern fighter aircraft are shown in Fig. 5.3.21.

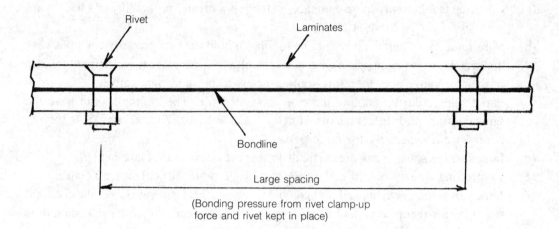

Fig. 5.3.20 Rivet-bond Method

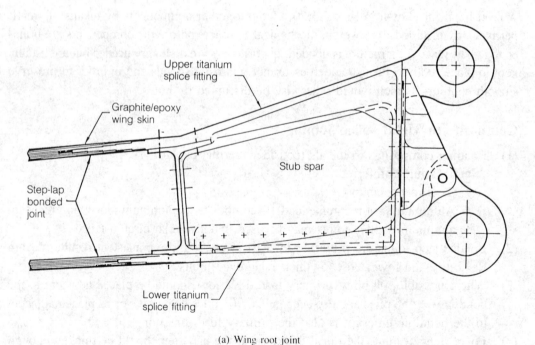

(a) Wing root joint

Fig. 5.3.21 Stepped-lap Joint Applications on Fighter Aircraft

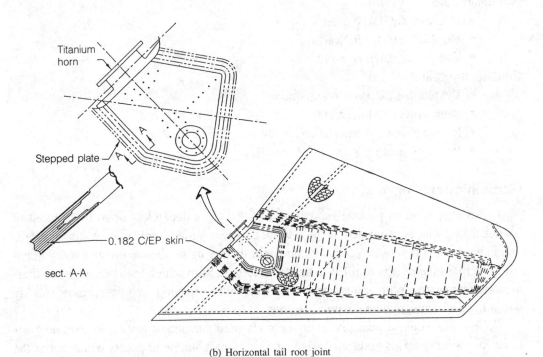

(b) Horizontal tail root joint

Fig. 5.3.21 Stepped-lap Joint Applications on Fighter Aircraft

5.4 WELDING

A composite welding process is possible only with thermoplastics and utilizes heat energy to melt the faying surface resin to accomplish the bonding process.

Resistance Welding

Resistance welding, as shown in Fig. 5.4.1, is a developmental joint technique which heat and melts the thermoplastic part at the faying surface by passing electric current through a resistive element which is placed at the interface. The welding process is described below:

(a) Place high resistance wires in the joint which act as heating elements when connected to a power supply (once the joint has consolidated, the wires remained trapped). The bond strength can be improved at the interface by placing a thermoplastic amorphous film which easily wets the bonding composite prepreg and produces a uniform bond line.

(b) Carbon fibers can replace the wires as the heating element to take advantage of the semi-conductive nature of carbon.

(c) Relatively simple and fast.

Advantages are:
- Potential for large area weld
- No thickness limitations
- Full faying surface weld

Considerations are:
- Developmental joint technique
- High power requirements
- Requires special surface treatment
- Requires heating element at interface

Ultrasonic Welding

This ultrasonic welding process, shown in Fig. 5.4.2, is dependent upon the amount of frictional heat that is generated at the interface of two joined laminates. Welded surface should have small bumps or points molded onto the faying surface to create energy intensifiers to focus the energy applied from the ultrasonic horn which melts through the thermoplastic laminates. With applied pressure on the molten material and the surfaces solidify into a high strength molecular bonded joint.

It is good to use a peel ply to create a textured surface as a surface treatment for ultrasonic welding. This texture consists of many small bumps or points which act as the energy directors or intensifiers required for ultrasonic welding. This treatment is necessary to achieve consistent high-quality welds and is only needed for one faying surface.

Ultrasonic welding has been used to produce high joint strength in some applications. However, ultrasonic welding does not lend itself well to continuous processing such as line welds. Coupling the ultrasonic energy into the composites while moving the ultrasonic head over the surface of the composites is very difficult to achieve. As such ultrasonic welding is limited to spot welding. Ultrasonic spot welding technology is well developed and has been used to assemble thermoplastic parts (see Fig. 4.6.7 in Chapter 4.0).

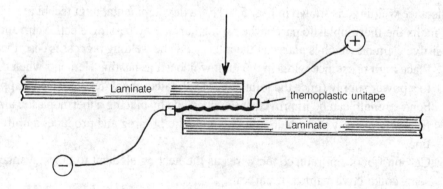

Fig. 5.4.1 Resistance Welding

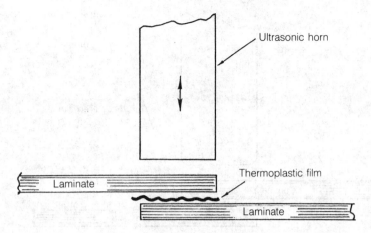

Fig. 5.4.2 Ultrasonic Welding

Advantages are:
- High strength properties
- Rapid cycles

Considerations are:
- Spot welding only
- Requires special surface treatment
- Must support back side
- Require locating tooling

Induction Welding

Induction heating, as shown in Fig. 5.4.3, is a unique process in that it is a non-contact form of heating that operates at a very high speed. Heating occurs when electrically conductive and magnetic materials are exposed to alternating magnetic fields.

Process considerations are:

(a) High temperatures can be achieved in the bondline in very short times and the area that is heated can be localized to the area of the joint
(b) Heating is a non-contact process which improves heating efficiency and repeatability since the heat transfer is not dependent on surface contact
(c) High speed temperature control is possible with induction heating. Since heating occurs immediately when the magnetic field is applied there is no thermal lag associated with induction heating.

Advantages are:
- Only single side access
- No problem on curved or complexly shaped surface
- Portable equipment available

Considerations are:
- Thickness limited
- Metallic susceptor needed in interface
- High risk

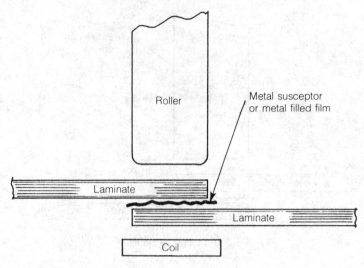

Fig. 5.4.3 Induction Welding

Vibration Welding

The vibration welding process (heat come from surface friction) is well developed for unreinforced thermoplastic materials and the technique has been used for welding reinforced thermoplastic composites. This method is fast and produces stronger lap shear joints. This method works well for unidirectional laminates, but cross-ply laminates are more difficult to weld since the fibers in adjacent plies may cut each other when they vibrate, decreasing the strength of the joint.

Vibration welding is well suited to situations where two small parts are being welded together or when a small part is being welded to a large part. The large part would remain stationary in the fixture while the small part is vibrated to make the weld. Advantages are:
- Rapid cycles
- Full faying surface weld

Considerations are:
- Requires locating tooling
- Equipment size limitation
- Both sides access required
- Requires special surface treatment
- Part size limitation — not suitable for large parts

Focused Infrared Heat Welding

Infrared heating is due to electromagnetic waves being absorbed and dissipated by the material surface. In this method, the interface to be welded is sufficiently heated through exposure to an intense narrow beam line of infrared heat without thermally damaging the welded thermoplastic laminates. Once they are molten, the laminates are pressed together until cooled down. This process focuses heat only at the bondline and has the capability to be automated in production assembly.

Advantages are:
- High strength properties
- Rapid cycles
- Full faying surface weld

Considerations are:
- Elaborate setup
- High risk

5.5 SEWING AND WET-CRUSH-RIVET

Sewing is an in-process assembly method is in its infancy. The process has already resulted in greater structural values at stiffener-to-skin joints (see aslo Fig. 2.5.8, in Chapter 2) and even on primary joints for business and smaller aircraft airframes. A typical sewing joint method is shown in Fig. 5.5.1.

Fig. 5.5.2 illustrates a wet-crush-rivet joint method which uses a segment cut from a long bundle of fibers wetted with thermoset resin (e.g., epoxy).

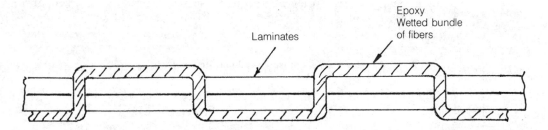

(a) Sew wetted bundle into joined laminates

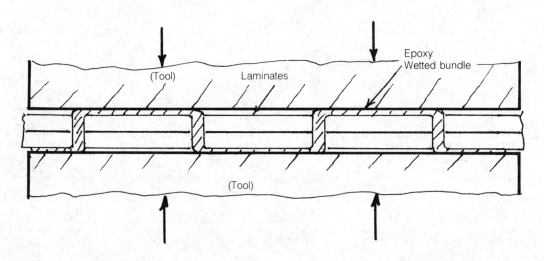

(b) Apply pressure with tools (heat optional)

Fig. 5.5.1 Hand Sewing Joint Method

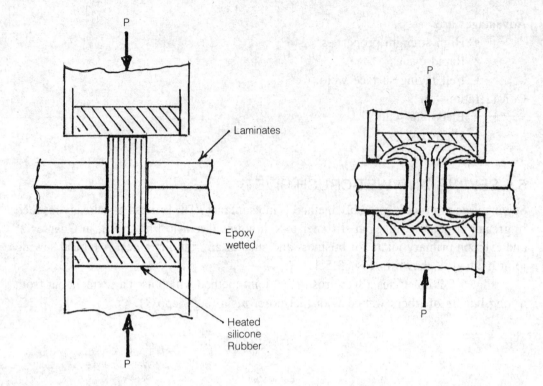

Fig. 5.5.2 Wet-crush-rivet Joint Method

References

5.1 Cosenza, F., "Mechanical Fasteners for Composites", MATERIALS ENGINEERING, Aug. 1987. pp. 33-37.

5.2 Phillips, J.L., "Manual and Automated Fastening of Composites Structures", SME paper No. AD85-07. 1985.

5.3 Anon., "The Government/Industry Bridfing on Manufacturing Technology for Low Cost Composite Fasteners" AFML Contract No. F33615-77-C-5050. 1979.

5.4 Hart-Smith, L. J., "Mechanically-fastenered Joints for Advanced Composites-Phenomenological Considerations and Simple Analysis", (Douglas Paper No. 6748A) 4th Conference on Fibrous Composites in Structrual Design, San Diego, CA. 14-17 Nov., 1978.

5.5 Phillips J. K., "Fastening Composite Structures with Huck Fasteners-Part 1", AIRCRAFT ENGINEERING, Nov., 1986. pp. 2-6.

5.6 Phillips, J. L., "Fastening Composite Structures with Huck Fasteners-Part 2", AIRCRAFT ENGINEERING, Dec., 1986. pp. 10-15.

5.7 Marsh, G., "Getting It Together", AEROSPACE COMPOSITES & MATERIALS, 1990.

5.8 Maguire, D. M., "Joining Thermolplastic Composites", SAMPE Journal, Vol. 25, No. 1, Jan/Feb, 1989 pp. 11-14.

5.9 Lewis, C. F., "Getting a Quick Fix From "Adhesives". Materials Engineering. Oct., 1988. pp. 57-60.

5.10 Anon., "ADHESIVES AGE", monthly published by Communication Channels, Inc.. 6255 Barfield Road, Atlanta, GA 30328.

5.11 Hart-Smith L. J., "Adhesive-Bonded Double Lap Joints". NASA CR112235, 1973.

5.12 Hart-Smith, L. J., "Adhesive-Bonded Single Lap Joints". NASA CR112236, 1973.

5.13 Hart-Smith, L. J., "Adhesive-Bonded Scarf Lap Joints". NASA CR112237, 1973.

5.14 Klein, A. J., "Cocuring Composites", ADVANCED COMPOSITES, Jan/Feb, 1988. pp. 43-53.

5.15 Anon., "Will the Joint Hold?", AEROSPACE ENGINEERING, Jan., 1987. pp. 10-14.

5.16 Petrie, E. M., "Adhesively Bonding Plastics: Meeting an Industry Challenge", ADHESIVE AGE, May 15, 1989. pp. 613.

5.17 Ziesman, W. A., "Influence of Constitution on Adhesion", HANDBOOK OF ADHESIVES, Van Nostrand Reinhold Co., 1977.

5.18 Landrock, A. H., "Adhesives Technology Handbook", Noyes Publications, Park Ridge, NJ 1985.

5.19 Hodges, W. T., "Bonding and Nondestructive Evaluation of Graphite/PEEK Compsite and Titanium Adherends with Thermoplastic Adhesives", SME Paper No. MF85-511, 1985.

5.20 Holmquist, H. W. and Lantz, R. G., "Advanced Composites Surface Preparation", AIRLINER, Apr-Jun, 1986. pp. 13-17.

5.21 Allbee, N., "Adhesives for Structural Applications", ADVANCED COMPOSITES, Nov/Dec, 1989. pp. 42-50.

5.22 Dreger, D. R., "Design Guidelines for Joining Advanced Composites", MACHINE DESIGN, May 8, 1980. pp. 89-93.

5.23 Anon., "Joining of Advanced Composites", ENGINEERING DESIGN HANDBOOK published by Army Material Development and Readiness Command, Alexandria, VA

5.24 Hart-Smith, L. J., "Specimen Shows Thicknesses Where Adhesive will Not Fail", ADHESIVE AGE, April, 1987. pp. 28-32.

5.25 Silverman, E. M. and Griese, R. A., "Joining Methods for Graphite/PEEK Thermoplastic Composites", SAMPE Journal, Vol. 25, NO. 5. Sept/Oct, 1989. pp. 34-38.

5.26 Cogswell, F. N., et. al., "Thermoplastic Interlayer Bonding", 34th International SAMPE Symposium, May, 1989. p. 2315.

5.27 Wu, S. Y., "Dual Resin Bonding of Thermoplastic Composites", 36th International SAMPE Symposium, April 1991.

5.28 Anon., "Moldable Shim Material", Dynamold, Inc., 2905 Shamrock Ave. Fort Worth, TX 76107.

5.29 Anon., "EA9394 and EA9396 Data Package — Structural Paste Adhesives", HYSOL Aerospace & Industrial Products Division, Pittsburg, CA. 94565, 1988.

5.30 Niu, C. Y., "Airframe Structural Design', Conmilit Press Ltd., 101 King's Road, North Point, Hong Kong. 1988.

5.31 Landrock, A. H., "ADHESIVES TECHNOLOGY HANDBOOK", published by Noyes Publications. 1985.

5.32 Anon., "ADHESIVES — ADHESIVES, SEALENTS & PRIMERS", co-published by D.A.T.A. Inc. and The International Plastics Selector Inc. 1986.

5.33 Benatar, A. and Gutowski, T. G., "A Review of Methods for Fusion Bonding Thermoplastic Composites", SAMPE Journal, Jan/Feb, 1987. pp. 33-38.

5.34 Niu, C. Y., "L-1011 Fastener Handbook", Internal Lockheed publication No. CER51-013, Lockheed Aeronautical Systems Co. 1973.

5.35 Stevens, T., "Joining Advanced Thermoplastic Composites", MATERIALS ENGINEERING, Mar., 1990. pp. 41-45.

5.36 Anon. "Engineering Materials Handbook, Vol. 1 — Composites", ASM International, Metals Park, Ohio 44073. pp. 681-728. 1987

Chapter 6.0

ENVIRONMENTS

6.1 INTRODUCTION

Composites for airframe structures must be designed to withstand the great diversity of terrestrial environments encountered in a variety of operations. Environmental effects, including combinations of heat, cold, moisture, lightning strikes, ultraviolet (UV) radiation, fluids and fuels, can reduce mechanical properties to varying degrees, depending on the composite system.

A typical environmental effect is the hot/wet condition in which the elasticity and strength design allowables of the composite may reduce as much as 10% to 20% (50% reduction occur in some wet-layup materials). The composite matrix generally is the component most vulnerable to environmental effects. Consequently matrix dominated properties such as loss of compressive strength are of greatest concern.

Another environmental concern is lightning strike (which occur once per year on the average for commercial aircraft; less frequently for military aircraft) — catastrophic failure of the aircraft structures can result.

The most common environmental hazards are listed below:
- Exposure to moisture
- Elevated temperatures
- Cold temperatures
- Lightning strikes and Electromagnetic interference (EMI)
- Sunlight — ultraviolet radiation
- Erosion of material
- Solvent and chemicals
- Ballistic impact
- Acoustic exposure
- Nuclear exposure

This chapter describes the main environmental hazards and problems peculiar to composite airframe structures, and recommends protective measures.

6.2 WEATHERING EFFECTS

In the terrestrial environment the combined effects of temperature and humidity must be considered when assessing long-term structural integrity.

Moisture/Temperature

The amount of moisture that composite materials absorb is a function of matrix and fiber type, time, component geometry, temperature, relative humidity, and exposure conditions. The absorption of moisture by the organic matrix of composites is an important factor at high temperatures. The moisture diffuses into the matrix, in both laminates and adhesives, causing them to swell and acts as a plasticizer or softener. The latter phenomenon is by far the most important for airframe applications and results in a decrease of the matrix glass transition temperature (Tg). This manifests itself in a decrease of matrix-dominated properties at high temperatures. For example, compression strength is clearly reduced, but tension behavior is relatively unaffected, at the upper temperature ranges of the material's usefulness.

Composite mechanical properties are dependent upon the amount of water (moisture) in the matrix and the temperature. The phenomenon may include some or all of the following:
- Composite matrix absorbs moisture until an equilibrium (or saturation) point is attained (see Fig. 6.2.1)
- Moisture lowers the glass transition temperature (Tg)
- Generally,
  - 350°F composites absorb more moisture than 250°F composites
  - Thermosets absorb 1% to 2% moisture
  - Thermoplastics absorb 0.1% to 0.3% moisture
- Hot/wet conditions cause the matrix to become more plastic
- Cold/dry conditions cause the matrix to become more brittle
- Moisture desorption gradients may induce microcracking as the surface desorbs and shrinks, putting the surface in tension. If the residual tension stress at the surface is beyond the strength of the matrix, cracks occur
- Cyclical swelling and contracting due to recurring moisture exposure may lead to joint loosening (a problem which is more critical with thermosets than thermoplastics)
- Lower design allowables or a protection system are required for composites which will be exposed to, if temperatures higher than normal design service temperatures since the material mechanical properties degrade after a very short time.
- To minimize moisture ingress:
  - Seal machined edges and surfaces of laminates
  - Provide surface paint or protective coating. However, paint will not prevent moisture from diffusing into the composite matrix

Some composite laminates may need to be protected in order to be resistant to fluids such as moisture, fuel, hydraulic fluid and anti-icing fluid. The following precautions should be taken for composite laminates:
- When a part is not painted, a thin film of thermoset or thermoplastic resin should be applied to the surface to reduce porosity (preventing ingress of moisture through surface cracks)
- Sealant should be applied when surfaces are exposed to weathering to reduce the effects of static rain, air oxidation, airblown sand, and ultraviolet radiation

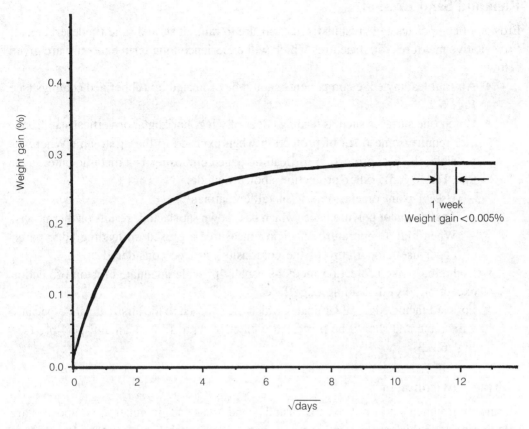

Fig. 6.2.1 Typical Equilibrium Moisture Content of a Thermoplastic

Ultraviolet Radiation

- Ultraviolet (UV) can deteriorate composite material integrity
- Carbon, glass and boron are impervious to UV, but Kevlar 49 is degraded by UV
- Presence of temperature/moisture with UV can magnifies degradation of material
- Conventional aircraft paint provides protection against UV

Hail and Foreign Objects Damage

The leading edge components of wings and empennage are vulnerable to rain, hail and foreign object damage. Complete impact protection is difficult to achieve and the equivalent thickness of composite part should be equal or near to of aluminium counterpart [e.g., 0.06 inch (1.52 mm) for large transport and will be less for small aircraft]. Fig. 6.2.2 shows the area on the leading edge (between points A and B) that is most susceptible to rain or hail damage.

Occasionally control surfaces such as ailerons, elevators, and rudders shows evidence of hailstorm including actual damage, puncture of the skin. The need for protection of control surfaces should not be overlooked.

Chapter 6.0

Rain and Sand Erosion

Erosion of composites edge results from exposure to rain, dust, and sand (in desert areas). Proventative measures for structures which will experience long term exposure are given below:

- Aircraft leading edge components should be protected by rubberized coatings (see Fig. 6.2.3)
- Composite surfaces such as leading edges of wing, horizontal or vertical stabilizers, etc. require some form of protection when exposed to the airstream. Wherever possible, the best protection for leading edge composites is a metallic outer layer (see Fig. 6.3.5); other protection choice include:
  — Use of paint or elastomeric/metallic coatings
  — Application of polyurethane (which has shown satisfactory results on thermosets)
  — When high temperature anti-icing measures are used on leading edge parts, upper use temperature of the composite must be considered
- Lightning strike protection methods usually provide adequate erosion protection except on aircraft leading edge faces
- Forward facing edges of laminates which are exposed to the airstream are especially vulnerable and should be protected with edge wrap film or an anti-peel ply (see Fig. 6.2.3)

Marine Environment

Marine environments occur at sea coasts and island areas and, in general, composites are able to withstand this extreme environment beyond the capability of metals. But this environment does attack areas of secondary adhesive bondlines (cold bonding) and mechanically metal fastened joints to weaken composite strength. The marine environment is composed of:

- High humidity
- Highly corrosive salty moisture

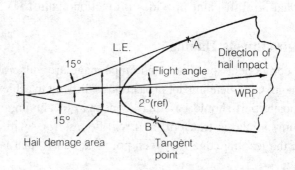

Fig. 6.2.2 Possible Hail Damage Area on The Leading Side Between A and B

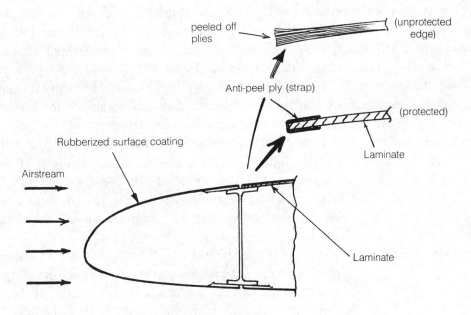

Fig. 6.2.3 Protection of Leading Edge Surfaces

6.3 LIGHTNING STRIKES

There is increased use of composite materials to replace aluminum in primary airframe structures. But composite materials are either not conductive at all or are significantly less conductive than aluminum. Unless protected, composites suffer more damage due to lightning strikes (see Fig. 6.3.1) than counterpart aluminum structures. Also, composite materials allow significant portions of lightning current to flow into onboard systems (e.g., electrical wiring, hydraulic lines, fuel and vent tubes, etc.) and provide less shielding of onboard electronic systems from lightning electromagnetic fields than to metal structures.

Airframe structures require lightning strike protection on exterior or mold line surfaces of the aircraft as defined in the specification listed in Fig. 6.3.2. Aircraft protruding tips, leading edges and trailing edges are the exterior mold line surfaces most likely to be primary lightning strike zones; other airfoil surfaces are secondary strike zones. Both must be conductive to facilitate lightning streamering and the dissipation of static electricity to ground or to static dischargers.

An idealized concept of aircraft lightning protection would be to have the entire exterior surface highly conductive and electrically continuous. This is impossible because of windshield transparency, radome, antenna, and other requirements. In addition, there are weight penalties associated with the idealized protection system. One compromise to idealized lightning protection is to have the exterior surface conductive to a degree that is consistent with system requirements and safe operation. The fuel system require special design attention.

Ref. 6.19 provides information and guidance concerning an acceptable means, but not the only means, of compliance with parts 23, 25, 27, and 29 of the U.S. Federal Aviation Regulations (FAR). Ref. 6.19 describes methods as applicable of preventing the hazardous effects due to lightning, from affecting electrical/electronic systems which perform critical/essential functions. In lieu of following the suggested methods, the applicant may elect to establish an alternative method of compliance that is acceptable to the FAA.

Terms commonly used to describe lightning strikes are given below:
- Attachment Point — A point of contact between the lightning flash and the aircraft.
- Lightning Flash — The total lightning event in which charge is transferred from one charge center to another. It may occur within a cloud, between clouds, or between a cloud and ground. It can consist of or more strikes, plus intermediate or continuing currents.
- Lightning Leader — The leader is the preliminary breakdown that forms an ionized path for charge to be channeled towards the opposite charge center. The "stepped" leader advances in a series of short, luminous steps prior to the first return stroke. The "dart" leader reionizes the return stroke path in one luminous step prior to each subsequent return stroke in the lightning flash.
- Lightning Strike — Any attachment of the lightning flash to the aircraft.
- Lightning Stroke (Return Stroke) — A lightning current surge that occurs when the lightning leader makes contact with the ground or another charge.
- Swept Stroke — A series of successive attachments due to sweeping of the flash across the surface of the airplane by the motion of the airplane.

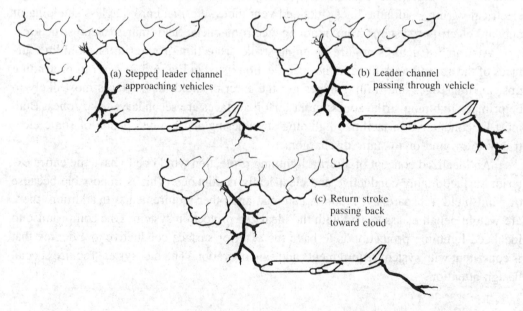

Fig. 6.3.1 Lighting Flash Striking on Aircraft

The aircraft can be divided into three lightning zones (Ref. 6.19) as shown in Fig. 6.3.3. These zones are defined as follows:
- Zone 1 — Surface of the vehicle for which there is a high probability of direct lightning flash attachment or exit
    - Zone 1A — Initial attachment point with low probability of flash hang-on, such as a nose
    - Zone 1B — Initial attachment point with high probability of flash hang-on, such as a tail cone
- Zone 2 — Surface of the vehicle across which there is a high probability of a lightning flash being swept by the airflow from a Zone 1 point of direct flash attachments
    - Zone 2A — A swept-stroke zone with low probability of flash hang-on, such as a wing mid-span
    - Zone 2B — A swept-stroke zone with high probability of flash hang-on, such as a wing trailing edges
- Zone 3 — Zone 3 includes all of the vehicle areas other than those covered by Zone 1 and Zone 2 regions. In Zone 3 there is a low probability of any direct attachment of the lightning flash arc, but Zone 3 areas may carry substantial amounts of electrical current by direct conduction between some pairs of direct or swept-stroke attachment points in other zones.

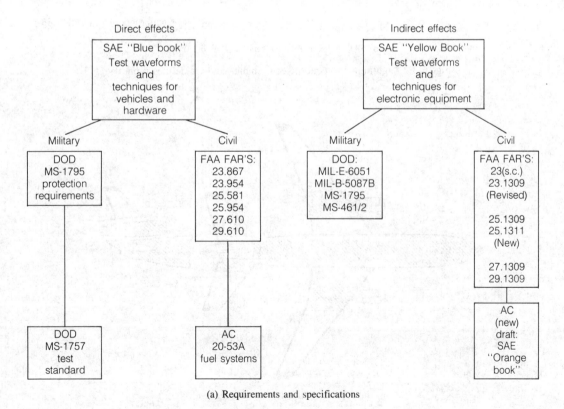

(a) Requirements and specifications

Fig. 6.3.2 Lightning Protection Requirements and Guidelines

Chapter 6.0

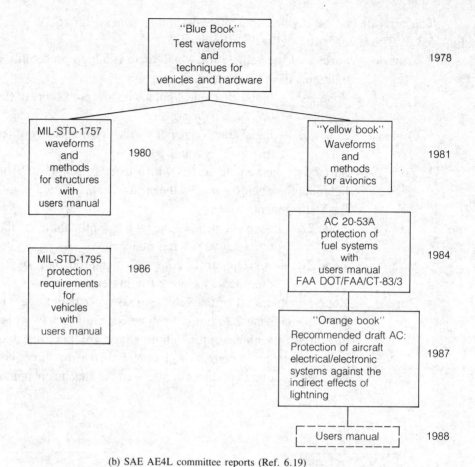

(b) SAE AE4L committee reports (Ref. 6.19)

Fig. 6.3.2 Lightning Protection Requirements and Guidelines (cont'd)

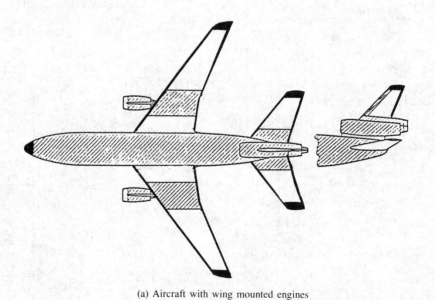

(a) Aircraft with wing mounted engines

Fig. 6.3.3 Lightning Strike Zones — Transport

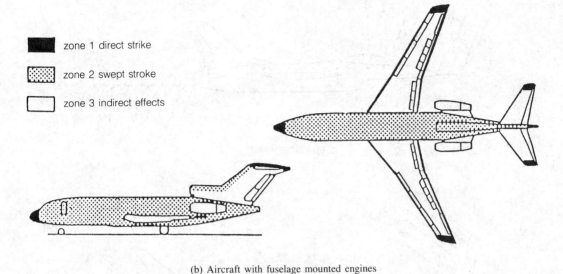

(b) Aircraft with fuselage mounted engines

Fig. 6.3.3 Lightning Strike Zones — Transports (cont'd)

Aircraft extremities, such as nose, tail cone, wing tips, empennage tips, and engine nacelles are susceptible to lightning attachment. Surfaces aft of the lightning strike attachment points may be contacted by swept strikes as shown in Fig. 6.3.4 and Fig. 6.3.5.

Lightning effects can be divided into direct effects and indirect effects:

- Direct Effects — Any physical damage to the aircraft and/or electrical/electronic systems due to the direct attachment of the lightning channel. This includes tearing, bending, burning, vaporization or blasting or aircraft surfaces/structures and damage to electrical/electronic systems.
- Indirect Effects — Voltage and/or current transients induced by lightning in aircraft electrical wiring which can produce upset and/or damage to components within electrical/electronic systems.

Different components often require different lightning protection methods depending on location of strike (the type of lightning strike zone), material, structural configuration, and interfaces. However, the extent and type of damage frequently can only be determined by testing.

The lightning protection system selected for use in composite applications must satisfy the following requirements:

(1) The system should withstand the mechanical forces involved in dissipating high electrical energy (lightning) loads and provide sufficient conductive surface to substructure continuity for safety-of-flight protection from the electrical wave forms.
(2) Neither the protective system nor its application process should detract from composite material properties.
(3) The system should permit the dissipation and flow of static electricity to a substructure ground or toward static discharges (pigtails, shown in Fig. 6.3.6) and should provide adequate shielding from electromagnetic interference (EMI).

Chapter 6.0

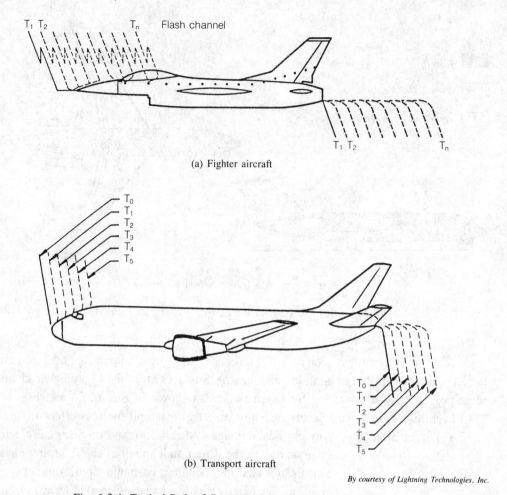

(a) Fighter aircraft

(b) Transport aircraft

By courtesy of Lightning Technologies, Inc.

Fig. 6.3.4. Typical Path of Swept-stroke with Attachment Points

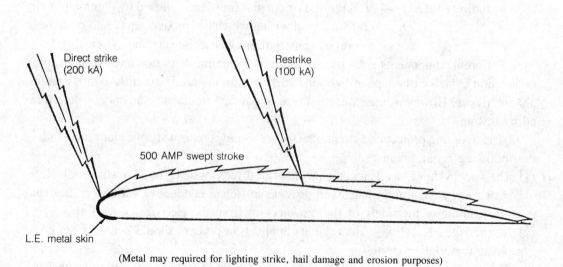

(Metal may required for lighting strike, hail damage and erosion purposes)

Fig. 6.3.5 Lightning Strike on Wing Surface

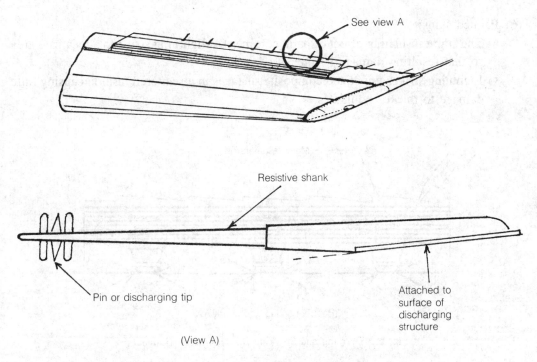

Fig. 6.3.6 Pigtail Precipation Static Discharge on Trailing Edge of Control Surface

(4) The system's conductivity characteristics and the electrical grounding joint should not significantly degrade with time or operational environmental exposure.
(5) The protective surface material system should be repairable, considering flight and ground service exposure conditions, and require a minimum of maintenance.

Areas which may require protection are:
(1) Non-conductive composites (e.g., fiberglass, Quartz, Kevlar, etc.):
   • Do not conduct electricity
   • Puncture danger when not protected
(2) Advanced composite skins and structures:
   • Generally non-conductive except for carbon reinforced composites
   • Carbon fiber laminates have some electrical conductivity, but still have puncture danger for skin thickness less than 0.15 inch (3.81 mm)
(3) Adhesive bonded joints:
   • Usually do not conduct electricity
   • Arcing of lightning in or around adhesive and resultant pressure may cause disbonding
(4) Anti-corrosion finishes:
   • Most of them are non-conductive
   • Alodine finishes, while less durable, do conduct electricity
(5) Fastened joints:
   • External fastener heads attract lightning
   • Usually the main path of lightning trasmission between components
   • Even use of primers and wet sealants will not prevent transfer of electric current from hardware to structure

(6) Painted skins:
- The slight insulating effect of paint confines the lightning strike to a localized area so that resulting damage is intensified
- Lightning strikes unpainted composite surfaces in a scattered fashion causing little damage to thicker laminates

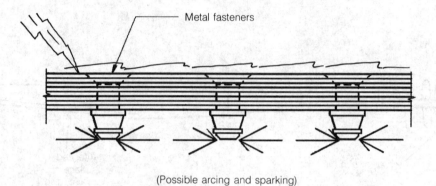

Fig. 6.3.7 Arcing and Sparking Between Metal Fasteners

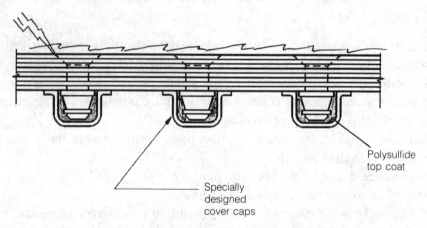

(a) Use plastic cover caps to prevent arcing

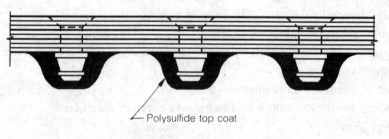

(b) Use top coat sealant

Fig. 6.3.8 Methods of Preventing Arcing and Sparking of Metal Fasteners in Carbon Composites

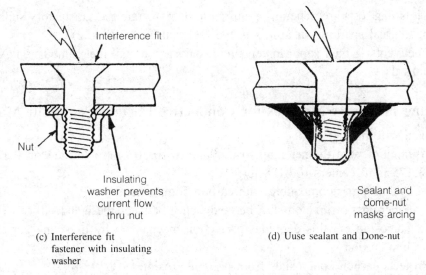

(c) Interference fit fastener with insulating washer

(d) Uuse sealant and Done-nut

Fig. 6.3.8 Methods of Preventing Arcing and Sparking of Metal Fasteners in Carbon Composites (cont'd)

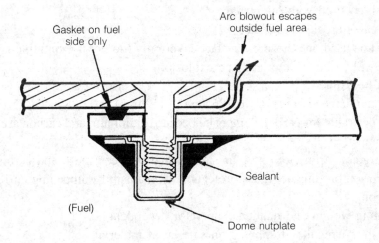

Fig. 6.3.9 Lightning Protection of Fuel Tank Access Door Seal

(7) Integral fuel tank:
- Dangers are melt-through of metal fasteners or arc plasma blow by between fasteners and the resulting combustion (see Fig. 6.3.7); some preventative measures are shown in Fig. 6.3.8
- Lightning protection for access doors is shown in Fig. 6.3.9
- Wing fuel tank skins will provide adequate protection against ignition of fuel vapors, if skin thicknesses are electrically equivalent to an aluminum thickness of 0.08 inch (2.032mm) (swept stroke studies indicate that 0.08 inch aluminum thickness would be adequate for conventional airframes)
- Consider use of bladders in aircraft wings which carry fuel to eliminate possibility of metal fastener arcing
- Use composite material fasteners

- Lightning tests are generally required to demonstrate the satisfactory suppression of internal sparking or arcing in the fuel tank area (Zone 2)
- Fuel vents — Fuel vapor in vents represents a potential fuel combustion explosion hazard

Lightning Protective Methods for Conductive Carbon Composite Materials

(1) Woven wire mesh:
  - Aluminum wire diameter up to 0.008 inch (0.203 mm) located on 0.125 inch (3.175 mm) centers (0.0128 lb/ft^2)
    — Aluminum incompatible with carbon composites
    — Fiberglass scrim cloth may be sandwiched between mesh and carbon to prevent corrosion but this degrades protection mechanism by impeding current flow into carbon
  - Protects carbon composites from particle erosion
  - Weight penalty
  - Difficult to install on compound contoured surfaces
  - Mesh cocured with laminate
  - Increases erosion resistance
  - Must be used on outermost surface layer only (do not use multiple layers because this will result in interlaminate explosion)
  - Can be painted

(2) Aluminum flame or arc spray [0.006 inch (0.152 mm) to 0.008 inch (0.203 mm) thick]:
  - Not desirable for carbon composite because aluminum and carbon are galvanically incompatible
  - Fiberglass scrim cloth may be sandwiched between mesh and carbon to prevent corrosion but this degrades protection mechanism by impeding current flow into carbon
  - Coating weight and quality is operator-dependent
  - Finish is irregular if sprayed on surface of material
  - Crazing may occur when applied to large areas
  - Can be painted

(3) Nickel coated fibers:
  - Nickel is electrodeposited on graphite filament as shown in Fig. 6.3.10
  - Lower conductivity than aluminum
  - Minimal weight increase
  - Substituted for one layer of material in layup
  - Can only be applied at surface layer
  - May be painted
  - Small loss of strength
  - Expensive

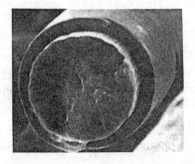

(a) 0.5 Micron thick nickel layer on about a 7 micron graphite fiber

(b) Filament-wound composite fuel tank is protected against lighting strikes by nickel-coated graphite fibers

By courtesy of American Cyanamid Co.

Fig. 6.3.10 Nickel Coated Fibers

(4) Aluminum interwoven wires [0.004 inch (0.102 mm) wires and 8 wires/inch]:
- Wires must be imbedded in matrix and kept dry, otherwise possible galvanic corrosion between wires and carbon composites
- Does not protect against indirect effects
- Outermost ply only
- Works well under painted surfaces

(5) Conductive paints:
   - High metal (copper or silver) content necessary for adequate conductivity
   - Cannot handle the higher levels of electric currrent
   - Works on the principle of diverting flash current over material surface
   - Cannot be used in proximity to other conductive elements
   - Can be painted
   - Least effective method
   - Should not be used for conductive composites (e.g., carbon, boron, etc.) because it causes greater damage
(6) Aluminum bar diverter with use fasteners:
   - Bar diverters may be used to transfer lightning current across Zone 3 regions and they must be properly located to avoid sharp bends of corners, which may be damaged by the magnetic forces during the transfer of lightning current. Fig. 6.3.11 shows a example of bar diverter application
   - Able to withstand first return stroke
   - Reduced efficiency when painted
(7) Conducting polymers (not commercialized) — Conducting polymers have been around since the mid-1970s and but are still in developmental stage. They have the potential to protect against lightning strikes and dissipate energy over the surface of the material, as shown in Fig. 6.3.12.

Lightning Protective Methods for Non-conductive Composite Materials

Non-conductive materials such as fiberglass, Kevlar, Quartz, etc. must be protected from electromagnetic waves (an indirect effect) produced by lightning strikes:

(1) For "transparent" applications (e.g., radome radar fairings, etc.) consider the following:
   (a) Aluminum bar diverter [diverter strips; 0.5 inch (12.7 mm) × 0.25 inch (6.35 mm)] and fasteners (#6-aluminum or #8-copper) with spacing 6 to 8 inch:
      - The nose radome (see Fig. 6.3.13) is not conductive and bar diverters used on the exterior surface of the radome tend to shield the interior components.
      - Reduced efficiency when painted
      - May hinder dispersion of electromagnetic waves, although weather radar is not affected
      - Commonly used on transport radomes
   (b) Segmented diverter (shown in Fig. 6.1.14)
      - Easily installed
      - Transparent to electromagnetic waves
      - Not as effective as aluminum bar diverter
      - Cannot be painted
      - Commonly used on radomes of fighter aircraft and some commercial aircraft use also
   (c) Dielectric film such as Lexan, Kapton, Ultem, etc.
      - Particle impact reduces insulating ability
      - Used on interior applications where not affected by particle impact

Environments

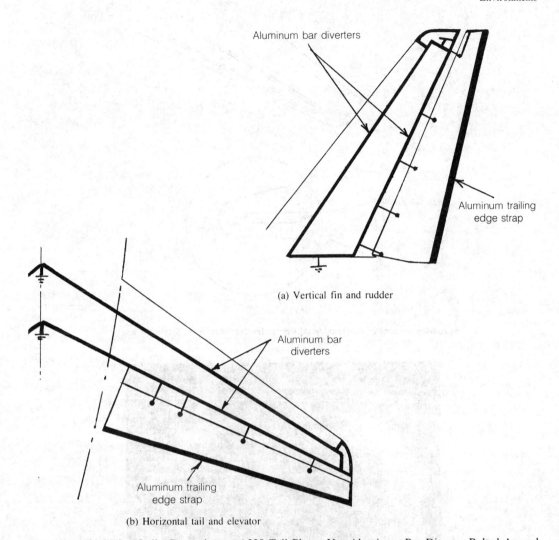

(a) Vertical fin and rudder

(b) Horizontal tail and elevator

Fig. 6.3.11 Lightning Strike Protection on A320 Tail Planes-Use Aluminum Bar Diverter Bolted Around The Composite Structures

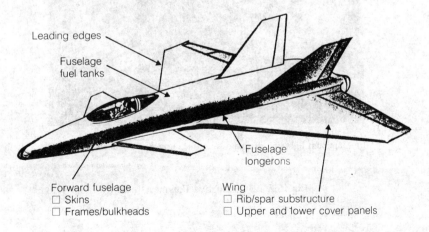

Fig. 6.3.12 Conductive Polymers

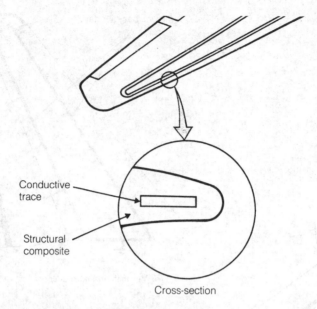

(a) Artist's concept of future advaced fighter features using conductive polymers

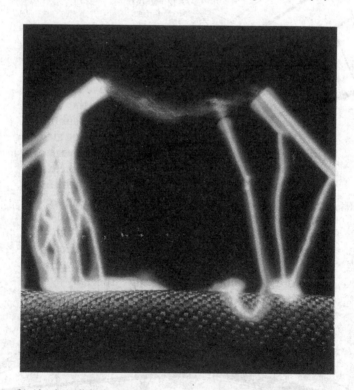

(b) Aluminum lightning dissipates over the surface of conductive polymers

By courtesy of Lockheed Aeronauticaly Systems Co.

Fig. 6.3.12 Conductive Polymers (cont'd)

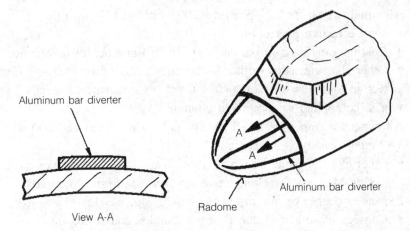

Fig. 6.3.13 Aluminum Bar Diverter Applied on Radome

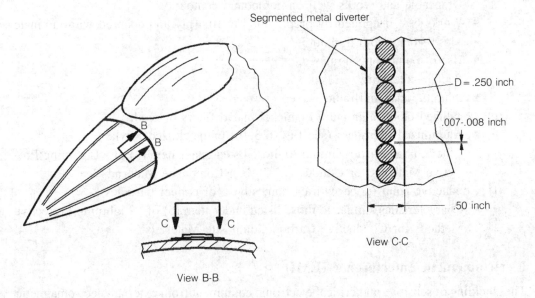

Fig. 6.3.14 Segmented Metal Diverter

(2) For "non-transparent" applications consider these protective measures:
   (a) Aluminum foil [0.005 inch (0.127 mm) thick]:
- Environmental seal of composite surface
- Uniform surface conductivity
- Surface material completely replaceable
- Foil stock width limiations
- Difficult to install on compound contours
- Poor repairability
- Poor handleability

(b) Aluminum flame or arc sprayed [0.005 inch (0.127 mm) thick]:
   - Very effective protection
   - Coating weight (approximately 0.1 lb/ft^2) and quality is operator-dependent
   - Other characteristics similar to those listed under item (2) of "Lightning Protective Methods for Conductive Carbon Composite Materials"
   - Can be applied to compound contours
(c) Aluminum woven wire mesh [0.004 inch (0.102 mm) and 200 wires/inch]:
   - Weight penalty is approximately 0.045 lb/ft^2
   - Other characteristics similar to those listed under item (1) of "Lightning Protective Methods for Conductive Carbon Composite Materials"
(d) Expanded copper or aluminum mesh (see Fig. 6.3.15):
   - Consists of solid foil that is slit, expanded and rolled flat
   - Due to long continuity, precludes sparking and gives much better conductivity than woven mesh
   - Drapeable and works well on compound contours
   - Works best if applied by prepregging to first ply and cocured with laminate
   - Mesh stock width limitations
   - Use of paint concentrates damage
   - Poor handleability
   - Inhibits radar and radio waves
(e) Metallized cloth (aluminum or nickel plated fibers or cloth):
   - Aluminized fiberglass (see Fig. 6.3.16) or metallized Kevlar
   - Other characteristics similar to those listed under item (3) of "Lightning Protective Methods for Conductive Carbon Composite Materials"
(f) Conductive paint (dope with carbon, silver, or copper particles):
   - Characteristics similar to those listed under item (5) of "Lightning Protective Methods for Conductive Carbon Composite Materials"

Electromagnetic Interference (EMI)

The shielding of sensitive and critical electronic equipment from external electromagnetic interference (EMI) is of vital importance in aerospace systems for safe flight operation (see Fig. 6.3.17). This is most directly accomplished by surrounding such equipment with an electrically conductive shell (metallic structure performs this function automatically).
- Organic composites (excluding metal matrix composites) are not adequate conductors or adequately grounded or interconnected to absorb by induction the incoming electromagnetic radiation
- Minimal conductive coatings, such as conductive paints, may not provide adequate EMI shielding
- Protective methods used to safeguard against lightning strikes usually serve as effective EMI shielding
- Two basic methods exist for protecting electronics from lightning:
  — Simply shield them with metal
  — Electrically isolate them from the induced charge

	Copper	Aluminum
Base material	110 Electrolytic Tough Pitch Annealed Copper	1145 Series Electrical Grade Annealed Aluminum
Conductivity vs. Silver Standard	101%	59%
Resistivity, micro ohm/cm	1.71	2.89
Minimum Elongation of Screen Material	100%	20%
Tensile Strength, ksi		20
Thickness, Inches Minimum Maximum	0.003 0.015	0.003 0.015
Minimum Weight lb/ft^2	0.03	0.015
Strand Width, Inches Minimum Maximum	0.003 0.020	0.003 0.020
Open Area, %	15-85	15-85
Roll Length, ft.	650	650
Std. Roll Widths, in.	13 & 31	13 & 31

(Source: Astroseal products Manufacturing Corp.)

Fig. 6.3.15 Lightning Strike Screen (non-woven) Materials

(unprotected)

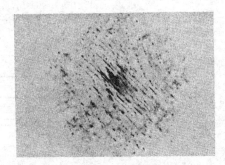

(protected)

Fig. 6.3.16 A Single Ply of Thorstrand Aluminized Glass Cloth Offers Significant Protection From Lightning

Chapter 6.0

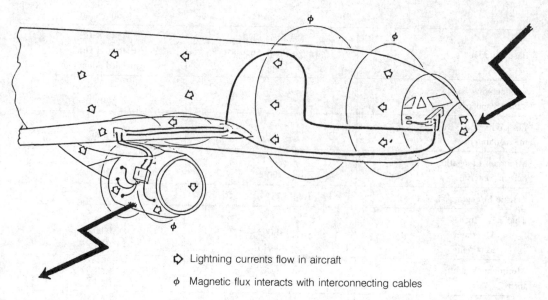

⇨ Lightning currents flow in aircraft
φ Magnetic flux interacts with interconnecting cables

By courtesy of Lightning Technologies, Inc.

Fig. 6.3.17 Lightning Intervaction with Electronic Systems

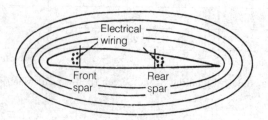

(a) Aluminum alloy leading edge & trailing edge

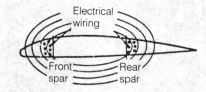

(b) Composite leading edge & trailing edge

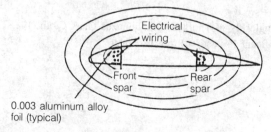

(c) Composite with aluminum alloy foil

Fig. 6.3.18 Composite Edge Panels Affect Electromagnetic Field

Composites (usually Fiberglass or Kevlar) are being used extensively for fixed leading and trailing edges (areas of minimum skin thickness). This presents a problem when electrical wiring needs to be installed in these areas. An electromagnetic field around a structure results, as shown in Fig. 6.3.18, from the flow of current from a lightning strike. If the leading and trailing edge are all-metal alloy, no problem exists. Lightning strike inducing higher voltage in the wiring may result in damage to engine instruments. Also, overloading wiring to fuel quantity probes in the integral tank may result in arcing to the structure. These problems can be solved by:
- Running wiring in conduits
- Putting fuel quantity probe wiring inside the integral tank
- Protecting the engine or control wiring by applying 0.003 inch (0.0762 mm) self-bonding aluminum alloy foil onto the composite skin panels to increase the electromagnetic field, as shown in Fig. 6.3.18(c).

6.4 GALVANIC CORROSION

The aircraft industry has vast amounts of data gained from history and research on corrosion to design out corrosion and to manufacture corrosion resistant airframes (see Ref. 6.9). This Section describes the most pronounced galvanic corrosion problems faced today and the means being employed to implement prevention into design and manufacturing practice.

Galvanic corrosion occurs where two materials from different groups in the galvanic series, as shown in Fig. 6.4.1, are in contact in the presence of moisture. This type of corrosion is usually accompanied by a buildup of corrosion products in the contact area. Corrosion progresses more rapidly the farther apart the materials are in the galvanic series. Material located toward the anodic end of the table will corrode sacrificially. For example, aluminum will corrode when in contact with carbon (or graphite) composites. This is especially a problem with the metallic fasteners generally used in assemblies (see Fig. 5.2.7 in Chapter 5).

Galvanic corrosion can be reduced by insulating the materials in the areas of contact and by applying protective coatings on both materials.
Recommendations are given below:
(1) Carbon composites in contact with metals
- Carbon will induce corrosion when in contact with aluminum when moisture exists, unless moisture intrusion is prevented
- Keep carbon composites out of electrical contact with any adjacent metals
- Carbon is highly cathodic and will severely attack aluminum and cadmium unless a protective ply of inert cloth (e.g., fiberglass, Kelvar, etc.) is used between them and/or protective paint or sealant is used
(2) If an aluminum part is to be used in contact with carbon components, all aluminum and carbon parts should be processed as follows prior to assembly:
  (a) Method A:
    - Anodize aluminum parts per MIL-A-8625, Type II
    - Finish external surfaces of both aluminum and carbon parts per MIL-F-18264:

— Two coats epoxy primer per MIL-P-85582
— Two coats white polyurethane enamel per MIL-C-83286
  (b) Method B:
  • Cure a layer of fiberglass ply to the composite interface
  • Seal edges of composites
(3) Boron composites in contact with metals
  • Corrosion does not occur when elemental Boron contacts metal
  • The Boron tungsten core fiber can be a corrosion cell if an electrical connection exists between the tungsten core and other metals such as a fasteners (anodic materials)
(4) Fasteners used with composites
  • Use fasteners of corrosive resistant materials (e.g., titanium, corrosion resistant steel, etc.) for carbon composites (see Fig. 5.2.7 in Chapter 5)
  • Do not use cadmium-plated fasteners in carbon composites
  • Install fasteners wet with corrosion inhibiting sealant

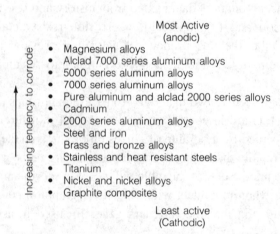

Fig. 6.4.1 Galvanic Series of Aircraft Materials

6.5 OTHER ENVIRONMENTAL CONCERNS

(1) Solvents and chemical resistance:
Composite properties may degraded by long time exposure to aircraft fluids/chemicals:
  • Application of flexible polyurethane paint is recommended
  • Aircraft fuel is absorbed by the composite matrix and causing weight gain and reduction of strength.
  • Hydraulic fluid does not usually affect most composites (Skydrol does)
  • Paint strippers will attack thermoset and thermoplastic composites
  • MEK cleaning fluid dissolves some thermoplastic composites
(2) Acoustic exposure:
When panel structures are located in the following areas, be aware that premature failure could be caused by exposure to acoustic environment:

- Cavity areas of bomb-bay, landing gear storage, etc.
- Near or close to engine areas
- Use sandwich or honeycomb panels
- Use bonded-joints instead of fasteners to eliminate stress concentration

(3) Nuclear exposure:

Protection from the following effects should be considered when aircraft structures are exposed to a nuclear explosion:
- Dynamic overpressure — a dynamic airload
- Thermal shock — could result in significant material degradation if panel surface has no protection or coating
- Nuclear radiation — in general, this does not cause the degradation of material for currently used composites; exposure to radiation should be considered in the development of future composite materials

References

6.1 Schweltzer, P.A., "CORROSION RESISTANCE TABLES", published by Marcel Dekker, Inc. 1986.

6.2 Boyer, H.E., "SELECTION OF MATERIALS FOR SERVICE ENVIRONMENTS", an ASM source book, published by ASM International, Metals Park, OH 44073. 1987.

6.3 Springer, G.S., "ENVIRONMENTAL EFFECTS ON COMPOSITE MATERIALS — Vol. 1", published by Technomic Publishing Co., Lancaster, PA. 1981.

6.4 Springer, G.S., "ENVIRONMENTAL EFFECTS ON COMPOSITE MATERIALS — Vol. 2", published by Technomic Publishing Co., Lancaster, PA. 1984.

6.5 Springer, G.S., "ENVIRONMENTAL EFFECTS ON COMPOSITE MATERIALS — Vol. 3", published by Technomic Publishing Co., Lancaster, PA. 1988.

6.6 Brick, R.O., "Multipath Lightning Protection for Composite Structure Integral Fuel Tank Design", presented at the 10th International Aerospace and Ground Conference on Lightning and Static Electricity, Paris, 1985.

6.7 Dexter, H.B., "Long-Term Environmental Effects and Flight Service Evaluation of Composite Materials", NASA TM-89067, National Aeronautics and Space Administration, Jan 1987.

6.8 Anon., "Conducting Polymers Open New Worlds", DESIGN NEWS, Jan 21, 1991. pp. 60-66.

6.9 NAVWEPS 01-1A-509, "CORROSION CONTROL FOR AIRCRAFT", published by bureau of Naval Weapons, USA.

6.10 Anon., "Protecting Electronics Against Lightning Gets Harder", AEROSPACE AMERICA, May 1986. pp. 30-34.

6.11 DeMels, R., "Lightning Protection for Aircraft Composites", AEROSPACE AMERICA, Oct 1984. pp. 62-65.

6.12 Kung, J.T., and Amason, M.P., "Lightning Conductive Characteristics of Graphite Composite Structures", 23rd National SAMPE Symposium and Exhibition, Vol. 23, May 2-4, 1978. pp. 1039-1053.

6.13 Clifford, F.L., "Materials keep a Low Profile", MATERIALS ENGINEERING, June 1988. pp. 37-41.

6.14 Anon., "Composite Materials in the Airbus", AIRCRAFT ENGINEERING, Dec 1989. pp. 20-29.

6.15 Fisher, F.A. and Plumer, J.A., "Lightning Protection of Aircraft", NASA RP 1008, Oct 1977. p. 53.

6.16 DOD-STD-1795 (USAF), "Military Standard Lightning Protection of Aerospace Vehicles and Hardware", May 1986.

6.17 MIL-STD-1757A, "Lightning Qualification Test Techniques for Aerospace Vehicles and Hardware", June 17, 1980.

6.18 AC 20-53, "Protection of Aircraft Fuel Systems Against Lightning", Federal Aviation Agency Advisory Circular, FAA, Dept. of Transportation, Washington, D.C. Oct 6, 1967. pp. 2-3.

6.19 SAE Committee AE4L (Orange Book), "Recommended Draft Advisory Circular-Protection of Aircraft Electrical/Electronic Systems Against the Indirect Effects of Lightning", Feb 4, 1987.

6.20 Luxon, B.A., "Metal Coated Graphite Fibers for Conductive Composites" Proceedings of the SPE 44th Annual Technical Conference & Exhibit, 1986.

6.21 Plumer, J.A., "Lightning Protection of Advanced Avionics Systems", paper from Lightning Technologies, Inc., 10 Downing Parkway, Piffsfield, MA 01201.

6.22 Plumer, J.A., "Protection of Aircraft Avionics from Lightning Indirect Effects", paper from Lightning Technologies, Inc., 10 Downing Parkway, Piffsfield, MA 01201.

Chapter 7.0

LAMINATE DESIGN PRACTICES

7.1 INTRODUCTION

Composite materials are uniaxial in their single-ply site state as shown in Fig. 7.1.1, having very high mechanical properties along their longitudinal axis, and low properties along their transverse axis (tape only). This is the primary difference, from a structural design and analysis standpoint, between composites and metals (see Fig. 7.1.2 and Fig. 7.1.3).

Metals are nearly homogeneous and isotropic in nature, and their reaction to an applied load can be defined by knowing two of the three basic elastic constants, the modulus of elasticity (E), the modulus of rigidity (G) and the Poisson's ratio (ν). On the other hand, a basic unidirectional lamina, or any balanced (implies that for every 45° ply, there exists a $-45°$ ply in the laminate) and symmetric (requires mirror image ply stacking about the midplane) laminate, is orthotropic in nature, having three mutually perpendicular planes of elastic symmetry. For planar applications, these types of materials can be defined by four of five basic elastic constants for orthotropic materials. It should be noted that there are twice as many independent planar elastic constants for orthotropic materials as there are for isotropic materials because of the different properties in the planes of symmetry (see Fig. 1.3.1).

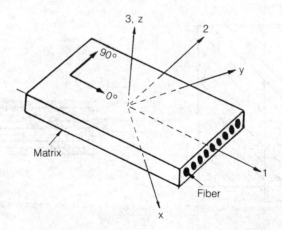

(Note: 1, 2, 3 are Lamina coordinates; x, y, z are laminate coordinates)

Fig. 7.1.1 Lamina Axes Rotation

Chapter 7.0

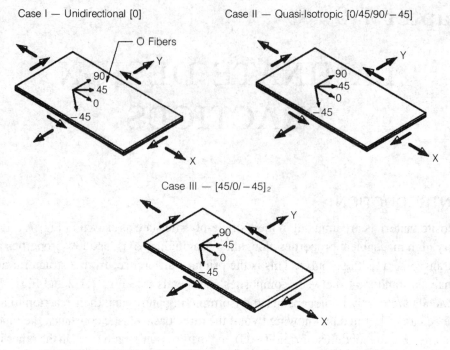

	F_X (Ksi)	E_X (Msi)	F_Y (Ksi)	E_Y (Msi)	F_{XY} (Ksi)	G_{XY} (Msi)
CASE I	1.0	1.0	1.0	1.0	1.0	1.0
CASE II	.38	.4	6.8	5.26	3	2.67
CASE III	.6	.58	4.5	2.3	4	3.6

Fig. 7.1.2 Laminate Strength Variation Versus Ply Angle Orientation

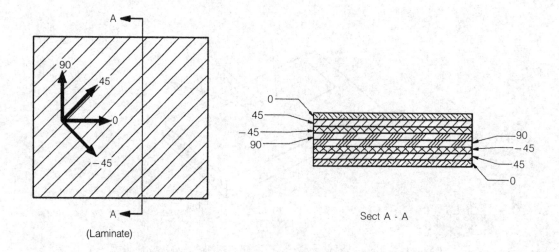

Fig. 7.1.3 Effects of Stacking Sequence

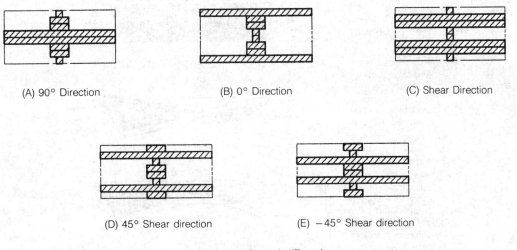

(A) 90° Direction   (B) 0° Direction   (C) Shear Direction

(D) 45° Shear direction   (E) −45° Shear direction

(Equivalent flexural stiffness)

Fig. 7.1.3 Effects of Stacking Sequence (cont'd)

For many composite applications (e.g., the forward swept wing of X-29A, shown in Fig. 1.1.8), the laminate is not even orthotropic, but is anisotropic, as shown in Fig. 7.1.4. This occurs when an orthotropic laminate is loaded in a direction which does not coincide with one of the principal axes, or when the laminate layup is symmetric but not balanced about a principal reference axis. A $[0/\pm 45/90]_s$ laminate is balanced and quasi isotropic, while a $[0/\pm 45]_s$ laminate is orthotropic. This type of laminate requires six elastic coefficients for definition. As a general rule, all laminates should be symmetrically laid up about their midplane; coupled laminates should be avoided. Fig. 7.1.5 shows the difference between the different types of laminates, including the differences a symmetric laminate and one which is not symmetric.

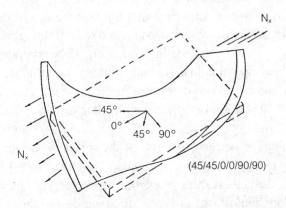

(Applying a load to an unsymmetrically laminated plate causes coupling between extension, shear, bending, and twisting)

Fig. 7.1.4 Anisotropic Laminate Behavior

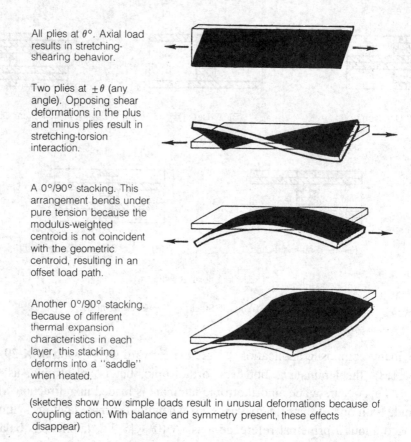

All plies at $\theta°$. Axial load results in stretching-shearing behavior.

Two plies at $\pm \theta$ (any angle). Opposing shear deformations in the plus and minus plies result in stretching-torsion interaction.

A 0°/90° stacking. This arrangement bends under pure tension because the modulus-weighted centroid is not coincident with the geometric centroid, resulting in an offset load path.

Another 0°/90° stacking. Because of different thermal expansion characteristics in each layer, this stacking deforms into a "saddle" when heated.

(sketches show how simple loads result in unusual deformations because of coupling action. With balance and symmetry present, these effects disappear)

Fig. 7.1.5 Effective of Stacking Sequence on Deflection of Laminates

The directional nature of composite laminae provides the ability to construct a material which can meet specific loads and/or stiffness requirements without wasting material by providing strength and stiffness where they are not needed. If the design requirement is simply to provide axial strength or stiffness, a high percentage of the material should be unidirectionally oriented. If this material is adhered to restraining members, theoretically all of the fibers may be so oriented [see Fig. 7.1.6(a)]. If the composite material is unconfined, it is wise to provide a nominal amount of transverse reinforcement to account for any off-axis (refers to rotation about any one of the laminate axes) loading that may occur, whether during fabrication or by induced loading and to reduce the Poisson's ratio effects.

If shear loading or shear stiffness is the primary design consideration [see Fig. 7.1.6(b)], then most of the material should be oriented at $\pm 45°$ to the longitudinal axis, as this provides the highest shear properties. However, care must be taken to evaluate any loading in the longitudinal or transverse directions, since the strength in these directions is quite low. Conditions which would not normally be critical in metal design may approach or exceed the strengths available in these directions when using a pure [$\pm 45°$] layup. This consideration may necessitate the inclusion of at least a minimal number of 0° and/or 90° laminae.

An example of an unrestrained application would be any plate or skin, whether or not it is an elastic base, or a stiffener application. A rule-of-thumb is to make the number of laminae in the transverse direction equal to at least 10% of the total number of laminae in the laminate part.

In addition to the pure axial or pure shear case just mentioned, there are many applications which require the ability to withstand a combination of loadings [see Fig. 7.1.6(c)]. Although it is possible to determine an optimum orientation sequence for any given loading condition, it is more practical from a fabrication standpoint to limit the number of orientations (see Fig. 7.1.7) to a few specific families (e.g., $0°$, $\pm 45°$ and $90°$) which can then be characterized by tests.

The anisotropic nature of composite materials, while allowing the engineer to tailor material more closely to the design requirements, imposes the problem of selecting the proper orientation for the application. This is a consideration which does not arise in metal design, and the engineer must be aware that the traditional methods of design and analysis have to be developed to higher orders of refinement for anisotropic materials, not only to provide a basis for selecting the proper orientation, but even for defining stresses and margins of safety (see Fig. 7.1.8).

One of the major differences in composite analysis, as opposed to metal, is that strain is the major concern and not stress. Composite structures are made up of different plies, and each ply will be stressed at a different level because the ply's elastic modulus is dependent upon the ply orientation. It is engineer's responsibility to determine the best ply orientation for the various loading conditions (see Fig. 7.1.9) following basic laminate requirements.

In the design of large structures, one of the basic ground rules is to establish the ultimate gross area cut-off strain (or stress), as shown in Fig. 7.1.10(a), when designing in tension and compression-critical areas. This cut-off automatically covers many design considerations, such as high local strain areas, joints of various kinds, and structural integrity in terms of impact damage and fracture. In the application of composite material to structures, the allowable levels (expressed in strain instead of stress) are low because of the following limits:
- Tolerance for impact damage (this is the dominant failure mode in compression for some materials)
- Flaw growth resistance
- Stress concentration associated with cutouts, fastened joints, etc.
- Reduced strength in hot/wet conditions

Currently, these factors restrict design ultimate gross area strains to about 50% of the unnotched, undamaged composite material failure strain, depending on loading and laminate orientation. Fig. 7.1.11 lists the primary factors which govern the design strains of composite materials and also shows the relative effect of the environment on laminate structures. To summarize:
- Under tension loading the governing consideration is notches (holes)
- Compression loading takes over when laminate damage occurs as the result of impact, or notches

Chapter 7.0

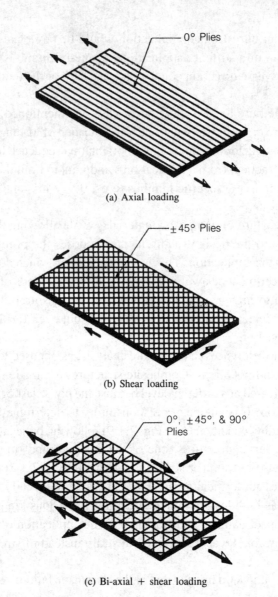

(a) Axial loading

(b) Shear loading

(c) Bi-axial + shear loading

Fig. 7.1.6 Tailoring Ply Orientation to Meet Loading Requirements

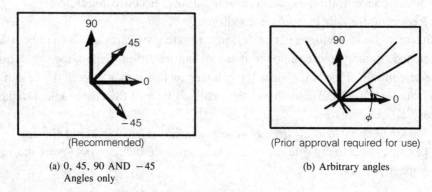

(Recommended)

(a) 0, 45, 90 AND −45 Angles only

(Prior approval required for use)

(b) Arbitrary angles

Fig. 7.1.7 Selection of Ply Angles

Laminate Design Practices

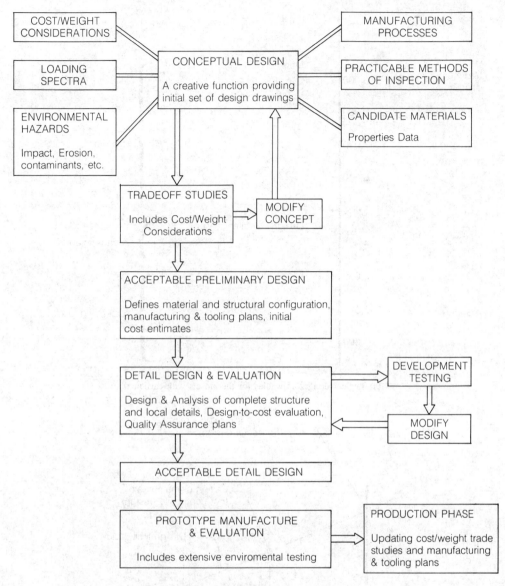

Fig. 7.1.8 Composite Design Methodology

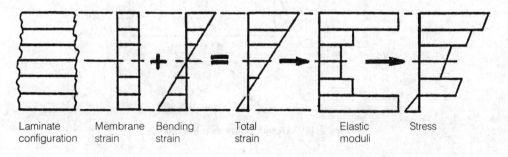

Fig. 7.1.9 Representative Strain, Moduli and Stress Distribution Across the Laminate Thickness

Chapter 7.0

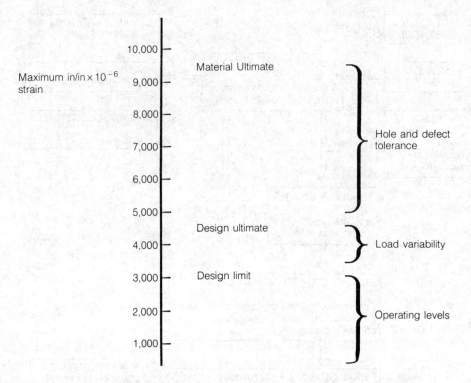

(a) Typical design allowables for thermosets with carbon fibers

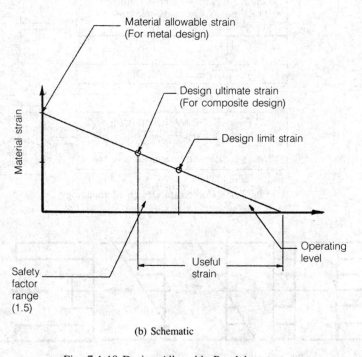

(b) Schematic

Fig. 7.1.10 Design Allowable Breakdown

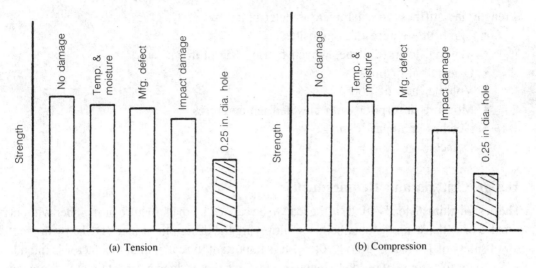

Fig. 7.1.11 Factors Affecting Design Strength

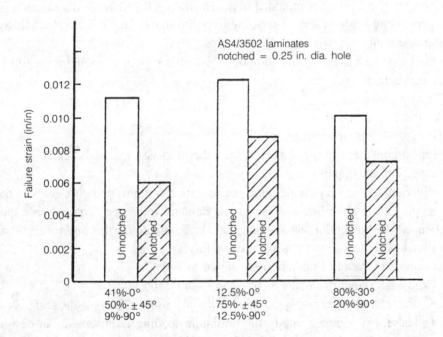

Fig. 7.1.12 Notch Sensitivity Versus Percentage of ±45° Plies

For composite laminates with a given flaw or hole size, strength retention in general increases as the percentage of ±45° plies present in an orientation is increased (see Fig. 7.1.12). However, total load carrying capacity of the laminate generally decrease due to reduced moduli. For example, the stress concentration factor for a drilled hole is more than three times higher in unidirectional carbon/epoxy than in a laminate consisting of only ±45° plies.

Strength and stiffness in a laminate is determined by:
- The brittle nature of composites
- Material form (unitape, woven fabric, 3-D preform, etc.)
- Orientation of plies
- Volume ratio of fibers
- Moisture absorption with elevated temperatures
- Notches or impact damage
- Defects

Design Criteria and Requirements

The design ultimate load (DUL) is the load to be carried by the structure or member without rupture or collapse and is obtained by multiplying the design limit load (DLL) by the factor of safety of 1.5 [see Fig. 7.1.10(b)]. It is important to realize that this factor is intended to cover for accuracy of load, analysis, etc. This factor does not contain any provision for problems of fabrication or processing of composite parts, nor for impact damage, environment, etc. The allowance included in the allowables for material degradation caused by local processing inadequacies is separate from the statistical variations that always occur in the determination of material allowables.

(1) A sound approach to designing composite airframe structures includes the following:
- Analysis methods
- Testing approach
- Material system(s)
- Design concepts (innovative)
- Hybrid construction (e.g., thermosets, thermoplastics, metals, etc.)
- Inspection and repair

   In view of the very challenging performance, weight, and cost goals of the airframe project, such influential factors as damage tolerance requirements must be challenged. It is essential that adequate data be available for this and for trade studies in order to produce the best mix of capability and cost.

(2) Selection of ply angle orientation as shown in Fig. 7.1.7:
   (a) Case I — Standard angles of $0°$, $45°$, $-45°$ and $90°$:
   - These are the basic ply angles commonly used in composite design
   - These 4 ply angles satisfy the minimum loading requirements in design and their use is strongly recommended to simplify analysis and manufacturing
   (b) Case II — Arbitrary angles:
   - Use of limited numbers of non-standard ply angles is allowed only when such use is critical to structural weight reduction and/or a special design
   - The number of different ply angles used should be kept to a minimum

(3) Common material thicknesses currently used in composite design:
- Unitape — Approximately 0.005 to 0.0075 inch (0.127 to 0.191 mm)
- Unitape — 0.002 inch (0.0508 mm) (very expensive and only used on space applications)
- Fabric — 0.010 to 0.015 inch (0.25 to 0.381 mm)

7.2 MATERIAL ALLOWABLE DETERMINATION

In the certification process for civil or military aircraft the use of statistically based material properties is required for composite structural analysis. Unidirectional and multidirectional properties of high-strength and high-modulus composites such as carbon/epoxy are shown in Fig. 2.5.5. The unidirectional strength values are not very useful for the design process. Since most structures are subjected to combined loading, it is necessary to orient fibers or individual plies of collimated fibers at specific angles to absorb these loads.

A description and example of the statistical procedures used to derive material design allowables under basic headings of 'A' and 'B' is provided in Ref. 7.12:

   (a) 'A' basis — The mechanical property value indicated is the value above which at least 99% of the population of values is expected to fall with a confidence of 95%. This value is used to design a single member whose loading is such that its failure would result in loss of structural integrity

   (b) 'B' basis — The mechanical property value indicated is the value above which at least 90% of the population of values is expected to fall with a confidence of 95%. This value is used on redundant or fail-safe structure analysis, where the loads may be safely distributed to other members.

Materials allowables must be obtained by testing (coupon or element tests) as described in Chapter 8.
Statistical analysis of the data has led to the formulation of 'A' and 'B' basis design allowables. 'B' basis design allowables are those mechanical properties that are most commonly used in composite structural analysis.

Composite structures are vulnerable to impact damage and coupon test data indicates that impact damage that is not visible may seriously degrade the compression strength of a laminate. Compression tests should be conducted on elements (later full scale component tests are also required to verify the coupon test data) containing both non-visible and visible impact damage to compare with a laminate containing a given fastener hole diameter (notched). If the compression impact damage is a lower valve than that of notched strength, the compression impact damage strength will be used as the design allowable or maximum design cutoff strain (e.g., 5000 micro in/in for example) for all environmental conditions.

Usually tension allowables are based on a 0.25 inch (6.35 mm) diameter hole, filled or unfilled depending on which is the lower value. Compression allowables are based on a 0.25 inch diameter hole or the damage tolerance requirement. If the hole has a larger diameter or is countersunk, a reduced allowable must be used as shown in Fig. 7.2.1.

Preliminary Design Allowables

Most of the time, a new project will need a group of preliminary allowables to do the initial sizing and analysis before the development of final allowables. Final allowables usually take more than a year to develop during which more than 4000 coupons are tested to generate complete allowable data for use in the certification program. Two methods are described below:

Chapter 7.0

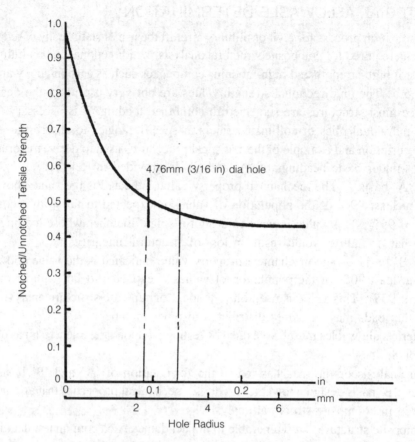

Fig. 7.2.1 Hole Radius Effects on Tensile Strength

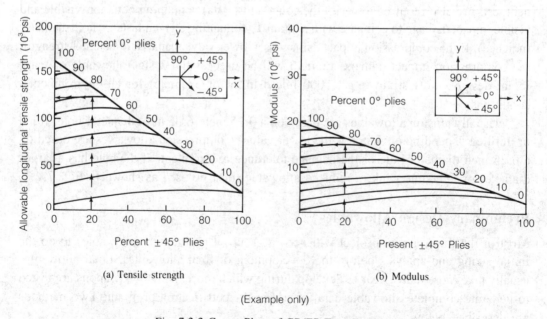

(a) Tensile strength

(b) Modulus

(Example only)

Fig. 7.2.2 Carpet Plots of GR/EP Tape Families

(1) Carpet Plot Method (See Coupon Tests in Section 8.3 of Chapter 8):
Carpet plot represent various combinations of symmetric and balanced laminates that contain 0°, 90° and ±45° plies. The ply data may be used to find:
- Properties for a given laminate
- Various laminates that satisfy a particular property requirement

In Fig. 7.2.2 shows two typical carpet plots (modulus and strength values) from a set of design curves for carbon/epoxy unitape (similar plots can be made of woven fabrics). Similar design curves can be found in Ref. 7.13. Such design data curves are only used for the (0/±45/90) family of composite and are only applicable for a particular material system. Consider a carpet plot for each of the following:
- Notched tensile strength
- Compression strength (notched)
- Notched shear strength
- Bearing strength
- Tension, Compression and shear modulus of elasticity
- Poisson's ratio

The notched stresses listed above are gross area and are commonly based on a 0.1785 inch (4.76 mm) or 0.25 inch (6.35 mm) diameter fastener hole with a spacing of 4 times fastener diameter. Each set of the design strain allowable data mentioned above should include following conditions:
- Room temperature dry (RTD)
- Hot/wet
- Cold/dry
- Impact damage

The failure criterion used for predicting strength is the maximum strain criterion with the following assumptions:
- For laminates with fibers in the direction of loading, failure is assumed when laminate strain exceeds the ply-level 0° ply direction failure strain
- For laminates which contain only ±45°/90° plies and which are loaded in tension in the 0° direction, failure is conservatively assumed to occur when the 90° tensile failure strain is exceeded
- When ±45°/90° laminates are loaded in compression, failure is conservatively assumed to occur when the shear strain in the ±45° plies exceeds the ply-level shear strain

(2) AML (Angle Minus Longitudinal) Plots Method (See Coupon Tests in Section 8.3 of Chapter 8):
It has been found in some cases that the failure strain level for stress concentrations varies according to the relative proportions of the various angle plies. Laminates with the highest proportions of 0° plies have the lowest failure strains, and laminates with the highest proportions of ±45° plies have the highest failure strains. Consequently, a current method plots failure strain against 'AML' (Angle Minus Longitudinal, or the % of ±45° plies minus the % of 0° plies; nothing that the term 'angle' refers only to the ±45° plies). The extremes are AML = -100 for all 0° plies and AML = $+100$ for all ±45° plies.

Laminates of different layups, which provide a range of AML values likely to be used are drilled or impacted and tested to failure. The allowable strain is derived from the test data and plotted as a function of the AML parameter. This method is based on a series of coupon tests (approximately 300 coupons of 16 to 24 ply laminates under room temperature, hot/wet, and cold/dry conditions) including tension tests (open and filled hole) and compression tests including open hole (either a 0.1875 inch or 0.25 inch diameter fastener hole size is commonly selected for the project requirement) and post-impact tests. An AML plot is shown in Fig. 7.2.3. Where holes larger than 0.25 inch diameter exist, further reductions are required locally based on the effect of the largest hole (see Fig. 7.2.1). This method neither accounts for the 90° plies or the stacking sequence, both of which affect the failure strain and it is generally conservative. However, it generally works well for laminates with less than 25% 90° plies. The application of this approach to strength analysis is illustrated in Section 7.6 in this Chapter.

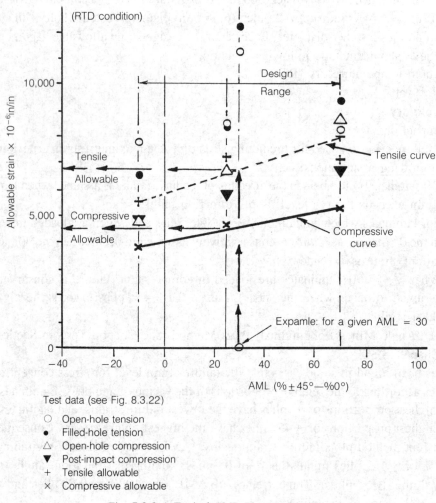

Test data (see Fig. 8.3.22)
○ Open-hole tension
● Filled-hole tension
△ Open-hole compression
▼ Post-impact compression
+ Tensile allowable
× Compressive allowable

Fig. 7.2.3 A Typical AML Plot (Carbon Fiber)

7.3 LAMINATE STRENGTH ANALYSIS

This section contains a brief discussion of the fundamental principles of strain (or stress) at a point in order to form a firm base for the analytical development for composites which follows.

The basic lamination theory is the stress-strain or constitutive relation and failure criteria for the individual laminae. Each lamina consists of a layer of unidirectional fibers or woven cloth, impregnated and fully surrounded by matrix material. The strong, stiff fibers provide the primary load carrying capability while the matrix protects and supports the fibers and transfers the load between them. The terminlogies of the theory of laminated plates are described below:

(a) Micromechanics — The study of the interaction between fiber and matrix in a lamina such that the mechanical behavior of the lamina can be predicted from the known behavior of the constituents:
  - Micromechanics establishes the relationship between the properties of the constituents (the fiber and matrix) and those of the unit composite ply
  - Complex formulations relating the shape, array and interactions have proven no better than a simple rule-of-mixtures approach based on volume fractions of the constituent
  - All approaches require a correction factor to correlate with measured ply level or laminate tests
  - All approaches suffer from the problem of measuring the properties of the constituents
    For these reasons, micromechanics is seldom used for aircraft strength or stiffness design. However, it is used for certain physical properties, e.g., density, fiber volume, etc., as will be discussed later.

(b) Macromechanics — The study of the mechanical behavior at any point in a laminate based on the known behavior of the laminae.
  - Macromechanics establishes the relationship between the properties of the plies and those of the resultant laminate
  - It is based on continuum mechanics which models each ply as homogeneous and orthotropic, ignoring fiber/matrix interface
  - Lamination theory is the principle mathematical tool for determining the property relationship between ply and laminate
  - The ply properties are determined from ply level or laminae tests

(c) Lamination theory — A mathematical formulation for predicting the macromechanical behavior of a laminate based on an arbitrary assembly of homogeneous orthotropic laminae. Two dimensional theory is most common, and three dimensional theory is most complex.

Laminate strength analysis:
  - A ply (or lamina) has five potential modes of failure as shown in Fig. 7.3.1.
  - A laminate has additional modes of failure:
    — Delamination and sublaminate buckling
    — Interlaminar shear
    — Interlaminar tension

- A laminate may have failed laminae and continue to carry load
- Strength and stiffness are directional

Poisson's ratio effects:
- Poisson's ratio (v) for composite materials is directional and varies with relative orientations of the individual plies (see Fog. 7.3.2)
- Composite parts may induce large loads to adjacent structures because of their relatively high Poisson's ratio
- The Poisson's ratio of composites is significantly higher than the usual value of metals
- Special care should be taken at the boundaries between composites and metals where induced stresses due to high Poisson's ratio are greatest

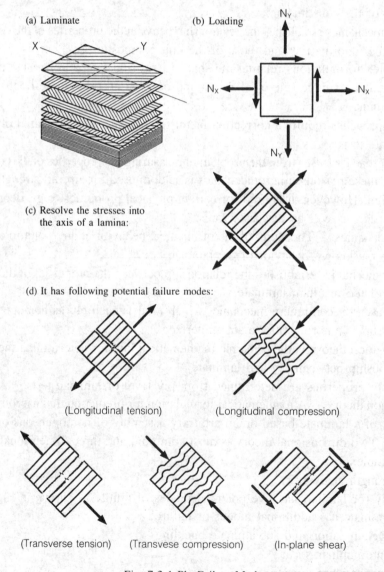

Fig. 7.3.1 Ply Failure Modes

- The top ply has the most x-direction stress because it is stiffer than the bottom ply in that direction
- The x-direction displacements are identical, but
  Without bonding:

  $\triangle_y^1 > \triangle_y^2$ (because $\nu_{12} > \nu_{21}$)

  With bonding:
  — The top ply must get wider (σ_{y_T} = tension)
  — And bottom ply must get narrower (σ_{y_B} = compression)
  — From equilibrium
    $\sigma_{y_T} t_T + \sigma_{y_B} t_B = 0$

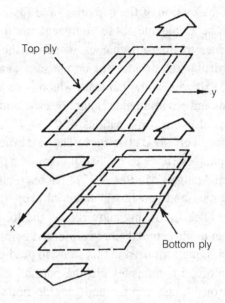

Fig. 7.3.2 Poisson's Ratio Mismatch of Two Plies

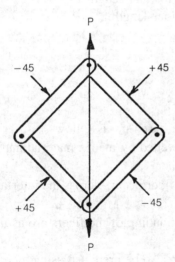

±45 Laminate:

Laminate relies on matrix for much of its rigidity

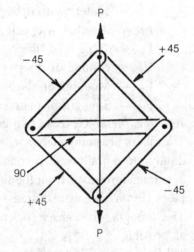

±45, 90 Laminate:

Like a truss, the addition of transverse fibers stiffen the longitudinal direction

Fig. 7.3.3 Lamina Interaction

With the advent of composite materials, there is an opportunity for materials design to be integrated into structural design as an added dimension. A basic understanding of the interaction of fiber, matrix, and fiber-matrix interface in composites (see Fig. 7.3.3) will be a valuable aid to engineers and to materials and structural analysis. The modern science of micromechanics, which is the study of structural material interactions, is of particular importance for composites analysis.

However, the laminate which is required to withstand a number of differing conditions and environments cannot be efficiently designed with micromechanics alone. Techniques of using combinations of 0°, ±45°, and 90° laminae can produce a simple design of relatively good efficiency. The best efficiencies, however, are achieved by computerized techniques (refer to the list of computer programs from ''Computer Programs for Structural Analysis'' in Ref. 7.17). These computerized techniques are based on both empirical test data and constituent material properties along with micromechanics.

The following study relates the longitudinal and transverse mechanical properties of an unidirectional fiber reinforced composite to the properties of the constituents. The basic equations used in this study to predict the longitudinal modulus and strength of fiber reinforced composites are the parallel element mixture equations. These are based on reasonable assumption and they do not violate theories of elasticity.

$$E_{11} = E_f V_f + E_m(1 - V_f) \tag{7.3.1}$$

$$F_{11} = F_f V_f + F_m(1 - V_f) \text{ (generally in tension)} \tag{7.3.2}$$

where:
E_{11} — Modulus of elasticity parallel to fiber length
F_{11} — Material strength parallel to fiber length
E_f — Modulus of elasticity of fiber
F_f — Strength of fiber
V_f — Volume of fiber
E_m — Modulus of elasticity of matrix
F_m — Strength of matrix

The assumptions on which these two equations are based on as follows:
- The fibers are completely sorrounded and wetted by matrix material and accordingly are not allowed to contact one another
- The transfer of load from the matrix to the fiber occurs across the interface surfaces comprising wetted areas
- The strength, size, shape, orientation, and bonding of the fibers are as uniform as possible

The overall Poisson's ravio (ν) of the composite can also be predicted using an equation of similar form.

$$\nu_{12} = \nu_f V_f + \nu_m(1 - V_f) \tag{7.3.3}$$

where ν_{12} — Poisson's ratio of the composite material
ν_f — Poisson's ratio of the fiber
ν_m — Poisson's ratio of the matrix

Predictions of the tensile modulus normal to the fiber direction and the shear modulus in a unidirectional composite are difficult to make because of their sensitivity to voids, and their dependence on accurate knowledge of both the matrix modulus and the details of the fiber-matrix packing. The two equations given below are used to approximate the transverse tensile modulus and shear modulus of unidirectional fiber reinforced composite materials.

$$E_{22} = 1/[(V_f/E_f) + (1 - V_f)/E_m] \tag{7.3.4}$$
$$G_{12} = 1/[(V_f/G_f) + (1 - V_f)/G_m] \tag{7.3.5}$$

where E_m — Modulus of Elasticity of the matrix
  G_m — Modulus of Rigidity of the matrix

Note that Equations (7.3.1) through (7.3.5) represent but a few of the many micromechanical expressions developed to predict the behavior of fiber reinforced composite materials (for further information, see Ref. 7.3).

Lamination Theory

When the plane stress assumption is valid in lamination theory for a thin plate which is subjected only to inplane or membrane loads that do not cause any instability, the stress-strain behavior become relatively simple. Lamina axes are defined by the numbers 1, 2, and 3 and the laminate is defined by x, y and z, as shown in Fig. 7.3.4.

Basic assumptions of lamination theory:
- The structure is restricted to be a thin plate or shell consisting of an arbitrary combination of laminae (plies) cured or consolidated into a single laminated plate
- The assumptions of plate and shell theory hold; i.e., the through plate thickness direct stress (σ_z) is zero and only plane stresses (σ_{xy}) exists
- The two transverse shear stresses (τ_{xz}, τ_{yz}) are neglected to meet the classic thin plate and shell theory [see Fig. 7.3.4(b)]
- The theory is a 'point' analysis in an effectively infinitely large plate and shell, completely ignoring the effects of neighboring edges, stiffeners, holes, cutouts or any other discontinuities
- The loading is assumed to be inplane memberane stress and moment resultants.

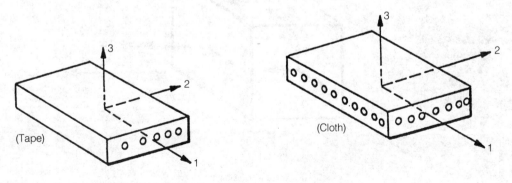

(a) Lamina or material axes for tape and cloth

Fig. 7.3.4 Definition of Composites Coordinates

Chapter 7.0

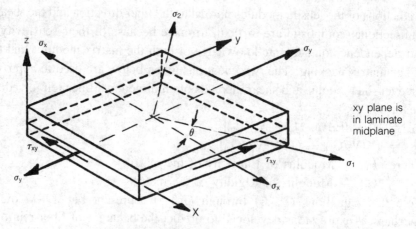

(b) Laminae and laminate coordinates and stresses

Fig. 7.3.4 Definition of Composites Coordinates (cont'd)

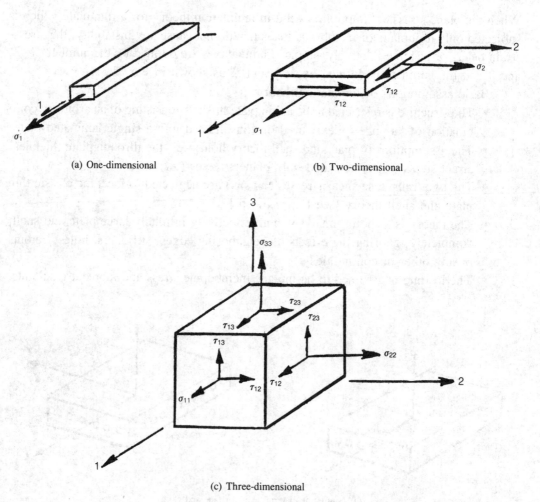

(a) One-dimensional

(b) Two-dimensional

(c) Three-dimensional

Fig. 7.3.5 Hooke's Law Theory

Hooke's Law Theory

(1) For a homogeneous isotropic material in a one dimensional stress state, [see Fig. 7.3.5(a)] the Hooke's law relationship is:

$$\sigma = E \times \epsilon \tag{7.3.6}$$

(where the proportionality constant, E, is Young's modulus or the modulus of elasticity and, ϵ, is strain)

(2) For a homogeneous isotropic material in a two dimensional stress state [plane stress, see Fig. 7.3.5(b)], the Hooke's law relationship become:

$$\sigma_1 = (\epsilon_1 + \nu\epsilon_2)\frac{E}{1-\nu^2} \tag{7.3.7}$$

$$\sigma_2 = (\epsilon_2 + \nu\epsilon_1)\frac{E}{1-\nu^2} \tag{7.3.8}$$

$$\tau_{12} = (\gamma_{12})\frac{E}{2(1+\nu)} \tag{7.3.9}$$

or the equations in matrix form are:

$$\begin{bmatrix} \sigma_1 \\ \sigma_2 \\ \tau_{12} \end{bmatrix} = \begin{bmatrix} C_{11} & C_{12} & 0 \\ C_{12} & C_{22} & 0 \\ 0 & 0 & C_{66} \end{bmatrix} \begin{bmatrix} \epsilon_1 \\ \epsilon_2 \\ \gamma_{12} \end{bmatrix} \tag{7.3.10}$$

where,

$$C_{11} = C_{22} = E(1-\nu^2) \tag{7.3.11}$$

$$C_{21} = C_{12} = \nu E/(1-\nu^2) \tag{7.3.12}$$

$$C_{66} = E/2(1+\nu) = G \tag{7.3.13}$$

It is evident that two independent elastic constants appear in Eq. (7.3.10). E and ν. The third elastic constant, shear modulus, G. is a function of E and ν. The relationship of stress and strain is shown in Fig. 7.3.6 and

$$G = E/2(1+\nu) \tag{7.3.14}$$

Therefore, for isotropic materials only two elastic constants are necessary to write the Hooke's law relationships for two or three dimensional applications.

(3) The Hooke's law relationships can be generalized for three dimensional anisotropic [see Fig. 7.3.5(c)] materials.
   (a) It requires twenty-one independent elastic constants for the Hooke's law relationships for an anisotropic material in three dimensions.
   (b) For an orthotopic material in three dimensions, only nine independent elastic constants are necessary.
   (c) For the case which is of the most practical interest, an orthotropic material in a two dimensional application, there are four independent elastic constants required to specify a Hooke's law relationship.

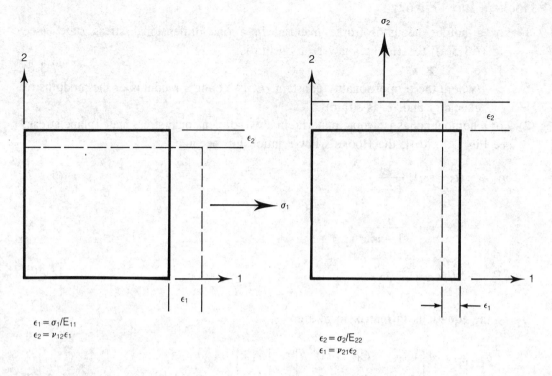

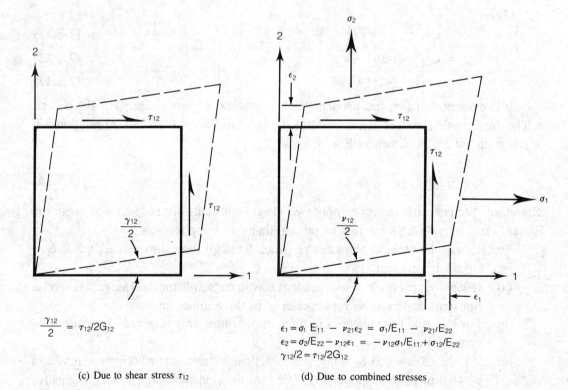

Fig. 7.3.6 Two-dimensional/Stress and Strain Relationship

Case (c) is of particular interest because the individual lamina of a filamentary composite may thus be modeled as an orthotropic material in a state of plane stress.

Generalized Hooke's law now can be written in matrix from (eighty-one elastic constants):

$$\begin{bmatrix} \sigma_{11} \\ \sigma_{22} \\ \sigma_{33} \\ \tau_{23} \\ \tau_{13} \\ \tau_{12} \end{bmatrix} = \begin{bmatrix} C_{11} & C_{12} & C_{13} & C_{14} & C_{15} & C_{16} \\ C_{12} & C_{22} & C_{23} & C_{24} & C_{25} & C_{26} \\ C_{13} & C_{23} & C_{33} & C_{34} & C_{35} & C_{36} \\ C_{14} & C_{24} & C_{34} & C_{44} & C_{45} & C_{46} \\ C_{15} & C_{25} & C_{35} & C_{45} & C_{55} & C_{56} \\ C_{16} & C_{26} & C_{36} & C_{46} & C_{56} & C_{66} \end{bmatrix} \begin{bmatrix} \epsilon_{11} \\ \epsilon_{22} \\ \epsilon_{33} \\ \gamma_{23} \\ \gamma_{13} \\ \gamma_{12} \end{bmatrix} \quad (7.3.15)$$

(where $[C_{ij}]$ is the stiffness matrix)

For a state of plane stress, the equation above reduces to:

$$\begin{bmatrix} \sigma_{11} \\ \sigma_{22} \\ \tau_{12} \end{bmatrix} = \begin{bmatrix} C_{11} & C_{12} & 0 \\ C_{12} & C_{22} & 0 \\ 0 & 0 & C_{66} \end{bmatrix} \begin{bmatrix} \epsilon_{11} \\ \epsilon_{22} \\ \gamma_{12} \end{bmatrix} \quad (7.3.16)$$

Where only 4 independent elastic constants exist.

$$C_{11} = \frac{E_{11}}{(1 - \nu_{12}\nu_{21})} \quad (7.3.17)$$

$$C_{12} = \nu_{12} C_{21} \quad (7.3.18)$$

$$C_{22} = \frac{E_{22}}{(1 - \nu_{12}\nu_{21})} \quad (7.3.19)$$

$$C_{66} = G_{12} \quad (7.3.20)$$

The following equation is used to transform the elastic constants of an orthotropic material (ply) in two dimensions to laminate axes, x and y (see Fig. 7.3.4). Without going into the derivation, the transformation relationships for the elastic constants for an orthotropic material in a plane stress state and the transformation matrix $[Q_{ij}]$ are as follows:

$$\begin{bmatrix} \sigma_x \\ \sigma_y \\ \tau_{xy} \end{bmatrix} = \begin{bmatrix} Q_{11} & Q_{12} & Q_{16} \\ Q_{21} & Q_{22} & Q_{26} \\ Q_{61} & Q_{62} & Q_{66} \end{bmatrix} \begin{bmatrix} \epsilon_x \\ \epsilon_y \\ \gamma_{xy} \end{bmatrix} \quad (7.3.21)$$

Where:

$$Q_{11} = C_{11} \cos^4\theta + 2(C_{12} + 2C_{66}) \sin^2\theta \cos^2\theta + C_{22} \sin^4\theta \quad (7.3.22)$$

$$Q_{22} = C_{11} \sin^4\theta + 2(C_{12} + 2c_{66}) \sin^2\theta \cos^2\theta + C_{22} \cos^4\theta \quad (7.3.23)$$

$$Q_{12} = (C_{11} + C_{22} - 4C_{66}) \sin^2\theta \cos^2\theta + C_{12} (\sin^4\theta + \cos^4\theta) \quad (7.3.24)$$

$$Q_{66} = (C_{11} + C_{22} - 2C_{12} - 2C_{66}) \sin^2\theta \cos^2\theta + C_{66} (\sin^4\theta + \cos^4\theta) \quad (7.3.25)$$

$$Q_{16} = (C_{11} - C_{12} - 2C_{66}) \sin\theta \cos^3\theta + (C_{12} - C_{22} + 2C_{66}) \sin^3\theta \cos\theta \quad (7.3.26)$$

$$Q_{26} = (C_{11} - C_{12} - 2C_{66}) \sin^3\theta \cos\theta + (C_{12} - C_{22} + 2C_{66}) \sin\theta \cos^3\theta \quad (7.3.27)$$

Q's elastic constants refer to the laminate x and y coordinates as shown in Fig. 7.3.4

General Constitutive Equation

The constitutive equation for a thin laminated anisotropic plate can be written (not including the term for thermomechanical properties):

$$\begin{bmatrix} N \\ M \end{bmatrix} = \begin{bmatrix} A & B \\ B & D \end{bmatrix} \begin{bmatrix} \epsilon \\ \varkappa \end{bmatrix} \quad (7.3.28)$$

or by the following form:

$$\begin{bmatrix} N_X \\ N_Y \\ N_{XY} \\ M_X \\ M_Y \\ M_{XY} \end{bmatrix} = \begin{bmatrix} A_{11} & A_{12} & A_{16} \\ A_{21} & A_{22} & A_{26} \\ A_{61} & A_{62} & A_{66} \\ B_{11} & B_{12} & B_{16} \\ B_{21} & B_{22} & B_{26} \\ B_{61} & B_{62} & B_{66} \end{bmatrix} \begin{bmatrix} \epsilon_X \\ \epsilon_Y \\ \epsilon_{XY} \end{bmatrix} + \begin{bmatrix} B_{11} & B_{12} & B_{16} \\ B_{21} & B_{22} & B_{26} \\ B_{61} & B_{62} & B_{66} \\ D_{11} & D_{12} & D_{16} \\ D_{21} & D_{22} & D_{26} \\ D_{61} & D_{62} & D_{66} \end{bmatrix} \begin{bmatrix} \varkappa_X \\ \varkappa_Y \\ \varkappa_{XY} \end{bmatrix} \quad (7.3.29)$$

where

$$A_{ij} = \sum_{k=1}^{n} (Q_{ij})_k (h_k - h_{k-1}) \quad (7.3.30)$$

$$B_{ij} = \sum_{k=1}^{n} (Q_{ij})_k (h_k^2 - h_{k-1}^2) / 2 \quad (7.3.31)$$

$$D_{ij} = \sum_{k=1}^{n} (Q_{ij})_k (h_k^3 - h_{k-1}^3) / 3 \quad (7.3.32)$$

(A_{ij} is the extensional or membrane stiffness. D_{ij} is the flexural or bending stiffness while B_{ij} is responsible for the coupling between membrane and bending behavior; see Fig. 7.3.7 for notations)

Eq. (7.3.28) is the general constitutive equation for laminated composites (plates and shells). In this form, the significant point is that there is coupling between extensional (membrane) deformation and bending deformation caused by the existence of the [B] matrix. In other words, even within the limits of small deflection theory, forced curvature, [$\varkappa$], within the laminate induces in-plane loads, [N], through this type of coupling. Also, in-plane strains, [ϵ], would induce curvatures, [$\varkappa$], in the laminate. This coupling is caused by the neutral axis and the midplane of the laminate not being coincident.

The procedures for simplifying the general equation for practical use are:

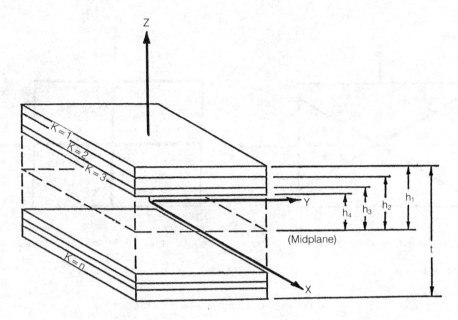

Fig. 7.3.7 Laminate Ply Notation

(a) Eliminate coupling matrix [B] — This can be accomplished by fabricating the laminate symmetrically about the midplane. In other words, an equal number of identical plies (thickness and orientation) are located at the same distance above the midplane as are located below the midplane. With this restriction, the constitutive equation reduces to:

$$[N] = [A] [\epsilon] \qquad (7.3.33)$$
$$[M] = -[D] [\varkappa] \qquad (7.3.34)$$

Hence, a laminate which is symmetrically laminated about the midplane is often referred to as "homogeneous" anisotropic.

(b) If the laminate is constructed with equal numbers of pairs of laminae with symmetry about the coordinate (x, y) axes (angle ply laminate, see Fig. 7.3.8), the [A] matrix becomes orthotropic in nature ($A_{16} = A_{26} = 0$) as

$$[A] = \begin{bmatrix} A_{11} & A_{12} & 0 \\ A_{12} & A_{22} & 0 \\ 0 & 0 & A_{66} \end{bmatrix} \qquad (7.3.35)$$

The [D] matrix remains fully populated and anisotropic in nature.

(c) If the laminate is constructed with equal numbers of pairs of laminae at angles of 0° and 90° (cross-ply laminate, see Fig. 7.3.9) to the x-y axes, the [D] matrix will become orthotropic in nature.

Finally, the practical equation may be written

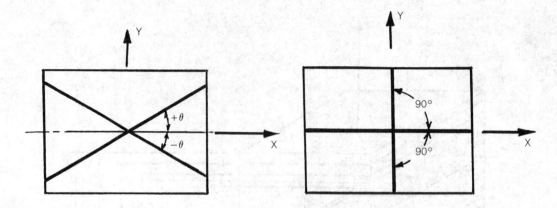

Fig. 7.3.8 Symmetric-ply Laminate   Fig. 7.3.9 Cross-ply Laminate

$$\begin{bmatrix} N_x \\ N_y \\ N_{xy} \end{bmatrix} = [A] \begin{bmatrix} \epsilon_x \\ e_y \\ \gamma_{xy} \end{bmatrix} \quad (7.3.36)$$

Dividing both sides by the total laminate thickness, t, yields

$$\begin{bmatrix} \bar{\sigma}_x \\ \bar{\sigma}_y \\ \bar{\tau}_{xy} \end{bmatrix} = \frac{1}{t}[A] \begin{bmatrix} \epsilon_x \\ \epsilon_y \\ \gamma_{xy} \end{bmatrix} = [\bar{A}] \begin{bmatrix} \epsilon_x \\ \epsilon_y \\ \gamma_{xy} \end{bmatrix} \quad (7.3.37)$$

This practical equation can be used for the following reasons:
- The majority of the laminates which will be utilized for design will necessarily be symmetrical about the midplane or nearly so because of the warpage which will occur during the fabrication process when symmetry does not exist.
- The majority of the applications for advanced composites in the past have been largely confined to structures where membrane loading exists due to the relatively low transverse shear strength of the laminates.

From Eq. (7.3.37), the σ's are the average laminate stresses. The [A] matrix is the laminate stiffness matrix just as the [C] and [Q] matrices, were the lamina stiffness matrices. The equation may be inverted and yield:

$$\begin{bmatrix} \epsilon_x \\ \epsilon_y \\ \gamma_{xy} \end{bmatrix} = [\bar{A}^*] \begin{bmatrix} \bar{\sigma}_x \\ \bar{\sigma}_y \\ \bar{\tau}_{xy} \end{bmatrix} \quad (7.3.38)$$

(where the laminate compliance matrix is given by $[\bar{A}^*]=[\bar{A}]^{-1}$)

The gross or average laminate elastic moduli may be obtained from the components of the laminate compliance matrix $[\bar{A}^*]$. The gross laminate elastic constants are given by the following:

$$E_{xx} = \frac{1}{\bar{A}^*_{11}} = \frac{A_{11}A_{22} - A^2_{12}}{A_{22} t} \tag{7.3.39}$$

$$E_{yy} = \frac{1}{\bar{A}^*_{22}} = \frac{A_{11}A_{22} - A^2_{12}}{A_{11} t} \tag{7.3.40}$$

$$G_{xy} = \frac{1}{\bar{A}^*_{66}} = \frac{A_{66}}{t} \tag{7.3.41}$$

$$\nu_{xy} = \frac{-\bar{A}^*_{12}}{\bar{A}^*_{11}} = \frac{A_{12}}{A_{22}} \tag{7.3.42}$$

$$\nu_{yx} = \frac{-\bar{A}^*_{12}}{\bar{A}^*_{11}} = \frac{A_{12}}{A_{11}} \tag{7.3.43}$$

By referring to Fig. 7.3.4(b) and Fig. 7.3.10 and using the right-hand-rule sign convention, the laminate stress resultants and stress couples may be determined as follows:

$$N_x = \int_{-t/2}^{t/2} \sigma_x \, dz \tag{7.3.44}$$

$$N_y = \int_{-t/2}^{t/2} \sigma_y \, dz \tag{7.3.45}$$

$$N_{xy} = \int_{-t/2}^{t/2} \tau_{xy} \, dz \tag{7.3.46}$$

$$M_x = \int_{-t/2}^{t/2} Z \, \sigma_x \, dz \tag{7.3.47}$$

$$M_y = \int_{-t/2}^{t/2} Z \, \sigma_y \, dz \tag{7.3.48}$$

$$M_{xy} = \int_{-t/2}^{t/2} Z \, \tau_{xy} \, dz \tag{7.3.49}$$

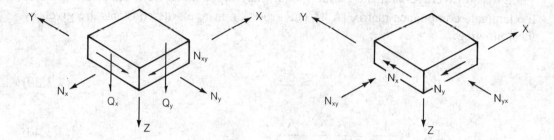

Fig. 7.3.10 Laminate Load Sign Convention

The macromechanics approach to fibrous composites is to model the individual ply or lamina as a homogeneous orthotropic medium subjected to plane stress. Laminate properties (stiffness) have been very successfully predicted from the individual ply properties and good agreement has been obtained with empirical data. Therefore, this technique of utilizing the individual lamina or ply properties to predict laminate properties seems to be the most advantageous approach to the mechanics of laminated composites.

Examples:

Example I:

Material Properties (Lamina or ply):

$E_{11} = 18.2 \times 10^6$ psi
$E_{22} = 1.82 \times 10^6$ psi
$G_{12} = 1 \times 10^6$ psi
$\nu_{12} = 0.30, \; \nu_{21} = 0.03, \; \nu_{21}E_{11} = \nu_{12}E_{22}$

(a) Lamina Stiffness Matrix (from Eq. (7.3.16) through Eq. (7.3.20))

$$[C] = \begin{bmatrix} C_{11} & C_{12} & 0 \\ C_{12} & C_{22} & 0 \\ 0 & 0 & C_{66} \end{bmatrix}$$

$$C_{11} = \frac{E_{11}}{1 - \nu_{12}\nu_{21}} = \frac{18.2 \times 10^6}{1 - 0.3 \times 0.03} = 20 \times 10^6$$

$$C_{12} = \frac{\nu_{12}E_{22}}{1 - \nu_{12}\nu_{21}} = \frac{0.3 \times 1.82 \times 10^6}{1 - 0.3 \times 0.03} = 0.6 \times 10^6$$

$$C_{22} = \frac{E_{22}}{1 - \nu_{12}\nu_{21}} = \frac{1.82 \times 10^6}{1 - 0.3 \times 0.3} = 2 \times 10^6$$

$$C_{66} = G_{12} = 1 \times 10^6$$

(b) Transform stiffness matrix from lamina to laminate axes (lamina axes 1 & 2 to laminate axes x & y coordinates):

$$[Q] = \begin{bmatrix} Q_{11} & Q_{12} & Q_{16} \\ Q_{12} & Q_{22} & Q_{26} \\ Q_{16} & Q_{26} & Q_{66} \end{bmatrix}$$

Let $\cos \theta = m$ and $\sin \theta = n$

Now from Eq. 7.3.22 through Eq. 7.3.27:

$Q_{11} = C_{11}m^4 + 2(C_{12} + 2C_{66})n^2m^2 + C_{22}n^4$

$Q_{22} = C_{11}n^4 + 2(C_{12} + 2C_{66})n^2m^2 + C_{22}m^4$

$Q_{12} = (C_{11} + C_{22} - 4C_{66})n^2m^2 + C_{12}(n^4 + m^4)$

$Q_{66} = (C_{11} + C_{22} - 2C_{12} - 2C_{66})n^2m^2 + C_{66}(n^4 + m^4)$

$Q_{16} = (C_{11} - C_{12} - 2C_{66})n\,m^3 + (C_{12} - C_{22} + 2C_{66})n^3\,m$

$Q_{26} = (C_{11} - C_{12} - 2C_{66})n^3\,m + (C_{12} - C_{22} + 2C_{66})n\,m^3$

For 0° Ply $m = 1; n = 0$

$$[Q] = \begin{bmatrix} 20 & .6 & 0 \\ .6 & 2 & 0 \\ 0 & 0 & 1 \end{bmatrix} \times 10^6$$

For 90° Ply $m = 0; n = 1$

$$[Q] = \begin{bmatrix} 2 & .6 & 0 \\ .6 & 20 & 0 \\ 0 & 0 & 1 \end{bmatrix} \times 10^6$$

For 45° Ply $m = 1/\sqrt{2}; n = 1/\sqrt{2}$

$$[Q] = \begin{bmatrix} 6.8 & 4.8 & 4.5 \\ 4.8 & 6.8 & 4.5 \\ 4.5 & 4.5 & 5.2 \end{bmatrix} \times 10^6$$

For $-45°$ Ply $m = 1/\sqrt{2}; n = -1/\sqrt{2}$

$$[Q] = \begin{bmatrix} 6.8 & 4.8 & -4.5 \\ 4.8 & 6.8 & -4.5 \\ -4.5 & -4.5 & 5.2 \end{bmatrix} \times 10^6$$

Example II

(A) Consider a 4 ply laminate (see Fig. 7.3.11):

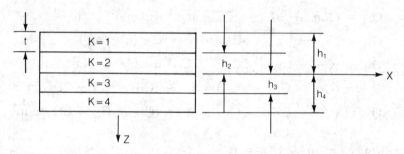

Fig. 7.3.11 Laminate Geometry (4 Plies)

Thickness per ply = t

$h_1 = -2t$

$h_2 = -t$

$h_3 = t$

$h_4 = 2t$

Membrance stiffness [(Eq. (7.3.30)]:

$A_{ij} = Q_{ij}^{(1)}(-t - (-2t)) + Q_{ij}^{(2)}(0 - (-t)) + Q_{ij}^{(3)}(t - 0) + Q_{ij}^{(4)}(2t - t)$

$= Q_{ij}^{(1)} t + Q_{ij}^{(2)} t + Q_{ij}^{(3)} t + Q_{ij}^{(4)} t$

Coupling (Eq. 7.3.31]:

$B_{ij} = \frac{1}{2}\left[Q_{ij}^{(1)}((-t)^2 - (-2t)^2) + Q_{ij}^{(2)}(0 - (-t)^2) + Q_{ij}^{(3)}(t^2 - 0) \right.$

$\left. + Q_{ij}^{(4)}((2t)^2 - t^2) \right]$

$= \frac{1}{2}\left[Q_{ij}^{(1)}(t^2 - 4t^2) + Q_{ij}^{(2)}(-t^2) + Q_{ij}^{(3)}(t^2) + Q_{ij}^{(4)}(4t^2 - t^2) \right]$

$= Q_{ij}^{(1)}\left(-\frac{3}{2}t^2\right) + Q_{ij}^{(2)}\left(-\frac{t^2}{2}\right) + Q_{ij}^{(3)}\frac{t^2}{2} + Q_{ij}^{(4)}\left(\frac{3t^2}{2}\right)$

Flexural stiffness (Eq. (7.3.32)):

$$D_{ij} = \frac{1}{3}\left[Q_{ij}^{(1)}\left((-t)^3 - (-2t)^3\right) + Q_{ij}^{(2)}\left(0 - (-t)^3\right) + Q_{ij}^{(3)}\left(t^3 - 0\right)\right.$$
$$\left. + Q_{ij}^{(4)}\left((2t)^3 - t^3\right)\right]$$
$$= \left[\frac{1}{3}Q_{ij}^{(1)}(-t^3 + 8t^3) + Q_{ij}^{(2)}(t^3) + Q_{ij}^{(3)}(t^3) + Q_{ij}^{(4)}(8t^3 - t^3)\right]$$
$$= Q_{ij}^{(1)}\frac{7}{3}t^3 + Q_{ij}^{(2)}\frac{1}{3}t^3 + Q_{ij}^{(3)}\frac{1}{3}t^3 + Q_{ij}^{(4)}\frac{7}{3}t^3$$

Values of A_{ij}, B_{ij} and D_{ij} are tabulated below:

	$Q_{ij}^{(1)}$	$Q_{ij}^{(2)}$	$Q_{ij}^{(3)}$	$Q_{ij}^{(4)}$
A_{ij}	t	t	t	t
B_{ij}	$-3/2\ t^2$	$-1/2\ t^2$	$1/2\ t^2$	$3/2\ t^2$
D_{ij}	$7/3\ t^3$	$1/3\ t^3$	$1/3\ t^3$	$7/3\ t^3$

Matrix [A] is independent of stacking sequence. 0/90/90/0, 90/0/90/0 and 0/0/90/90 have identical Matrix [A].

Matrix [B] is only zero if laminate is balanced. 0/90/90/0, $\theta/-\theta/-\theta/\theta$.
$\theta/-\theta/\theta/-\theta$ will not have zero Matrix [B]; e.g., $45/-45/45/-45$.

$$B_{16}\text{ and }B_{26} = \left[4.5\left(-\frac{3t^2}{2}\right) - 4.5\left(-\frac{t^2}{2}\right) + 4.5\left(\frac{t^2}{2}\right) - 4.5\left(\frac{3t^2}{2}\right)\right]10^6$$
$$= [9t^2]\,10^6$$

Matrix [D] is dependent on stacking sequence.

Note: D_{16} and D_{26} are zero for $+\theta/-\theta/+\theta/-\theta$ but not for $+\theta/-\theta/-\theta/+\theta$;
e.g., $+45/-45/+45/-45$ and D_{16} value is:

$$D_{16} = \left[4.5\left(\frac{7t^3}{3}\right) - 4.5\left(\frac{t^3}{3}\right) + 4.5\left(\frac{t^3}{3}\right) - 4.5\left(\frac{7t^3}{3}\right)\right]10^6 = 0 = D_{26}$$

Under layup of $+45/-45/-45/+45$ and D_{16} value is:

$$D_{16} = \left[4.5\left(\frac{7t^3}{3}\right) - 4.5\left(\frac{t^3}{3}\right) - 4.5\left(\frac{t^3}{3}\right) + 4.5\left(\frac{7t^3}{3}\right)\right]10^6$$
$$= [18t^3]\,10^6 = D_{26}$$

(B) Consider an 8 ply laminate.

	$Q_{ij}^{(1)}$	$Q_{ij}^{(2)}$	$Q_{ij}^{(3)}$	$Q_{ij}^{(4)}$	$Q_{ij}^{(5)}$	$Q_{ij}^{(6)}$	$Q_{ij}^{(7)}$	$Q_{ij}^{(8)}$
A_{ij}	t	t	t	t	t	t	t	t
B_{ij}	$-7/2\ t^2$	$-5/2\ t^2$	$-3/2\ t^2$	$-1/2\ t^2$	$1/2\ t^2$	$3/2\ t^2$	$5/2\ t^2$	$7/2\ t^2$
D_{ij}	$37/3\ t^3$	$19/3\ t^3$	$7/3\ t^3$	$1/3\ t^3$	$1/3\ t^3$	$7/3\ t^3$	$19/3\ t^3$	$37/3\ t^3$

Matrix [A] [B] and [D] are summarized below:

Layup sequence	$[90/0/45/-45]_s$	$[90/-45/0/45]_s$	$[0/45/-45/90]_s$
Matrix [A] =	$\begin{bmatrix} 71.2 & 21.6 & 0 \\ 21.6 & 71.2 & 0 \\ 0 & 0 & 24.8 \end{bmatrix} 10^6 t$	$\begin{bmatrix} 71.2 & 21.6 & 0 \\ 21.6 & 71.2 & 0 \\ 0 & 0 & 24.8 \end{bmatrix} 10^6 t$	$\begin{bmatrix} 71.2 & 21.6 & 0 \\ 21.6 & 71.2 & 0 \\ 0 & 0 & 24.8 \end{bmatrix} 10^6 t$
Matrix [B] =	$\begin{bmatrix} 0 & 0 & 0 \\ 0 & 0 & 0 \\ 0 & 0 & 0 \end{bmatrix} 10^6 t^2$	$\begin{bmatrix} 0 & 0 & 0 \\ 0 & 0 & 0 \\ 0 & 0 & 0 \end{bmatrix} 10^6 t^2$	$\begin{bmatrix} 0 & 0 & 0 \\ 0 & 0 & 0 \\ 0 & 0 & 0 \end{bmatrix} 10^6 t^2$
Matrix [D] =	$\begin{bmatrix} 339 & 48 & 18 \\ 48 & 555 & 18 \\ 18 & 18 & 65 \end{bmatrix} 10^6 t^3$	$\begin{bmatrix} 233 & 81 & -54 \\ 81 & 593 & -54 \\ -54 & -54 & 98 \end{bmatrix} 10^6 t^3$	$\begin{bmatrix} 613 & 99 & 36 \\ 99 & 301 & 36 \\ 36 & 36 & 115 \end{bmatrix} 10^6 t^3$

Design Failure Criteria

Determining whether a design adequately meets design criteria is more difficult with composites than with conventional metal materials because failure criteria are more complex in composites, which can be delaminate or have either matrix or fiber failure. Therefore, most engineers use conservative approaches which needlessly limit where composites can be applied.

Since composite failure modes often operate concurrently, interactively, sequentially with the different modes of failure of a unidirectional fiber composite laminate. The most
- Establish low design strain level from testing
- No matrix failure at limit load
- No fiber failure at ultimate load

Fig. 7.3.12 shows a reasonable set of criteria that satisfy the desire to deal separately with the different modes of failure of a unidirectional fiber composite laminate. The most commonly used are the maximum strain, the maximum stress, and the quadratic criteria which is an empirical description of the failure of a composite component subjected to complex states of stress or strains.

(a) Maximum strain theory:
- No interaction between strains
- Failure occurs whenever:

$\epsilon_1 = \epsilon_L^{UT}$ or ϵ_L^{UC}

$\epsilon_2 = \epsilon_T^{UT}$ or ϵ_T^{UC}

$\gamma_{12} = \gamma_{LT}^{SU}$

where: UT — Ultimate tension strain or stress
UC — Ultimate compression strain or stress
SU — Ultimate shear strain or stress
Y — Yield stress or strain
T — Longitudinal direction
L — Transverse direction
LT — Long transverse

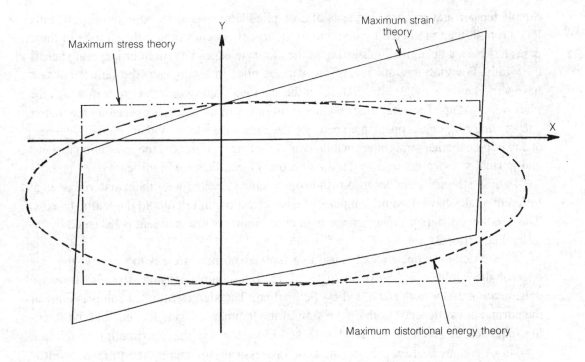

Fig. 7.3.12 Design Failure Strength Envelope Comparison (Ref. 7.16)

(b) Maximum stress theory:
- No interaction between stresses
- Failure occurs whenever:
  $F_1 = F_L^{UT}$ or F_L^{UC}
  $F_2 = F_T^{UT}$ or F_L^{UC}
  $F_{12} = F_{LT}^{SU}$

(c) Hill's (Azzi-Tsai) maximum distortional energy theory:
- Based on metallic yielding theory, interaction between stresses
- Elliptical curve-fit
- Yielding occurs whenever:
  $$(F_1^2 - F_1 F_2)/(F_L^Y)^2 + F_2^2/(F_T^Y)^2 + F_{12}^2/(F_{LT}^{SU})^2 = 1$$

(d) Tsai-Wu strength tensor theory:
- Tensorial so that it is subject to transformation
- Interaction between stresses
- Failure occurs whenever:
  $K_i F_i + K_{ij} F_{ij} = 1$ (i,j = 1, 2, 6)
  Where K_i, K_{ij} — strength tensors (inverse) that require off-axis tensile tests to evaluate

Free Edge Effects

Simple tension and compression tests of crossplied laminates have exhibited significantly less strength than predicted by lamination theory. Frequently, the failure mode in these cases has been a delamination starting at the laminate edges and often propagating across the width. For the same amount of the various plies in a laminate, the failure stresses have varied quite strongly, depending on the stacking sequence or the order in which the plies were laid up. These failures resulted from interlaminar stresses exceeding the matrix strength and edge cracks opening up between the plies at the edges. Within laminate thickness of any edge, whether straight or on a hole or cutout circumference, the interlaminar shear and normal stresses ignored by lamination theory can become significant.

These stresses arise because orthotropic lamina stacked with their material axes at different angles have dissimilar inplane Poisson's ratios, in relation to the laminate axes. This produces different inplane stresses in each lamina whose moment is balanced by interlaminar stresses.

The stacking sequence can cause interlaminar normal stresses to occur at the free edge of the laminate as shown in Fig. 7.3.13. Interlaminar tension stresses can cause delamination under both static and cyclic loading. The sign (tension or compression) of the normal stress depends both on the sign of the laminate inplane loading and the stacking sequence. A given ply set can be stacked in such a way that maximum or minimum tension or compression σ_z can be obtained. In classical lamination theory, no account is taken of interlaminar stress such as σ_z, σ_{xz} and σ_{yz}. Accordingly, classical lamination theory is incapable of providing predictions of some of the stresses that actually cause failure in a composite material. Interlaminar stresses are one of the failure mechanisms uniquely characteristic to composite materials. Moreover, classic lamination theory implies values of τ_{xy} where it cannot possible exist, namely at the edge of a laminate. Physical grounds can be used to establish that:

- At the edges of a laminate, (or hole), the interlaminar shearing stress is very high (perhaps even singular) and would therefore cause the debonding that has been observed in these areas
- Layer stacking sequence changes produce differences in the tensile strength of a laminate even though the orientations of each layer do not change (in classical lamination theory, such changes would not effect stiffness). Interlaminar normal stress (σ_z) changes near the laminate boundaries are believed to provide the answer to such strength differences

The effect of stacking sequence on the interlaminar stress is shown in Fig. 7.3.14.

It is therefore important to consider the stacking sequence in selected orientations to minimize interlaminar stresses. To prevent free edge effects:

- It is good practice to minimize the angle between adjacent tape plies
- The ply stacking sequence should be as homogeneous as possible in order to reduce the effects of interlaminar stress
- Drop-off plies of the same orientation should be dispersed as uniformly as possible
- Keep to a minimum the grouping of plies of the same orientation
- Unitapes are expected to be more sensitive to edge effects than cloth plies

Laminate Design Practices

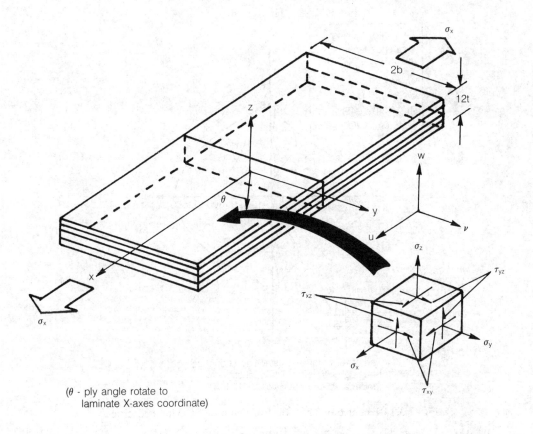
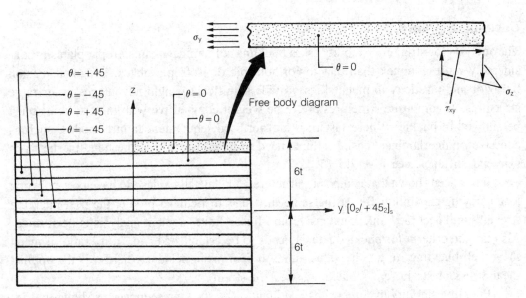

Fig. 7.3.13 Interlaminar Geometry and Stresses

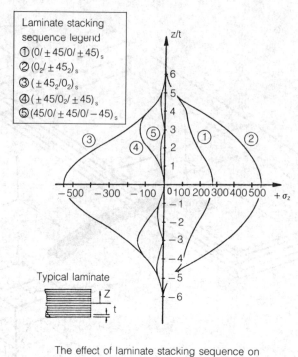

The effect of laminate stacking sequence on free-edge interlaminar normal stresses (under axial compressive $\sigma_x = 48$ ksi).

Fig. 7.3.14 Example of The Effect of Laminate Stacking Sequence

Buckling Stability

The analysis for structural stability (skin buckling) of laminated anisotropic plates is considerably more complex than that for orthotropic or isotropic plates. Diagonal tension behavior and plasticity in metal skins provide significant weight advantages in airframe structures. If composite structures are to be weight competitive with metals, skins must be allowed to buckle. Where fasteners are not used, equivalent failure modes, such as adhesive or interlaminar tensile failure, need to be considered at the bondline between skin and stiffener; see Fig. 7.3.15.

It has been shown that composite structures, like metallic ones, do have post-buckling load carrying capability. But at the same time it is difficult to predict the post-buckling strength with accuracy unless extensive non-linear finite element analysis is performed. Also fatigue criteria for buckled panels needs to be established, i.e., q/q_{cr} ratio of metal shear web buckling. (e.g., the value equal to 5 in metal design; where q is the applied shear stress or strain; q_{cr} is buckling stress or strain)

Buckling stability in composites is a function of stacking sequence as shown in Fig. 7.3.16 and Fig. 7.3.17, and the following design considerations improve this capability:
- Buckling strength can be improved by placing $\pm 45°$ plies on the outer surfaces. The maximum shear buckling strength depends on the direction of th shear load. The direction which produces compression in the outermost ply will have a greater buckling load than if the outermost ply is in tension as shown in Fig. 7.3.18.

- Increase stiffener torsional stiffness to increase the panel critical buckling strength
- Two laminates that differ only in stacking sequence will have the same membrane stiffness but different shear stress distributions (see Fig. 7.3.18)
- Column buckling of plates can be improved by placing the 0° plies farthest from the midplane. Sometimes, depending on plate geometry, it may be beneficial to put ±45° plies on outside
- Shear buckling of plates can be improved by placing the ±45° plies farthest from the midplane

This problem will not be discussed further in this chapter; the references listed at back of this chapter will allow the reader to pursue this subject.

Although such analysis data are usually company proprietary information, some public domain data are available for use — refer to the list of computer programs from "Computer Programs for Structural Analysis" in Ref. 7.17.

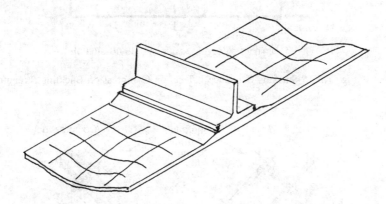

(a) Design which prevents separation

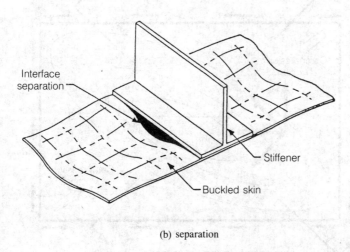

(b) separation

Fig. 7.3.15 Skin-Stiffener Interface After Skin Buckling

Chapter 7.0

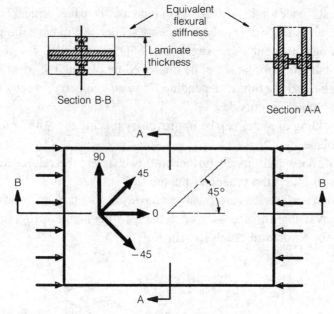

Fig. 7.3.16 Stacking Sequence Effects on Compression Buckling Strength

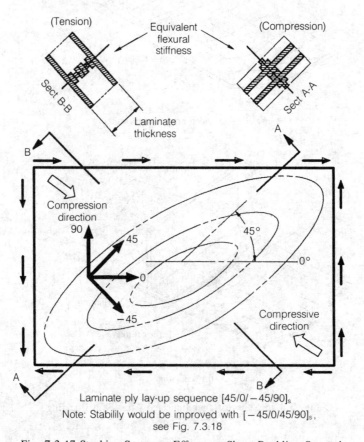

Note: Stability would be improved with $[-45/0/45/90]_s$, see Fig. 7.3.18

Fig. 7.3.17 Stacking Sequence Effects on Shear Buckling Strength

Laminate Design Practices

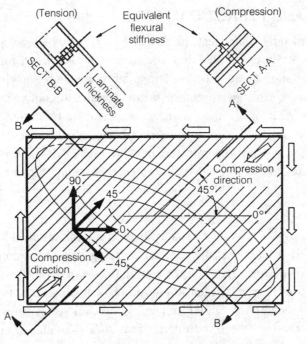

(a) Laminate ply lay-up sequence $[45/0/-45/90]_s$

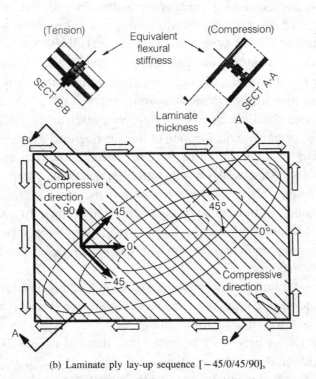

(b) Laminate ply lay-up sequence $[-45/0/45/90]_s$

Fig. 7.3.18 Use Outer Plies of 45° or −45 in Compression Due to Shear Loading to Increase Flexural Stiffness

421

7.4 FATIGUE AND IMPACT DAMAGE

Design of metal structural components for fatigue loading may be accomplished by relatively straightforward procedures. With composite structures, however, methodology is complicated by consideration of the effects of temperature, moisture, notches, impact damage, etc., and a variety of other performance factors unique to composites.

In composite structure, impact damage can override the fatigue problem in structural design because impacted compressive strength generally becomes the design allowable. Therefore, attention should generally be paid to impact damage rather than fatigue. Fig. 7.4.1 compares the differences between composite and metal structures.

Fatigue

Metal fatigue typically occurs by mechanisms of crack initiation and propagation. Final fatigue results from reduction of the net load-carrying area to the point at which the applied stress exceeds the material's ultimate strength. In general, the fatigue behavior of composites is superior to metal for tension-tension type loadings (see Fig. 1.3.5). But most structures are not loaded purely in tension and some composites show significant fatigue weakness in compression, shear and interlaminar shear loadings after impact, as shown in Fig. 7.4.2. Fatigue analysis of metallic structure involves many empirical equations derived from tests and experience. Such analysis data does not yet exist for composites because of the lack of database. Currently failures induced by flaws are analyzed using failure criteria (see Fig. 7.3.12) rather than the classic metallic fracture toughness approach. The following fatigue characteristics are unique to composites:

(a) Tensile fatigue — This includes crack initiation and growth as is typical with metals. However, instead of a single crack expanding, many micro cracks develop and grow. These could lead to failure by joining with each other to the extent that the matrix becomes unable to perform its function of transferring load from one fiber to another.
(b) Crack initiation — Cracks can result from stress concentrations, inclusions, or voids. With composites cracks usually result from filler bebonding, voids, fiber debonding, fiber discontinuities, or other stress concentrations (see Fig. 7.4.3).
(c) Crack propagation — In metals cracking is typically limited to a single crack which grows in an unimpeded manner with each cycle. But with composites many matrix cracks, which grow simultaneously, may propagate through the matrix or move along a fiber/matrix interface.
(d) Final failure — Failure in metals usually results in a relatively clean fracture surface. With composites failure is usually a complex combination of matrix and/or fiber failure or fiber pull-out, or sublaminate buckling.
(e) Because fiber fatigue properties are better than those of the matrix, it is desirable to have fiber domination in fatigue whenever possible. This is normally accomplished by using continuous or highly oriented fibers in the composite's principal stress direction. Generally, high modulus fibers such as carbon exhibit outstanding fatigue properties in unidirectional composites. This is because matrix strains in these composites are very low, resulting in fatigue performance almost totally dominated by the fibers.

Characteristics	Metal	Composites
Stress-strain behavior	Low strain to failure	High strain to failure
Notch sensitivity	$\dfrac{\text{Notched strength}}{\text{Unnotched strength}} > 1.0$	40-60% reduction in tension strength
Environmental affect on properties	Relatively insensitive	Affected by moisture and temperature
Causes of damage	Fatigue; corrosion; stress corrosion	Foreign object damage and fabrication damage
Type of critical damage	Cracks	Delamination
Critical damage tolerance loading condition	Tension	Compression
Detectability of damage prior to failure	Generally visual detectable	NDI detectable (may not be visually detectable)
Prediction capability	Good	Poor to nonexistent
Service experience	Extensive	Very limited

Fig. 7.4.1 Comparison of Metal and Composite Characteristics Affecting Fatigue and Damage Tolerance

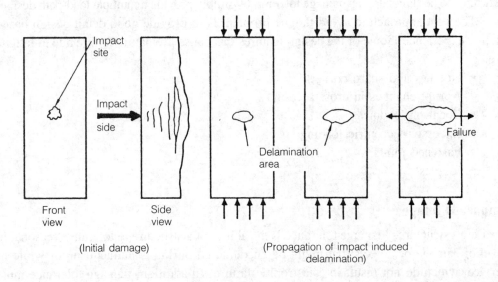

Fig. 7.4.2 Propagation of Delamination Due to Impact Under Cyclic Compression Loading

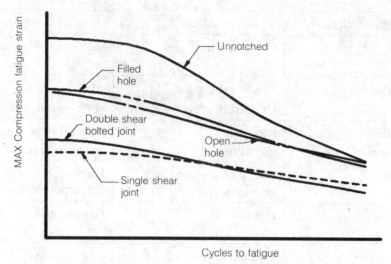

Fig. 7.4.3 Fatigue Effects on Notched Laminates

(f) Debonding and cracking of composites can occur relatively early in fatigue life, with microscopic cracks existing for most of its lifetime. For this reason, fatigue failure of composites is considered to occur gradually.
(g) Peel stresses in composite laminates are detrimental to fatigue life.
(h) The environment can significantly affect fatigue properties of adhesively bonded joints (hot/wet provide the worst case)

It should be noted that for orthotropic laminated composites, fatigue and fracture cracks are stable, and progress at a significantly slower rate than do cracks in isotropic metal which are unstable. Another safeguard against fatigue is to design the composite structures to be 'fail-safe' or 'damage tolerant' by simply provide a multiple load-path design.

The best approach to avoid fatigue problem is to provide good detail design based on past experience. Some of the design features that cause premature failure due to fatigue are:

- Notches and sharp corners
- Abrupt changes in cross-section
- Local ply padups
- Excessive eccentric loading
- Fastened joints
- Joggles

Impact Damage

In aircraft structures, damage tolerance is the ability of a structure to tolerate a reasonable level of damage or defects that might be encountered during manufacturing or while in service, which do not result in catastrophic failure. In addition, damage tolerance must be achieved in conjunction with maximum structural efficiency, minimum weight and minimum manufacturing, maintenance, repair, and supportability costs.

The damage tolerance requirements for both U.S. civil and military aircraft are described below:

(a) Civil aircraft — The requirements are established in U.S. Federal Aviation Regulation (FAR) 23 (Ref. 1.10) and (FAR) 25 (Ref. 1.11) and in the Advisory Circular AC-107 (Ref. 7.39). The latter is not a mandatory requirement, but provides both industry and the Federal Aviation Administration (FAA) with guidelines for composite structure compliance with FARs.
- Impact damage must be considered as part of the static ultimate allowables, up to the level of detectability
- Large impacts are considered under damage tolerance requirements

(b) Military aircraft — The U.S. Air Force delineates damage tolerance criteria for metal structures in MIL-A-83444 (Ref. 7.24). But the composite structural damage tolerance requirements are contained in MIL-A-87221 (Ref. 1.8) which covers composite durability and damage tolerance, and defines nondetectable flaws and damage, and corresponding design and load requirements.

Impact damage in composites, as shown in Fig. 7.4.4, is of paramount importance because of the tendency toward delamination, even when the impactor has low kinetic energy and does not appear to cause any damage. This includes incidents where small tools or hard packages are dropped on the composite surface. In all such situations, the first concern is detecting the degree of damage inflicted. It is well established that impacts which produce little or no surface damage detectable to the eye can cause severe internal damage, usually in the form of delamination as shown in Fig. 7.4.5. Sometimes such degradations do not markedly influence tension strength; sometimes they do. However, the resulting compressive capability can be lower than half the undamaged strength, as shown in Fig. 7.4.6. Once the damage becomes visible, the loss of compressive strength can be even greater.

Causes of damage include:
- Dropped tools
- Runway debris
- Damage by failed engine turbine blade(s)
- Hail and bird impact damage
- Mishandling
- Ballistics

Initial flaw/damage criteria are listed below:
- Scratches — Surface scratch 4 inches in length and 0.02 inches deep
- Delamination — Interply delamination equivalent to a 2 inch diameter circle (may be more or less critical depending on location)
- Impact damage — Damage from a 1.0 inch diameter hemispherical impactor with 100 ft-Lbs of kinetic energy or with that kinetic energy required to cause a dent 0.1 inches deep, whichever is less

Low energy impact (tool drop) zones are shown in Fig. 7.4.7.

It is impossible to design a structure with complete protection from all damage but use the following guidelines:
- Avoid designing with minimum gage whenever possible
- Design for repairability
- Design for replaceability

Chapter 7.0

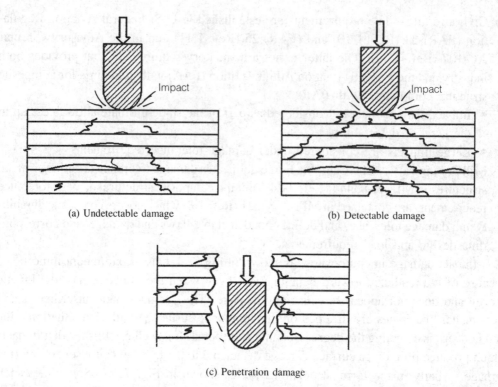

(a) Undetectable damage

(b) Detectable damage

(c) Penetration damage

Fig. 7.4.4 Types of Damage

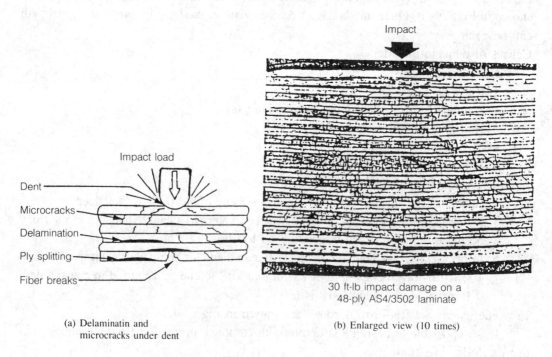

(a) Delaminatin and microcracks under dent

(b) Enlarged view (10 times)

30 ft-lb impact damage on a 48-ply AS4/3502 laminate

Fig. 7.4.5 Delamination After Impact

426

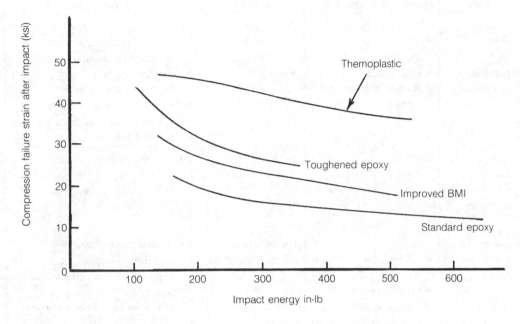

Fig. 7.4.6 Comparison of Compression Strength After Impact (Graphite Laminates)

Improve damage tolerance by:
- Minimizing the grouping of too many plies of same orientation
- Adding hybrid materials such as Kevlar, fiberglass, etc. to the basic carbon laminate to improve impact resistance, as shown in Fig. 7.4.8 (caution, it may cause thermally induced microcracking)
- Using 'soft skin design' as shown on curve ① in Fig. 7.4.8 by increasing the percentage of ±45° plies in a laminate (This is may not be a practical design for structures which must be bolted to ribs, spars, fittings, etc.)
- Using ±45° plies at the outer surface of the laminate to increase damage tolerance
- Using fabric plies at the outer surface to increase damage tolerance
- Using thermoplastic systems which are considered to be more damage tolerant than thermosets (see Fig. 7.4.6)

In any situation involving damage don't overlook these considerations:
(a) Parameters influencing degree and type of damage — Composite response to damage, including the various degrees of damage and its influence on residual strength cannot yet be computed with any high degree of accuracy. Hence, design guidance relies very heavily on experiments. It is obvious that laminate thickness is a very important parameter, and energy absorbed at a prescribed level seems to be roughly linearly dependent on part thickness, as shown in Fig. 7.4.9. Impact data is therefore frequently presented in terms of impact energy per laminate thickness.
(b) Effect of quantifiable damage on strength — It is important to note that the thicker the laminate, the greater the internal damage (such as delamination) is likely to be for a given amount of front face damage.

Zone	Damage source	Damage level	Requirements
Zone 1 - High probability of impact	(Tool drop) • 0.5 inch dia. impactor • Low velocity • Normal to horizontal surface	• Visible (0.1 inch deep) • 6 ft-lbs	• No functional impairment or structural repair required for two design life times and no water intrusion after field repair • No visible damage from a single 4 ft-lbs impact
Zone 2 - Low probability of impact	Same as Zone 1	Same as Zone 1	• No functional impairment after two design life times and no water intrusion after field repair
All vertical and upward facing horizontal surfaces	(Hail) • 0.8 inch dia • 90 ft./sec • Specific Gravity = 0.9 • Normal to horizontal surfaces • 45° angle to vertical surfaces	Uniform density 0.8 inch on center	• No functional impairment or structural repair required for two design lifetimes and no water instrusion after field repair
Structure in path of debris	(Runway debris) • 0.5 inch dia. • Specific gravity = 3 • Velocity approx. to airplane		• No functional impairment for two design lifetimes and no water intrusion after field repair

Fig. 7.4.7 Low Energy Impact Zones (Ref. 7.34)

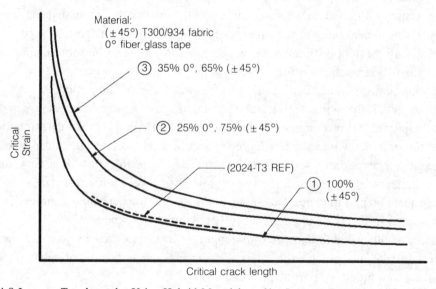

Fig. 7.4.8 Increase Toughness by Using Hybrid Materials and/or Increase Percentage of ±45° Plies

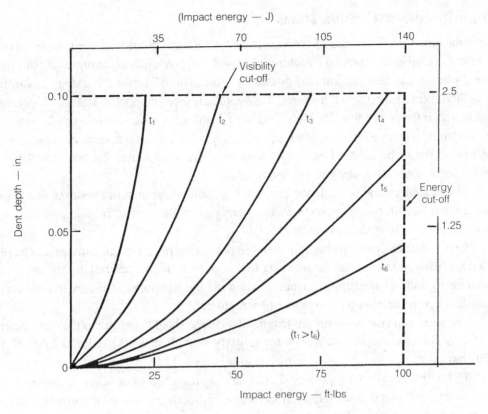

Fig. 7.4.9 High Energy Impact Criteria (Ref. 7.3.5)

(c) Diagnostic techniques whereby the damage is quantified — This is the province of non-destructive inspection (NDI) and will not be discussed in this chapter; see Chapter 10.

Variables that may influence impact damage:
- Impact velocity and location (bonded stringers are especially sensitive to this) are important
- Of particular importance is the part played in load transmission by the structure surrounding the impact point (one extreme is thin laminate; this is the hard point of stiffener locations)
- Delaminations caused by impacts that do not penetrate may have a more serious effect on compression strength than a clean hole of the same surface area.

Since damage due to impact is likely to occur during the lifetime of a structure, the design ultimate strain has to be restricted. The types of damage are:
- Type 1 Damage — Non-visible damage that will not grow to degrade the strength to less than design ultimate load
- Type 2 Damage — Visually detectable damage that must be repaired before residual strength degrades to less than design limit load

Chapter 7.0

Crash Resistance Considerations

Airframe structure fabricated of composite materials must provide at least the same level of safety as conventional metal structures. Fig. 7.4.10 shows that aluminium sustains more than 24 times the deformation and possesses more than 65 times the energy absorption capability of carbon composite material. Therefore, advanced composite material is generally considered inferior because its high degree of brittleness makes it less crashworthy. If an aluminium structure (especially a fuselage) is replaced with a composite material, energy absorption would be reduced and more structural break-up would be expected to occur unless some innovative designs are incorporated.

There is encouraging evidence that in a fire, composite material systems would provide greater burn-through protection; but pyrolysis of the matrix may impose an additional threat due to toxic gas release.

Experience has proven that considerable protection from critical damage can be provided for fuselage frames and center-wing fuel tanks of transport aircraft during survivable accidents by appropriate support structure allowing the composite structure to provide the same level of protection as conventional structures.

The structural energy-absorbing featurs of airframe should conform to the requirements of MIL-S-58095A and MIL-S-81771 for military aircraft and FAR 23 and FAR 25 for civil transports.

(1) Crashworthiness — Composites should have at least the same level of protection as conventional metal (aluminium) structures — difficult because of the brittle nature of the material, except when innovative design concepts are used. The crashworthiness of an airframe structure is measured by three major capabilities:
- The reduction of mechanical forces upon impact with the ground, debris or other objects
- The capability of the structure to remain intact and provide the occupants with protection in the event of a post-crash fire
- The maintenance of fuel tank integrity in a crash

Fig. 7.4.11 and Fig. 7.4.12 shows a few design concepts which improve energy-absorbing capability.

(2) Ballistic impact:

When a composite structure is subjected to ballistic impact damage caused by a projectile (e.g., 30-50 mm caliber canon, see Fig. 7.4.13) the damage generally consists of:
- A roughly circular hole
- The production of internal hydro-dynamic ram pressure [e.g., 30-50 psi (207-345 kPa)] in the structural box

The requirement for military aircraft such as fighters is that the design enables the aircraft to withstand the effects of damage caused by ballistic impact.

Design considerations are:
- Innovative design concepts
- Material selection, e.g., Spectra, Kevlar, etc.
- Material forms, e.g., 3-D preform (see Chapter 2.5)

Laminate Design Practices

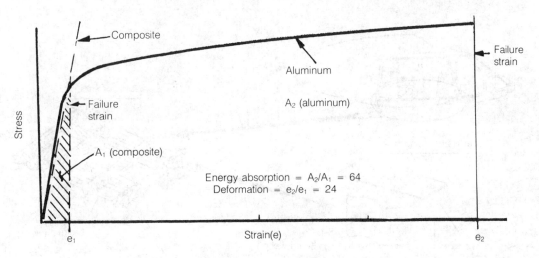

Fig. 7.4.10 Stress and Strain Relationship of Graphite/epoxy and Aluminum Material.

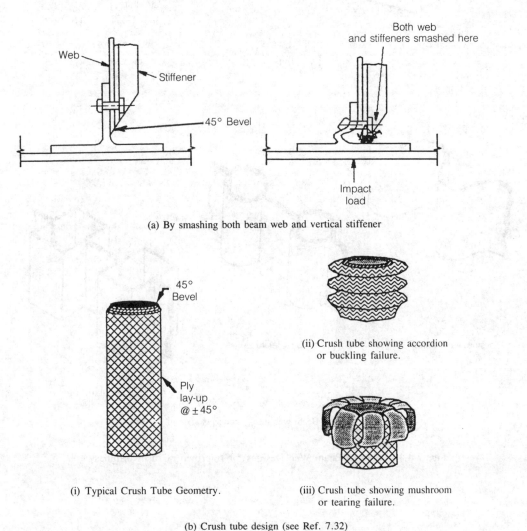

(a) By smashing both beam web and vertical stiffener

(b) Crush tube design (see Ref. 7.32)

Fig. 7.4.11 Examples of Design which Improves Crashworthiness

431

Chapter 7.0

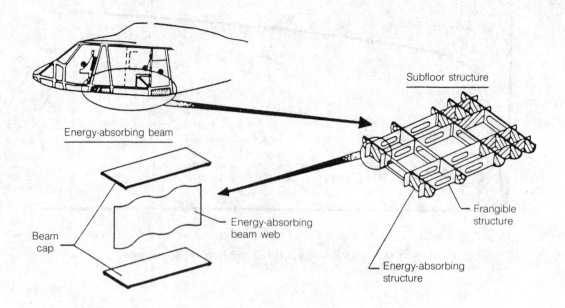

(a) Helicopter subfloor structure

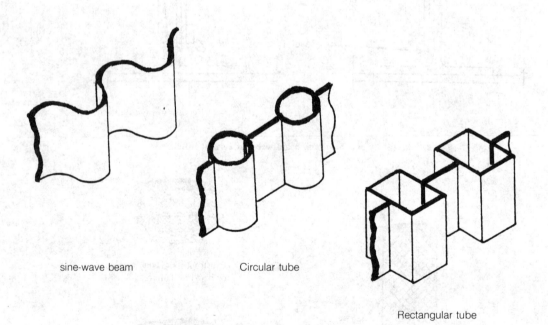

(b) Beam web design concepts

Fig. 7.4.12 Potential Beam Web Designs for Improving Crushworthiness

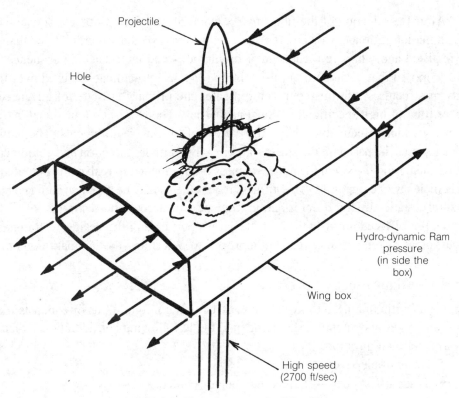

Fig. 7.4.13 Ballistic Damage of a Wing Box

7.5 DESIGN PRACTICES

It is recognized that one of the outstanding features of composite structures is their ability to be tailored, through crossplying, to match individual loading or stiffness requirements. This being so, it follows that large numbers of different crossplied laminates are likely to be encountered from one application to another. Each of these laminates is unique in its properties and characteristics and, hence, must be distinctly indentified whenever it is to be associated with specific quantitative or numerical data.

A laminate ply and orientation code has been defined in Fig. 1.7.4 (refer to the shorthand laminate orientation code in Chapter 1) which provides both concise reference and positive identification for a laminate. For example, $(0_5/90_2)$ indicates 5 plies in the 0° direction and 2 plies in the 90° direction.

A distinction is made between $(+)$ and $(-)$ angles; however, when there are the same number of $(+)$ and $(-)$ angles of the same magnitude, they may be combined into a $(\pm)$ type of notation in the listing; $(0_5/\pm45_3/90_2)_2$ indicates 10 plies in the 0° direction, 6 plies in the +45° direction, 6 plies in the −45° direction and 4 plies in the 90° direction.

It should be noted that these codes only signify the number of plies in each direction. They do indicate the stacking sequence of the plies within the laminate part. At the back of this book, Appendix B gives an engineering drafting practice which shows a method of indicating the stacking sequence of plies.

After the selection of the fiber/matrix material, the design process concentrates on the lamination rationale: for a particular cross-section of the structure, how many plies are required and what are their angular orientations and their stacking sequence? Some have proposed that a standard quasi-isotropic $(0/\pm 45/90)$ laminate be used over the majority of structures, but this approach mitigates the primary attribute of advanced composites, that of high specific directional properties. There exists no universal lamination geometry which effectively satisfies the variety of loading requirements. The lamination geometry must be based on the stress state (i.e., magnitude, direction and combined biaxial and shear) and the strength and/or stiffness requirements, to realize the potential structural efficiency of composites. The stiffness requirements may be based on laminate buckling (flexural), static structural deformation (inplane), or aeroelastic restraints, e.g., flutter. The degree-of-freedom of both the material properties and the design restraints should be sufficient to require a systematic approach to determine the lamination geometry.

Panel Configurations

There is a continuing need to explore new structural configurations or concepts for composite airframes and attention should be paid to the following considerations when selecting a panel configuration:
- Design objectives
- Trade-off studies — reduce number of choice
- Type of stiffening
- Type and arrangement of substructures
- Fabrication methods (cost)

Types of application on composite structure:
- (a) Reinforcement of metal structure
  - Selectively reinforced metals (see Fig. 2.3.4)
  - Weight reduction: small
- (b) Replacement or material substitution
  - Thickness change only — no change in geometry
  - Weight reduction: nominal
- (c) Redesign/no resizing
  - The same external geometry and interfaces
  - Optimize internal geometry for weight and/or cost
  - Good weight reduction
- (d) Redesign/resizing
  - Resize all geometry and interfaces for optimum weight
  - Maximum weight reduction

The following are skin panel design considerations:
- Structural efficiency
- Ease of manufacture
- Load transfer between stringer and skin or web
- Ease of repair
- Splice or joint compatibility

- Damage tolerance capability
- Compatibility of attachment to either frames (fuselage) or ribs (wing)
- Structure must be designed to support foot loads of 300 lbs distributed over a 3 inch (7.62 cm) length. This load applies in all directions and includes assembly and inspection activity
- Contour compatibility

Several currently used composite panel stiffening methods with their pros and cons are described below:

(1) The following panels are primarily used for axially loaded panels:
  (a) Blade stiffened panel [see Fig. 7.5.1(a)]:
   Pros
   - Simple to fabricate
   - Easy to stabilize at frame (fuselage) or rib (wing or empennage)
   - Simple tie-in at frame and rib
   Cons
   - Inefficient in bending
   - Edge wrapping may be needed to prevent delamination
   - Marginal torsional stability under axial loading
  (b) Bulb panel [see Fig. 7.5.1(b)(c)]:
   Pros
   - Fair in bending
   - Fair tie-in at frame or rib
   Cons
   - Difficult to splice
   - Difficult to fabricate because of possible compaction of bulb
  (c) Hat- (or bead) stiffened panel [see Fig. 7.5.1(d)(e)]:
   Pros
   - Torsionally stable
   - Double skin flanges permit wider spacing between stiffeners
   Cons
   - Difficult tie-in (requires blind fasteners or bonding) at frame or rib
   - Very difficult to splice
   - Entraps bilge fluid inside of the hats
  (d) J-stiffened panel [see Fig. 7.5.1(f)]:
   Pros
   - Simple tie-in at frame or rib
   - Double skin flange improves peel and post-buckling strength
   Cons
   - Torsionally unstable
   - Difficult to fabricate compared to blade
  (e) I-stiffened panel [see Fig. 7.5.1(g)]:
   Pros

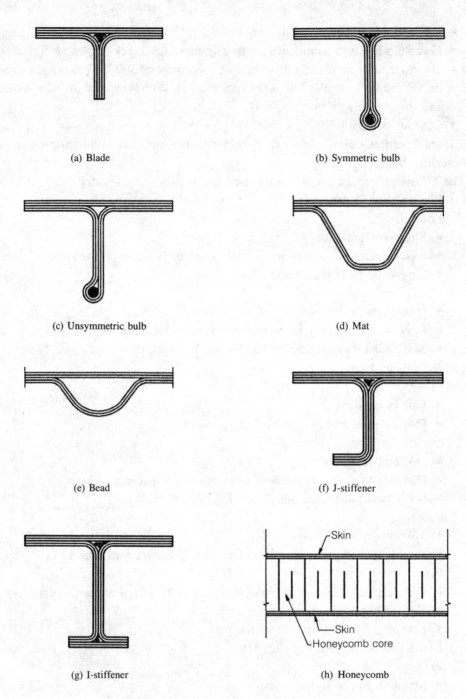

Fig. 7.5.1 Composite Panels

- Symmetric cross-section improves torsional stability
- Double flanges improve peel and post-buckling strength

Cons
- More difficult to fabricate
- Splicing difficult because of narrow flanges
- Difficult tie-in at frame or rib

(f) Honeycomb panel [see Fig. 7.5.1(h)]:

Pros
- Light weight
- Good stability
- Low cost
- Good accoustic fatigue resistance

Cons
- Moisture ingestion
- Closeout at edge and cutouts
- Poor damage tolerance

(g) Orthogrid panel (grids are composed of narrow strips of composite and syntactic materials as shown in Fig. 7.5.2):

Pros
- Structural continuity in frame and stringer
- Permits automated fabrication at low cost

Cons
- Difficult to splice
- Careful design needed to bond skin and grids
- Difficult to inspect (quality assurance and/or delamination damage)

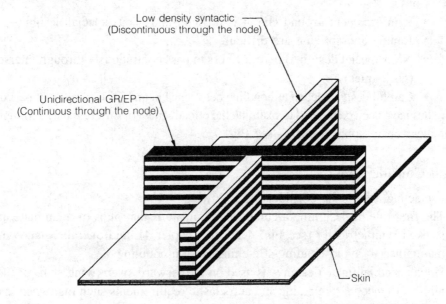

By courtesy of Lockheed Aeronautical Systems Co.

Fig. 7.5.2 Orthogrid Panel Concept

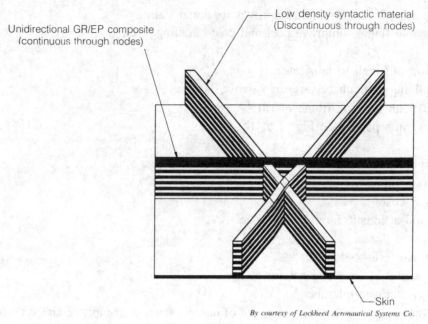

Fig. 7.5.3 Isogrid Panel Concept

(h) Isogrid panel (grids are composed of narrow strips of composite and syntactic as shown in Fig. 7.5.3):

Pros
- Excellent damage tolerance and fail-safe capability
- Structural continuity
- Permits automated fabrication at low cost

Cons
- Reinforcement around cutout is difficult and needs special design
- Joining and splicing are difficult
- Need special design to permit fibers to pass continuously through intersections (no joggles)
- Careful design needed at bondline between skin and grids to resist impact damage

(2) Studies have been conducted to examine the effectiveness of different panel configuration in shear beam applications — see Fig. 7.5.4.

Design Considerations

When selecting a laminate consider the following:
(1) The first rule of designing a composite laminate is that plies of a laminate must be stacked symmetrically (see Fig. 4.2.8 in Chapter 4), and overall balance must be maintained to avoid bending-stretching-torsion coupling.
(2) Unbalanced laminate design was used on the forward swept wing skins of the X-29 aircraft to meet aerodynamic and aeroelastic requirements such that when loads are applied to the wing it will twist in a nose-down position to eliminate aerodynamic divergency.

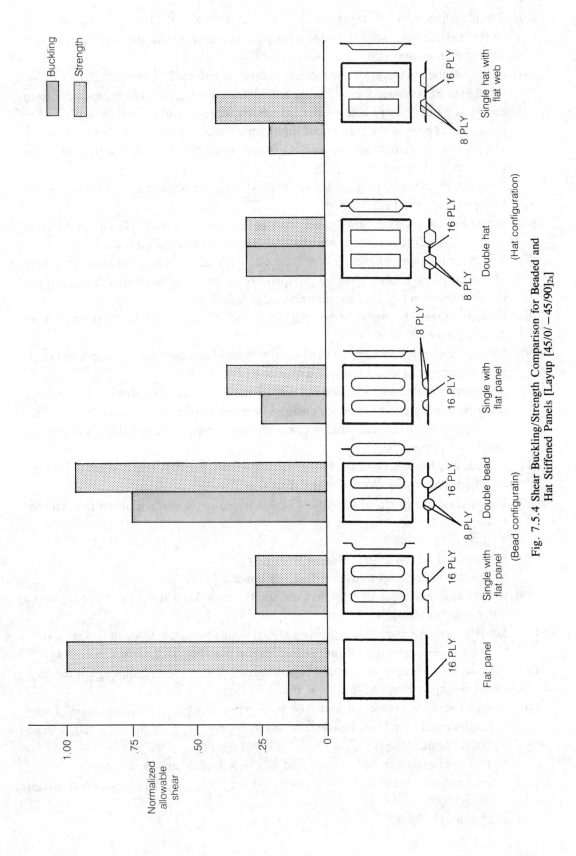

Fig. 7.5.4 Shear Buckling/Strength Comparison for Beaded and Hat Stiffened Panels [Layup $[45/0/-45/90]_{2s}$]

(3) The ply adjacent to a bonded joint should be oriented with the fibers parallel or ±45° to the direction of loading. Joints with plies oriented 90° to the loading direction have minimum strength.

(4) Adjacent plies (except woven fabrics) should be oriented (when possible) with no more than 60° between them. Studies have shown that microcracking can occur from curing stresses if there is a greater than 60° differences in the adjacent plies. The same rule applies to the transfer of interlaminar shear stresses. While not normally affecting static strength, this can affect fatigue strength. This rule can be waived for laminates with less than 16 plies.

(5) Avoid stress concentration design in areas where there is the potential for delamination problem, see Fig. 7.5.5.

(6) If possible, avoid grouping 90° plies; separate them by 0° or ±45° plies (0° is direction of critical load) to minimize interlaminar shear and normal stresses.

(7) Wherever possible maintain a homogeneous stacking sequence and avoid grouping of similar plies. If plies must be grouped, avoid grouping more than 5 plies of the same orientation together to minimize edge splitting.

(8) Exterior surface plies should be continuous and 45° or −45° to the primary load direction (not 0° or 90°)

(9) Hybrid laminates, consisting of two or more types of materials in the composite layup, may experience internal thermal expansion effects.

(10) Symmetry should be maintained not only within a laminate but within an assembly, if the elements are to be bonded, cocured or co-consolidated. (A possible exception to this rule is when local build-up or a doubler is required, but the doubler itself should by symmetric).

(11) Provide at 10% of each of the four ply orientations (the 10% rule design) to prevent direct loading of the matrix in any direction.

(12) Use at least one group of 45/90/−45 plies (tape) at the surface of the laminate to:
 • Improve damage tolerance
 • Improve stability
 • Reduce machining difficulties
 • Provide better load transition through bonded joints

(13) In the location of mechanically fastened joints, use at least 40% ±45° plies to maximize bearing strength.

(14) Local beef-up plies should always be tapered with a moderate slope of less than 10°, to prevent the occurence of fatigue separation (see Fig. 4.2.17 in Chapter 4).

(15) Honeycomb core cell sizes should be small enough to prevent 'dimpling' of the face sheets during curing under normal pressure loads.

(16) Aluminum core is not recommended because of potential galvanic corrosion problems:
 • If aluminum core must be used for strength reasons, it should be isolated from carbon laminates
 • This application should not be used where lightning strikes will occur

(17) Careful designs needed at locations prone to delamination (shown in Fig. 7.5.6) which is caused by:
 • Manufacturing defects

Laminate Design Practices

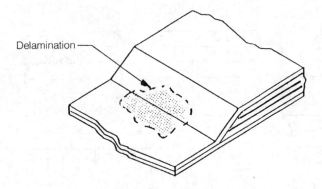

(a) High interlaminar load transfer

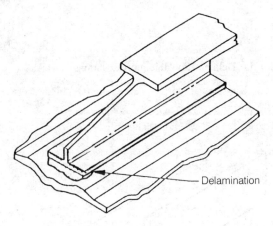

(b) Stringer run out

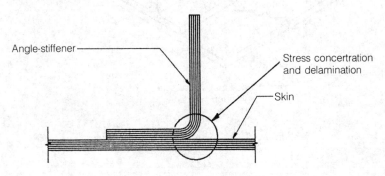

(This configuratin is not allowed)

(c) Bonded angle striffener to skin

Fig. 7.5.5 Potential Delamination Critical Locations

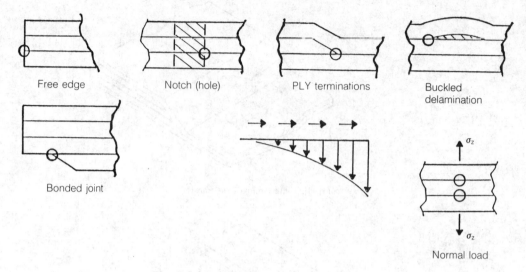

Fig. 7.5.6 Delamination Initiation Locations

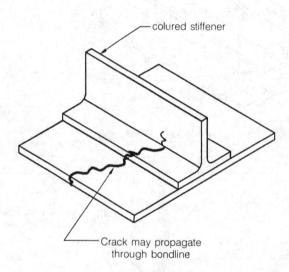

Fig. 7.5.7 Crack Propagation at Bondlines

- Out of plane loads
  — Applied loads giving rise to σ_z, τ_{xz} stresses
  — Ignoring the rules of laminate geometry
(18) Bonded structures are prone to cracking which passes across the bondline into adjacent structure (see Fig. 7.5.7)
(19) The stress concentration factor for a hole drilled in all 0° unidirectional carbon/epoxy is approximately 7, and reduces to 2 for a hole drilled in a laminate consisting of only ±45° plies
(20) Avoid dropping a 0° ply adjacent to a 90° ply.

(21) Flatwise tensile stresses are induced by moment in curved members or small angle bend radii (see Fig. 7.5.8).
(22) No liquid shim material should be used in integral fuel tank boundary structure because it degrades too easily.
(23) Design to reduce the Poisson's ratio, shown in Fig. 7.5.9.
 • Consider the use of 90° plies to reduce the Poisson's ratio effect
 • Reduce the percentage of 0° plies in a laminate
 • Reduction of Poisson's ratio is critical in bonded parts as shown in Fig. 7.5.9(a).
(24) Local beef-up thickness will increase fastener bearing strength (see Fig. 7.5.10)
(25) Edge bonding support of a rigid panel (e.g., honeycomb panel) should be soft in bending stiffness to avoid debonding, as shown in Fig. 7.5.11.
(26) Laminate thickness transition is always a problem area and careful design should be used as shown in Fig. 7.5.12.

For other design considerations refer to Practices in Section of 4.2 of Chapter 4.

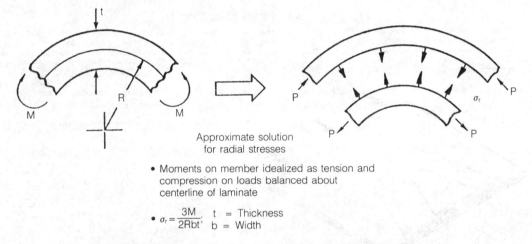

Approximate solution for radial stresses

• Moments on member idealized as tension and compression on loads balanced about centerline of laminate

• $\sigma_r = \dfrac{3M}{2Rbt}$; t = Thickness b = Width

Fig. 7.5.8 Flatwise Tensile Stresses Are Induced by Moments in Curved Parts

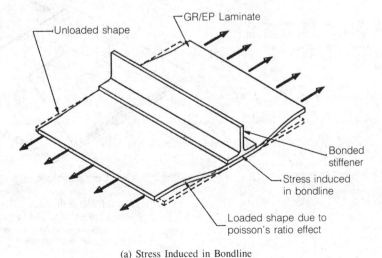

(a) Stress Induced in Bondline

Fig. 7.5.9 Poisson's Ratio Effect

Chapter 7.0

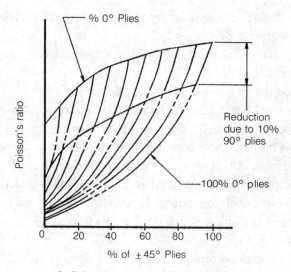

(b) Poisson's ratio versus % of ±45° plies

Fig. 7.5.9 Poisson's Ratio Effect (cont'd)

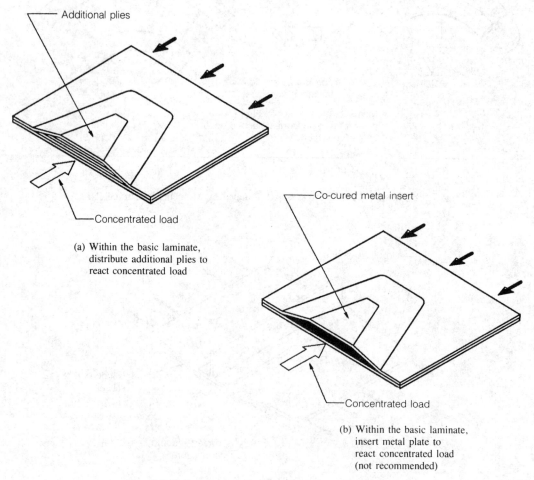

Fig. 7.5.10 Laminate under An Edge Concentrated Load

444

Laminate Design Practices

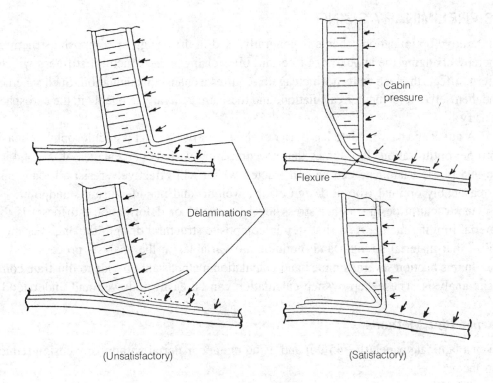

Fig. 7.5.11 Honeycomb Bulkhead Edge Supports

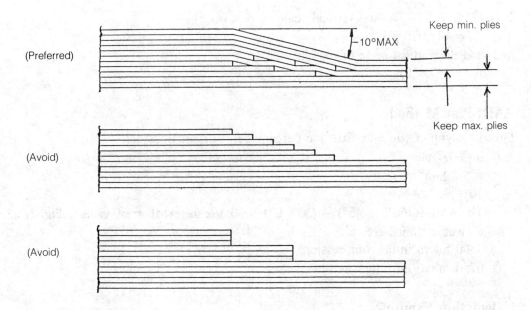

Fig. 7.5.12 Laminate Thickness Transitions (Symmetric and Balanced Laminate)

445

Chapter 7.0

7.6 PRELIMINARY SIZING

An iterative optimization scheme is generally used in the design of composite structures whereby changing the angularity of certain plies changes the structural stiffness which, in turn, affects the load pattern which again requires a change in angulation. Such schemes lend themselves to computer calculation, and these are generally available in the aerospace industry.

A quick and easy method has been developed for design that is a philosophy of quasi-isotropic optimization. It assumes that by combining various percentages of $0°$, $\pm 45°$, and $90°$ plies, a laminate can be constructed which will effectively resist all loads and be reasonably optimal from a design effort, weight, and manufacturing standpoint.

In structural design, either stress level (strength) or deformation (stiffness) is the material limiting factor. As a first step in composite structural design, the engineer must realize that material strength and modulus are variables in the analysis process.

In this Section several simple hand-calculation methods are provide to illustrate composite analysis. These step-by-step calculation can be easily followed and understood.

Carpet Plot Methods

Given a laminate panel $(0/\pm 45/90)$ and using unitape material without any design reduction factor:
 70% $0°$ plies
 20% $\pm 45°$
 10% $90°$ plies
From Fig. 7.2.2. read the vertical scale of plot (a),
 $F = 125,000$ psi
and read the vertical scale of plot (b),
 $E = 15 \times 10^6$ psi

AML Plot Method

Given a layup of $(0_3/\pm 45_3/90)$, the percentage of orientations are:
 30% $0°$ plies
 60% $\pm 45°$ plies
 10% $90°$ plies
The AML is $(60\% \pm 45°) - (30\% \ 0°) = 30$; use this AML = 30 value in Fig. 7.2.3. The allowable strains are:
 4500 micro in/in (compression)
 6750 micro in/in (tension)

Calculation Examples

Example 1:
To demonstrate some of the difference between designing with metals and composites, consider the case of a cylindrical pipe loaded by internal pressure [see Fig. 7.6.1(a)]. Material strength theory gives:

circumferential (y) stress: longitudinal (x) stress:
$$f_y = (pr)/t \qquad f_x = f_y/2 = (pr)/2t$$
where r is the average pipe radius and t is the pipe wall thickness (assume the thickness is reasonably small compared with the pipe radius). Therefore, for any given pressure and radius, the required wall thickness can be directly calculated for any given material.

For composites, however, the engineer can determine an optimum orientation so that the pipe is stronger in the direction of maximum load (f_y) than in the direction of the lesser load ($f_x = f_y/2$). It is apparent that, since shear strength is not a requirement, some combination of 0° plies (along the length of the pipe) and 90° plies (circumferentially around the pipe) will provide the optimum structure.

The solution to the design is shown in Fig. 7.6.1(b). This verifies what would be intuitive for this type of loading case: that 67%, or two-thirds, of the laminae should be oriented in the 90° direction, since the stress in that direction is twice the stress in the 0° direction. The required wall thickness can then be calculated.

Thus, the composite pipe is designed in a more optimal manner than the metal pipe, with the biaxial strength of the pipe matched in the type of loading applied.

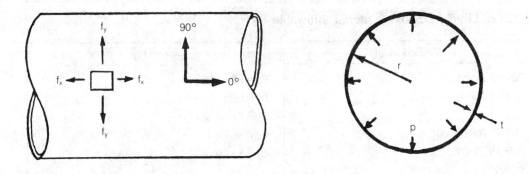

(a) Pipe geometry

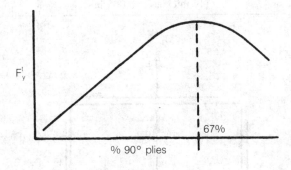

(b) F_y^t versus percent 90° plies for $f_y/f_x = 2$.

Fig. 7.6.1 Cylindrical Pipe with Internal Pressure

Chapter 7.0

Example II:

(a) Given load: $N_x = 7000$ lb/in; $N_y = 0$;
$N_{xy} = 4500$ lb/in

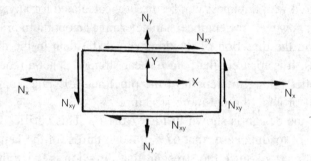

Requirements: $GT = 600000$ lb/in
$ET = 1.5 \times 10^6$ lb/in
where T = total thickness of laminate

(b) Given Graphite/Epoxy material allowables:

Unidirectional Ply	±45° Ply
$F_{11}^t = 98$ ksi $F_{11}^c = 74$ ksi $E_{11}^t = 20.5 \times 10^3$ ksi $E_{11}^c = 18 \times 10^3$ ksi $F_{22} = 0$ $E_{22} = 0$	$F_{11}^{45} = F_{22}^{45} = 15$ ksi $E_{11}^{45} = E_{22}^{45}$ $\quad = 3 \times 10^3$ ksi $F_{12} = 38$ ksi $G_{12} = 5 \times 10^6$ ksi
Ignore the strength from epoxy contribution	
Ply thickness = 0.005 in	

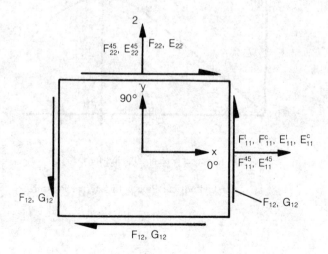

(c) Define the number of plies in each direction and the stacking sequence

   (i) Find number of $\pm 45°$ plies:
   Based on shear requirement,
   $N_{xy} = 4500$ lb/in

   $$\frac{N_{xy}}{tF_{12}} = \frac{4500}{0.005\,(38000)} = 23.68 \text{ plies}$$

   Based on shear stiffness requirement, $GT = 600000$ lb/in

   $$\frac{GT}{tG_{12}} = \frac{600000}{0.005\,(5)(10^6)} = 24 \text{ plies}$$

   A minimum total of 24 $\pm 45°$ ($+45°$ and $-45°$) plies are required to satisfy the above requirement and the total thickness $t_{45} = 0.005\,(24) = 0.12$ in.

   (ii) Find the number of 0° plies:
   Based on axial load requirement, $N_x = 7000$ lb/in

   $$\frac{N_x - F_{11}^{45} t_{45}}{tF_{11}^t} = \frac{7000 - (15000)(0.12)}{(0.005)(98000)}$$
   $$= 10.6 \text{ plies}$$

   Based on flexural stiffness requirement, $ET = 1.5(10^6)$ lb/in

   $$\frac{ET - E_{11}^{45} t_{45}}{tE_{11}^t} = \frac{1.5(10^6) - 3(10^6)(0.12)}{0.005\,(20.5)(10^6)}$$
   $$= 11.17 \text{ plies}$$

   A minimum total of 12 0° plies are required to satisfy the above requirement and the total thickness $t_0 = 0.005\,(12) = 0.06$ in.

   (iii) Find the number of 90° plies:

   Given $N_y = 0$

   Four 90° plies are arbitrarily added (about 10%) for lateral stability, crack propogation inhibition as well as to relieve the poisson's ratio effect.

   (iv) Total number of plies and stacking of the laminate:

   Total number of plies $= 0_{12}/\pm 45_{12}/90_4$ (40 plies)

   Laminate thickness $= 0.005\,(40) = 0.200$ in

   Recommended stacking: $[\pm 45_2/0_2/\pm 45_2/0_2/\pm 45_2/0_2/90_2]_2$

(d) Find allowable strength in y-direction:
From 90° plies:

   $(t_{90})\,(F_{11}^t) = [4(0.005)](98000) = 1960$ lb/in

From $\pm 45°$ plies:

   $(t_{45})(F_{22}^{45}) = (0.12)(15000) = 1950$ lb/in

From 0° plies:

$(t_0)(F_{22}) = (0.06)(0) = 0$

The allowable strength:

$N_y = 1960 + 1950 + 0 = 3910$ lb/in

or $F_y = \dfrac{3910}{0.200} = 19550$ psi

References

7.1 Brostow, W. and Corneliussen, R., "FAILURE OF PLASTICS", published by Hanser Publishers. 1986.

7.2 Wang, A.S.D., et. al., "FAILURE ANALYSIS OF COMPOSITE LAMINATES", published by Technomic Publishing Co., Inc. 1985.

7.3 Tsai, S.W., "COMPOSITE DESIGN", published by Think Composites, 3033 Locust Camp Road, Dayton, OH 45419. 1987.

7.4 Agarwal, B.D. and Broutman L.J., "ANALYSIS AND PERFORMANCE OF FIBER COMPOSITES", published by Wiley Interscience, New York, NY. 1980.

7.5 Sendeckyi, G.P., "COMPOSITE MATERIALS, VOL. 2 — MECHANICS OF COMPOSITE MATERIALS", published by Academic Press, New York, NY. 1974.

7.6 Broutman, L.J., "COMPOSITE MATERIALS, VOL. 5 — FRACTURE AND FATIGUE MATERIALS", published by Academic Press, New York, NY. 1974.

7.7 Chamis, C.C., "COMPOSITE MATERIALS, VOL. 7 — STRUCTURAL DESIGN AND ANALYSIS — PART I", published by Academic Press, New York, NY. 1974.

7.8 Chamis, C.C., "COMPOSITE MATERIALS, VOL. 7 — STRUCTURAL DESIGN AND ANALYSIS — PART II", published by Academic Press, New York, NY. 1974.

7.9 Dvorak, P.J., "Designing with Composites", MACHINE DESIGN, Nov. 26, 1987.

7.10 Rouse, N.E., "Building Composites with Computers", MACHINE DESIGN, Nov. 12, 1987.

7.11 Rouse, N.E., "Optimizing Composite Design", MACHINE DESIGN, Feb. 25, 1988.

7.12 MIL-HDBK-5D, "Metallic Materials and Elements for Flight Vehicle Structures", U. S. Government Printing Office, Washinton, D.C. 1986.

7.13 Anon., "Advanced Composites Design Guide, Vol. II", Air Force Materials Laboratory, Wright-Patterson AFB, OH. 1983.

7.14 Jones, R.M., "MECHANICS OF COMPOSITE MATERIALS", McGraw Hill Inc., 1975.

7.15 Ashton, J.E., Holpin, J.C. and Petit, P.H., "PRIMER ON COMPOSITE MATERIALS: ANALYSIS", published by Technomic Publishing Co., Inc. 1969.

7.16 Kaminski, B.E. and Lantz, R.B., "Strength Theories for Failure and Anisotropic Materials", Composite Materials: Testing and Design, ASTM STP 460, American Society for Testing and Materials. 1969. pp. 160-169.

7.17 Anon., "ENGINEERING MATERIALS HANDBOOK, VOL. I — COMPOSITES", ASM International, Metals Park, Ohio 44073, 1987. pp. 173-282 and pp. 417-495.

7.18 Leissa, A.W., "Buckling of Laminated Composite Plates and Shell Panels", AFWAL-TR-85-3069, Air Force Wright Aeronautical Laboratories, 1985.

7.19 Anbartsumyan, S.A., "THEORY OF ANISOTROPIC PLATES", published by Technomic Publishing Co., Inc. 1970.

7.20 Ashton, J.E. and Whitney, J.M., "THEORY OF LAMINATED PLATES" published by Technomic Publishing Co., Inc. 1970.

7.21 Vinson, J.R. and Chou, T.W., "COMPOSITE MATERIALS AND THEIR USE IN STRUCTURES", Applied Science Publishers, 1975.

7.22 Calcote, L.R., "THE ANALYSIS OF LAMINATED COMPOSITE STRUCTURES", Van Nostrand Reinhold. 1969.

7.23 Christensen, R.M., "MECHANICS OF COMPOSITE MATERIALS", John Wiley & Sons. 1979.

7.24 MIL-A-83444, "Airplane Damage Tolerance Requirements", U. S. Air Force, July, 1974.

7.25 Starnes, J.H. and Rouse, M., "Post-buckling and Failure Characteristics of Selected Flat Rectangular Graphite/Epoxy Plates Loaded in Compression", AIAA Paper No. 81-0543. 1981.

7.26 Smillie, D.G., "The Impact of Composite Technology on Commercial Transport Aircraft", AIRCRAFT ENGINEERING, May, 1983. pp. 2-10.

7.27 Mills, A.L. and Tanis C., "Advanced Composites Manufacturing As Influenced by Design", SAMPE Quarterly, Jan. 1977. pp. 22-29.

7.28 Duthie, T., "Developing New Design Methods", ADVANCED COMPOSITES ENGINEERING, Sept. 1986, pp. 10-13.

7.29 Anon., "FRP Composite Designs Involve Special Fatigue Considerations", Vol. 90, No. 5, Society of Automotive Engineers, Inc., 1982. pp. 51-57.

7.30 Kliger, H.S., "Simplified Design and with Graphite-Reinforced Plastics", MECAHNICAL ENGINEERING. June 1978. pp. 38-43.

7.31 Palmer, R.J., "Investigation of The Effect of Resin Material on Impact Damage to Graphite/Epoxy Composites", NASA CR 16577, 1981.

7.32 Egerton, M.W. and Gruber, M.B., "Thermoplastic Filament Wound Parts Demonstrating Properties in Crush Tube and Torque Tube Application", 34th International SAMPE Symposium, May 8-11, 1989. pp. 159-169.

7.33 Anon., "Predicting Performance of Advanced Plastic Composites", DESIGN ENGINEERING. May, 1980. pp. 39-44.

7.34 AFGS-87221A, "General Specification for Aircraft Structures", Air Force Guide Specification. June, 1990.

7.35 Whitehead, R.S., "Lessons Learned for Composite Aircraft Structures Qualification", Proceedings of the 1987 Aircraft/Engine Structural Integrity Program (ASIP/ENSIP) Conference, AFWAL-TR-88-4128. June, 1988. pp. 8-43.

7.36 Horton, R.E., Whitehead, R.S. and et. al., "Damage Tolerance Composites", AFWAL-TR-87-3030, three volumes. July, 1988.

7.37 Gallagher, J.P., Giessler, F.J., Berens, A.P. and Engle Jr, R.M., "USAF Damage Tolerance Design Handbook: Guideline for the Analysis and Design of Damage Tolerance Aircraft Structures", AFWAL-TR-82-3073. May, 1984.

7.38 O'Brien, T.K., "Fatigue Delamination Behavior of PEEK Thermoplastic Composite Laminates", J. Reinforced Plastics and Composites, Vol. 7, July, 1988. pp. 341-359.

7.39 Anon., "Composite Aircraft Structure Advisory Circular AC-107", U. S. Department of Transport, Federal Aviation Administration.

7.40 Pafitis, D.G., and Hull, D., "Design of Fiber Composite Conical Components for Energy Absorbing Structures", SAMPE Journal, Vol. 27, No. 3, May/June, 1991.

Chapter 8.0

STRUCTURAL TESTING

8.1 INTRODUCTION

Airframe structure design requires a continuing assessment of structure function to determine whether or not the requirements have been satisfied. The expected service performance must be assessed before the structure enters the service environment. This assessment is the structural testing which will ensure and substantiate structural integrity per certification criteria for either civil or military requirements. The basic "building block" approach, shown in Fig. 8.1.1, for testing of anisotropic laminate structures should be established in the early stages of development because the validation process for composite structures is very dependent on testing of all levels of the fabrication process.

Composite structural testing is similar to most metal structural testing (the majority of metal testing procedures are applicable to composite structures) in that it requires knowledge of design and analysis. The difference is that composites behave anisotropically and need thorough experimental testing, not only of the structure as a whole, but also of test specimens at the coupon, element, and component levels.

Design with composite materials requires a knowledge of lamination theory and appropriate failure criteria (see Chapter 7) as well as related analyses. These analyses must deal with the new set of material properties that result from the making of a laminate. Laminate properties test results are not useful to the engineer until the data is reduced, translated into design allowable, and then reported in a standard format that is easy understand.

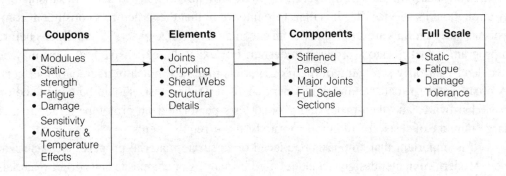

(a) Building block test sequence

Fig. 8.1.1 Building Block Testing Approach

Chapter 8.0

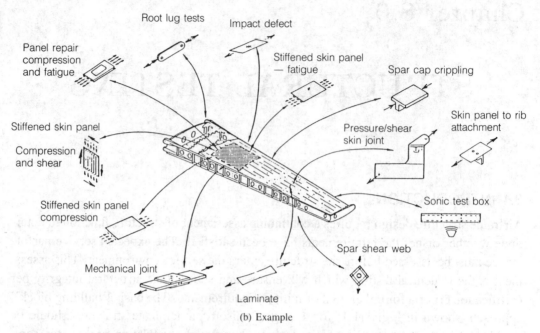

(b) Example

Fig. 8.1.1 Building Block Testing Approach (cont'd)

This chapter presents a comprehensive overview of the structural testing of composite materials and hardware to assist in validating design predictions and verifying methods of analysis. It is not possible to foresee and describe all the situations which might arise in a given experimental project. This chapter will not provide detailed description of those tests which are performed on a routine and highly repetitive basis. The engineer will instead be directed to the references which list appropriate published specifications and standard testing technical reports. These references will better serve to alert the engineer to potential difficulties, give deepening understanding of their purpose and limitations, and explain how the data must be reduced.

The purpose of a structural test program is to establish failure modes, demonstrate compliance with criteria, and correlate test results with theoretical predictions and thus assure confidence in the part or overall airframe structure that it will perform satisfactorily throughout its service life. It is therefore important that the structure's constituent components and materials are studied under an encompassing range of service conditions before a program is locked into a production design. For example, expensive redesigns may be avoided by an early screening of matrix materials to assess moisture degradation effects. A broad range of material and component characterization tests should be completed on basic hardware, and then proof and ground tests performed on prototype structures. A large number of tests are required to satisfy these requirements.

It is important that emphasis be placed on accurate material property characterization. Modern computer design techniques used in analysis of composite anisotropic materials are extremely dependent on, and sensitive to, the quality of the material property parameters which are furnished by the testing results.

General Considerations

Test engineers must constantly bear in mind the characteristics of composites and must be aware of the following test problems which are unique to composites and cannot be ignored:
- The variations in composite material strengths can lead to increased scatter in test results
- The very low interlaminar shear and transverse tension strengths of laminates, relative to their inplane strength, lead to difficulties in load introduction.
- Special precautions must be taken to diffuse the applied load into composites.
- The free edge effects causes some cross-plied test laminates to exhibit a tendency to delaminate, a failure mode fundamentally different from virtually all isotropic metal structures
- Composite materials are generally brittle, which tends to influence testing as follows:
  — Response is often linear up to failure, thereby easing the measurement problem
  — Small damage resulting from specimen preparation or pretest handling may result in high scatter, i.e., low test values
  — A small misalignment of fibers relative to the test direction may also cause significant data scatter
  — Lack of ductility around fastener holes means extra care must be taken in creating joints which diffuse the load into the structure.
- Testing to assess moisture-induced degradation is of course unique to composite and adhesive joints
- Non-visible damage that results from impact, in-process procedures, NDI limitations, etc. must be addressed in the material properties

Testing Standards

Most of the effort devoted to materials testing is directed toward the testing of small samples in order to evaluate the properties required for use in lamination theory analysis. While new improved methods are always being evaluated, much of this type of testing follows the standard procedures described in Ref. 8.1 and includes:

(1) The measurement of unidirectional coupons to determine the material moduli, Poisson's ratio, and failure stresses and strains in the direction of the two primary axes (0° and 90°). This refers to the fiber direction and normal to the fibers in tape while for cloth it is the warp and fill directions. This can only be accomplished when the test specimens have all plies laid up in the same direction and considerable effort is devoted to providing a uniform strain in the test section.

(2) Relative to testing to obtain the strength of cross plied laminates, using lamination theory and the failure criteria derived from the above unidirectional materials testing, a good estimation can be made of laminate failure. However, the testing of typical laminates is required to evaluate the accuracy of such predictions. Both material and laminate tests are repeated often enough to provide sufficient data to compute the "B" basis values of strength, stiffness, etc. The combined effects of moisture and temperature must be included in these tests. The degradation of matrix-dominated properties such as compression strength, due to moisture absorption is very important.

It must be remembered that laminate theory is not valid in the immediate neighborhood of joints, cutouts, free edges, simulated damage, and locations where concentrated loads are applied. Thus the material properties derived for use with lamination theory are an insufficient description of material behavior. To account for the behavior of the composite in the areas mentioned, it is necessary to test specimens containing the relevant source of stress concentration. This type of testing must be done on all structural elements tested.

The testing of fiber reinforced composites is so complex that often a new test concept can be evaluated more efficiently with the aid of a computer than by design and fabrication of expensive 'proof-of-method' test hardware which an analysis may indicate would not function suitably.

Types of Structural Testing

(1) Material Qualification Testing
 (a) Its purpose is to determine minimum mechanical properties and manufacturing processing requirements for qualification (see Fig. 8.1.2) to a process specification for:
   - New materials
   - Second source materials (requires approximately 70 coupons for same family materials)
   - Alternate source materials

Requirement	Specification
Basic resin type	Toughened thermosetting
Cure temperature	350°F
Service temperature	−65°F to 200°F
Uncured resin content	35±2% by weight
Volatile content	0.4% weight
Tack	Within specification
Areal weight	145±5 gm/m²
Storage life at 0°F	6 months
Out-time at 70% RH and temperature noted	10 days at 80°F max 15 days at 70°F max
Chemical characterization High pressure Liquid chromatography (HPLC) High resolution Infrared spectrophotometry Viscosity profile	 Information only Information only Information only

Fig. 8.1.2 Example of Matrix/Prepreg Property Requirements Tests (NASA/Industry Standard)

(b) Mechanical property tests (includes environmental effects):
- Strength requirements
- Stiffness requirements

(c) Manufacturing processing requirements:
- Cure and post-cure cycles
- Raw ingredients
- Resin flow and content
- Tack
- Out and storage time
- Drape
- Volatile content
- Fiber areal weight
- Temperature

It is recommended that a chart or plan be developed which summarizes the material qualification tests to be conducted for each new tape or bidirectional fabric composite material that is tested. These tests include the type of tests defined by standard testing methods and specified in each project. Specimen configurations should be defined. The numbers and types of tests for fabric laminates are the same as for tape laminates except for the number of plies used in the specimens. For fabric laminates each ply consists of fibers oriented at 0°/90° or ±45°. Fabric plies are thicker than tape plies and therefore fewer plies are required for equivalent laminate thickness.

The level of moisture conditioning will be determined on hole notched specimens from each batch (e.g., minimum 3 to 5 specimens per batch) used for the moisture conditioning tests. The fabricated specimens will be soaked in 160°F demineralized water up to 12 weeks. Each week the specimens will be removed and weighed. A plot of the percentage of moisture gain versus exposure time will be made to determine the percentage weight gain corresponding to approximately the 2/3 (67% is the real world case and should be used) saturation level for the wet tests. The same weight gain should be used for the moisture (wet) condition tests discussed in the subsequent sections of this chapter.

There are other recommended method for moisture conditioning which is accomplished in a moisture controlled chamber under 160°F temperature with 85% relative humidity (RH) condition.

(2) Coupon tests (static/fatigue-durability):
- Establish lamina material properties
- Establish laminate design allowable (design criterion varies for particular applications)

(3) Element and component tests (static/fatigue-durability) — Used to verify the structural adequacy of a particular structural configuration (e.g., co-cured panel, sine-wave web, beaded web, etc.)

(4) Full-scale tests (static/fatigue-durability) — Full-scale airframe, empennage stabilizer, canard, control surface, pylon, etc., is tested.

(5) Retrofit Test Program

A retrofit program is one in which the design objective is to substitute a composite part or component(s) for existing metal counterpart(s). In material substitution, the composite laminate thickness and stiffness will closely duplicate those of the metal part being replaced. A large number of tests should be performed to satisfy design requirements, but the expense will preclude such tests on most retrofit programs. Standard coupon testing is usually performed to either verify the allowable with minimum testing or to develop an allowable with substantial testing. The selected retrofit part should be designed and tested as a quick way to start verifying:
- The structural design without incurring costly design risk
- The manufacturing approaches and structural integrity

(6) Other tests:
- Environment (part of Material Qualification Testing)
- Lightning strike (see Section 6.3 in Chapter 6)
- Damage tolerance (part of Material Qualification Testing)
- Hydraulic ram (see Fig. 8.1.3)
- Ballistic impact
- Fluid resistance test (functional test)

Moisture and Temperature Effects

Composites are sensitive to both temperature and moisture absorption from service environment exposure which must be accounted for by means of the analysis and tests required for certification. The induced stress-strain reductions are caused by the structural response at the following levels:

(a) At the ply-level (or coupon level, see Fig. 7.2.1) — The need for testing at this level is due to the physical compliance of the laminate which is usually accounted for by ply-level properties. These moisture and temperature effects are accounted for in the stress analysis of the structure by using a reduced allowable.

(b) At the structural element or component level — The actual allowable strength of the composite element results from the use of different ply orientations, and the inherent structural redundancy of a high performance complex airframe structure. This is verified by testing.

The selection of environmental conditions, the relative humidity and temperature, to be used for the conditioning process is a function of the exposure conditions of the actual hardware and the time available for the test program.

Environmental conditions to be tested are:
- Temperature — cold, room and high
- Moisture content — 67% saturation and dry

Data Documentation

Test data involves identifying and tabulating the appropriate:
- Specifications
- Materials and processing records

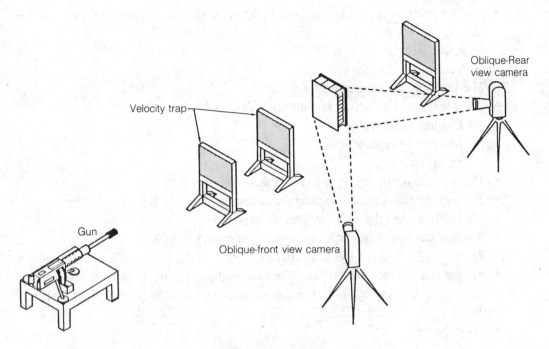

Fig. 8.1.3 Hydraulic Ram Testing Setup

- Inspection records
- Specimen preparation records
- Test data logs
- Stress-strain curves

Complete traceability is necessary for individual specimens regarding their:
  - Locations in the laminate
  - Inspection and processing records
  - Material identification — prepreg batch number and fiber lot number

Obtaining this level of traceability requires proper organization of the test program from the beginning, proper monitoring in progress, and collection of all documentation for data analysis. Controlling and monitoring all these steps virtually assures that most of the test data will be acceptable, thus minimizing retest requirements. Correlation of experimental and analytical data is one of the basic parts of data analysis.

It is easier to record test data and information than to reconstruct it later; the test report should include:
  - A complete description of the material, including:
    — Code numbers
    — Lot numbers
    — Run numbers
    — Any physical test data

- A description of the technique used to fabricate the parts especially noting layup room:
  — Out-time
  — Moisture conditions
- The number and orientation of plies
- The specimen fabrication techniques including:
  — Cutting methods
  — Moisture conditions
  — Storage times
- The measured dimensions of each specimen
- The instrumentation and methods of attachment, including storage time
- All information relating to deliberate exposures
- The test environment, including temperature and humidity
- The type of machine, the grip, and the speed of the test
- All the calculated values of modulus, strength, strain-to- failure, and Poisson's ratio, including any test points that were discarded and the reasons for discarding them
- The type of failure

8.2 DURABILITY AND DAMAGE TOLERANCE TESTS

An airframe must satisfy static, durability, and damage tolerance requirements to achieve structural integrity and reliability. Test data reveals that:
- A laminate can lose approximate 60% of its undamaged static strength with impact damage that is essentially non-visible
- The loss of strength in the cyclic-load durability test is also greatest from impact

Durability (Fatigue) Testing

Durability testing is generally understood to be fatigue testing, either constant cycle or longtime-cycle spectrum load testing. However, with composite materials, the effects of environmental exposure on static and dynamic behaviors must be considered. Thus, durability testing for composite structures becomes a function of load cycling and environmental exposure. Airframe durability testing, as shown in Fig. 8.2.1, is accomplished in a complicated and sophisticated manner, using a flight-by-flight real-time loading spectrum related to aircraft lifetime and, concurrently, environmental exposure based on flight temperatures and ground-based moisture environments. In addition, accelerated flight spectrum loading and accelerated moisture/temperature environments have been used to simulate real-time testing, with some success and some failure in correlation.

For fighter airframes, accelerated testing to 4 to 5 lifetimes under the worst environmental conditions would simulate one lifetime of real-time testing. Design requirements for fighter airframes generally state that the durability testing must simulate 4 real lifetimes; therefore, accelerated testing would require at least 16 lifetimes.

Structural Testing

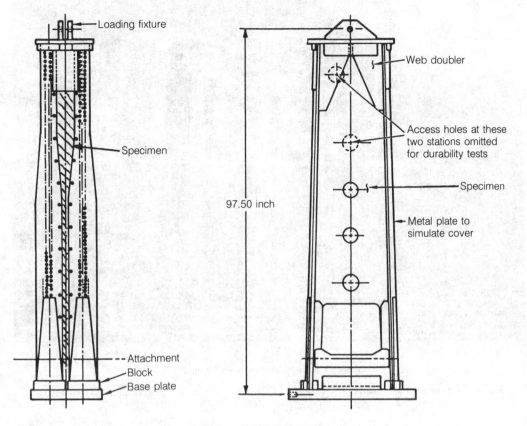

(a) Spar durability test specimen

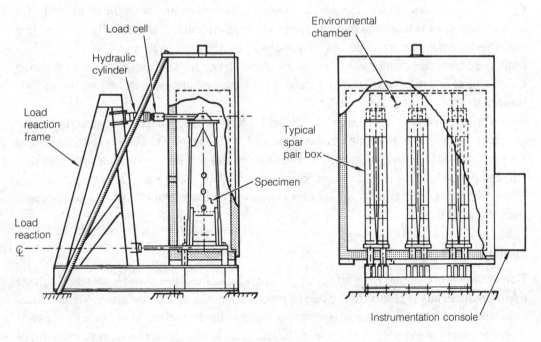

(b) Spar durability test setup

Fig. 8.2.1 Spar Segment Durability Test For L-1011 Vertical Fin Box (Graphite/Epoxy)

Chapter 8.0

By courtesy of Lockheed Aeronautical System Co.

(c) Specimen in the environmental chamber

Fig. 8.2.1 Spar Segment Durability Test of L-1011 Vertical Fin Box (Graphite/Epoxy) (Cont'd)

Fiber-dominated laminates are considerably more efficient in load-carrying ability than are matrix-dominated laminates; however the latter are sometimes needed, for multidirectional loadings and damage tolerance requirements. It is generally assumed that matrix-dominated laminate design is governed by durability strength, whereas fiber-dominated laminate design is governed by static strength. Therefore, durability testing for structural integrity verification of matrix-dominated laminates must necessarily include bonded joints.

The effects of cyclic loading on "current" carbon/epoxy composite structures have generally been shown to be non-critical due to low sensitivity to stress concentration at the low level fatigue load spectrum. The load threshold at which composites become sensitive to cyclic loading is a very high percentage of their static load. This threshold load is very high and most aircraft do not experience cyclic loads that even begin to approach their ultimate loads.

Damage Tolerance Testing

Damage tolerance testing is significantly different for composites than for metals. Damage tolerance in metals is related to the rate of propagation of a crack of a given size and location, whereas damage tolerance in composites is primarily dependent on resistance to impact.

Structures made with composite materials must be designed to support design loads after impact that has a reasonable probability of occurring during fabrication or service life. To define a strain allowable to account for impact damage compression stress is somewhat akin to defining fatigue allowable for metal structure (tensile stress is critical). The fatigue allowable are selected based on limited tests and previous design experience. However, final fatigue substantiation is based on durability (fatigue) tests conducted on full-scale components or the complete airframe. Compression tests are conducted on impact damaged coupons to select preliminary compression design stress allowable, and then compression tests of impact damaged structural panels and subcomponents are conducted to substantiate the design allowable.

To define design allowable for impact damage, tests are conducted on flat laminates loaded in compression. These may have varying amounts of impact damage, dependent on the panel thicknesses and the damage tolerance requirements for damage visibility and maximum impact energy. The panels must be large enough to nullify size effects, e.g., 10 inches (25.4 cm) × 12 inches (30.48 cm). The results are representative of damage to the areas of the structure between reinforcements (e.g., stiffeners). The effect of impact damage where reinforcements are attached to the skin or the effect on the reinforcements themselves is evaluated based on tests of reinforced panels.

Since strength and damage sustained may be a function of layup configuration, several variations of each laminate should be tested. The effect of environment will also need to be evaluated with tests for given conditions. Some tests should be conducted with higher impact energies to determine the trend of data for wider damage widths. It will also be necessary to conduct sufficient cyclic tests to ensure that no detrimental damage growth will occur during the expected service life.

Impact damage testing of a laminate is done on laminate panels, as shown in Fig. 8.2.2, using a support fixture with either simple or fixed edge support. Fig. 8.2.3 shows an example of integrally-stiffened panel in a test fixture. The damage requirements vary considerably, depending on mission and lifetime requirements. The requirements for a typical military composite structure are as follows (Ref. 7.34 and also see Section of 7.4 in Chapter 7):

(a) Low level impact damage:
- An impact of 6 ft-lbs (8.4 J) from an impactor with a 0.5 inch (12.7 mm) diameter hemispherical head
- The damaged laminate should have the capability of carrying static ultimate loads.

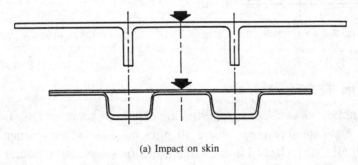

(a) Impact on skin

Fig. 8.2.2 Impact Locations On Test Panels Prior To Post-Impact Compression Test

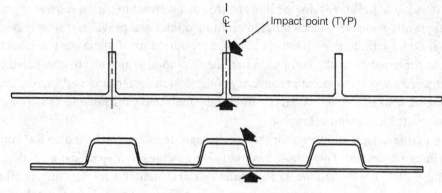

(b) Impact on stiffener and on skin-opposite-stiffener

Fig. 8.2.2 Impact Locations On Test Panels Prior To Post-Impact Compression Test (cont'd)

(b) High level impact damage tests:
- An impact of 100 ft-lbs (140 J) from an impactor with a 1.0 inch (25.4 mm) diameter hemispherical head; or an impactor which would not cause a dent deeper than 0.1 inch (2.54 mm)
- The damaged laminate should have the capability of carrying static limit load

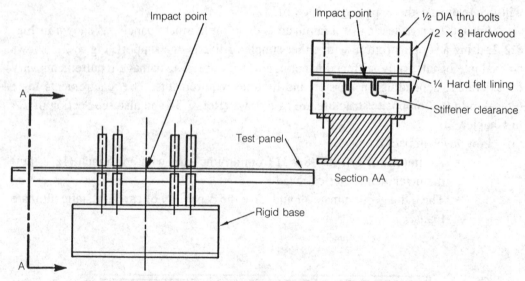

Fig. 8.2.3 Arrangement For Impact Tests Of An Integrally-Stiffened Panel

8.3 COUPON TESTS

Single ply (lamina; tape or fabric) properties are obtained experimentally from multi-ply, unidirectional laminate specimens where all plies have the same orientation. For tape laminates with all fibers aligned in the same direction (also may consider to test cross-plied laminates to obtain unidirectional properties), the ply properties needed for design (see Fig. 8.3.1) are:

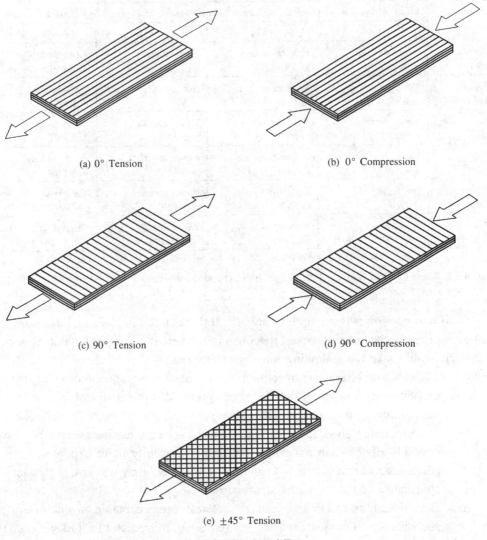

(a) 0° Tension

(b) 0° Compression

(c) 90° Tension

(d) 90° Compression

(e) ±45° Tension

Fig. 8.3.1 Ply Level Tests

- Ultimate strength values
- Elastic constants
- Poisson's ratio values

Test coupons that are designated to be weighed during the conditioning process should be weighed immediately after fabrication. All of the coupons are then stored in a dry, desiccated chamber prior to conditioning. It is important that the fiber volume (see Fig. 8.3.2) and void content of each coupon be known. Moisture is absorbed by the matrix, so the percentage of matrix in a given coupon will affect the amount of moisture absorbed. The size and concentration of voids present in the coupon must also be known. The relative humidity in the controlled chamber will determine the maximum moisture content of the conditioned test specimens (see curve shown in Fig. 6.2.1).

Property	Effect of fiber volume percentage (FVP) on the mechanical properties of the following laminates.*			
	$[0]_{nt}$	$[90]_{nt}$	$[\pm 45]_{ns}$	$[(\pm 45)_5/0_{16}/90_4]_c$
Ultimate Strength	Varies directly with FVP	Not sensitive to FVP	Varies directly with FVP	varies directly with FVP
Ultimate Strain	Not sensitive to FVP	Not sensitive to FVP	Not sensitive to FVP	Not sensitive to FVP
Prop. Limit Stress	Varies directly with FVP	Not senstive to FVP	Varies directly with FVP	Varies directly with FVP
Prop. Limit Strain	Varies directly with FVP	Not senstive to FVP	Varies directly with FVP	Varies directly with FVP
Poisson's Ratio	Not sensitive to FVP	Not sensitive to FVP	Varies Inversely with FVP	Not sensitive to FVP
Modulus of Elasticity	Varies directly with FVP	Varies directly with FVP	Varies directly with FVP	Varies directly with FVP

*The above deductions are valid for both tensile and compressive properties.

Fig. 8.3.2 Effect of Fiber Volume Percentage (FVP) on the Mechanical Properties of Test Laminates

Load introduction tabs are used to hold the test specimen in place and to insure that local stresses remain low so that the test specimen fails where it is designed to fail. Specimen tab design should take the following into considerations:

(a) Tabs for graphite/epoxy specimens are made of woven glass/epoxy or suitable graphite/epoxy prepregs that have been premolded under suitable pressures and temperatures into solid laminate sheets.
(b) Unidirectional glass/epoxy and graphite/epoxy tape laminates have been successfully used as tab materials, but they are usually more expensive.
(c) Metal tab materials such as titanium or steel may be used where higher tab-to-specimen bonded strengths are required.
(d) Both aluminum and magnesium tab materials are acceptable on small elements such as coupon or small articles where a small mismatch of CTEs between the tab and test specimen is acceptable.

A table should be made which summarizes the tests to be conducted to determine the lamina properties for tape and bidirectional fabric laminates. The specimens can be cut from the same laminate plate, but oriented so the loading direction is in the 0° direction and 90° direction as required.

Fig. 8.3.3 shows test fixtures for composites. The strain measurement can be obtained by using either strain gages and/or extensometers (see Fig. 8.3.4). Testing at elevated temperature requires certain modifications from standard practice because of the increased drying rate caused by heating. Ideally, testing in an environmental chamber (see Fig. 8.3.5) would eliminate the requirement to make undesired compromises by maintaining the moisture in the saturated specimen.

Structural Testing

(a) IITRI compression fixture has a unique design that reduces friction for more accurate test results.

(b) Boeing compression-after-impact fixture allows a reduction in material allocation.

(c) Split disc tensile fixture is used for testing reinforced plastic rings.

(d) Two-rail shear fixture is recognized industry-wide for consistent shear property determination.

(e) Iosipescu shear fixture allows strength and modulus to be obtained in a uniform shear stress state.

(f) Short beam shear fixture tests shear strength of composite materials. Meets the requirements of ASTM D2344.

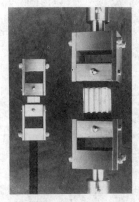

(g) Flat wise tension fixture is for transverse tension testing of composite sections.

(h) Climbing drum peel test fixture measures peel resistance of adhesive bonds.

By courtesy of Instron Corp.

Fig. 8.3.3 Various Testing Fixtures For Composites

Chapter 8.0

By courtesy of Optra Inc.

Fig. 8.3.4 Laser Extensometer

By courtesy of Instron Corp.

Fig. 8.3.5 Environmental Chamber System For Coupon Tests [Temperatures Range From −100°F (−73°C) to 400°F (200°C)]

Test results from composite materials tend to exhibit relatively large scatter and thus statistical confidence values are particularly influenced by sample population size; test matrices should be planned to allow for a statistically appropriate number of tests. An approximately 4000 coupons are needed to generate a final set of allowable for both "A" and "B" values.

There are several basic coupon tests which are routinely performed on composite materials. Two of these tests, tensile and compression, require the specimen to have special load introduction tabs (use on unidirectional specimens for modulus tests only; multi-directional specimens do not use tabs), furnish design data. The short beam shear and flexural tests, provide qualitative data which is more appropriate for acceptance evaluation and assessing processing adequacy. Two other coupon tests are the inplane shear and interlaminar shear tests.

(1) Tensile Test:

The strain measurement of tensile composite specimens can be sensitive to the specimen configuration (configuration does not affect such testing on metals). For unidirectional tape specimens, the recommended configuration is shown in Fig. 8.3.6 which is described in detail in Ref. 8.7

(2) Compression Test:

The measured strain for compression specimens is sensitive to specimen configuration and the fixture used for loading. The specimen must first be constrained from buckling as shown in Fig. 8.3.7; the recommended test fixtures, shown in Fig. 8.3.8 and Fig. 8.3.9, were developed for unidirectional composite test specimens (Ref. 8.8).

(3) Shear Test:

There are numerous shear test methods; some ASTM methods are listed below:

ASTM	Type	Description
D2344	Interlaminar	Short beam shear (3-point) (Ref. 8.9)
D3846	Interlaminar	Short beam shear (4-point) (Ref. 8.10)
D3518	Inplane	$\pm 45°$ tensile test (Ref. 8.11)
D4255	Inplane	Rail shear (Ref. 8.12)

The rail shear method shown in Fig. 8.3.10 tests inplane shear:
   (a) Test area of laminate is about 0.5 inch (1.27 cm) wide by 6.0 inches (15.24 cm) long; width may be reduced to avoid buckling
   (b) Rails may be bonded to specimen or faced with abrasive paper to avoid failure through bolt holes
   (c) Knife-edged bars may replace roller spacers to avoid induced transverse tension under larger shear deflections

Fig. 8.3.11 shows a method of testing interlaminar shear:
   (a) Two parallel cuts on opposite faces must both sewer the center lamina at the thickness midplane
   (b) Self-aligning grips are used to permit the load to pass through the centerline of the specimen

Chapter 8.0

Tensile test specimen. (1) Fiberglass tabs shall be positioned, both sides, two places; (2) Tabs shall be bonded with adhesives to graphite, dependent upon test temperature; (3) Specimen thickness shall not vary more than 0.076 mm. (0.003 in.) from nominal; (4) Specimen edges shall be parallel to 0.076 mm (0.003 in.); (5) Specimen shall be used for unidirectional (0°) and bidirectional (0°/90°), graphite tape laminates.

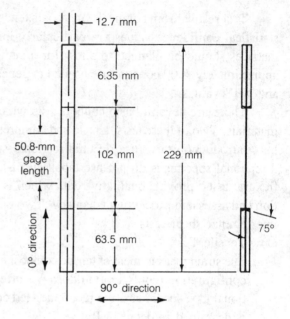

Fig. 8.3.6 Example Of Tension Test Specimen (GR/EP) (Ref. 8.7)

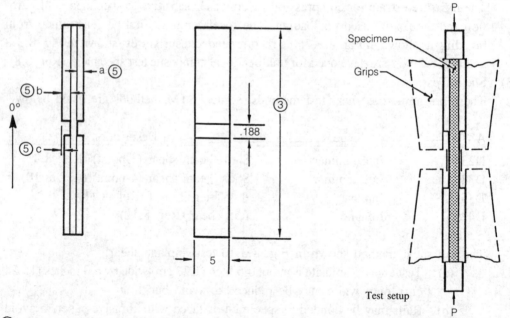

① Test panel layup same as for tensile test specimen
② Support tabs shall be fabricated from a unidirectional graphite material with 0 degree direction in the longitudinal direction of the specimen. Tabs shall be made of 12 plies of grade 190, or equivalent thickness when using other grades.
③ Specimen length is 3.150 ±0.005 inch.
④ Prepare the specimen prior to bonding the tabs by hand sanding the bonding area of the test panel with 150 grit sandpaper and sandblasting or sanding the tab to remove all surface gloss and cleaning thoroughly with acetone or MEK
⑤ A must equal b to within 0.010 inch and c must be less than 0.0025 inch. Rework of the sides of the tabbed specimen may be necessary.

Fig. 8.3.7 Example Of Compression Test Specimen (GR/EP)

Structural Testing

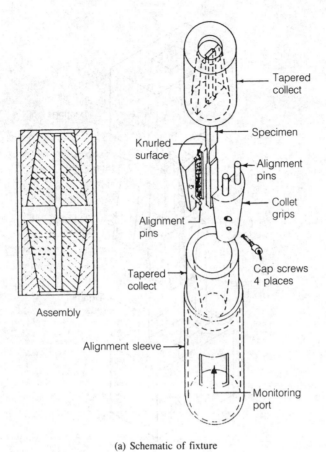

(a) Schematic of fixture

(b) Fixture parts   (c) Wyoming modified celanese compression test fixture

Fig. 8.3.8 Celanese Compression Test Fixture

471

Chapter 8.0

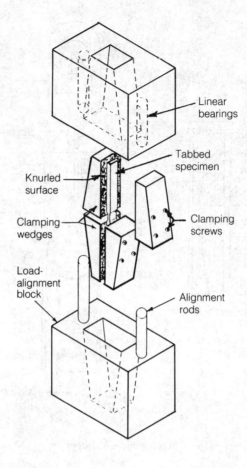

(a) Test fixture

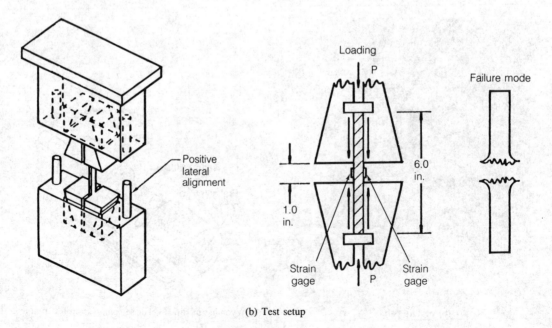

(b) Test setup

Fig. 8.3.9 IITRI Compression Test Fixture

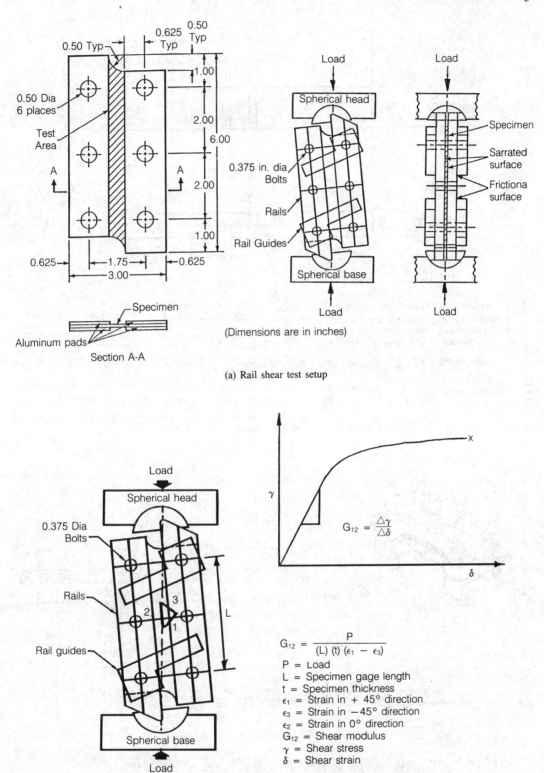

(a) Rail shear test setup

(b) Experimental determination of lamina stiffness shear modulus, G_{12}

Fig. 8.3.10 Rail Shear Test

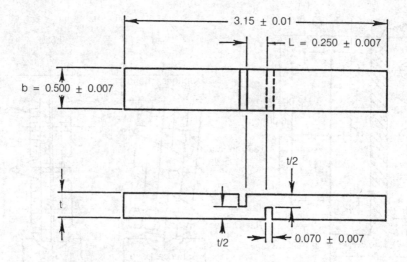

Fabrication
1. Laminate orientation [45/0/−45/90]$_{6s}$
2. Specimen edge parallel and end perpendicular requirement ± 1°
3. Edge finish shall be 32√ in accordance with MIL-STD-IOA
4. Cut specimen notches with an abrasive wheel such that:

Notch depth = t/2 $^{+\,0.010}_{-\,0.000}$ and notch penetrates centerply of laminate

Notch corner radius = 0.005 $^{+\,0.001}_{-\,0.000}$

(a) Example specimen (all dimensions are in inches)

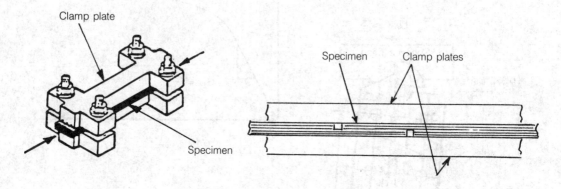

(b) Method "A" test (with clamp plates)

(c) Method "B" test (without clamp plates)

Fig. 8.3.11 Interlaminar Shear Tests Using Compression Test Method

(c) Method (A) employs metal plates held snugly against the faces of the test specimen by clamps
(d) Method (B) is loaded to failure without plates
(e) Distance between notches, the laminate thickness and longitudinal modulus affect the test result
(f) ASTM D2345-65T standard:
   - Sawcuts to be parallel within 0.030 inch (0.76 mm)
   - Depth of sawcut to be equal to:
     — Half the laminate thickness plus one ply, or
     — Half the laminate thickness plus 0.005 inch if number of plies or ply thickness is unknown

The 3-point (Ref. 8.9) or 4-point (Ref. 8.10) short beam test method is a standard test because it is a good and simple measurement and so much data is based on it. The Iosipescu test method [see Fig. 8.3.3(e) and Ref. 8.19] is the most commonly used for testing shear specimens; this method measures both modulus and strength.

Generally, the extensometer is not a reliable instrument to measure inplane shear strain, so strain gages are used.

(4) Flexural Test:

Flexural strength is not considered an intrinsic property, but the test is inexpensive to run and is considered a good quality control test. Ref. 8.13 is the source which is most quoted for flexural test methods; the test arrangement is shown in Fig. 8.3.12.

(5) Short Beam Tests:

This test, as shown in Fig. 8.3.13, is useful for evaluating the interlaminar shear behavior of the laminate matrix. Processing variables which will affect the test results are:
- Elevated temperature
- Matrix moisture content
- General matrix condition

This test is generally accepted as a method for obtaining a qualitative measure of the laminate matrix condition rather than as a procedure for generating valid design data.

Notched Effect Tests

The notched tension and compression tests are conducted to determine the most damaging combinations of temperature and moisture The effect of hole diameter size (fasteners) on residual tensile and compression strength are evaluated based on the given tests requirements. Tests are conducted for several hole sizes larger than the baseline hole e.g. 0.25 inch (6.35 mm) is the most common baseline with a w/d of 6. The tests are conducted for the most critical environmental condition determined from the tension and compression tests (e.g., see specimen example shown in Fig. 8.3.14). The ratio of hole diameter to specimen width will be the same for all specimens. A reduction factor as a function of increasing hole diameter is determined from these tests.

Chapter 8.0

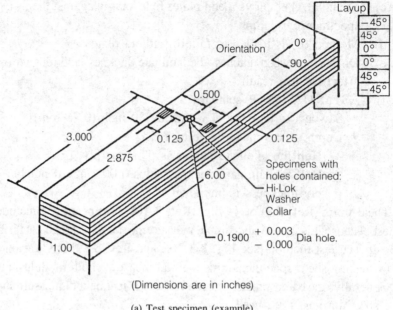

(a) Test specimen (example)

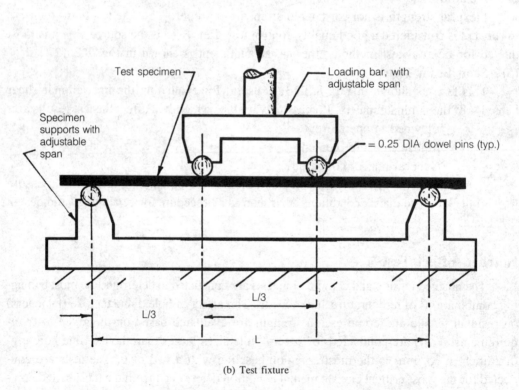

(b) Test fixture

Fig. 8.3.12 The 4-Point Flexural Test

Structural Testing

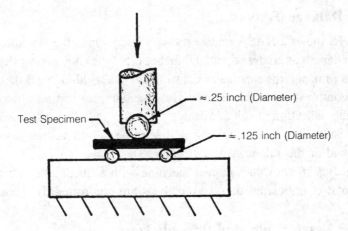

Fig. 8.3.13 The Short Beam Shear Test

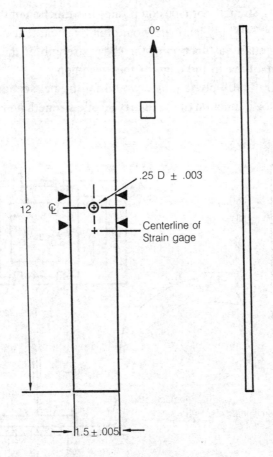

1. ▲ AND ▲ Indicate extensometer placement centered relative to the hole. 1-inch extensometer only.
2. Unless otherwise indicated, dimensional tolerances are ± 0.010 inch.
3. Edge roughness in accordance with MIL-STD-10A.
4. Dimensions are in inches.

Fig. 8.3.14 Example Of Open Hole Tension And Compression Specimen

Impact Damage Tests

Fig. 8.3.15 shows a NASA impact test specimen which has a width of 7.0 inches (17.78 cm) and a length of not less than 10.0 inches (25.4 cm) nor greater than 12.0 inches (30.48 cm). After impact, the specimens are trimmed to a width of 5 ± 0.03 inches (7.54 ± 0.0762 cm) for compression test to failure. However, impact testing on coupons does not give an accurate indication of suitable design properties and is merely used to compare material characteristics. The impact testing to determine design values should be done on components and/or the full-scale test structure.

Fig. 8.3.16 shows an impact machine with a pneumatic clamping fixture and environmental chamber; all data and analysis are controlled by a computer system.

Fastener Bearing and Pull-through Tests

(1) Fastener Bearing tests:

Fastener bearing strength for tape composites is a function of the layup configuration:
- The 100% 0° ply laminate would fail by shear tear out, and strength would be essentially a function of the shear strength of the matrix and the cross-sectional area to the edge of the specimen
- The 100% 90° ply laminate would fail by net section tension and strength would be a function of the matrix tensile strength and the net cross-sectional area

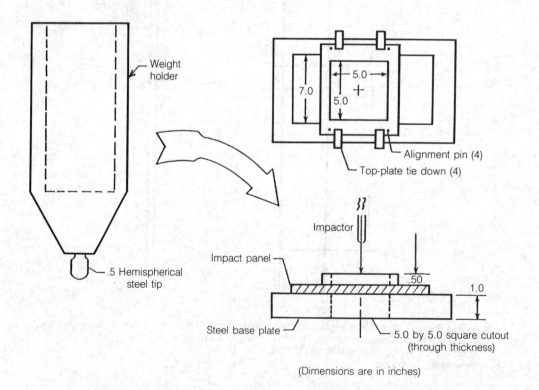

(Dimensions are in inches)

Fig. 8.3.15 NASA Impact Test Apparatus (Ref. 8.14)

Structural Testing

By courtesy of dynatup General Research Corp.

8.3.16 Impact Test Machine (Model 8250) With Pneumatic Clamping Fixture And Environmental Chamber

- Both of these are comparatively weak failure modes, multi-directional reinforcement is required if appreciable bearing strength is to be obtained

The bearing strength allowable can be determined from the following equation:

$$F_{bru} = F_{br} \times K_e \times K_{csk} \tag{8.3.1}$$

where: F_{br} — "B" allowable for room temperature dry (RTD) bearing strength of non-countersunk holes

K_e — Environmental correction factor

K_{csk} — Countersunk correction factor

The "B" bearing allowable for a non-countersunk hole and RTD condition is modified by K_e and K_{csk} to account for environment, t/d. w/d, single lap shear, etc. and countersunk thickness for flush fasteners. To determine the F_{br} allowable, tests are conducted on double lap shear specimens, as shown in Fig. 8.3.17(a), made from several different laminate thicknesses. These tests results are pooled to establish "B" bearing allowable. The bearing allowable for a countersunk hole can be tested by using the test fixture shown in Fig. 8.3.17(b).

479

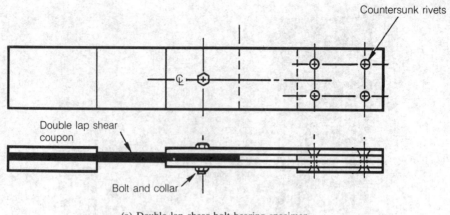

(a) Double lap shear bolt bearing specimen

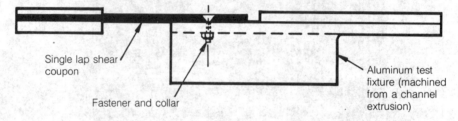

(b) Single lap shear countersunk screw bearing specimen

Fig. 8.3.17 Examples Of Fastener Bearing Tests

The test for extreme environmental conditions are used to provide the K_e environmental correction factors. Each specimens used to test environmental conditions should be cut from the same location in the laminates. Then the ratio of the environmental bearing strength to the RTD bearing strength can be used for statistical analysis.

The bearing tests for fabric laminates are similar to the tests for tape laminates.

(2) Pull-Through Tests:

The pull-through test, as shown in Fig. 8.3.18, is conducted to determine the load required to pull fasteners through composite laminates. This property is important for structures subjected to internal pressure loads or to permit buckling of skins (or webs) without failure at fastener attachments. The majority of failures of secondary supports (e.g., secondary tension due to diagonal tension shear buckling effect) are in the form of fasteners pulling through laminates, particularly countersunk fastener heads. Composites are generally weak in pull-through strength.

Pull-through strength is a function of:
- Laminate thickness
- Fastener diameter
- Configuration of fastener head type
- Laminate deflection

Therefore, tests are conducted for several thicknesses, fastener diameters, and fastener head types. Laminate support should supply rigidity to that expected for the structure.

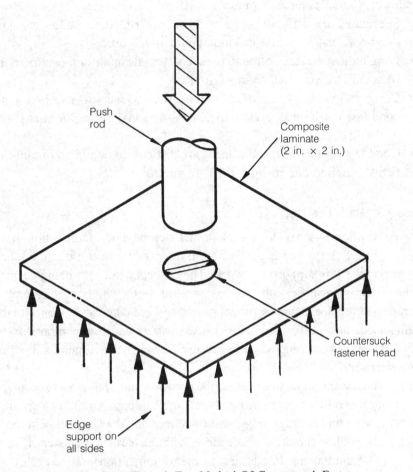

Fig. 8.3.18 Use Push-Through Test Method Of Countersunk Fastener

Material qualification tests

```
┌─────────────┐    ┌──────────────────┐    ┌─────────────┐
│ Candidate   │    │ Test services lab│    │ Engineering │
│ material    │───▶│ Qualification test│──▶│ Test results│
│ Submitted   │    │ performed        │    │ reviewed    │
└─────────────┘    └──────────────────┘    └─────────────┘
```

Batch acceptance and tag end tests

```
┌─────────────┐    ┌──────────────────┐    ┌─────────────┐
│ Material    │    │ Test services lab│    │ Production  │
│ lot         │───▶│ Batch acceptance │──▶ │ Parts and tag│
│ received    │    │ tests performed  │    │ ends layed up│
└─────────────┘    └──────────────────┘    └─────────────┘

┌─────────────┐    ┌──────────────────┐
│ Production  │    │ Test services lab│
│ OK to continue│◀─│ Tag end tests    │◀──
│ fabrication │    │ performed        │
└─────────────┘    └──────────────────┘
```

Fig. 8.3.19 Material Qualification Tests Vs. Batch Acceptance And Tag-End Tests

Chapter 8.0

Testing of Aramid Materials

The testing of Aramid materials requires special attention and care for the following reasons:
- Specimens are difficult to machine (fibers rip loose at the edges) because of toughness and a special machine cutter is required
- Longitudinal end tabs often debond before tension failure occurs in the specimen (Aramid has low surface bond strength)
- Compression tests are critical because Aramid has low compression values; for good test results use thicker specimens and make sure there is good alignment of the fibers
- Good test results on inplane shear are difficult to obtain and failure mode is difficult to define due to high fiber toughness

Process Control Testing

Process control testing provides an additional step beyond qualification and acceptance testing (see Fig. 8.3.19) and is derived from the need to ensure that the fabrication process is working as it is supposed to work. The concept came into being during metal bonding when adhesive qualification coupons were run along with bonded assemblies. As they were processed together, successful coupon tests were taken as evidence that the process was being adequately performed. This concept was extended to composites (tag-end tests are usually called out on composite engineering drawings; see Appendix B) with the following considerations:

(a) The test specimen must reflect the process (cure cycle) that is occurring at critical locations (often more than one on large complex parts). This means the significant factors in the process must be reflected in the fabrication of the test coupon.

(b) The coupon should be made along with the part (same material, same exposure to contamination in addition to the aforementioned cure cycle)

(c) The coupon should be a standardized type so that acceptance criteria can be established with statistical relevance

(d) The method should be simple enough to be systemized to the point that only manufacturing quality control interface is involved. The engineer should only become involved during set-up and when deviations to the requirements occur

Tests must satisfy items (a) and (b) above and should be conducted on pieces trimmed from production parts. However, this approach is contrary to items (c) and (d) requirements and therefore needs a trade-off study for the most cost effectivity.

Carpet Plots (Design Curves)

Tests are conducted on different laminates which contain varying percentages of $0°$, $\pm 45°$ and $90°$ plies and which cover a range of layup configurations to be used for structural applications. Fig. 8.3.20 is a carpet plot showing the general relationship between strength and various layups of composite tape. Points Ⓐ, Ⓑ and Ⓒ are laminates used to define the lamina stiffness and Poisson's ratio properties of the material for predicting the strength and stiffness of various layups. Laminates Ⓐ, Ⓑ and Ⓒ are not used by themselves in structural applications.

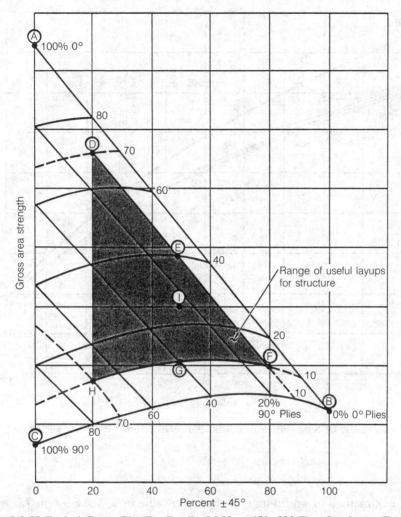

Fig. 8.3.20 Typical Carpet Plot For Family Of 0°, ±45°, 90° Tape Laminates (Example)

The crosshatched area indicates the range of laminates that are most useful for structural design applications. Laminates generally used in structures contain a minimum of 10% 0°, 10% 90°, and 20% ±45°, plies. Data for laminates Ⓔ and Ⓖ, and Ⓓ and Ⓗ can be obtained from one laminate tested in two different directions. Laminate Ⓘ is an isotropic laminate with equal amounts of material in four directions. Tests conducted on laminates Ⓓ through Ⓘ will be used to characterize the tape composite strength properties for the whole crosshatched region.

The properties for different percentages of 0°, ±45°, 90°, bidirectional fabric plies in laminates do not vary as much as do tape laminates. Each ply consists of fibers oriented at 0°/90° or at ±45°. An illustration of the variation in strength based on the percentage of 0°/90°, 90°/0° and ±45° plies is shown in Fig. 8.3.21. The strength is slightly higher in the 0° (warp) direction than in the 90° (fill) direction because the percentage of fibers is slightly higher in the 0° direction. The laminates Ⓐ, Ⓑ and Ⓒ, shown in Fig. 8.3.21, are not used in structures, but the properties of these laminates are used for predicting the properties of other combinations of laminates.

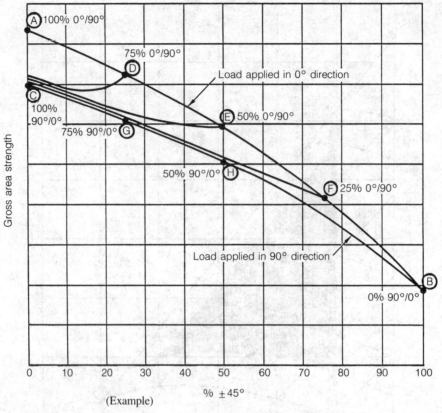

Fig. 8.3.21 Schematic Of Strength Characteristics For 0°/90°/±45° Bidirectional Fabric Laminates

AML Plots

This is a quick method of preliminary design allowable (also see Section 7.2 in Chapter 7). Since controversy exist over the use of this method, here, the text merely shows the procedures of creating AML plots. It can be obtained from the laminate-level coupon (3 specimens are used per batch and approximately 300 – 400 coupons are required) tests by the following steps:

(1) Laminate-level tests should conducted on the following configurations:
 - (50/40/10) — i.e., % of plies (% of 0°/ % of ±45° /% of 90°)
 - (25/50/25)
 - (10/80/10)

These three configurations have AML's (percentage ±45° minus percentage 0° plies) of −10, 25 and 70, respectively.

(2) Unnotched tension and compression tests are performed on the above three laminate configurations at RTD (room temperature dry) and on the (25/50/25) laminate configuration at ETW (elevated temperature wet). These specimens contain measuring instruments which determine modulus as well as failure stress and strain.

(3) Specimens with an open hole are tested in both tension for OHT (open hole tension) and compression for OHC (open hole compression) at RTD, and a lower percentage are tested at ETW and CTD (cold temperature dry).

(4) Specimens with fasteners are tested for FHT (filled hole tension) at RTD with some at ETW and CTD.
(5) In addition to the tests of OHC and FHT described above, OHC and FHT specimens are machined from a (25/50/25) panel and tested at RTD to determine statistical factors.
(6) Specimens are impacted at specified "A" impact level and tested for compression at RTD and ETW.
(7) Quasi-isotropic (25/50/25) specimens are tested for PIC (post-impact compression) at RTD after impact at various specified impact levels.
(8) Specimens are tested in single-shear bearing using specified diameter (e.g. 0.25 inch) protruding head and/or countersunk fasteners to determine bearing strengths at RTD and at ETW.
(9) Pull-through tests are conducted using countersunk head fasteners (e.g. 0.25 inch).
(10) Picture frame shear tests are conducted on (10/80/10) laminates to determine initial buckling and the effect of specified "A" impact level [see item (6) above] on shear strength.

The test data (an example of test data for RTD is shown in Fig. 8.3.22) should be carefully reviewed and an appropriate "knock down" factor, e.g., 80% to 90% be determined.

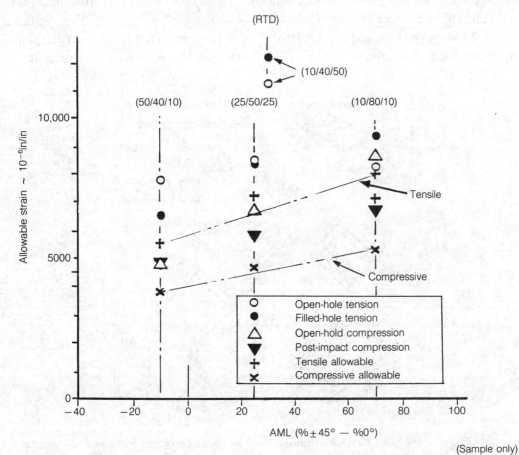

(Sample only)

Fig. 8.3.22 AML Design Curves (Knock Down Factor Included)

Chapter 8.0

8.4 ELEMENT AND COMPONENTS TESTS

The design of composite structures is often verified by testing actual structural components, as shown in Fig. 8.4.1 and Fig. 8.4.2; tests are used for allowable verification and for fulfillment of structural integrity requirements. These specimens occasionally contain holes, notches, stringer run-outs, joggles, etc., and typically have more instrumentation and require more effort to both load introduction and test fixtures, as shown in Figures. 8.4.3 through 8.4.5, because few standard methods are available at the component level.

Load-carrying structures usually contain sections which require intensive detailed analytical and experimental examination to ascertain their effects upon the performance of the total structure.

For example:
- An access hole through a skin structure may drastically alter the stress concentration and redistribution in the surrounding area
- A fastened, bonded and/or fastened joint likewise can produce significant stress perturbations in its immediate vicinity

These sections of components may induce large stress perturbations in the constitutive material and induce failure modes very different from those predicted by lamination theory.

In addition to inplane axial and shear loads, concentrated normal tension load on a composite integrally stiffened panel tests the flatwise tension and peel strength between skin and stiffener which are much lower than inplane laminate strengths. Thus, stiffener pull-off strength tests should be conducted. Fig. 8.4.6 shows a test in the form of a simple method which allows the pull-off results to translated into a running-load strength.

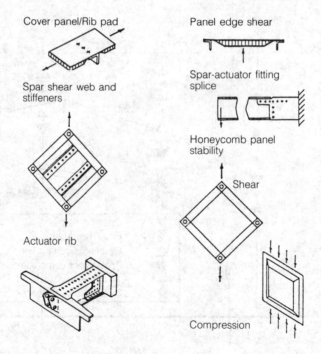

Fig. 8.4.1 Key Subcomponent Tests For B727 Composite Elevator

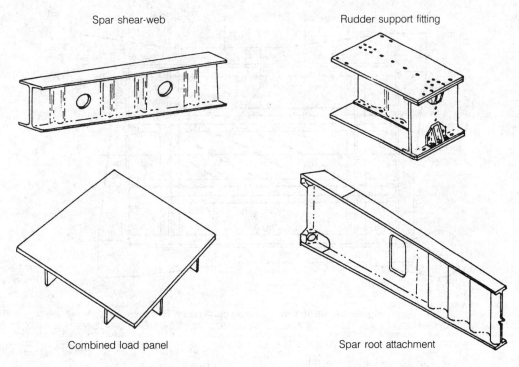

Fig. 8.4.2 Component Tests (DC-10 Vertical Composite Fin Box)

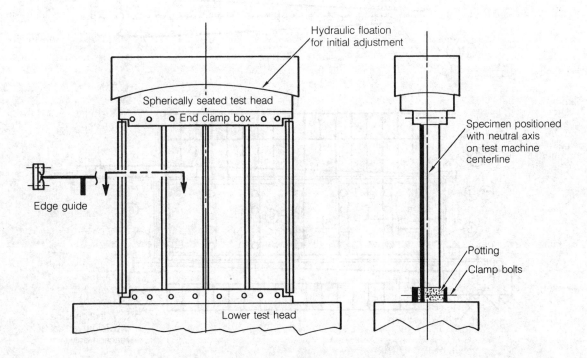

Fig. 8.4.3 Compression Test Panel Arrangement

Chapter 8.0

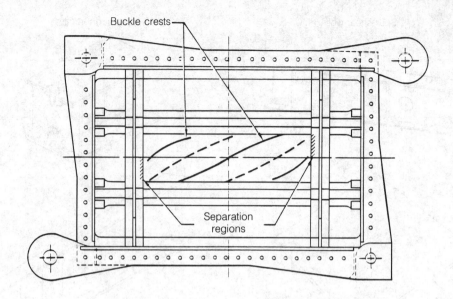

Fig. 8.4.4 Shear Panel Test Setup (Picture Frame)

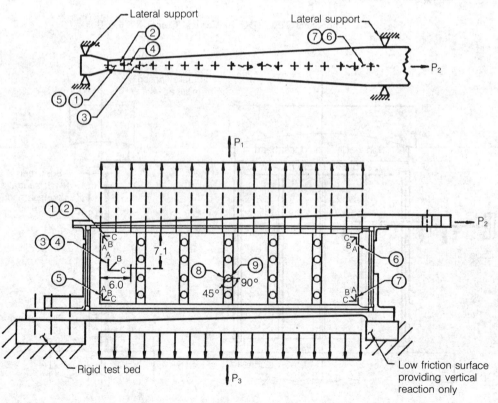

Notes:
1. Strain gage rosettes ①, ②, ⑤, ⑥ and ⑦ are named in corners or web, reinforcement intersection.
2. Strain gage rosettes ③ and ④ are centrally located on list as shown.
3. Strain gages ⑧ and ⑨ are inside hole as shown.

Fig. 8.4.5 Sine Wave Spar Combined Shear And Tension Test Setup

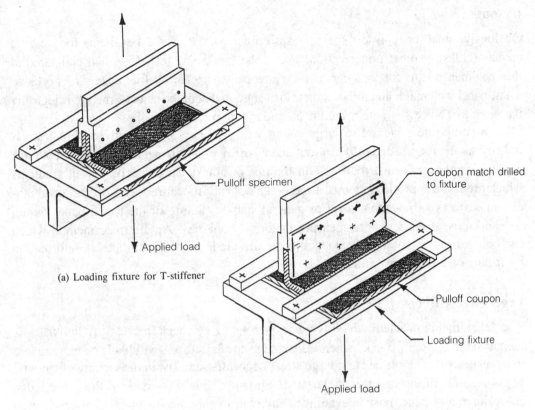

Fig. 8.4.6 Stiffener Pull-Off Strength Tests

Joint Design Between Specimen and Fixture

One of the primary challenges related to joint testing is introducing load into the joint in a fashion which is representative of the boundary condition of a test component specimen. For example, it may be difficult or virtually impossible to determine, much less duplicate in a test, the stiffness boundary conditions which are present at the joint in actual service. The choice of boundary conditions which are readily reproducible in most tests are either free or fixed supports. The engineer may be able to furnish information as to the testing procedure and gripping hardware which would be most appropriate for approximating in situ conditions. The service stress distribution in the components which border on the joint must be predicted by the engineer. Then it is possible to approximate the same stress proportions by using boundary control techniques which are related to an active feedback signal from the component under test. Such a test may be expensive, but the application might be critical enough to warrant resorting to such a technique.

It is often necessary to measure the deflection of the joint as test loads are applied. Load and deflection data can be combined to provide a measure of the stiffness of the joint. Obviously, it is difficult to use a strain gage to measure the sliding motions which occur in most joints. A simple deflection sensor is an ideal transducer for measuring joint deflections.

Cutouts

Obviously, small coupon test specimens are inappropriate for experiments focusing on cutouts (holes) or other imperfections unless the flaw being tested is small compared to the specimen width; coupons can also be adversely affected by free edge stress effects. Thus, panel with major and minor dimensions close to that of the actual structure is logically the obvious choice for notch, cutout or imperfection tests.

In composite structure, a large cutout will present a significantly different stress redistribution around edge of the cutout and an array of strain gages may be adequate for quantifying the strain distribution. But the tips of cracks cause steeper strain gradients which are more appropriately measured by an optical procedure (e.g., Moire' or photo-elastic coatings). Preventing local or general stability failure of the tested panel should be considered if the test is to measure material properties. An alternate method which can be used to prevent a general stability failure mode is to restrain lateral deformation by means of a reaction fixture.

Free Edge Effects

The delamination problem which is associated with free edges in cross-ply laminates is mentioned in Section 7.3 in Chapter 7. Free edge normal stresses will likely be more severe in laminates with cutouts because large stress concentration exist in the vicinity of cutouts. Measurements of through-the-thickness deformation should be made at the cutout edge since this may be the most relevant measurement to support analytical characterization studies. In addition, strain gages, displacement sensors and optical methods all have potential application for delamination strain characterization.

8.5 VERIFICATION FULL-SCALE TESTS

Full-scale testing (FST) of the completed airframe, as shown in Fig. 8.5.1 and Fig. 8.5.2, or testing of a large segment as a single unit (see Fig. 8.5.3), is the major test in an airframe structural test program. FST is one of the primary methods of demonstrating that an airframe can meet the structural performance requirements and it is extremely important because it tests all the related structures in the most realistic manner.

Typical FST includes static, durability (fatigue), and damage tolerance (may not be required in FST). The use of FST must take into account the unique characteristics of composite structures and their response to the expected service conditions as simulated by the test.

FST is a necessary check in the process of developing satisfactory structural systems. Analysis techniques have significantly improved in recent years, particularly with the advent of computers and the now possible finite element analysis; however, the complexity of composite structural systems still requires FST verification programs.

Test requirements such as limit and ultimate loads are often established on the basis of material test scatter derived from metals. Since composites usually exhibit higher scatter, problems may develop. Also, composite laminates exhibit relative brittleness, low interlaminar strength and a difference in CTE (in contact with metal parts) and all these factors present serious problems for the FST programs.

Structural Testing

By courtesy of McDonnel-Douglas Corp.

Fig. 8.5.1 Full-Scale Test Of AV-8B VTOL Aircraft (Tension Patches Method)

Full-scale ground test setup

Fig. 8.5.2 Full-Scale Test of B737 Composite Horizontal Stabilizer Box Structure (Ref. 8.17 And 8.18)

Chapter 8.0

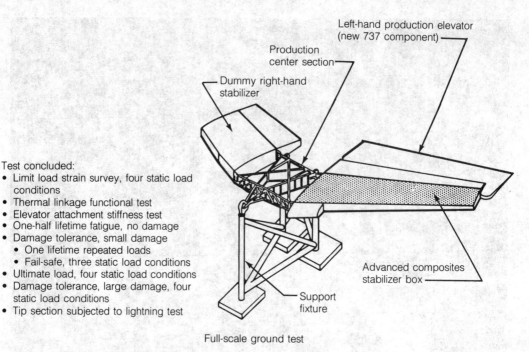

Test concluded:
- Limit load strain survey, four static load conditions
- Thermal linkage functional test
- Elevator attachment stiffness test
- One-half lifetime fatigue, no damage
- Damage tolerance, small damage
  - One lifetime repeated loads
  - Fail-safe, three static load conditions
- Ultimate load, four static load conditions
- Damage tolerance, large damage, four static load conditions
- Tip section subjected to lightning test

Full-scale ground test

Fig. 8.5.2 Full-Scale Test of B737 Composite Horizontal Stabilizer Box Structure (Ref. 8.17 And 8.18) (cont'd)

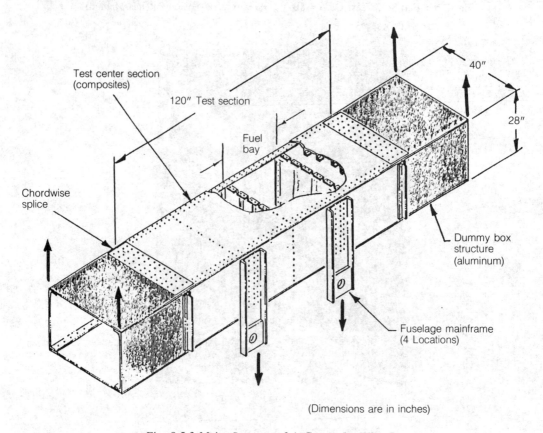

(Dimensions are in inches)

Fig. 8.5.3 Major Segment of A Composite Wing Box

There are three considerations when choosing the size of the test article:
- It must be large enough allows for proper complex loading and also for load interactions at interfaces that would otherwise be difficult to simulate
- If the component is small enough it is less costly to use a FST environmental test to certify the structure
- Structural configuration also play a role in the environmental condition test:
  — Primary or secondary structure
  — Type and complexity of loading

The purpose of FST is addressed below:
- To verify analysis with actual internal load distribution
- To see if any unexpected discrepancies occur
- To evaluate whether durability and damage tolerance have been adequately assessed
- To evaluate the durability of combination of composites and metals, particularly in interface areas (e.g., thermal expansion problems)

Instrumentation (all data is electrically recorded and controlled by computer) used on FST structures includes:
- Strain gages (most common)
- Deflection indicators
- Accelerometers
- Stress coatings
- Acoustic emission detectors
- Evener systems

Pre-test predictions of FST structure failure loads, locations and mechanisms are important to the test plan. These are based on minimum margin of safety calculations and the known statistical variations of the material allowable developed from coupon tests and used in analysis.

Appropriate "knock-down" factors are applied to the test margins after completion of the mechanical property (moisture) is testing. These results must be verified with a long-term aging study where the structure is subjected to real-life environmental conditions and tested at various intervals throughout the duration of the test program. Perfect duplication of temperature/moisture/time histories is not possible on complex FST structures and even attempting it can lead to unacceptably high test costs.

If FST must include environmental effects. It can be anticipated that FST will have to be wet-conditioned, damaged and tested in an environmental chamber having the capability of qualifying structures at the extreme temperatures specified for the design. The problem is workable and not as formidable as it sounds, but development work must to be done to accurately assess the costs and additional program risks associated with these requirements.

When composite mating structures are required for load introduction extra care must be taken to diffuse loaded into the structures. Generally:
(a) In tension tests — The mating structures must be sufficiently strong that they will not fail before the structure being tested.

(b) In compression tests — The mating structure must be simulated and the loads applied to it such that the rotational characteristics are approximated. This subjects components which are buckling critical to appropriate end-fixity conditions and ensures adequate load diffusion into the tested structure.

Static Test

The FST static test is the most important test in qualifying composite airframes because of their brittleness and sensitivity to stress concentrations compared to metal counterparts. The ability of the test data to meet certification requirements must be inherent in each of the FST static test requirements.

(1) The parameters to consider for the static test are:
- Type of test structure
- Type and number of load conditions
- Usage environment to be simulated
- Type and quantity of data to be obtained

It is difficult to conduct a FST under both temperature and moisture conditions but these environmental effects are addressed at the analysis, coupon, structural element, and component level (e.g., see Fig. 8.2.1) in the "building block" testing approach. The sums of these tests must be consolidated in such a way as to validate the consideration of environmental effects.

(2) The subject of loading methods on a FST composite structure needs careful consideration due to the composite's weak through-the-thickness strength (tension) and sensitivity to stress concentrations; possible test methods (e.g., on wing surfaces) are listed below:
  (a) Tension-patches method (see Fig. 8.5.4):
   - Offers uniform load distribution with closer representation of the real structural load (see Fig. 8.5.1); a more costly method
   - Involves a more complex test set-up (higher cost)
   - Introduction of load directly to the composite bonded surface must be done more carefully than with metal bonds because of through-the-thickness weakness
  (b) Loading frame method (see Fig. 8.5.5):
   - Less complex loading set-up and the least costly method
   - All loads are converted into numerous compressive concentrated loads; this is not as effective as the tension-patches method but it is acceptable
   - The attachment of substructures such as spars, ribs etc., at locations of concentrated loads needs careful investigation to make sure there is sufficient strength

(3) The recommended FST test sequence is as follows:
  (a) Check the test set-up, which involves functional testing of:
   - Loading jacks and evener system
   - Instrumentation
   - Data recording

Structural Testing

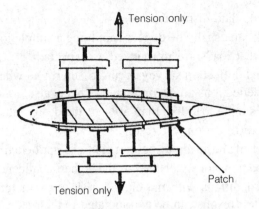

Fig. 8.5.4 Tension-Patches Method

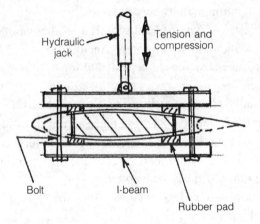

(a) Loading frame arrangement

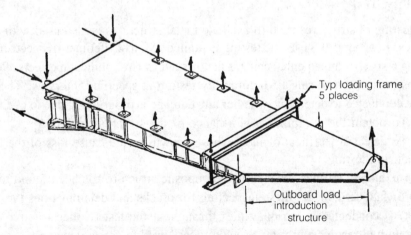

(b) Test setup example

Fig. 8.5.5 Loading Frame Method

495

- Real-time data displacements

This check can usually be done by applying a simple loading case at low levels to ensure that loads are introduced as expected.

(b) A strain and deflection survey is run to determine whether the strain distributions and deflections are as predicted. This is also done by applying low level loading (a simple load less than 50% of the design ultimate load) that will not affect the certification test results.

(c) The lowest of the loads to be certified are applied first:
- The conditions for which there is the highest confidence of success are run first
- Those conditions with the highest risk of premature failure are run last

(d) The early test results can be extrapolated to the predicted design ultimate load level

(e) If there is some risk of failure before reaching the required design load, a careful review and investigation should be conducted.

(4) Ultimate load requirement — Type of load required by the qualifying or certifying agencies to meet their validation requirements includes:

(a) U.S. FAA requires the structure to be tested to the limit load (same as conventional metal structures)

(b) U.S. military usually requires testing the ultimate load:

After completing the required FST, whether to limit or ultimate load, the airframe manufacturer usually chooses the most critical load condition to test to failure. If the destruction failure load exceeds the design ultimate load, the airframe future growth margin is warranted.

(5) The final step is a review of the data obtained from the test and evaluation of its correlation with the stress analysis.

Durability Test

Cyclic testing of structures used to evaluate metal structures is also used with composite structures. In general, FST cyclic testing is limited to 2 to 4 lifetimes of spectrum loading, including a spectrum load enhancement factor such as environmental effects. Periodic inspections must occur during FST durability testing at specified intervals. These inspections are conducted to determine whether any damage is progressing due to cyclic loading:

- To obtain the durability performance of the structural details
- To detect any critical damage whose growth would cause loss of the test article during testing

For example, stiffness change in a composite structure has been found to be an indication of fatigue damage; these inspecting for cracks and delaminations (very difficult to detect) is conducted at various times throughout the test. Nondestructive inspection methods with ultrasonic and x-ray are commonly used to detect damage.

Finally, a post-test inspection of the test article offer the FST durability test is very important to ensure that no damage has occured which would threaten the structural integrity of the composite design.

Damage Tolerance Tests

Testing composite FST structures for damage tolerance is especially important because it addresses the concerns associated with both the static and durability tests. The damage tolerance test, like the static test, is a qualification requirement of both civil (e.g., FAA) and military authorities. The load specified by civil and military requirements varies (both specify a residual strength requirement) and the requirements also vary depending on:
- Ability to inspect damage
- Type of service inspection used
- Type of aircraft

As in the durability test, the critical flaw or damage may be associated with either its initial state or its growth after cyclic loading. The environmental effect during the cyclic test is not easily defined but the load enhancement of the spectrum as recommended for the durability test is the best option. Because the FST damage tolerance test has many similarities to the static and durability tests, all the testing considerations which apply to them are also applicable to this test.

If the residual strength test is successfully passed, the structure can then be loaded to failure to further evaluate or integrate its damage tolerance capability.

References

8.1 ASTM committee D-30, Subcommittee on "High Modulus Fibers and Their Composites", American Society for Testing and Materials.

8.2 Daniel, I. M., "COMPOSISTE MATERIALS; Testing and Design", (ASTM STP 787) 6th ASTM Conference, Philadelphia, 1982.

8.3 Chamis, C. C., "TEST METHODS AND DESIGN ALLOWABLES FOR FIBROUS COMPOSITES", ASTM Special Technical Publication 734. 1981.

8.4 Anon. "ENGINEERING MATERIALS HANDBOOK, VOL. 1 — COMPOSITES", ASM International, Metals Park, Ohio 44073. 1987. pp.283-351.

8.5 Tuppeny, W. H. and Kobayashi, A. S., "Manual on Experimental Stress Analysis", Society of Experimenatal Stress Analysis. 1965.

8.6 Read, B. E. and Dean, G. D., "Experimental Methods for Composite Structures", AGARD L.S. 55. 1976.

8.7 Anon., "Standard Test Method for Tensile Properties of Fiber-Resin Composites", D 3039, Annual Book of ASTM Standards, American Society for Testing and Materials.

8.8 Anon., "Standard Test Method for Compression Properties of Unidirectional or Crossply Fiber-Resin Composites", D 3410, Annual Book of ASTM Standards, American Society for Testing and Materials.

8.9 Anon., "Standard Test Method for Apparent Interlaminar Shear Strength of Parallel Fiber Composites by Short-Beam Method", D 2344, Annual Book of ASTM Standards, American Society for Testing and Materials.

8.10 Anon., "Standard Test Method for Inplane Shear Strength of Reinforced Plastics", D 3846, Annual Book of ASTM Standards, American Society for Testing and Materials.

8.11 Anon., "Standard Practice for Inplane Shear Stress-Strain Response of Unidirectional Reinforced Plastics", D 3518, Annual Book of ASTM Standards, American Society for Testing and Materials.

8.12 Anon., "Standard Guide for Testing Inplane Shear Properties of Composite Laminates", D 4255, Annual Book of ASTM Standards, American Society for Testing and Materials.

8.13 Anon., "Standard Test Method for Flexural Properties of Unreinforced and Reinforced Plastics and Electrical Insulating Materials, D 790, Annual Book of ASTM Standards, American Society for Testing and Materials.

8.14 Anon. "Standard Tests for Toughened Resin Composites", NASA RP 1092, National Aeronautics and Space Administration. 1983.

8.15 Anon., "Standard Test Method for Apparent Tensile Strength or Ring or Tubular Plastics and Reinforced Plastics by Split Disk Method", D 2290, Annual Book of ASTM Standards, American Society for Testing and Materials.

8.16 Anon., "Standard Method of Shear Test in Flatwise Plane of Flat Sandwich Constructions or Sandwich Cores", C273, Annual Book of ASTM Standards, American Society for Testing and Materials.

8.17 Johnson, R. W., McCarty, J. E. and Wilson, D. R., "Damage Tolerance Testing for the Boeing 737 Graphite/Epoxy Horizontal Stabilizer", Paper presented at The Fifth Conference on Fibrous Composites in Structural Design, Department of Defense/National Aeronautics and Space Administration. Jan., 1981

8.18 McCarty, J. E. and Wilson, D. R., "Advanced Composite Stabilizer for Boeing 737 Aircraft", Paper presented at The Sixth Conference on Fibrous Composites in Structural Design, Department of Defense/National Aeronautics and Space Administration. Jan., 1983.

8.19 Walrath, D. and Adams, D. F. "Iosipescu Shear Properties of Graphite Fabric/Epoxy Composite Laminates", UWME-DR-501-103-1, University of Wyoming

8.20 Anon. "Standard Tests for Toughened Resin Composites", NASA RP 1142, National Aeronautics and Space Administration. 1985.

8.21 MIL-STD-10A, "Military Standard — Surface Roughness, waviness and Lay", Jan. 3, 1966.

Chapter 9.0

QUALITY ASSURANCE

9.1 INTRODUCTION

Non-destructive inspection (NDI) generally refers to the examination of a part or assembly in such a manner that the test article is not affected in any way. The methods used in determining composite quality include visual, Eddy-current, ultrasonic, radiographic (X-ray), and holographic inspection. With the small design margins used for airframe structures, NDI is essential to insure that composite structures are free from anomalies and can provide necessary performance.

The original impetus to develop a database on aging materials came from the nuclear power industry, which required more than 40 years of service life/safety from materials that would be subjected to critical wear and exposure during the life of a nuclear reactor. A materials aging database helps engineers to determine the proper materials to use for specific airframe applications. Once the material is chosen, it becomes the job of maintenance to ensure that the material will stand up to the stresses they are designed to withstand.

Composites are prone to manufacturing defects that compromise structural integrity and when they do occur they are generally not visible to the eye. While composites have been routinely inspected for a long time, two factors combine to increase the need for better methods of evaluating structural integrity:
- Composite materials are beginning to be used in structurally critical areas such as empennage stabilizers, fuselages, fighter aircraft wings, etc.
- The growing popularity of using composite materials has increased production rates and has forced manufacturers to look for smaller flaws and do it faster than ever before.

No matter how well a composite component is made, close examination inevitably reveals flaws; typically, large flaws in high-stressed regions are of the most concern. Deciding which flaws are critical, the size of flaw that can be ignored, and locations where such flaws can be tolerated or are unacceptable must be established in the design process. Acceptance criteria can vary widely depending on the application. An airframe primary structure obviously would have more stringent requirements than would a secondary structure.

Composite structures are difficult to design for visual inspection. Internal damage to a laminate skin is rarely apparent from the outside. Delamination or debonding of honeycomb core, caused by a dropped tool, frozen moisture or lightning, can be seen by X-ray; but because even a few square inches of internal damage can be unacceptable, a labor-intensive NDI scan of the whole component becomes necessary. Unlike the structure of most metallic materials, the built-up nature of composite laminates and honeycomb panels makes it easy for defects to occur. When a metal skin is hit it visibly dents, but when composites are hit nothing can be seen on the outside surface which unacceptable damage has occured to the internal structure.

It should be noted that, to keep controls adequate and cost-effective, many of the control functions cannot be the responsibility of Q. A. alone. Q. A. naturally retains the overall responsibility for structural product integrity, but certain areas of control must be delegated to Engineering, Planning, and Manufacturing. The task of quality control is to assure that composite parts represent a level of quality consistent with design requirements. This can be accomplished by planning, in cooperation with other disciplines, to develop the necessary documentation and specifications to facilitate the early detection of defects.

NDI is a test method which does not affect serviceability of the test article and its advantages are:
- Failure prevention
- Cost reduction
- Customer satisfaction
- Manufacturing process control
- Quality maintenance

In order to insure the proper quality assurance standard for a particular layup, the engineer must add a note to the drawing that calls out the controlling specification or document. Inspection criteria must stated and other requirement noted, such as those calling out NDI or testing requirements, on the composite engineering drawing.

There are three terms which generally need a full explanation to prevent misunderstanding:
- Non-destructive Testing (NDT) — This refers to the development and application of NDT methods and is the most general term
- Non-destructive Inspection (NDI) — This refers to the performance of inspections to established specifications and procedures using NDT methods to detect discrete anomalies
- Non-destructive Evaluation (NDE) — This refers to the capability to assess the state of a material, a component, or a structural form from a set of quantitative NDT measurements and to predict the serviceability of the item in question from these measurements when evaluated in the context of appropriate failure modes

Of these three terms this chapter is primarily concerned with non-destructive inspection (NDI) because the procedures and specifications which are implemented are used to detect defined anomalies, but in neither a quantitative nor serviceability predictive sense.

Function of Quality Assurance (Q.A.):

(a) Production quality analysis:
- Support team function with estimates (based on past experience) of defect and rejection rates (a percentage of parts inspected) for different types of manufacturing operations
- Recommend procedures for tracking part rejections and part variability during manufacturing
- Establish and implement procedures to identify the causes of high scrap/rejection rates or high variability in parts

(b) Quality assurance technology:
- Recommend Q. A./inspection procedures to aid Manufacturing in improving quality and reducing part variability
- Identify needed development of in-process inspection technology for error-avoidance and improvement of controlled-tolerance manufacturing
- Recommend procedures to integrate Q.A./inspection technology development with manufacturing technology and automation

In order to achieve an economical product which meets or exceeds requirements, it is essential that quality be considered as a goal throughout the design process. The engineer has the responsibility of designing a product which can be produced consistently with good quality and whatever tools are necessary must be employed to achieve that quality. Typical Q. A. involvement in the verification of a design is shown in Fig. 9.1.1.

Fabrication Level Inspection

(a) Fabrication inspection will include general surveillance to verify that:
- Manufacturing tooling has been conditionally accepted pending tool-try part acceptance
- Material and material utilization is controlled
- All measuring and test equipment calibrations are current
- Process specifications and engineering drawings are maintained in compliance with the manufacturing plan

(b) Incremental buy-offs during the fabrication process will verify:
- Tool preparation
- Number of plies
- Ply orientations after layup
- Consistency of ply trimming operations
- Proper layup in the tool

(c) Tool assembly, and bagging and leak check prior to cure will also be verified.

(d) During the cure, pressure, vacuum, temperature, and dwell time will be monitored, with a final buy-off recorded on the process control chart

(e) After cure, visual, dimensional, and non-destructive inspections will be performed in accordance with the engineering drawing and process specifications.

(f) Any non-conformances will be identified and the item will be held for engineering disposition.

(g) Process control specimens will be evaluated by the Q. A. laboratory and the results documented.
(h) The flow diagram shown in Fig. 9.1.2 gives the typical incremental inspections to be performed.

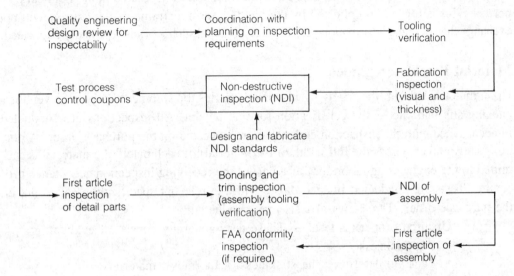

Fig. 9.1.1 Quality Assurance Involvement in the Verification of The Design

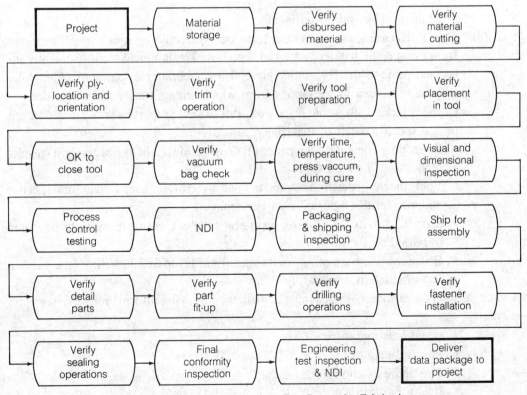

Fig. 9.1.2 Inspection Operations For Composite Fabrication

9.2 MATERIAL QUALIFICATION

Unlike the metallurgy industry, manufacturers of composite materials generally arc unwilling to let airframe manufacturers know the chemical compositions of their materials. This has forced many airframe manufactures to hire their own chemists to analyze the materials received from the materials companies to better understand their chemical properties. This is further complicated by the fact that most airframe manufacturers will not exchange the chemical information they have gained, and such work is repeatedly duplicated.

Material Batch Acceptance

Upon material receipt, the receiving inspector verifies the shipment against the vendor's packing slip and checks the certification for that receipt. All inspections are conducted in accordance with the inspection codes that are indicated on the purchase order. A portion (a strip taken across the full width of the material) of each batch of a material is submitted to the engineering laboratory as a part of the receiving inspection acceptance procedure. Tests are conducted in accordance with the relevant specification called out on the purchase order. Two types of tests may conducted:

(a) Uncured properties tests — These tests generally consist of evalutating
  - Ply thickness
  - Tack (to determine the stickiness of thermoset materials)
  - Volatile content
  - Resin content
  - Areal weight
  - Moisture absorption (optional)

(b) Cured properties tests — These tests are conducted on a laminate fabricated from the specific batch of material received. The laminate is ultrasonically inspected prior to the laboratory tag-end test (see later discussion). The panel may be machined into coupons from which strength tests are performed.
Cured laminate (ultrasonic inspected prior to test) properties tests are conducted on the specific batch of material received:
  - Resin/fiber content — determines ratio of resin weight to fiber weight (results reported as a percentage)
  - Void content — determines ratio of voids to resin and fiber weight for a specific volume (results reported as a percentage)
  - Specific gravity — ratio of resin and fiber weight to the weight of water (reported as a ratio)
  - Grind-down or burn-off — used to determine layup (results indicate the number and orientation of each ply)

A material safety data sheet (MSDS) must contain the following information (also see Ref. 1.2):
  - Hazardous ingredients
  - Physical data
  - Fire and explosion hazard data
  - Health hazard data

- Reactivity data
- Spill or leak procedures
- Special protection requirements
- Special precautions
- Method of transportation

Tag-end Tests

A Typical tag-end testing includes the following:
- Grind-down — ply orientation must conform to drawing
- resin content — by weight (e.g., $32 \pm 2\%$)
- Fiber volume — for information and reference only
- Void volume — for information and reference only
- Thickness — for information and reference only
- Laminate density — by lb/in^3 (or g/cc)
- Micrographs perpendicular to 0° — ply stacking sequence, porosity, and delamination
- Micrographs perpendicular to 90° — ply stacking sequence, porosity, and delamination

The tag-end example shown in Fig. 9.2.1 can be used to the above grind-down process listed above.

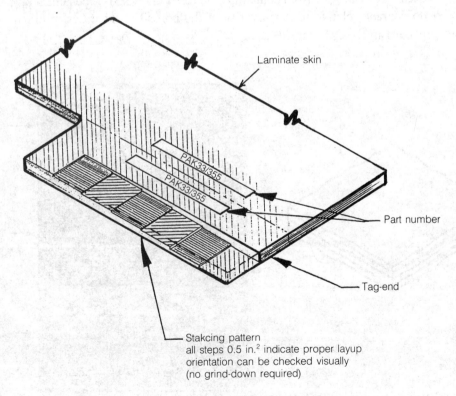

Fig. 9.2.1 Use Stacked Pattern To Avoid Grind-down Process

9.3 TYPES OF DEFECTS (FLAWS)

In composites, the many variables in the manufacturing process make many types of defects or flaws possible. Fortunately, most flaws occur infrequently, are easily prevented, or are not particularly detrimental to product performance. Defects that generally are of most concern are delamination and voids. These are gaps in the structure that, depending on location, can dramatically reduce strength.

(a) Delamination — Separation of plies within a laminate as a result of gas pockets or contamination. Delamination is caused by improper surface preparation, inclusion of foreign matter, or impact damage during shipping or handling.(see Fig. 9.3.1)

(b) Voids — Small clusters of air or gas micro-bubbles which have a tendency to collect at plies. Voids (see Fig. 9.3.2) are usually due to one of two factors:
- A mismatch in tooling during the cure cycle which results in unequal pressure distribution through the part, forming gaps that resin cannot fill.
- Resin with high volatile content combined with a short cure cycle. If volatiles cannot escape before the resin sets, voids will form. Occasionally, voids form when moisture is absorbed by the resin. Proper storage and process control will prevent this.

(c) Porosity — A condition of air or gas micro-bubbles in a given area within solid material. Caused by the incomplete flow of resin during cure and localized excessive heating or resin contamination (see Fig. 9.3.3). Mechanical properties do degrade relative to the severity of the porosity, but it has minimal effect if porosity is less than 3%.

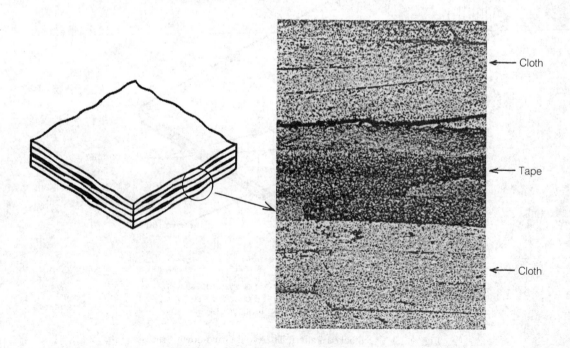

Fig. 9.3.1 Delamination

Quality Assurance

Fig. 9.3.2 Voids and Porosity

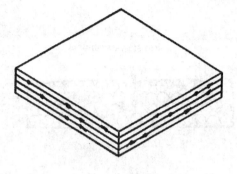

(See photo in Fig. 9.3.2 and Fig. 9.3.6)

Fig. 9.3.3 Porosity

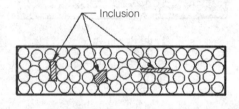

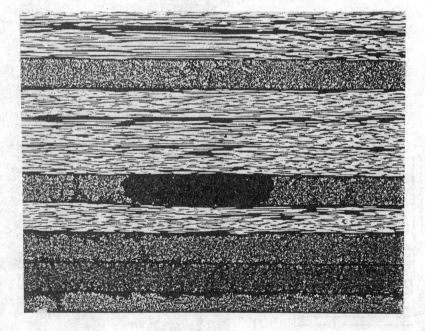

Fig. 9.3.4 Inclusions

Chapter 9.0

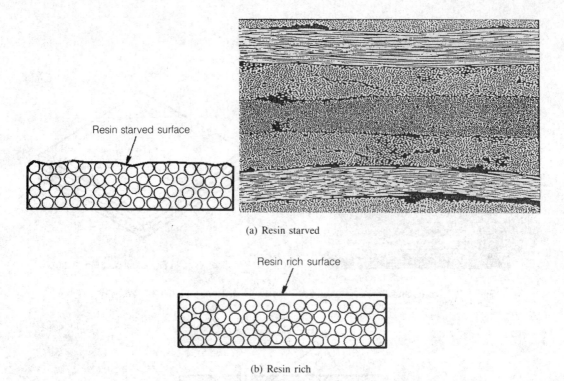

(a) Resin starved

(b) Resin rich

Fig. 9.3.5 Resin Variations

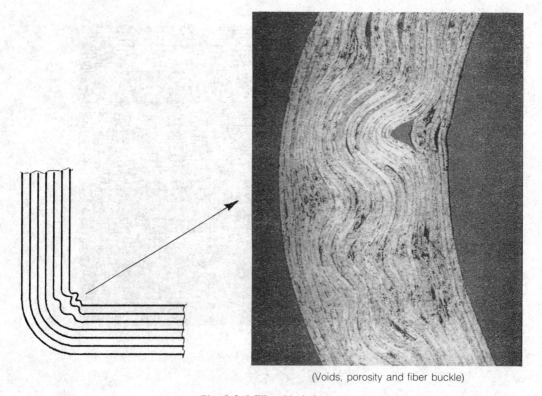

(Voids, porosity and fiber buckle)

Fig. 9.3.6 Fiber Variations

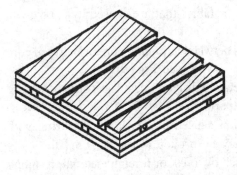

Fig. 9.3.7 Gaps at Splices

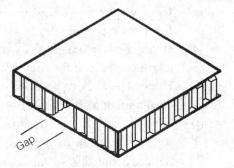

Fig. 9.3.8 Honeycomb Splice Gaps

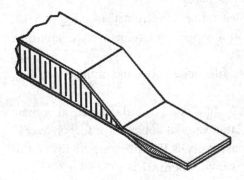

Fig. 9.3.9 Disbond

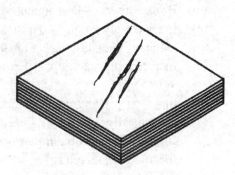

Fig. 9.3.10 Surface Scratch

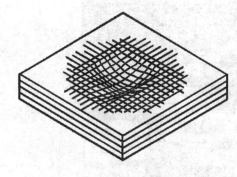

Fig. 9.3.11 Surface Depression

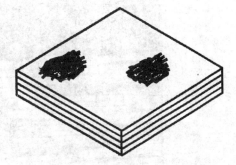

Fig. 9.3.12 Frayed Or Burned Material

(d) Inclusions — Foreign material inadvertently embedded in a layup such as chips, backing paper, tape, and other contaminants (see Fig. 9.3.4).

(e) Resin variations — Resin-rich or resin-starved areas may occur when laminate is improperly compacted or bleeding is improperly controlled during cure (see Fig. 9.3.5).

(f) Fiber variations — Abrupt changes in fiber orientation appearing as wrinkles, distortion, gaps, etc., resulting from problems in layup or cure (see Fig. 9.3.6)

(g) Gaps at splices — Gaps between tapes or fabric that are caused by poor layup practices (see Fig. 9.3.7).

(h) Honeycomb splice gaps — Open spaces between adjoining core or between core and face sheet are caused by improper layup (see Fig. 9.3.8).

(i) Disbond — Lack of bond joining between two different details because of surface contamination; bad fit between detail parts (see Fig. 9.3.9).

(j) Surface scratch — A narrow depression caused by marking or tearing with something rough or pointed; fibers may be broken (see Fig. 9.3.10).

(k) Surface depression — An indention caused by tools, or foreign material, resulting in deformed, not broken fibers (see Fig. 9.3.11)

(l) Frayed or burned material — Caused by extreme heat during cure and marked by noticeable color change in material (see Fig. 9.12)

(m) Broken fibers — Discontinuous or misplaced fibers in laminate (see Fig. 9.3.13)

(n) Incorrect ply count — Too many or too few plies in laminate, may not affect laminate thickness (see Fig. 9.3.14).

(o) Incorrect ply stack-up — Error in order of stacked plies in laminate (see Fig. 9.3.15).

(p) Incorrect ply alignment or orientation — In a composite laminate, alignment can typically vary by $\pm 2°$ in either direction without noticeable effect on overall strength. One problem that occurs occasionally is that plies are totally out of specified alignment, e.g., 45° or 90° is used where 0° is called for (see Fig. 9.3.16).

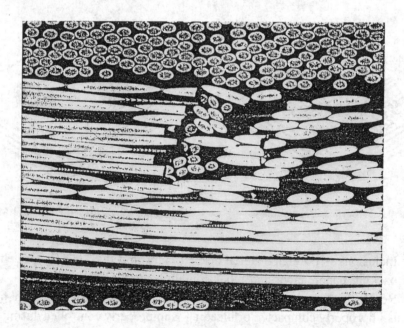

Fig. 9.3.13 Broken Fibers

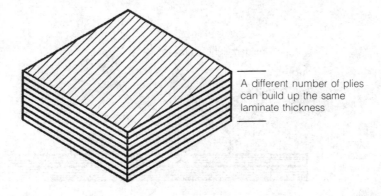

Fig. 9.3.14 Incorrect Ply Count

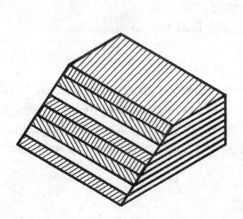

Fig. 9.3.15 Incorrect Ply Stack-up Sequence

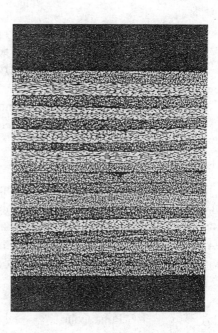

Fig. 9.3.16 Incorrect Ply Alignment

9.4 NDI METHODS

Visual Inspection

Visual inspection is the simplest and most economical inspection method and is routinely used for process control and final part inspection. Adequate, controlled lighting is necessary, and visual aids, such as mirrors, borescopes, magnifying glasses, microscopes, and other optical devices are used to inspect for defects and missing components. Fig. 9.4.1 illustrates the recommended angle to be used when visually inspecting with a light.

Visual inspection may be summarized as follows:

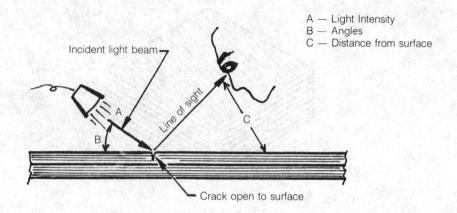

Fig. 9.4.1 Visual Inspection

- Requires adequate controlled lighting
- Utilizes visual aids (mirrors, borescopes, magnifying glasses, microscopes, etc.)
- Simplest and most economical method
- Routinely used for process control and final inspection

Ultrasonic Inspection

Ultrasonic inspection has become the most widely used method for detecting internal flaws in composite laminates and honeycomb assemblies. In this method, high frequency sound energy is introduced into the test part, and interpretation of the returned signals determines the presence of porosity, voids, delaminations, disdbonds, and other anomalies associated with composite materials (see Fig. 9.4.2).

Ultrasonic process:
- Methods:
  - Pulse-echo
  - Through-transmission
  - Reflected through-transmission
- Procedures:
  - Immersed (in water)
  - Contact (transducer on part)
  - Squirter (uses water stream)
- Displays:
  - A-scan (on CRT scope)
  - B-scan (on CRT scope and printout on x-Z recorder)
  - C-scan (on CRT scope and printout on x-y recorder)

Ultrasonic inspection:
- Relatively low cost inspection
- Is capable of providing permanent record for data retention in the form of hard copies and/or magnetic media

- High sensitivity to:
  - delaminations
  - voids
  - porosity
- Radiography complements the ultrasonic methods

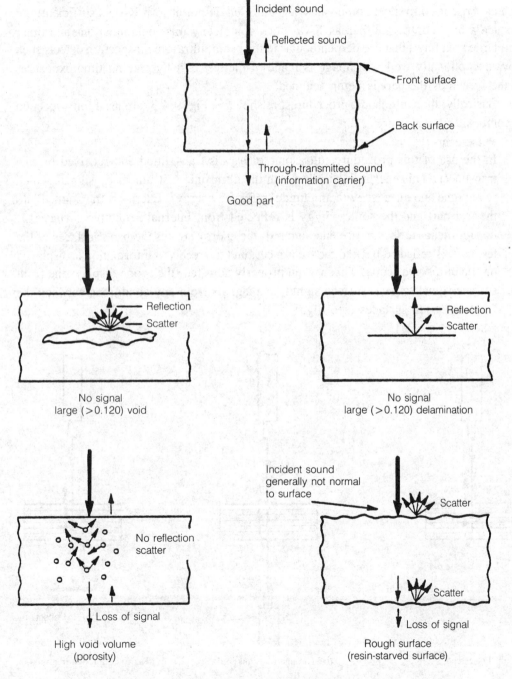

Fig. 9.4.2 Ultrasonic Wave Transmission Through Composite Laminate Panel

Disadvantages are:
- Relatively slow scanning method
- Difficulty imaging complex shapes
- Unbonded surfaces in intimate contact ("kissing") are not normally detectable (difficult to detect but can be detected)

Ultrasonic inspection is one of the few techniques capable of detecting flaws deep within the interior of a laminate. It is more sensitive to variations in organic materials, such as those used in most composites, which often do not absorb X-rays sufficiently to produce a high contrast. Whereas X-rays can completely miss delaminations and other thin, large area flaws that are perpendicular to the beam, ultrasonic inspection detects such flaws exceptionally well. An image is made of the test part by making time exposures on the screen as the part is being scanned.

Basically, three ultrasonic procedures, as shown in Fig. 9.4.3, are used for inspection of composites:

(1) Pulse-echo:
In the use of this procedure, ultrasonic energy is transmitted and received by one transducer. This energy is displayed on the ultrasonic instrument by signals which relate time and distance with amplitude (reflected energy). Energy is transmitted into the test part and the sonic energy is reflected from internal reflective surfaces. If no significant reflectors are encountered, the energy strikes the back surfaces of the test part, is reflected back to the transducer, and the received information is displayed on the ultrasonic scope. Thus, when properly adjusted, the scope displays the front surface signal, a back surface signal, and signals from any conditions between the front and back surfaces (see Fig. 9.4.4).

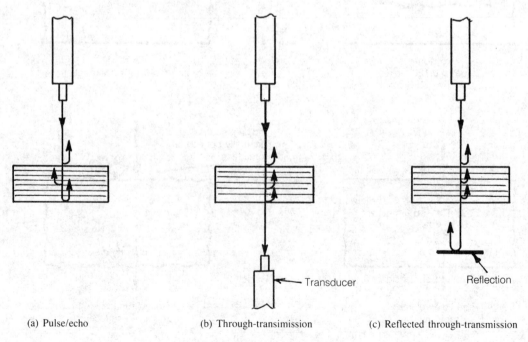

(a) Pulse/echo   (b) Through-transimission   (c) Reflected through-transmission

Fig. 9.4.3 Ultrasonic Procedures

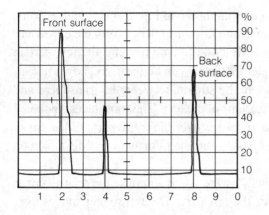

Fig. 9.4.4 Ultrasonic Scope Face

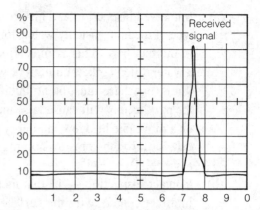

Fig. 9.4.5 Scope Showing Through-Transmission Signal

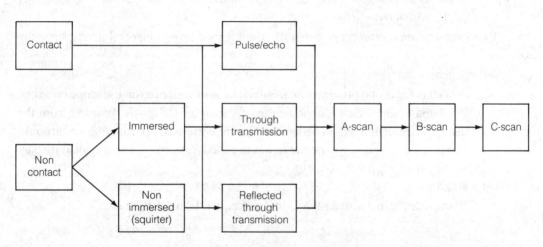

Fig. 9.4.6 Ultrasonic Inspection Procedures

(2) Through-transmission:

As the name suggests, ultrasonic energy is transmitted by one transducer and received by another, usually located on the opposite side of the test part. The ultrasonic scope displays a signal from the receiver as shown in Fig. 9.4.5. No reflections from the part interior are seen; the scope displays only the energy that passes through the part. If there is no flaw within the test part, the signal would remain at the calibrated height. If some anomaly interrupted the sound beam, then the received signal would be reduced. Honeycomb structures can only be inspected by through-transmission method (some pulse-echo can be done on honeycomb structure but is generally limited to laminate skin).

(3) Reflected-through-transmission:

This method is same as that of through-transmission except using reflector on the opposite side of the test part instead of transducer.

These procedures can be implemented in two ways (see Fig. 9.4.6):

(a) Immersed inspection — This procedure, as shown in Fig. 9.4.7, requires some type of medium, usually water, to transmit ultrasonic energy from the transducer to the test part. This is most frequently done by submerging the test part in a water-filled tank; a column of water (squirter) can be used, as shown in Fig. 9.4.8, but there must be no bubbles in the stream between the transducer and the part, as this will cause false indications of defects. The transducer, mounted on a moveable bridge scanning system, passes over the entire part. Either pulse-echo or through-transmission may be used and the information is displayed on a scope.

(b) Contact inspection — In this method, as the name implies, the transducer is in direct contact with the surface of the test part. The transducer moves over the surface to be inspected, as shown in Fig. 9.4.9. The transducer is coupled to the part by a thin non-toxic and non-corrosive sound conducting film, such as gelled water (tap water may be used as well). Either pulse-echo or through-transmission may be used, depending on the size of the part and accessibility of the inspected surface.

Two presentation methods are generally used for ultrasonic inspection of composite structures:

(a) A-scan:
A-scan is the scope presentation in which time or distance (on the scope baseline) is related to amplitude (scope height of the signal). Signals returned from the test part are displayed, whether contact or immersed, pulse-echo or through-transmission. The scope presentation is called "A-scan" and a typical display is shown in Fig. 9.4.10(a)(b).

(b) B-scan:
B-scan can also be used but is not a frequently used method.

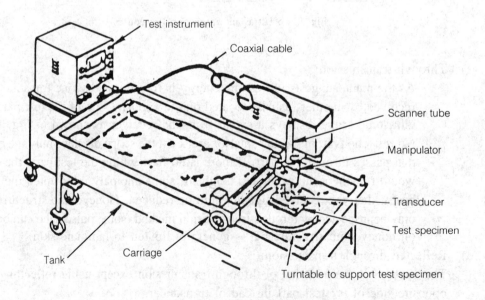

Fig. 9.4.7 Tank and Instrumentation for Immersed Testing

Quality Assurance

By courtesy of McDonnell Douglas Corp.

Fig. 9.4.8 McDonnell Aircraft AUSS-V System To Scan Contoured Composite Panels Inspected Using A Squirter

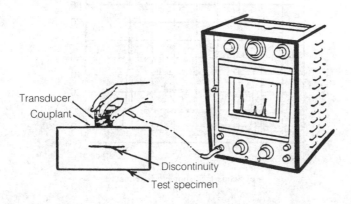

Fig. 9.4.9 Contact Ultrasonic Inspection

(c) C-scan:

C-scan is a two dimensional plan view recording of data and information taken from the "A-scan" when a part is inspected. All of the information shown on the "C-scan" is first displayed on the scope but is then plotted by a C-scan plotter. There are controls on the scope which allow the technician to determine which signals, and what degree of signal, will be converted into information regarding material quality to be shown on the C-scan's recording [see Fig. 9.4.10(a)(b)(c)]. Fig. 9.4.11 shows a large C-scan machine which uses several transducers, scanning simultaneously, and greatly reduces the time required for C-scan of large parts.

Chapter 9.0

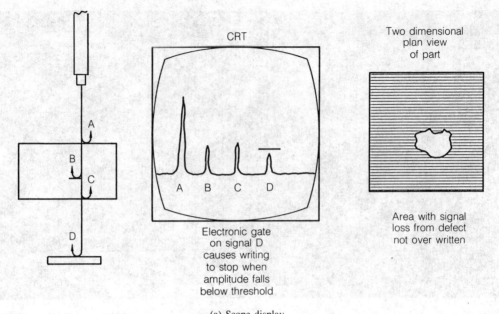

(a) Scope display

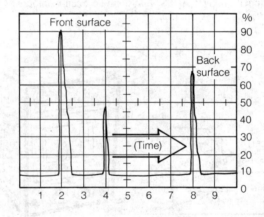

(b) Scope face showing baseline

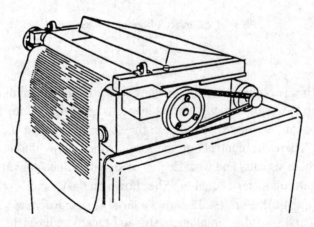

(c) Plotter (for C-scan only)

Fig. 9.4.10 Ultrasonic Inspection Methods

Quality Assurance

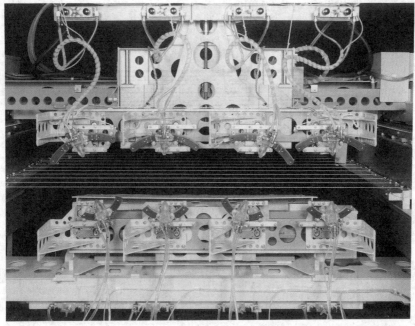

By courtesy of Custom Machine Inc.

Fig. 9.4.11 C-Scan Of Large Part, Using Several Transducers

It is imperative that a satisfactory reference standard is provided prior to any attempt at performing ultrasonic inspection. This standard must meet the fabrication guidelines and limitations for composite reference standard fabrication. The reference standard is fabricated from the same material, using the same layup method and cure procedures that are used for the test part. The reference laminate standard contains detectable discs (e.g., Grafoil discs with diameters of 0.5, 0.25, 0.125 , etc. inches) which simulate flaws. Honeycomb reference standards usually contain 0.75 or 0.5 inch inserts. The reference standard provides a means for the technician to calibrate the instrument that is used to detect anomalies, and when an anomaly is detected, allows the technician to compare the anomaly to a known flaw (see Fig. 9.4.12). The use of reference standards is summarized below:

- Requirements:
  — Is fabricated from same material and by same process as the part it will be used on
  — Porosity/void content is within specification limits
  — Utilizes Grafoil or Teflon discs in the layup to simulate unbonded areas
- Benefits:
  — Provides a means of sizing defects
  — Provides a means to standardize the calibration of ultrasonic equipment
  — Provides repeatability of calibration

Chapter 9.0

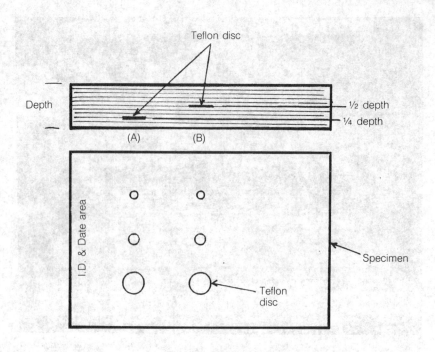

(a) Reference disc imbedded in laminate specimen

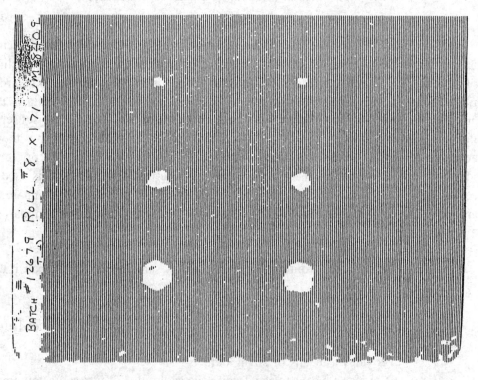

(b) C-scan of laminate specimen

Fig. 9.4.12 Example Of An Ultrasonic Reference Standards

(b) Example of ultrasonic evaluation criteria is given below:
- Single void area — 0.25 in². maximum
- Total accumulated area of any detectable voids — 1.0 in². in any 1.0 ft². area
- Single porous area — 1.0 in². maximum
- Total accumulated area of any detectable porosity — 4.0 in. in any 1.0 ft². area
- Distance between any detectable defects — 4.0 in. minimum
- Distance of detectable defect from finished edge — 1.0 in. maximum
- Delamination — None allowed

Ultrasonic inspection utilizes sound waves, usually between 500 kHz and 15 mHz. Within certain limits, the higher the frequency, the smaller the detectable flaw. Most inspection are conducted between 2 to 10 mHz.

The selection of an ultrasonic instrument, transducer, and method, etc. is usually followed to inspection specification which should be established in written procedures as to how parts will be scanned (This will help to standardize the inspections being performed). Some equipment is portable (see Fig. 9.4.13); C-scan equipment is not.

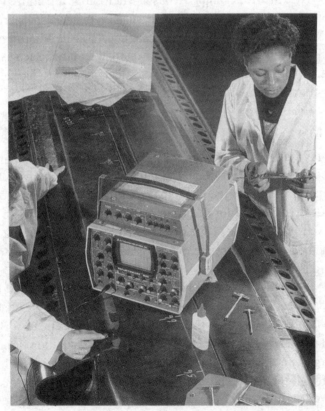

By courtesy of Sikorsky Aircraft

Fig. 9.4.13 Portable Ultrasonic A-Scanner Used On Composite Wing Structure

Acoustic Microscopy

The resolution of a C-scan is adequate for most composite structures; however, C-scans are limited by their relatively long wavelength. Acoustic microscopes operate over smaller areas, using frequencies several orders of magnitude higher than C-scan to obtain higher resolution (e.g., 1.0 μm). The higher resolution of acoustic microscopes is often needed when certain parameters, such as fiber population or orientation, fiber/matrix interface bonding [see Fig. 9.4.14(a)], or fiber breakage, are critical.

Acoustic microscopy (also known as Acoustic Micro Imaging — AMI) techniques are basically advanced forms of tradition Ultrasonic Testing techniques and provide greater sensitivity, more precise data gathering and higher resolution acoustic image with faster acquisition times than conventional ultrasonic C-scan systems. Different types of Acoustic Microscopes, SLAM, SAM and C-SAM as described below [see Fig. 9.4.14 (b)]:

(a) SLAM — The SLAM, Scanning Laser Acoustic Microscope, is a real-time, through transmission system and is able to detect defects or material variations throughout the entire thickness of a sample with one acoustic image. Its main advantage is that it can be used to qucikly screen a large number of samples to determine if they are of acceptable or rejectable quality.

(b) SAM — The SAM, Scanning Acoustic Microscope are utilized for the investigation of a sample's surface and near surface material properties.

(c) C-SAM — The C-SAM (C-Mode Scanning Acoustic Microscope) type systems are utilized to investigate a sample for internal features, such as bond or material interfaces, physical defects (cracks, voids, delaminations, inclusions, etc.), and/or microstructure at a specific depth or level.

Radiographic (X-ray) Inspection

Radiography (X-ray) is a useful tool for the inspection of composite parts. Radiography provides an excellent means to examine the interior of honeycomb parts for conditions such as misalignment, missing parts, core damage, inclusions, foam porosity, foreign objects, etc. One effective method for determining ply separations in a laminate is to apply a radio-opaque dye to the laminate edge or damaged area and perform a radiographic inspection. This method can disclose the extent of damage, and to some degree, the plies affected by separation or damage.

The cost of X-ray film is relatively high, but not prohibitive and the method provides the advantage of a permanent record. Portable inspection is suitable for field applications, but is limited because of the critical safety requirements.

The method uses an X-ray source to pass high energy radiation through the test part. Some of the energy is absorbed by the material under examination, while the reminder is transmitted through the part. The energy forms an image on the sensitive X-ray film which, when developed, displays the densities of the material being tested.

Radiographic inspection can be summarized as follows:

(a) X-ray passes high energy radiation through the test part and an image is formed on sensitive X-ray film, displaying the density of tested material

(b) Provides an excellent means to examine interior of honeycomb panels

Quality Assurance

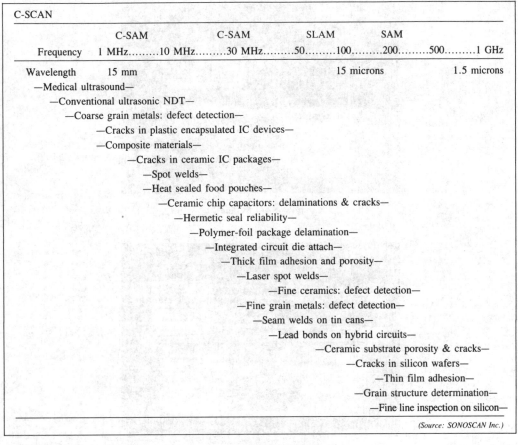

Note: SLAM and C-SAM acoustic microscopes made by SONONSCAN, Inc.

(a) Guide to Acoustic Micro Imaging

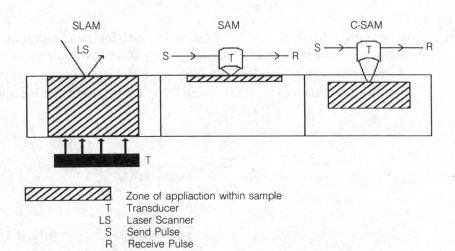

(b) Diagram indicating the various types of Acoustic Microscopy Technologies.

By courtesy of Sonoscan, Inc.

Fig. 9.4.14 Acoustic Microscopy Technology

Chapter 9.0

(c) C-SAM System

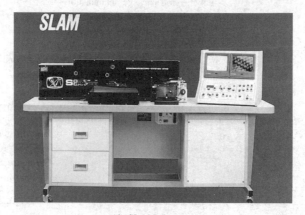

(d) SLAM System

By courtesy of SONOSCAN, Inc.

Fig. 9.4.14 Acoustic Microscopy Technology (cont'd)

(c) Laminate ply separations can be detected by applying radio-opaque dye to laminate edge or damaged area and performing X-ray inspection
(d) Provides permanent records on film (see Fig. 9.4.15)
(e) Special precautions must be taken because of potential radiation exposure

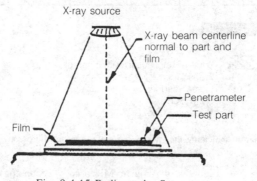

Fig. 9.4.15 Radiography Setup

524

The real-time radiography systems have been used on composite structure to eliminate slow setup and high film costs; while video-imaging systems are available to enhance the image. Low-energy X-rays improve contrast allowing better inspection of composite honeycomb panels and sandwich structuries, as shown in Fig. 9.4.16.

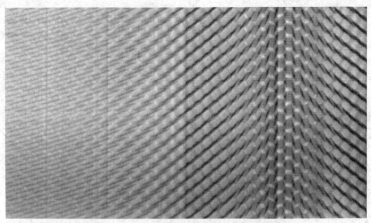

(a) Structure of a helicopter wing (honeycomb)

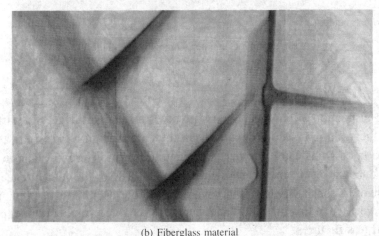

(b) Fiberglass material

By courtesy of Lixi Inc.

Fig. 9.4.16 Real-time Low-Energy X-Ray On A Composite Honeycomb Panel

Computer Tomography

Computer tomography is a means of retrieving data which is lost in standard X-ray imaging (which is produced by passing X-rays emitted from a single source only through the object in question). Computer tomography produces a slice of the area of interest, and can produce a 3-D image from a series of slices, as shown in Fig. 9.4.17. Computer tomography takes X-ray images from a number of different angles and processes this information electronically to produce a 3-D image. Theoretically, it cuts the material to examine the interior of the laminate without affecting part functionality.

This method is generally used to examine the matrix laminate, looking for resin-rich or resin-poor regions, or lack of bonding.

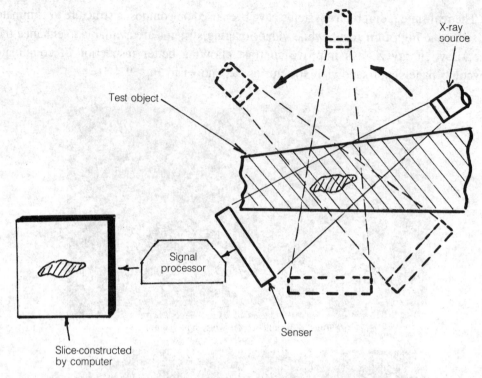

Fig. 9.4.17 Computer Tomography Method

Holography (Laser)

Holographic inspection of composite parts involves taking an image of a part while it is at rest, and then taking a second image when the part is slightly stressed. Defects such as delamination or voids cause the part surface to deform differently than in defect-free regions, produces a distortion, as shown in Fig. 9.4.18, in the interference pattern.

Because of its sensitivity to small movement, problems in fixturing and part stiffness occur. However, when this procedure can be used it is a very rapid and inexpensive way to inspect large structures.

This system allows real-time viewing of holographic images, or recording of permanent images on photographic film for future reference.

Shearography (Laser)

Shearography (see Fig. 9.4.19) is a laser-based interferometer that is used to detect impact damage, cracks in rivet holes, disbonded or overlapped joints, and to evaluate honeycomb repairs. Moreover, it allows on aircraft inspection without requiring the removal of the structure being tested. Benefits of shearography includes:
- Compared to Holography, no vibration isolation required
- Can be performed in ambient light
- Portablility allows use in field applications
- Real-time imaging
- Non-contact

Quality Assurance

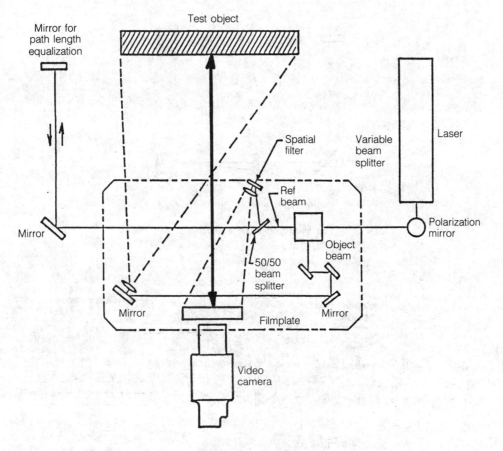

(a) Schematic illustration

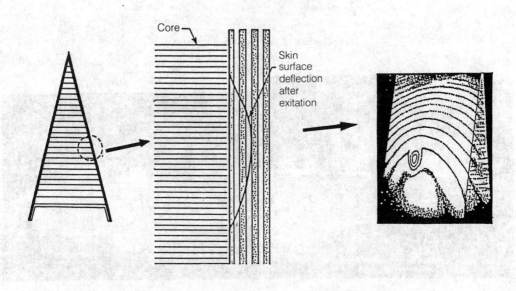

(b) Test specimen (honeycomb wedge)

Fig. 9.4.18 Holography Method

Chapter 9.0

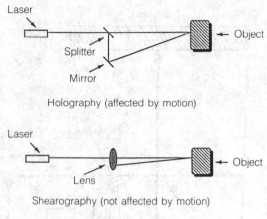
(a) Shearography vs. holography

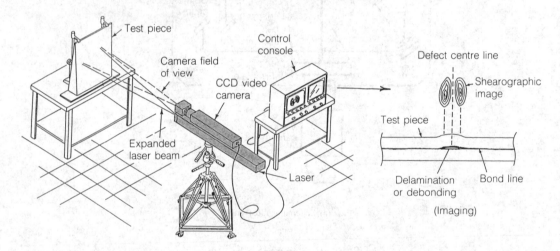
(b) Schematic of shearography arrangement

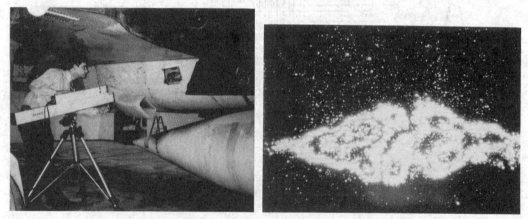

By courtesy of Laser Technology Inc. (PA)

(c) Shearography equipment to detact area of damage

Fig. 9.4.19 Shearography Inspection Method

Eddy-current

The Eddy-current technique is generally used to detect small surface cracks in metal structures, but this technique can also be used to detect damage in carbon-fiber composites. That is because carbon/fibers are relatively good electrical conductors; the flow of Eddy-current in carbon fibers is impeded where fibers are broken.

Thermography

Thermography is an infrared imaging system which can be used to detect pressurization leaks, thermal-duct leaks, and structural delaminations. It produces high-resolution, real-time thermal images and also has video-recording capability. This system only works on small parts, since providing a heat source for large parts is difficult and expensive.

A summary of NDI capabilities is shown in Fig. 9.4.20.

NDI method	Delaminations	Disbonds	Porosity	Resin variations	Fiber orientation	Core damage	Inclusions	Mislocated details	Fiber defects
Ultrasonics	A	A	A	B/A	C	B	B	C	C
Radiography	C	C	B	B	C	A	A	A	C
Thermography	B	B	B	C	C	C	B	A	C
Dye penetrants	A ⚠	B ⚠	B ⚠	—	—	—	—	—	—

Legend: A — Good capability
B — Fair capaibilty
C — Poor capability
Notes: — No capability
⚠ — Edges or surface only

Fig. 9.4.20 NDI Capability

9.5 DIFFICULT-TO-INSPECT DESIGNS

To avoid excessive costs and assure adequate inspection, the composite engineer must be aware of designs which are not readily inspectible by current NDI methods and attempt to avoid them whenever possible. The engineer should coordinate with quality assurance early in the design process to develop adequate inspection processes and standards for designs which are not readily inspectible.

Figures 9.5.1 through 9.5.8 show some difficult-to-inspect areas of which the engineer should be aware. This is by no means a complete list, but contains a few of the lessons learned by past composite programs. By having some knowledge of NDI techniques and their capabilities and limitations, the engineer may be able to prevent other difficult-to-inspect parts from being produced.

Chapter 9.0

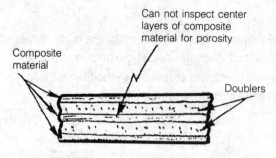

Fig. 9.5.1 Doublers

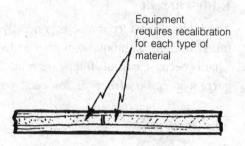

Fig. 9.5.2 Changes In Material Type

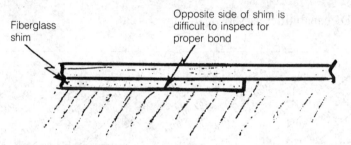

Fig. 9.5.3 Shims

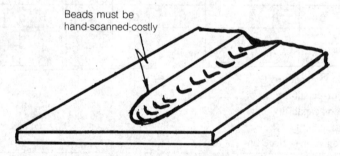

Fig. 9.5.4 Beaded Panels

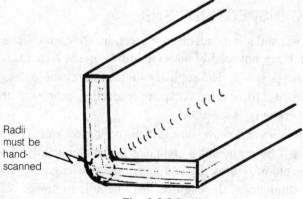

Fig. 9.5.5 Radii

Quality Assurance

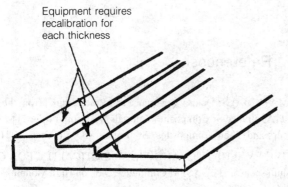

Equipment requires recalibration for each thickness

- It can be inspected for slight changes in thickness without recalibration
- It can be inspected for large thickness but is dependent on inspection system capabilities

Fig. 9.5.6 Steps/Changes In Part Thichness

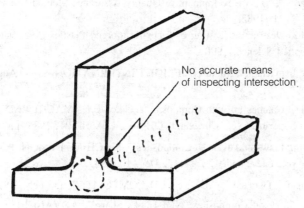

No accurate means of inspecting intersection

(It can be inspected but it is difficult)

Fig. 9.5.7. Tees

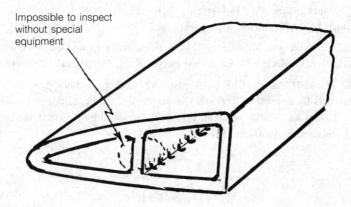

Impossible to inspect without special equipment

Fig. 9.5.8 Closed Out Areas

References

9.1 Browning, C. E., "COMPOSITE MATERIALS: Quality Assurance and Processing", from the Symposium on Producibility and Quality Assurance of Composite Materials, sponsored by ASTM Committee D-30 on High Modulus Fibers and Their Composites, St. Louis, MO, 20 Oct., 1981.

9.2 Pipes, R. B., "NONDESTRUCTIVE EVALUATION AND FLAW CRITICALITY FOR COMPOSITE MATERIALS", a symposium sponsored by ASTM Committee D-30 on High Modulus Fibers and Their Composites, PA, 10-11 Oct., 1978.

9.3 Korane, K. J., "Spotting Flaws in Advanced Composites", MACHINE DESIGN, Dec., 10, 1987.

9.4 Chen, J. S. and Hunter, A. B., "Development of Quality Assurance Methods for Epoxy Graphite Prepreg", NASA CR-3531. 1982.

9.5 Summerscales, J. "Non-Destructive Testing of Fiber-Reinforced Plastics Composites, Vol. 1 and Vol. 2", Elsevier Applied Science. 1990.

9.6 Ramsden, J. M. "The Inspectable Structure", FLIGHT INTERNATIONAL, 30 August, 1986. pp. 113-116.

9.7 Kulkarni, S. B., "NDT Catches up with Composite Technology", MACHINE DESIGN, Penton Publishing Inc., 1100 Superior Ave., Cleveland, OH 44114. April. 21, 1983. pp. 38-45.

9.8 Korane, K. J., "Spotting Flaws in Advanced Composites", MACHINE DESIGN, Penton Publishing Inc., 1100 Superior Ave., Cleveland, OH 44114. Dec. 10, 1987.

9.9 Leonard, L., "Testing for Damage", ADVANCED COMPOSITES, Jan/Feb., 1991. pp. 40-45.

9.10 Leonard, L., "Inside Story on Composites: NDI Looks Sharp", ADVANCED COMPOSITES, Jan/Feb., 1990. pp. 52-56.

9.11 Lewis, C. F., "Ultrasonics Reveal The Inside Picture", MATERIALS ENGINEERING, March, 1988. pp. 39-43.

9.12 Albugues, F. and LeFlock, C., "Holographic Techniques are Well-suited to Non- destructive Testing of Composites", ICAO Bulletin, May, 1981. pp.22-25.

9.13 Kessler, L. W., "Acoustic Microscopy", In Nondestructive Evaluation and Quality Control, Vol. 17, METALS HANDBOOK, 9th Ed., ASM International, Metals Park, OH. 44073. 1989. p, 465.

9.14 Kessler, L. W. AND Martell, S. R., "Acoustic Microscopy Technology (AMT) Analysis of Advanced Materials for Internal Defects and Discontinuities", Proceedings of The International Symposium for Testing and Failure Analysis (ISTFA), Published by ASM International, Metals Park, OH 44073. November, 1990. pp. 491-504.

Chapter 10.0

REPAIRS

10.1 INTRODUCTION

The increasing use of composites in both military and commercial aircraft mandates the development of proven repair methods that restore the integrity of the structure. Frequently these repairs must be applied in a timely manner with limited fabrication resources. The basic goal of repair is to restore a part's structural integrity in the shortest time span and at the least cost. The repair should match the strength and stiffness of the original part, while keeping any additional weight to a minimum. Depending on the part, often a trade-off must be made between aerodynamic smoothness and ease of repair. Continued and expanded use of composites critically depends upon the development of repair methods which are both structurally adequate and economically practical. See Fig.10.1.1 for commonly used methods of repair.

This chapter will provide essential information for a basic understanding of composite structures that is not only relevant to repair procedures (see Fig. 10.1.2) but also to the design of repairable composite structures. The design requirements and considerations discussed in Chapter 5 (Joining) are also applicable to repair designs.

Large part or large area repair often mandates replacement of the part, or its removal for remanufacturing. For damage over smaller areas, repair methods utilize prepregs and wet-layup techniques in conjuction with bonding, bolting and flush patching, etc. Two major repair techniques (see Fig. 10.1.3) can be categorized as follows:
- Bolted repair
- Bonded repair
  - Laminate repair
  - Honeycomb repair
  - Injection repair of delaminations

Repair categories are:
- Depot level (military) or maintenance base (commercial) —Performed at major maintenance bases or the manufacturer's facility
- Field level (military) or line station (commercial) — Performed at a forward operating base with limited facilities

Even with depot repair it may still be desirable to make the repair right on the aircraft to reduce downtime. This also lessens the chance of incurring further damage by removing the part.

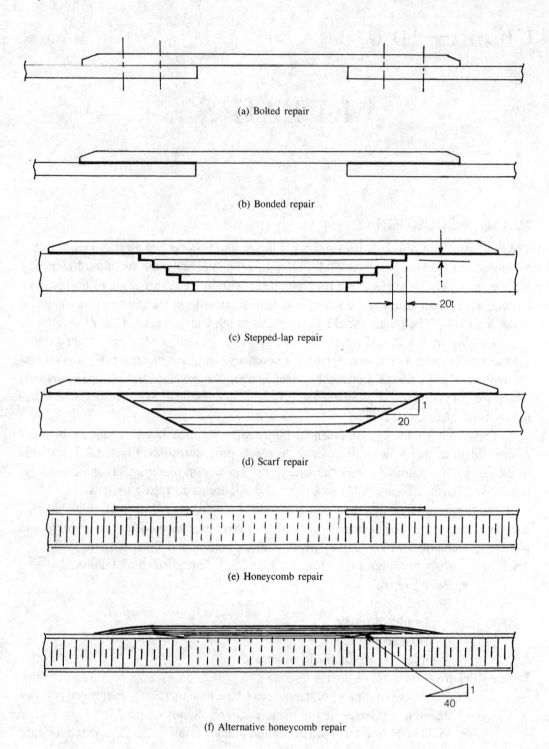

Fig. 10.1.1 The Most Common Methods Of Composite Repair

Repairs

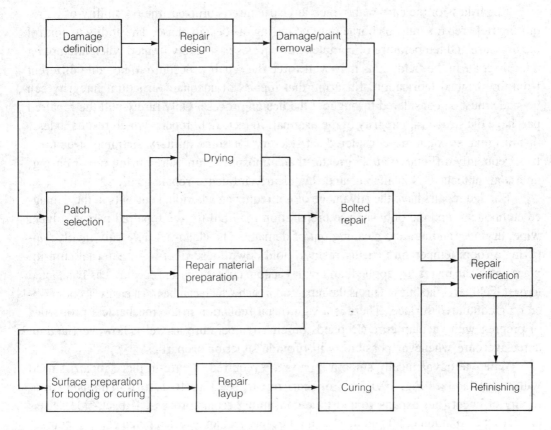

Fig. 10.1.2 Repair Process

Repair method	Pros	Cons
Adhesive repair	• Lighter weight • Distribution of load over wider area • Better for thin laminates	• Degradation in service due to temperature and humidity • Requires surface preparation • Difficult to inspect • More difficult to perform correctly
Mechanically fastened repair	• Not adversely affected by temperature and humidity • Less surface preparation required • Easy to repair and suitable for field repair • Generally visual inspection required, and sometimes may need Eddy-current to inspect holes	• Adds weight and bulk • Requires holes in weakened members • Produces stress concentration • Clearance between bolt and hole distributes load unevenly • Susceptible to delamination damage from driling/machining

Fig. 10.1.3 Bonded Vs. Bolted Repairs

Regardless of the care with which any structure is utilized, the possibility of its requiring field repairs must be borne in mind during the design phase. The ability to control the pressure and temperature of complex cure cycles is severely limited when attempting a local repair in the field. This may well force the repair procedure to be quite different from the original fabrication. Ensuring that repairs of adequate structural integrity can be made must be considered throughout the design process. Only rarely will the repaired part have the structural integrity of the original; in fact, field repairs which results in less-than-original strength are considered satisfactory for some military airframe structures. It is frequently difficult to apply greater than atmospheric pressure during repair curing, unless an autoclave is available, and this clearly influence repair quality.

Bonded repairs have the advantage of not requiring additional cut-outs in the damaged component, and the only surface preparation required for the repaired part is solvent wipe, drying, sanding, and cleaning out of damage. The clean-out may only involve applying a room temperature curing resin to hold down loose fibers. An external cure-in-place patch can be readily applied, and with vacuum cure this is a procedure easily adapted to most field level facilities. This is the repair approach which can be most easily accomplished on a contoured surface. There is a significant reduction in the mechanical properties of prepreg with vacuum cure (Depends on cure temperature as well), however, and an autoclave cure whenever possible, will provide superior properties.

When aerodynamically smooth repairs are required, a cure-in-place repair is used with a step-lap [see Fig. 10.1.1(c)] or scarf repair [see Fig. 10.1.1(d)]. This is a structurally efficient, but expensive and time-consuming repair process. If back-side access exists (not a common case) and there is no interference, an external patch can , of course, be applied to the inner surface thus preserving aerodynamic smoothness.

A problem encountered in many field repair situations is the lack of storage facilities for frozen composite materials. Use of C-staged prepreg, which can be stored at room temperature but still has flow characteristics necessary for proper cure, is one approach which looks promising. The use of two-part wet layup resins, analogous to current fiberglass repairs, would eliminate storage problems; but these systems are generally have high porosity with reduced durability compared to structural prepreg systems.

The effective composite repair requires knowledge of the following:
- Application and control of heat/pressure/vacuum systems
- Material behavior related to coefficient of thermal expansion
- The different characteristics exhibited by metals and composites
- Classes/types of adhesives/resins and their intended service environment
- Classes/types of fiber reinforcements and their use in prepregs and wet layup
- Characteristics of honeycomb core materials.
- Surface preparation techniques
- The creation of test specimens which are used to verify the strength of the repaired structure (see Fig. 10.1.4)

Criteria

The repair must restore the capability of the part to withstand design ultimate loads (without limitations unless otherwise specified) and must restore the full service life of the part. Design strength is based on the notched strength of composite in order to account for fastener holes and undetected damage. As part of the certification of a new aircraft, the FAA and military specifications require a satisfactory plan for accomplishing airframe repairs.

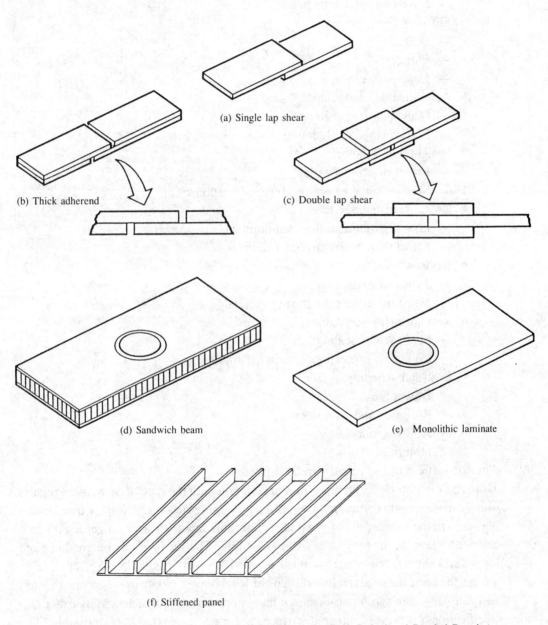

Fig. 10.1.4 Repair Test Specimen Configurations (For Both Bolted and Bonded Repairs)

(1) Common criteria for repairs include:
- Static strength and stability:
  — Full versus partial strength restoration
  — Stability requirement
- Durability for the life of the part:
  — Fatigue loading
  — Corrosion
  — Environmental effects
- Stiffness requirements:
  — Deflection
  — Flutter
  — Load path changes
- Aerodynamic smoothness:
  — Manufacturing criteria
  — Performance degradation
- Weight and balance:
  — Effect of size of repair
  — Mass balance effect on dynamic response
- Operational temperature:
  — Low/high temperature requirements
  — Effect of temperature on materials
- Environment:
  — Types of exposure
  — Moisture effects on repair materials
- Related airframe systems:
  — Fuel system sealing
  — Lightning protection system continuity
- Cost and schedule:
  — Downtime
  — Facilities and equipment
  — Skill personnel
  — Material handling

(2) Aircraft battle damage repair (or field repair):

This repair is a quick fix that has little in common with the type of permanent repairs usually made to both commercial or military airframe structures. Rather than restoring an airframe to meet the original condition, this repair focuses on getting a battle aircraft back in the air for one or two more flight. The repair might be good enough for several flights, but is not intended to last for years. Battle damage repair does not use the same materials, repair designs or level of complexity as peacetime repairs.

Aircraft battle damage repair does not have to restore the structure to its exact factory specifications; however, repaired surfaces are made as smooth as is possible. The aircraft may use more fuel than normal but at least it can be flown. A battle damage repair procedure or manual is used to control the repair procedures.

Classification of Damage

Most repairs relate to three types of damages:
- Manufacturing anomalies — Voids delaminations, surface depression, etc. during curing process
- Mishandling damage — Aircraft and parts are damaged on the ground (impact damage resulting from ground handling is the primary cause of damage to composite structures, see Fig. 10.1.4)
- Environmental damage — Hail, lightning strikes, bird strikes, debris (kicked up on takeoff or landing), etc.

Damage is defined as any deformation or reduction in cross section of a structural member or skin of the aircraft. Damage can consist of a hole, wrinkle, twist, nick, crack, scratch, delamination, or areas which are corroded (usually galvanic corrosion between metal and carbon composites) or burned. In general, damage to the structure is classified as either negligible, repairable, or necessitating replacement according to the following definitions:

(1) Negligible damage:

Damage to the airframe which does not affect the strength, rigidity, or function of the part involved is classified as negligible. Negligible damage is defined as the removal or distortion of material which can be permitted to exist as it is or which may be corrected by a simple procedure, such as the smoothing out of nicks or scratches.

(2) Repairable damage:

Repairable damage is divided into two basic categories as follows:

(a) Damage repairable by patching — Damage which can be repaired by installing a patch across the damaged portion of a part is defined as damage repairable by patching. The patch is attached to the undamaged portions (on all sides of the damaged portion) of the part to restore the full load-carrying characteristics and airworthiness of the airframe. Damage repairable by patching is specified, if practicable, for each member of an airframe on the material identification and repair index illustrations included in the structural repair manual.

(b) Damage repairable by insertions — Damage which has to be repaired by cutting entirely through a part and inserting a length of member identical in shape and material to the damaged part is defined as damage repairable by insertion. The member spliced into the space resulting from the removal of the damaged portion is identified as the insertion member. Splice connections or patches at each end of the insertion member provide load-transfer continuity between the existing part and the insertion member.

Generally, insertion repairs are used to simplify repair when the damaged portion is relatively long, when the damaged member has a complex shape, or when interference between the repair members and adjacent structural member is to be avoided.

	Surface impact
Dropped tools	Relatively common occurrence, although seldom specifically documented. Majority of hand tools weigh less than one pound although power tools of related equipment may be considerably heavier. Height of drop depends on height of work stands which may be used adjacent to the part impacted.
Dropped equipment	Although not specifically reported, dropping of mechanical, electrical or hydraulic equipment or access doors during installation or removal is a potential cause of impace damage. Weight and height as well as degree of risk depends on the specific aircraft.
Maintenance stands	Cases are reported in which a corner of a maintenance stand is pushed or falls against an aircraft component. Typical damage is a puncture or a surface gouge, although severity can vary widely.
Major damage	Several cases are reported in which a vehicle (truck, forklift, etc.) drove or backed into an aircraft. Since no reasonable degree of protection can be effective against this, ease or repair or replacement of susceptible components is necessary.
	Edge and corner impact
Dropped part	Frequently occurs for parts which are removeable from aircraft especially if the part is heavy or awkward to handle. Corners most easily damaged when striking flat surface such as concrete floor. Edges may be damaged but less frequently when striking other object, e.g., work bench corner, door frames, etc. Impact energy depends on wieght of component, height of drop, velocity at impact, incidence angle and resistance of object struck.
On-aircraft impacts	Impacts are reported due to work stands, hoisting cables or other equipment striking the exposed edges of fixed panels or opened doors. Appears to occur less frequently than dropping of removeable parts. Severity can vary widely from negligible to severe.
	Walking/Heel damage
Local pressure	Numerous cases are reported of dents believed to be caused by walking. Must frequent near intended walkways. Can result from heel presure or other foregin objects.
Walking	Disbonding of face sheets from honeycomb core has been observed and may be caused by walking, although this casue has not been definitely established.
	Fastener hole wear
Fastener shank abrasion	Retaining ring grooves in the shank of quick release fasteners have caused hole elongation. Condition is aggravated by a lateral force on the fastener during installation or removal, Force tends to be caused by dead weight of component, misalignment of fastener, etc. No hole elongation noted with smooth shank fasteners.
Pull through	Wear under countersunk head and local cracking and delamination can result if gasket or substructure configruation permits high tensile force in fastener.

Fig. 10.1.5 Causes of Damage (Ground)

(3) Damage necessitating replacement:
Damage which cannot be repaired is classified as damage necessitating replacement. Damage to parts which are of relatively short length, unless defined as negligible, should be considered as damage necessitating replacement because the repair of short members, generally, is impracticable. Some highly stressed members cannot be repaired because the repaired member would not have an adequate margin of safety. Such members, when damaged beyond negligible limits, must be replaced. Members with configurations not adaptable to practicable repair procedures must also be replaced when damaged.

10.2 BOLTED REPAIRS

The bolted repair method is not new and it is borrow from conventional metal sheet metal repair. Probably the quickest repair method is to bolt a patch over the damaged area. Plates of metal, such as aluminum (not to carbon composite skin) or titanium, or precured composite laminates can be bolted into place with little or no surface treatment.

Bolted repairs eliminate many of the facility problems and limitations of bonded repairs, but do require additional cut-outs which is structurally undesirable. The drilling operation, particularly where back-up cannot be provided, can result in additional damage or an oversize hole which will not pick up load properly. Special drills (see Fig. 10.2.1) are available, however, which minimize this problem. Bolted repairs are better adapted to field level facilities than bonded repairs, but field repairs must typically be accomplished without back-side access, requiring the use of blind fasteners. Improved blind fasteners have been developed for composites, and are currently being used. Fig. 10.2.2 shows both aluminum and titanium patches typically used to repair a 4.0 inch (10.16 cm) diameter damaged hole.

Repair patch material selection considerations:
- Titanium — Titanium sheet (6Al-4V) is a good patch material because of its corrosion resistance and high stiffness-to-weight ratio
- Aluminum — Aluminum sheet is generally more readily available under field repair conditions and is more easily machined than titanium, has a lower density, and can be protected against corrosion
- Stainless steel — Stainless sheet has corrosion resistance and high stiffness, but has a high density and is difficult to machine; it is generally not used
- Precured carbon/epoxy laminate — This laminates may be feasible for both bolted and bonded patches
- Woven fabric for wet layup process
  - Use 5 harness 12×12 in^2, weave (6,000 fibers/tow)
  - Impregnate on-site with resin or adhesive

The thickness requirements for bolted repair should take the following into consideration:
- Minimum thickness — No feather edges (knife edge) should be allowed if countersunk fasteners are used

Chapter 10.0

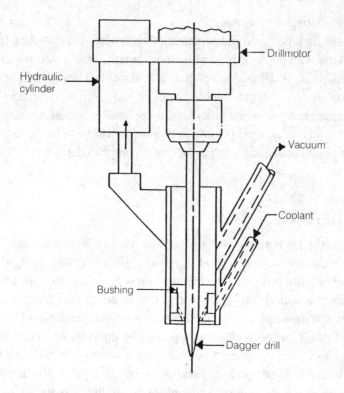

Fig. 10.2.1 Drill And Countersink System Schematic

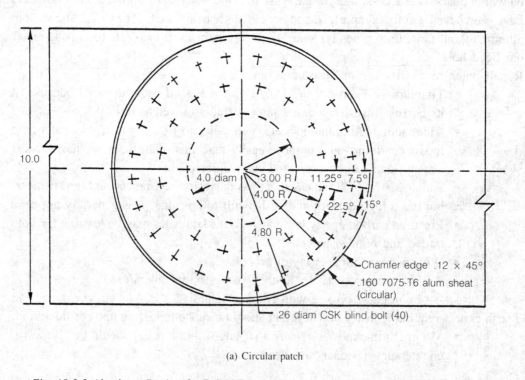

(a) Circular patch

Fig. 10.2.2 Aluminum Patches for Bolted Repair (Parent Laminate — 24 Plies Graphite/Epoxy)

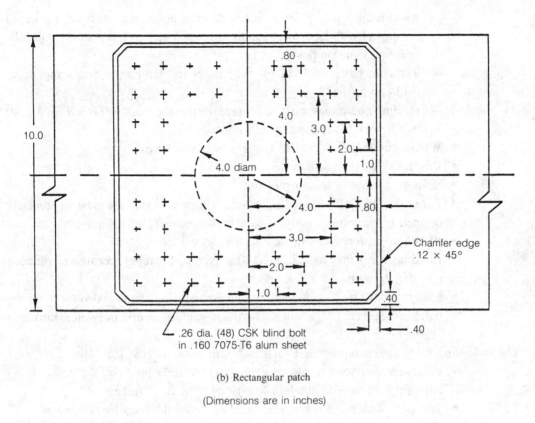

(b) Rectangular patch

(Dimensions are in inches)

Fig. 10.2.2 Aluminum Patches For Bolted Repair (Pavent Laminate — 24 Plies Graphite/Epoxy) (cont'd)

- Maximum thickness — Because of the requirement for surface aerodynamic smoothness, a maximum patch thickness [such as 0.16 inch (4.06 mm)] should not be exceeded in bolted repair (also bonded repair)
- Compexity of analysis with stepped straps or doublers; lots of field fasteners

There are drawback because this repair is not tailored to the needs of composite structures. Metal patches are usually heavier than the composite materials they are replacing, and they can alter the radar cross section.

Bolted repairs are simple because minimal equipment is needed to do the job and neither refrigeration nor heating blankets are required. Bolted repairs are the most common for battle-damage repairs because of reliability and relatively fast application to a damaged part. The design of bolted repair must consider the following:

- Patch material, thickness, and shape
- Fastener type, material, head configuration, and shank diameter
- Geometric arrangement of the fastener pattern
- In military aplications, no back-side access is assumed, so blind fasteners must be used (see Section 5.2 in Chapter 5)
- Recommended drilling procedures:
  — Wet installation of fasteners and use of faying surface sealant in carbon/aluminum interfaces is mandatory

- Aluminum patch — Drill undersize pilot holes in aluminum, locate on composite with clecos or clamps, and use patch as template to drill full-size holes in composite
- Titanium patch — Drill full size holes in titanium, use as template to drill composite
- Drilling operations must be separate or used special drill (see Fig. 5.2.40 in Chapter 5), if metal patch is used
* Where possible it is best to use existing bolt holes
* Bolt holes weaken structures
* Bolts in a blind hole can pull out
* For critical or highly loaded structures — fastener locations must be precisely defined by structural analysis to achieve recovery of design strength
* Effects of fastener clearances must be considered
* Blind fasteners may have limited effectiveness for repair because of reduced pull-off strength
* Fasteners must be wet installed for aluminum (also recommended for titanium) patches because of galvanic corrosion and moisture or water intrusion problems

The following rules are recommended for patch bolt patterns (see Fig. 10.2.2):
* Each bolt is spaced no closer than approximately four bolt diameters to an adjacent bolt three diameters to any edge of the laminate
* The edge distance for metallic patches is should keep two diameters
* The damage must be surrounded by at least two rows of bolts to protect against biaxial or off-axis load and inadvertent oversize bolt holes

10.3 BONDED REPAIRS

Bonding a patch usually is more reliable than bolted repair because bonding produces no holes, which are regions of increased stress. Both metal and composite patches can be bonded over a damaged area. External patches are made of either aluminum, titanium, precured laminate, wet, or prepreg layup. For metal patches, titanium is commonly used because of its non-galvanic reaction with the parent carbon/epoxy skin and its high strength and stiffness to weight ratio. The patch functions as a seal over the underlying repair and structurally carries some of the applied loading which the repaired part or component is expected to experience. Bonding methods require more rigorous control of surface treatments, storage of heat-sensitive bonding materials, and special equipment and is expensive.

Field repair environments generally restrict users to vacuum-bag cure techniques, and localized heat sources such as heat-blankets or pads, infrared lamps, or hot air guns, are generally used. Although epoxies are preferred by many for repair systems, their inherent reactive nature and potential storage problems make thermoplastics attractive as bonding agents. However, the disadvantage with thermoplastic is that they all require processing temperatures in excess of 500°F (260°C) and can not be used on themoset structures (250°F service temperature).

The design of a bonded repair must take the following into consideration:
- Patch material, thickness, shape and step taper
- Surface preparation
- Adhesive or resin requirements (see Fig. 10.3.1)
- Room temperature storage ability (because of the logistical problems of re-supply for field repair)
- Control of cure cycle
- Storage history of adhesive and treated patch materials
- Specialized skills and equipment required
- Vacuum pressure (see Fig. 10.3.2) adequate for bonded and cure- in-place repairs
- Cure-in-place repair (see Fig. 10.3.3) is advantageous for curved surfaces
- Structural advantages of bonded repair:
  — No cutout required
  — Can restore higher strength levels than bolted repair in thin laminate skin
- Flush aerodynamic bonded and cure-in-place repairs provide the most effective strength recovery; can restore design strength for thick high load laminates
- External bonded and cure-in-place repairs are adequate for restoration of design strength in lightly loaded structures
- Boron is more difficult to drill or cut then carbon, so bonded repairs are preferrable to bolted repairs

The alternative repair considered for the small patch include metal foils [titanium or aluminum (not used on carbon composite skin) with thickness of 0.008 inch (0.203 mm)], and several precured forms of composite sheets. Using a stack of progressively smaller foil sheets minimizes the peak adhesive shearing stresses and minimizes peeling stresses at the edge of the patch. Treatment of titanium or aluminum to provide a bondable surface is not feasible in shipboard environments. Etched and primed foils must be supplied in kit form for bonded repairs.

Fig. 10.3.3 shows a titanium patch which consists of several layers of precured composite plies and titanium foil, e.g., 0.008 inch (0.2 mm) thick or 0.016 inch (0.4 mm) thick, that are bonded to each other, and then to the damaged skin.

Care must be taken not to overheat the laminate being repaired. Overheating causes moisture to expand, forcing the plies to delaminate. Too much heat in one place can cause skin blisters. Constant temperature heat blankets provide a reliable cure, and virtually eliminate overheating problems when used correctly by trained operators. Heat is typically applied by means of flexible silicone heating pads laid directly on to the repair (see Fig. 10.3.4). These pads have sophisticated thermal controls for precise staging of the cure cycle. The time/temperature/pressure cure cycle can be monitored manually or automatically with a programmable repair console as shown in Fig. 10.3.5.

Fig. 10.3.6 shows a heating system (Moen Systems) which can deliver hot air to the repair area at temperatures up to 1500°F (816°C). This system expels high-velocity, low-pressure air to the repaired surface and also effectively dries parts.

Chapter 10.0

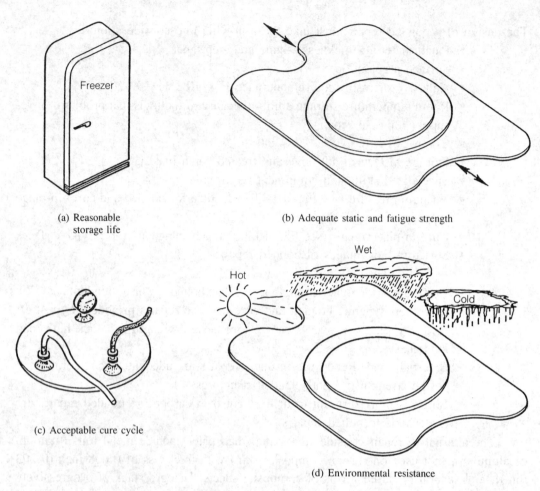

(a) Reasonable storage life

(b) Adequate static and fatigue strength

(c) Acceptable cure cycle

(d) Environmental resistance

Fig. 10.3.1 Adhesive and Resin Requirements

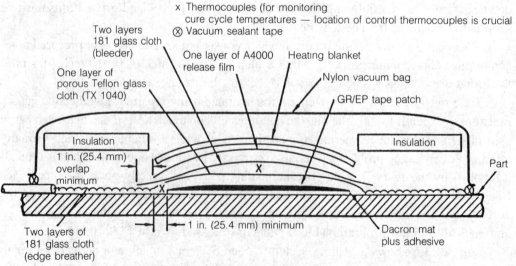

(Note: Insulation is placed and arranged over the vacuum bag such that the patch temperatures are within the specified cure temperature range)

Fig. 10.3.2 Typical Repair Bagging Arrangement

546

Bonded repair can be either cure-in-place or precured plies used with an adhesive. Sometimes metal and/or precured and cocured plies are combined in the same repair laminate:

(a) Cure-in-place — The entire patch is cocured to the part in one operation (see Fig. 10.3.7) or in thicker laminate or on vertical surfaces multiple cures may be performed; the part may not be able to withstand both the pressure and temperature of an autoclave, so lower pressure and temperature are applied. This method is the best one to use on curved or irregular surfaces.

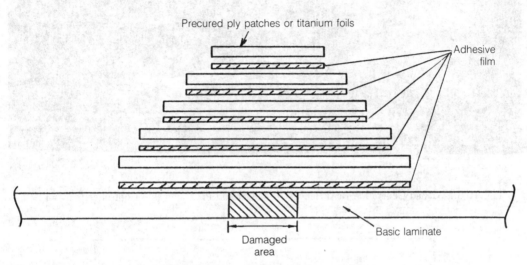

Notes: (1) Precured 3 or 6 ply patch layer
(2) Titanium foil — 0.008 inch or 0.016 inch thick

Fig. 10.3.3 Typical Metal Or Precured Patch Configuration

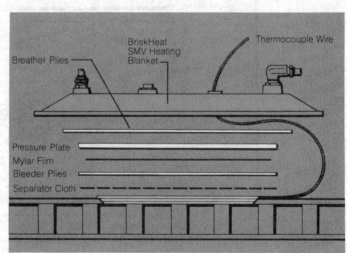

(a) First generation pad and controller circa 1986-87

(b) Typical cure layup

By courtesy of BriskHeat Corp.

Fig. 10.3.4 Flexible Silicone Heating Pad

547

Chapter 10.0

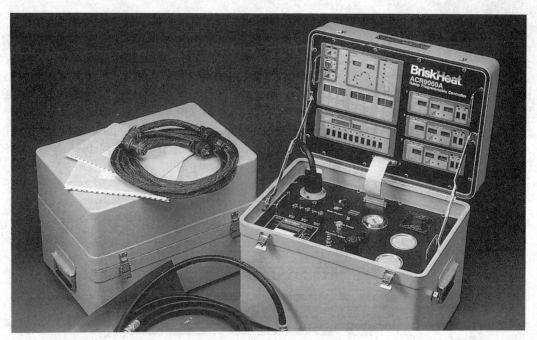

By courtesy of BriskHeat Corp.

Fig. 10.3.5 Second Generation BriskHeat ACR9000 Ramp Programmable Controller, Circa 1990

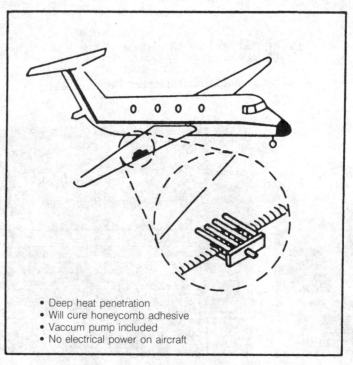

- Deep heat penetration
- Will cure honeycomb adhesive
- Vaccum pump included
- No electrical power on aircraft

By courtesy of Heat Transfer Technologies

Fig. 10.3.6 Moen Systems (Up To 1500°F)

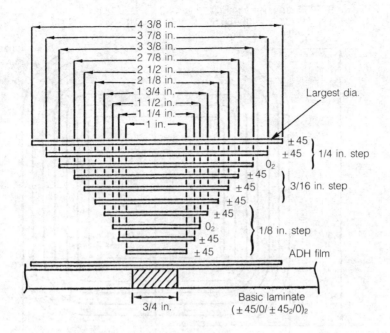

(a) Cut out damaged area (0.75 inch diameter)

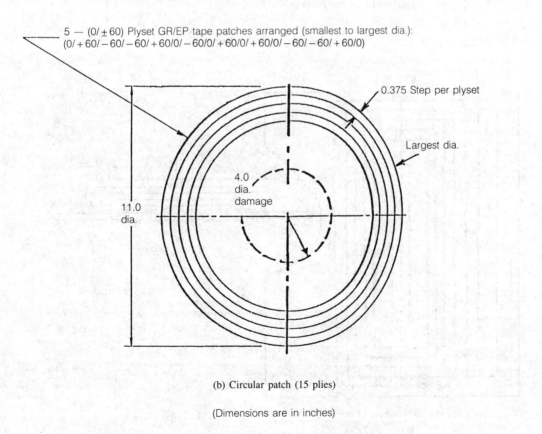

(b) Circular patch (15 plies)

(Dimensions are in inches)

Fig. 10.3.7 Samples Of Cure-In-Place Repair (Carbon/Epoxy Prepregs)

Chapter 10.0

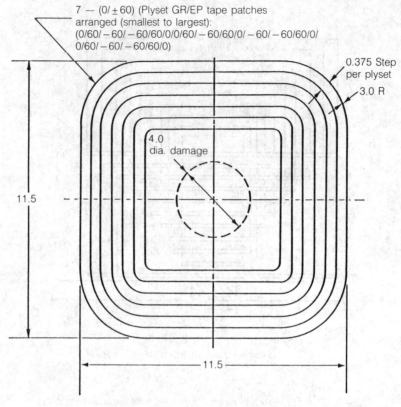

(a) Square patch (21 plies)

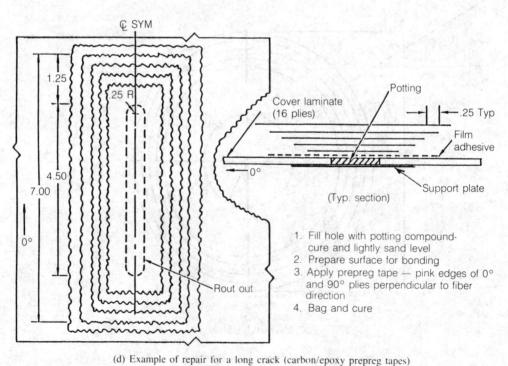

(d) Example of repair for a long crack (carbon/epoxy prepreg tapes)

(All dimensions are in inches)

Fig. 10.3.7 Samples of Cure-In-Place Repair (Carbon/Epoxy Prepregs) (cont'd)

- Fabric prepregs can be readily formed to curved repair surfaces
- Fabric prepreg plies can be readily stepped to form tapers
- Pre-staged prepreg (B-stage prepregs are available and some limited information on C-stage is also available) can be used; pre-staged material has long term storage ability and can still be formed to curved surfaces after storage
- Carefully control to obtain the low level of porosity in a patch when cured under vacuum pressure only

(b) Precuring — Precuring often achieves greater strength than cocuring, but requires more time. A replacement patch can be precured in an autoclave. This method tends to work better using several thin patches, e.g., 3 ply and 6 ply patches on curved surfaces or one thick patch for flat surface
- Patch can be multiple thin layers forming a step-lap or scarf repair (see Fig. 10.1.1) or a single splice plate (for simplicity) with chamfered edges
- This approach can readily utilize pre-kitted patch materials

(c) Bonded metal Patch
- One piece aluminum or titanium patch or titanium foil (see Fig. 10.3.3)
- Vacuum bonded or autoclave bonded
- Use of film adhesive or two-part system adhesive

Flush Tapered Bonded Patch

A flush tapered bond patch repair can restore full design strength. This type of repair costs more than other methods and is usually restricted to field level repairs. This method machines the laminate to a scarf surface for greater joining strength, as shown in Fig. 10.3.8; several external plies can be added which overlap onto the repair area. These edges of overlap plies (0° plies) are cut with standard pinking shears to produce serrations 0.125 inch (3.2 mm) deep, as shown in Fig. 10.3.9. The serrations have been shown to prevent peeling of the longer plies, thus resulting in a significant improvement in strength. This method can be used to repair a skin with one or more plies damaged and even through the thickness delamination. This is the most structurally efficient patch approach, but it is expensive and time-consuming.

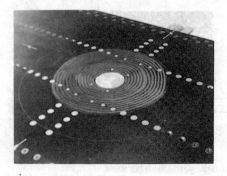
(a) Remove damaged material

(b) Flush repaired panel

By courtesy of McDonnell-Douglas Corp.

Fig. 10.3.8 Typical Flush Repair (Scarf)

Chapter 10.0

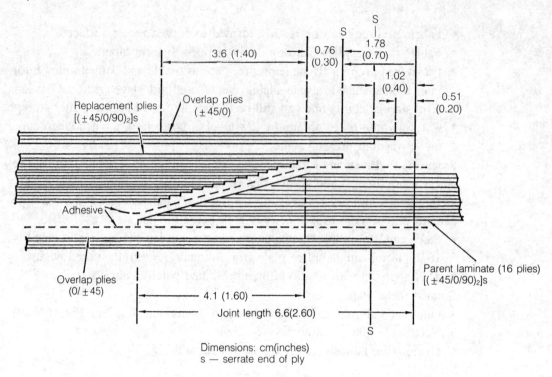

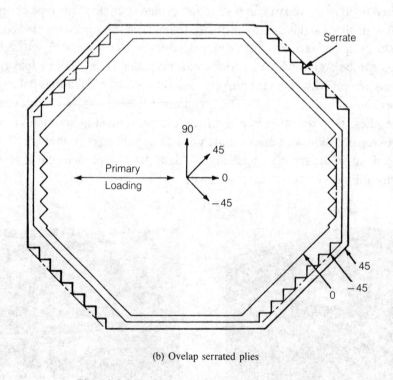

(b) Ovelap serrated plies

Fig. 10.3.9 Examples Of Flash Repair Configuration

Flush repair basically consists of removing the damaged area and replacing the plies:
- Inspect the part to determine the location and extent of the damage
- Remove the damaged area by cutting or sanding to form a scarf surface (sloping 1:30)
- Dry the surface with hot air, a heat gun, or in a oven, etc.
- Orient and stack the plies in the same sequence as the original laminate
- Place a vacuum bag over repair area
- Cure by heating with a heat blanket or similar heating system
- Inspect the repaired area, using ultrasonic inspection methods (see Chapter 9)
- Patch material is usually cure-in-place material with patch orientation matching the part orientation ply-for-ply

When the criteria cited for bolted repair for an external repair patch cannot be met, an alternative scarf or step-lap patch repair is considered. Testing has shown that a scarf repair [see Fig. 10.1.1(d)] is the most efficient joint that can be achieved, and that step-lap repair [see Fig. 10.1.1(c)] in the next most efficient method. However, a scarf joint is relatively easier to machine as opposed to accurately machining/cutting out a step-lap repair. Scarf or step-lap patches are generally considered to be flush patches and normally will have only a few doubler plies external to the moldline. These additional external doubler plies are designed to reduce the peel stresses that will build up at the end of the joint. This type of repair is expensive, requiring extra time and skill. The patches are uniquely designed, and fabricated from prepregs, then cocured and adhesively-bonded on in an autoclave or by the heating blanket curing process. Scarf and step-lapped patch repairs are applied only at the depot or maintenance base.

The most difficult part of the scarf operation is cutting the circular patches to the right size:
- Patch is too small leave gaps which are filled by resin-a weak area
- Patch is too big fold up at the edges

Wet or Prepreg Layup Repairs

Wet layup repairs usually are suitable for lightly loaded structuraes or crack damage and areas not subjected to high temperatures since cure occurs at room temperature. Wet layup patches are uniquely designed from dry woven cloth which has be wetted with resin by hand-impregnation.

These patches have the following characteristics:
- Any number of plies
- Any ply stacking sequence
- Shape specified by the repair engineer
- Surface may be contoured
- Lower strain level (e.g., 2000 μ in/in)

If damage is located in a region containing lower applied strain, multiple surface ply drop-offs, sharp contour breaks, or complex contours, the best repair is a precured "wet layup" patch. If high strain is required, a prepreg patch can be fabricated on a moldform tool or slave part. A vacuum bag is generally required because it allows vacuum and heat to be applied at the repair area.

10.4 HONEYCOMB REPAIRS

Honeycomb panel damage is usually found visually or during NDI. However, with honeycomb core damage it is often difficult to ascertain the full extent of the damage until the skin over the damaged area has been removed.

Repairing honeycomb panels frequently entails cutting out a new piece of core to replace the old and splicing it firmly in place with foaming adhesives. Exterior skin plies are then built up in configurations which match the original skin. For rapid repair with minimal equipment, simply filling in a small damaged area with a body filler or syntactic foam can return a part to its original aerodynamic surface profile frequently with little or no performance penalty, and only a small weight penalty.

The usual course of action for a very badly damaged honeycomb panel is to carefully remove the damaged portions of skin and core and sand away any paint and/or primer which may have been exposed. Replacement plies are cut out to exactly match the damaged area, and are laid up in the same orientation as the original prepregs. The repaired area is then covered with a vacuum-bag to apply pressure to the area during cure for maximum bond integrity and minimal voids.

A common repair problem with honeycomb panel structures is ingression of moisture into the honeycomb, usually as a result of microcracking in the composite skins. Skins must be peeled back, the honeycomb core dried out or replaced, and a repair patch applied. If moisture can get into the honeycomb core, it will become steam during the repair heating procedure, and can blow the part apart.

The type of repair used depends upon the type and extent of the damage, as well as on the loads in the area. The amount of repair weight which can be added to the area may also be a factor. Honeycomb panel can be repaired by following methods:

(a) Method A — For minimal surface damage:
 - Clean up the damage surface
 - Fill in the damaged area with filler or syntactic foam

(b) Method B — For intermediate damage:
 - Use potting compound to stabilize the parent core, if:
   - Damage area is small, and/or the loads are not high
   - The repair weight does not exceed the allowable weight limitation
 - If the limitations cited above are exceed then the following Methods C and D should be considered

(c) Method C — Use syntactic foam filler:
 - Remove damaged area
 - Fill in with syntactic foam or equivalent
 - Filler(s) are sanded flush with the surrounding skin
 - Bond or cocure the external skin over it

(d) Method D — Use honeycomb filler:
 - Remove damaged area
 - Fill in with new piece of honeycomb core
 - Use adhesive or syntactic foam to hold the new core in place
 - Bond or cocure the external skin over it

Fig. 10.4.2 shows how composite helicopter blades can be repaired in the field:
- A plug patch corrects skin and core damage [see Fig. 10.4.2(a)]
- A double-plug patch repairs holes [see Fig. 10.4.2(b)]
- A V-shaped patch is applied to the trailing edge [see Fig. 10.4.2 (c)]

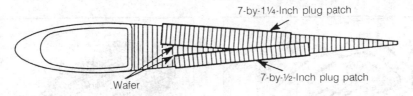

(a) Repair for through-the part damage

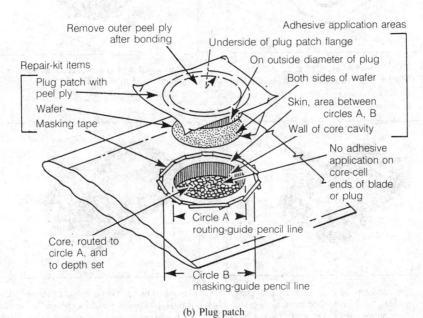

(b) Plug patch

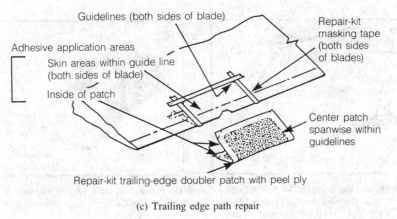

(c) Trailing edge path repair

By courtesy of Kaman Aerospace Corp.

Fig. 10.4.2 Honeycomb Blade Repair

Chapter 10.0

Example of Field Fixes for Beech Starship honeycomb panel

For the all-composite Beech Starship, a special repair method which utilize pre-cut prepreg patches and other special materials was formulated as the airframe was designed. It specifies all the necessary laminates, the ply sequences, and orientation of fibers.

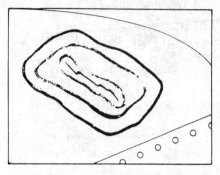

(a) Areas are marked to define actual damage and to locate area to be cleaned and prepared for patching process.

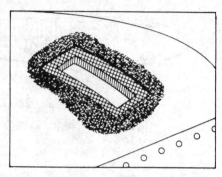

(b) Damagaed area is removed to evaluate inside damage and to make further preparations for patching.

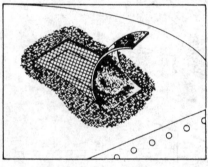

(c) The damaged core is replaced and epoxy painted graphite is applied in the same orientation and number of plies as the damaged skin which has been removed.

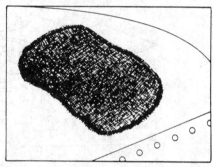

(d) One or two plies of graphite/epoxy are added on top of patch to make a "bridge" to undamaged area in order to accommodate designed load.

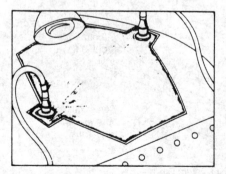

(e) A vacuum bag is applied to repaired area to assure good contact during 24 hours of curing. Then the repair is subjected to one hour of heat at 180 degrees F.

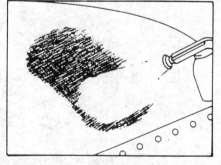

(f) Finally, the repaired area is sanded, primed and given a final coat of paint. It's as good as new!

By courtesy of Beech Aircraft Corp.

Fig. 10.4.3 Example of A Field Fix for A Beach Starship Skin

The repair of a Starship skin panel follows a controlled order which is specified in the detailed procedure. In the example shown in Fig. 10.4.3, the repair process involves a load bearing area of a wing skin punctured by a large tool which was dropped from a considerable distance. The drop splinters the fibers and penetrates the skin, crushing part of the Nomex core.

The major steps in the repair procedure are as follows:

(1) The first step is to inspect the extent of the damage. This is typically done by tapping the laminate and listening for resonant changes in the sound. A circle is then drawn around the area to define the actual damage.

   A second circle is drawn beyond that one to indicate the area which will help support the patch. And then a third circle on the outside of the second will define the total area to be cleaned and prepared for the patching process.

(2) The second step is to sand the surface and wipe it down with a special solvent to assure that it is bondable.

(3) The third step is to remove the face sheet over the damaged area and replace any damaged core.

(4) The fourth step is to lay on as many plies of graphite, painted with epoxy, as are necessary to fill the gap left from the removal of the damaged face sheet. This must be the exact number of plies as contained in the original skin and the plies must have the same orientations as the original skin.

(5) The fifth step is to build a bridge over the damaged spot to undamaged areas using woven graphite fiber soaked in epoxy. Enough layers are used to create a patch that restores the ability of the part to carry its designed load.

(6) The plies laid down are carefully, feathered on their outer edges to make a smoother patch.

(7) The final stage of the repair involves covering of area with a vacuum bag (a depressurized plastic sack) to assure good contact during curing.

(8) When the full epoxy reaction has taken place, the repaired area is ready for sanding, priming and painting.

(9) The whole process is simple and can usually be done overnight.

10.5 INJECTION REPAIRS

This method, as shown in Fig. 10.5.1, injects resin directly into the delaminated area, without removing the damaged materials. Resin injection can be performed quickly and right in the field. The resin injection method of repair is the primary approach for repairing delaminations and skin-to-core disbonds. Resin is injected into the damaged area at an edge, or through drilled holes/fastener holes, and re-glues the damaged area together. Some delaminations can be repaired by drilling two holes through the skin to the depth of the delamination, and then injecting resin into one hole until it flows freely from the other hole.

Chapter 10.0

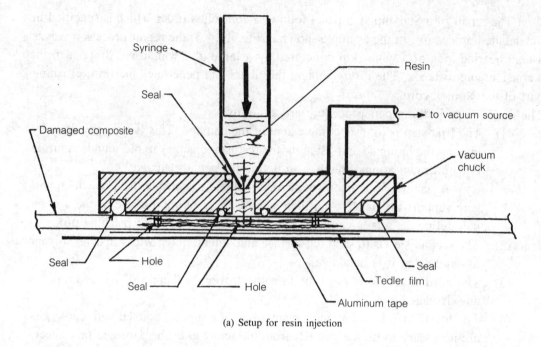

(a) Setup for resin injection

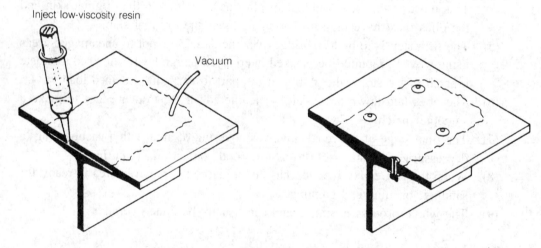

(b) Bonding pressure by using fastener's clamped up force

Fig. 10.5.1 Resin Injection Method

Injection repair for filling of internal delaminations identified by NDI is a uniquely composite repair, and is not analogous to any type of metal repair. This repair can be readily accomplished at depot level or even a field level facility, using a two-part low viscosity resin.

Generally, injection repair is not effective for filling internal delamination in GR/EP structure, because it will not flow enough to create a bond.

References

10.1 Brown, H., "Composite Repairs: SAMPE Monograph No. 1", published by SAMPE International Business Office, Covina, CA 91722. 1985.

10.2 Hamermesh, C. L., "Composite Repairs: SAMPE Monograph No. 2", published by SAMPE International Business Office, Covina, CA 91722. 1991.

10.3 McConnell, V. P., "In Need of Repair", ADVANCED COMPOSITES, May/June, 1989. pp.60-70.

10.4 English, L. K., "Field Repair of Composite Structures", MATERIAL ENGINEERING, Sept., 1988. pp.37-39.

10.5 Klein, A. J., "Repair of Composites", ADVANCED COMPOSITES, July/August, 1987. pp.50-62.

10.6 Lynch, T. P., "Composite Patches Reinforce Aircraft Structures", DESIGN NEWS, April 22, 1991. pp.116-117.

10.7 Brahney, J. H., "Composite Repair Techniques Examined", AEROSPACE ENGINEERING. May, 1986. pp.20-24.

10.8 Anon., "Field Fixes for Starship Skins are Simple, Logical and Mostly Overnight", Internal publication of Beech Aircraft Corp.

10.9 Watson, J. C., "Bolted Field Repair of Composite Structure", NADC-77109-30, Naval Air Development Center Report. March, 1979.

10.10 Bohlmann, R. E., Renieri, G. D. and Libeskind, M., "Bolted Field Repair of Graphite/Epoxy Wing Skin Laminates", Joining of Composite Materials, ASTM STP 749. American Society for Testing and Materials, Philadelphia, PA 1981. pp.97-116.

10.11 Stone, R. H., "Repair Techniques for Graphite/Epoxy Structures for Commercial Transport Applications", NASA Contractor Report No. 1590576. Jan., 1983.

10.12 Anon., "Advanced Composite Repair Guide", Contract No. F33615-79-3217, Air Force Wright Aeronautical Laboratories, Wright-Patterson Air Force Base, OH 45433. 1982.

Chapter 11.0

COMPOSITE APPLICATIONS

11.1 INTRODUCTION

Composites are now routinely used on aircraft and not only on control-surfaces and fairings but also on primary structures on both military and commercial aircraft. It can only be a matter of time before aluminum alloys are ousted from their position as the primary airframe material. Composites such as carbon fiber-, Kevlar-, and fiberglass-reinforced laminates have made progressive inroads into airframe designs and some manufacturers of business aircraft have embraced composites wholeheartedly. Much of the credit for persuading manufacturers to obtain composite experience goes to NASA (see Fig. 11.1.1), which in the 1970s spent more than $60 million on the Aircraft Energy Efficiency (ACEE) Program for the design, manufacturing, and testing of composites.

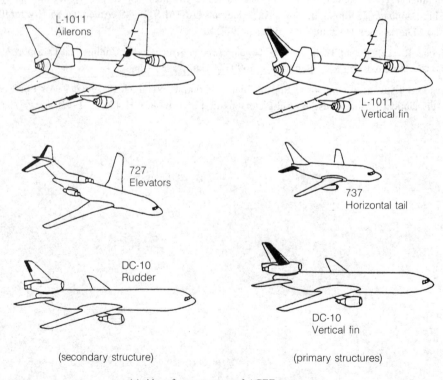

(a) Aircraft components of ACEE program

Fig. 11.1.1 NASA Aircraft Energy Efficiency (ACEE) Programm

Aircraft component	Planform area, m²	Chord		Span, m	Metal Design weight, kg	Composite Design weight, kg	Weight savings %
		root, m	tip, m				
Secondary Structures:							
727 Elevator	4.1	1.24	0.53	5.26	128.7	95.7	25.6
DC-10 Rudder	3.0	.97	.60	4.00	41.4	30.3	26.8
L-1011 Aileron	3.2	1.34	1.39	2.49	64.1	47.3	26.2
Primary Structures:							
737 Horizontal stabilizer	4.8	1.31	0.61	5.09	118.2	86.2	27.1
DC-10 Vertical fin	9.36	2.07	1.10	6.95	456.0	363.9	20.2
L-1011 Vertical fin	13.9	2.73	1.31	7.62	390	283.2	27.4

(b) Structural data and weight savings

Fig. 11.1.1 NASA Aircraft Energy Efficiency (ACEE) Program (cont'd)

Product	Component and/or producer	Composite
	PRODUCTION	
F-14	Horizontal stabilizer	Boron-epoxy
F-15	Tail section, cabin floor, and stabilator	Boron-epoxy
UTTAS Helicopter	Structural beam reinforcement	Boron-epoxy
F-111	Wing pivot doubler	Boron-epoxy
Mirage 2000	Rudder	Boron/graphite-epoxy
Space shuttle	Fuselage	Boron-aluminum tubes
	RESEARCH AND DEVELOPMENT	
F-14	Overwing and fairing	Boron + other filaments/fibers-epoxy
A-7	Outer wing	Boron/graphite-epoxy
C-130	Wing box	Boron-epoxy-reinforced aluminum
F-4	Rubber	Boron-epoxy
B707	Foreflap	Boron-epoxy
F-100	Engine fan-blades	Boron-aluminum
C-5A	Wing slat	Boron-epoxy
CH-54	Fuselage string and tail skid	Boron-epoxy
B-1	Horizontal and vertical stabilizers and wing slat	Boron/graphite-epoxy
F-111	Horizontal stabilizer	Boron-epoxy
CH-47	Rotor blade	Boron-epoxy

Fig. 11.1.2 Early Military Boron Composite Programs

With each new generation of military aircraft, the use of composite materials has increased. Combat aircraft entering service in the 1990s will be the first to be designed from the ground up to take full advantage of composites (see Fig. 11.1.2).

Currently, 10 percent of the structural weight of existing modern combat aircraft, such as the F-16 or F-18, is accounted for by composites. This proportion is only a few percent for the latest commercial transport, but the tendency is towards more widespread use of composites. The aim of second-generation composite development is to lower manufacturing cost while improving the acceptable stress limits, particularly in compression. Good damage tolerance properties and acceptable performance of the matrix-fiber combination in environmental conditions, such as those encountered in operation must be maintained.

The following composite applications are for reference only; some are in production and others are only in development. This chapter does not cover all composite applications but a few select components.

11.2 COMMERCIAL TRANSPORT AIRCRAFT

Application of composites in commercial transports has generally lagged behind military usage because:
- Cost is a more important consideration
- Safety is a more critical concern, both to the manufacturer and government certifying agencies
- A general conservatism because of past experiences with financial penalties from equipment down time

Composite materials are widely used, not only for exterior secondary components such as leading and trailing edges, fairings, radomes, landing gear doors, etc., but also for fuselage interiors on commercial transports. All interior materials must meet both flammability resistance (if applicable) and smoke and toxic-gas emission requirements. In general, the phenolic resin system is used for aircraft internal applications because of its excellent fire-resistant properties. The main considerations for interior panel application are:
- Impact resistance (mainly floor and side-wall panels)
- Stiffness
- Surface smoothness and appearance

NASA Aircraft Energy Efficiency (ACEE) Program

The ACEE program which started in 1975, greatly expanded the scope of commercial transport composites applications; it included three secondary and three primary structures:
(a) Secondary structures
- Lockheed L-1011 inboard aileron
- Boeing 727 elevator
- McDonnell-Douglas DC-10 rudder
(b) Primary structures
- Lockheed L-1011 vertical fin box
- Boeing 737 Horizontal stabilizer box
- McDonnell-Douglas DC-10 vertical fin box

The principal criteria applied to the design of the composite components in the ACEE program were:
- The part must meet the same design load requirement as the present metal part (in some cases, detail parts are being made to match the strength of the metal part even if it was originally over designed)
- The loads induced into adjoining structures must be the same as, or less severe, than the present loads (that is, no modifications of adjoining structure should be required)
- The interface should require little or no change and any change must accommodate a standard metal part as well as the composite part
- Handling characteristics of the aircraft should not be significantly altered (particularly, there should be no change in control response or adverse changes to the flutter envelope of the aircraft)

(1) Lockheed L-1011 Inboard Aileron:

The L-1011 aileron [see Fig. 1.1.3(a)] is a two-spar, five rib-stiffened structure with sandwich skin construction. Because the aileron is located just aft of the wing engines, the aileron must withstand acoustic loadings.
- Skins are made of graphite/epoxy tape facesheets and syntactic foam cores. The syntactic core or Syncore (see Fig. 2.6.14), consists of epoxy resin filled with small glass microballoons
- Graphite/epoxy tape is used for the main ribs; graphite/epoxy cloth for the intermediate ribs
- Graphite/epoxy tape and graphite/epoxy cloth are used for the front spar

(2) Boeing 727 Elevator:
- The Boeing 727 elevators (see Fig. 11.2.1) are constructed using Nomex honeycomb sandwich panels with graphite/epoxy facesheets for surface panels and four ribs.
- Use of sandwich covers allows elimination of most ribs.
- The ribs have honeycomb stabilized webs, and the spars are solid laminates.
- Skin panel facesheets have a layer of graphite fabric oriented at 45° and a single layer of unidirectional tape at 90°. The tape is used as the outer layer of the exterior facesheet to provide a smooth, nonporous surface.
- The outer layer of the inner facesheet is fabric, because it is more resistant to fiber breakout during drilling. Only fabric is used for the ribs
- Due to the weight savings on the elevator, removal of mass from the balance weight contributes to an overall weight savings of 29 percent (25.6% structural savings, see Fig. 11.1.1)

(3) McDonnell-Douglas DC-10 Rudder:

The all-graphite/epoxy structural box (see Fig. 11.2.2), providing a weight savings of 33% (including balance weight), is a two-spar multi-rib construction with solid skins. Skins are made from unidirectional broadgoods and substructures from fabric. Leading and trailing edges are fiberglass. The cost-effectiveness of the design results from following factors:

Chapter 11.0

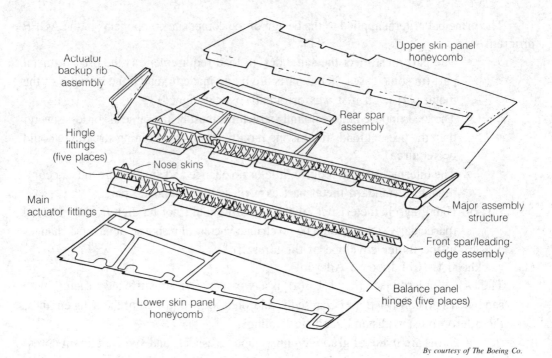

Fig. 11.2.1 Boeing 727 Composite Elevator-Structural Arrangement

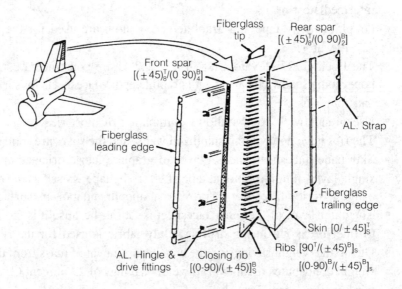

(a) Construction of DC-10 composite rudder. Superscript T denotes uniweave, wide tape and B denotes biwoven fabric. Numerical subscripts denote the number of plies and subscript S denotes symmetry.

Fig. 11.2.2 DC-10 Graphite/Epoxy Composite Rudder

(b) DC-10 Composite rudder

By courtesy of McDonnell-Douglas Corp.

Fig. 11.2.2 DC-10 Graphite/Epoxy Composite Rudder (cont'd)

- Use of broadgoods and fabrics, which reduces layup time
- The structural box is made as a single cocured unit using the beanbag approach which uses hollow aluminum beads in an inflatable bag. When a vacuum was pulled on the bag, it forms a solid mandrel which could be used as a form-block for rib channel members
- After assembling the bags within the specified steel curing tool, heat is applied to the bags and the total structure is cured
- Curing is done in an oven which is less expensive to use than an autoclave

(4) Lockheed L-1011 Vertical Fin Box

The metal version of the L-1011 vertical fin box, as shown in Fig. 11.2.3, is composed of solid skins over a substructure consisting of two spars and seventeen ribs. Unidirectional tape was used for the skins on the composite version because of its better mechanical properties and suitability for use with automatic layup processes. The number of ribs was reduced from 17 to 11 when composites were used [see Fig. 1.1.3(b)].

- The three upper ribs were solid graphite/epoxy laminates with integrally molded caps and bead stiffeners
- The eight lower ribs combine graphite/epoxy caps with extruded aluminum truss webs.

- The covers are solid laminates with integral, cocured hat stiffeners which is lighter than a honeycomb design
- The closed hat section was preferred over open channels or I-sections because it does not require tie clips at the ribs to prevent twisting instability

By courtesy of Lockheed Aeronautical Systems Co.

Fig. 11.2.3 L-1011 Composite Vertical Fin Box

Composite Applications

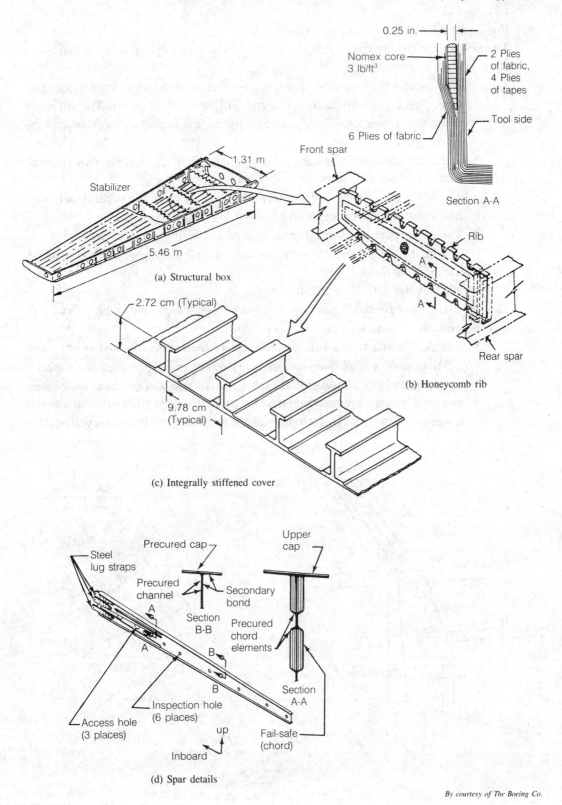

Fig. 11.2.4 Boeing B737 Composite Horizontal Stabilizer

(5) Boeing 737 Horizontal Stabilizer Box:
The Boeing 737 Graphite/epoxy Horizontal Stabilizer (see Fig. 11.2.4) is a two spar, eight rib box construction.
- The cover skin is a solid laminate, I-stiffened skin made from fabric and tape. The cover skin and its integral stiffeners are cocured. The stiffened composite cover skin is only slightly lighter than its unstiffened aluminum counterpart.
- The front and rear spars are solid laminates that are made in two channel sections which are subsequently bonded back-to-back
- The ribs are of a honeycomb sandwich construction in the vertical web portion only; the upper and lower flanges are solid laminates
- The cover panels are mechanically fastened to the substructure
- The major weight reduction comes from the rib and spar, and the reduced numbers of ribs

(6) McDonnell-Douglas DC-10 Vertical Fin Box:
The DC-10 vertical fin box consists of honeycomb sandwich skins with a four spar and thirteen rib substructure (see Fig. 11.2.5).
- The skins consists of fabric graphite/epoxy facesheets oriented at 45° over a Nomex core; two plies on the outer surface and one ply on the inner surface.
- The spars and ribs all use graphite fabric/epoxy sine-wave webs in most areas
- Spar and rib caps are built into the skin sandwich by a gridwork of unidirectional tapes in a quasi-isotropic pattern to locally replace honeycomb.

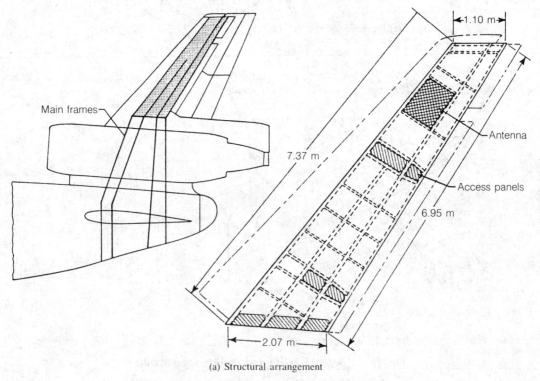

(a) Structural arrangement

Fig. 11.2.5 DC-10 Composite Vertical Stabilizer

By courtesy of McDonnell-Douglas Corp.

(b) DC-10 Vertical stabilizer

Fig. 11.2.5 DC-10 Composite Vertical Stabilizer (cont'd)

B747, B757, B767 and B777 Transports

The experience gained from the ACEE programs has resulted in increased composites usage on the next generation of commercial transports, such as the Boeing B757 and B767. In general, with the exception of small, detail parts, most composite components are of honeycomb sandwich construction. Some examples include:

(1) B747 — 6 ft-high winglet, carbon-fiber front and rear spars covered with carbon-fiber epoxy honeycomb sandwich skin panels
(2) B757 — Aileron, inboard spoiler, outboard spoiler, rudder, elevator, and inboard trailing edge flap (see Fig. 11.2.6) are made of carbon/epoxy materials.
(3) B767 — Ailerons, spoilers, rudder, and elevators
   - Outboard aileron is a full-depth honeycomb construction, as shown in Fig. 11.2.7
   - Rudder cover panels were split for manufacturing reasons. Basic design is a two spar multi-rib box using honeycomb sandwich ribs, spars and cover panels (see Fig. 11.2.8)
(4) B777 — Besides incorporation of composite components used on previous Boeing aircraft [see Fig. 1.1.9(a)], this aircraft has a composite empennage.

Chapter 11.0

By courtesy of The Boeing Co.

Fig. 11.2.6 Boeing 757 Composite Inboard Trailing-edge Flap

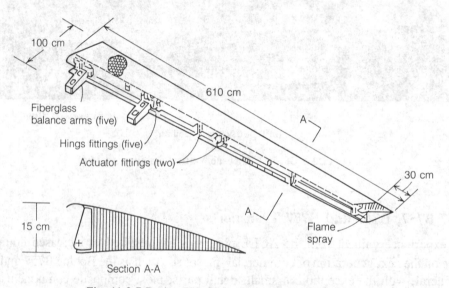

Fig. 11.2.7 Boeing 767 Composite Outboard Aileron

By courtesy of Aeritalia Societa Aerospaziale Italiana S.P.A.

(a) Composite rudder

Fig. 11.2.8 Boeing 767 Composite Rudder

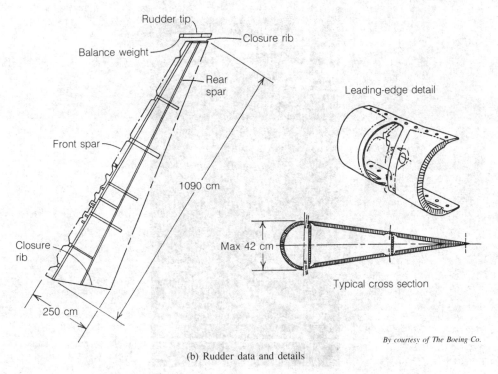

(b) Rudder data and details

Fig. 11.2.8 Boeing 767 Composite Rudder (cont'd)

Airbus Transports

In the early design stages of the original A300 transport, the increasing cost of fuel fixed economy of operation by weight reduction as a vital parameter. To further reduce cost of ownership, production costs and maintenance expenditure had to be reduced to the minimum. In 1985, Airbus became the first airframe manufacturer to use composite materials for a series production of primary structures when it began to assemble the A310 with fins built of carbon/epoxy. The use of composite fins reportedly has resulted in cost and weight savings and the elimination of corrosion.

The progress of Airbus composite development is as follows:
- 1972: A300 early design of fiberglass fin leading edge, fiberglass fairings
- 1978-1979: CFRP spoilers, air brakes, landing gear doors, and rudder
- 1980-1985: A300/A310 vertical fin (see Fig. 11.2.9 and Fig. 11.2.10)
- 1985-1987: A320 horizontal tail (see Fig. 11.2.11) and vertical fin using same design approach as A310. One piece module fabrication of A320 flap (see Fig. 11.2.12)
- 1987: Development of new tooling and manufacturing methods in the design of the A330 and A340 vertical and horizontal tail

The A300/A310 vertical fin is a most impressive composite structure. It is 27 ft 3 inches (8.3 m) high and 25 ft 7 inches (7.8 m) wide at the base end represents a weight savings of about 22% compared to its aluminum counterpart. In addition, it consists of only 95 parts compared with 2076 parts in the previous aluminum box structure, insuring a reduction of assembly costs.

Chapter 11.0

By courtesy of MBB GMBH

Fig. 11.2.9 Airbus A310-300 Composite Fin In Final Assembly

Production costs are reduced by use of automatic processes for the vertical and horizontal tail boxes. A module concept (Ref. 11.10) evolved and patented by MBB (Germany) is described below:
- Prepreg (usually are $\pm 45°$, plies) is wrapped around three-piece- aluminum modules [see Fig. 11.2.10(c)]: this will constitute the web of the stringers and the rib shear tie
- Prepreg is laid into a mold; this will constitute the external skin and stringer bottom flange
- The aluminum modules are assembled into the mold
- Unidirectional tape is laid on the stringer (spanwise) to form the upper flanges
- During the cure pressure is applied to the skin and stringer flanges by the autoclave and to the stringer webs by thermal expansion of the aluminum modules
- A large number of similar modules are set in a fixed grid.
- The complete assembly, under vacuum produced pressure, is cured at 250-350 °F (121-176 °C) and 100 psi (0.69 Mpa) in an autoclave.

P-180 Avanti

The P-180 designs utilizes composites on the nose cone, forward wing (canard), nacelles, wing trailing edge, empennage (see Fig. 11.2.13), and control surfaces. Composites make up 20% of the total weight.

Composite Applications

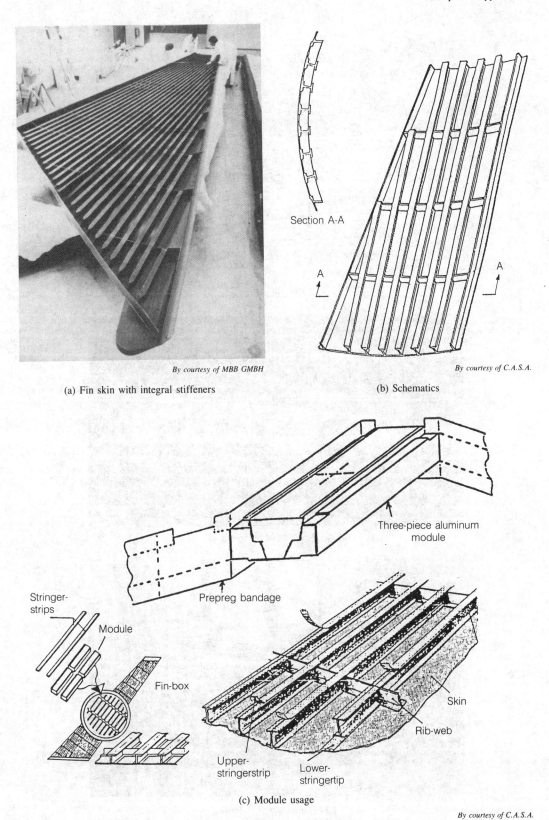

(a) Fin skin with integral stiffeners
By courtesy of MBB GMBH

(b) Schematics
By courtesy of C.A.S.A.

(c) Module usage
By courtesy of C.A.S.A.

Fig. 11.2.10 Airbus A310-300 and A300-600 Composite Vertical Fin (Skin with Integral Stringers, Spar Caps and Shear Ties)

Chapter 11.0

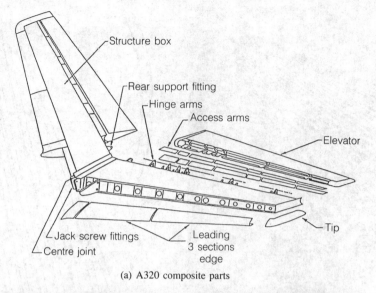

(a) A320 composite parts

(b) A320 composite horizontal tail

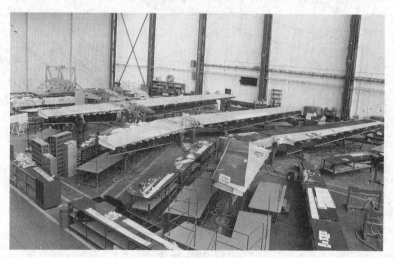

(c) A340 composite horizontal tail

By courtesy of C.A.S.A.

Fig. 11.2.11 Airbus A320 Composite Horizontal Tail

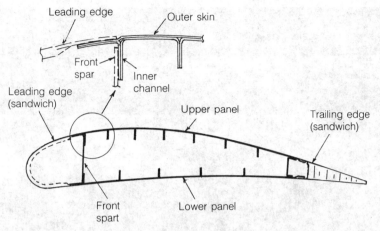

Fig. 11.2.12 Airbus A320 Composite Flap

(a) P-180 aircraft

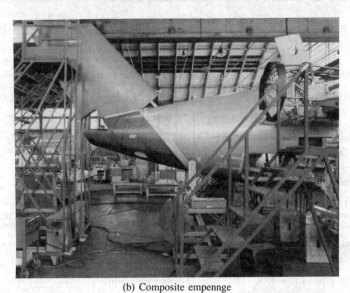

(b) Composite empennge

By courtesy of Rinaldo Piaggio S.P.A.

Fig. 11.2.13 All-Composite Tail of The P-180 Aircraft

Chapter 11.0

By courtesy of Rockwell International

Fig. 11.2.14 Space Shuttle Payload Doors On Assembly Fixture (60ft × 15ft)

Space Shuttle Orbitor

The upper cargo doors, shown in Fig. 11.2.14, are fabricated from graphite/epoxy materials. The double doors (two each side) each have a total length of 60 ft (18.3 m) and are 15 ft (4.56 m) wide. Honeycomb material is used in the door panels while the frames are made of solid laminate.

11.3 MILITARY AIRCRAFT

(1) F-4 (U.S.A.) — Rudder manufactured by using boron filament-reinforced epoxy skins over full-depth aluminum honeycomb

(2) F-14A (U.S.A) — Uses boron/epoxy (B/EP) skins on the horizontal stabilizer box, as shown in Fig. 11.3.1, a full depth honeycomb core and internal rib and spar structure. The stabilizer was designed so that no mechanical fasteners penetrate the boron skins

Composite Applications

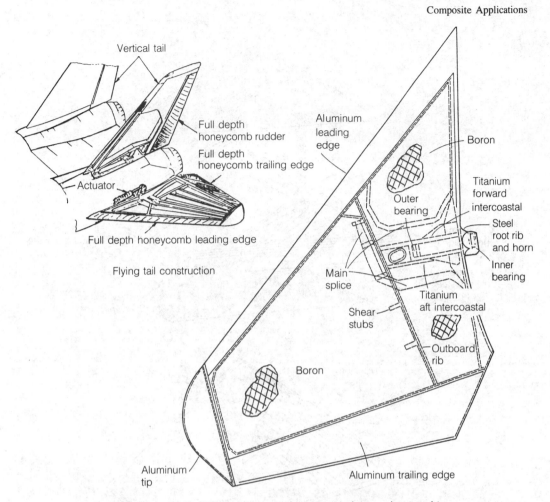

(a) Stabilizer details (full depth honeycomb core except between intercostals)

(b) Composite tail

By courtesy of Grumman Corp.

Fig. 11.3.1 F-14A Boron/Epoxy Composite Horizontal Tail

Chapter 11.0

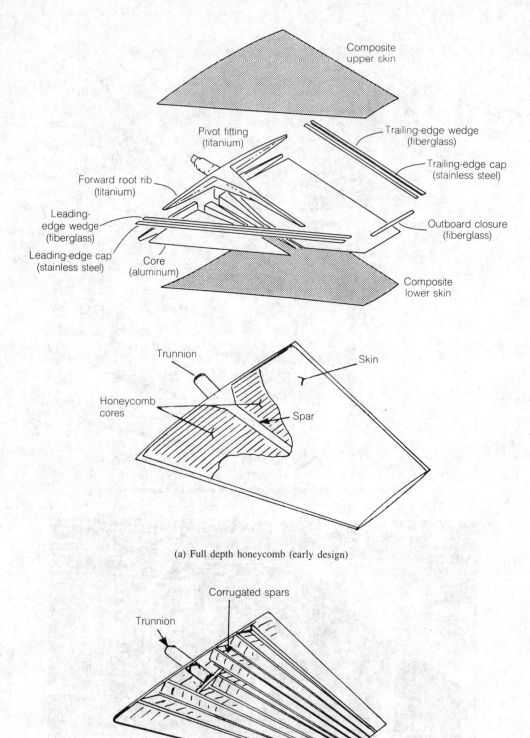

(a) Full depth honeycomb (early design)

(b) Corrugated spar (aluminum) with composite skin (later design)

Fig. 11.3.2 F-16 Composite Horizontal Stabilizer

(3) F-15 (U.S.A) — Vertical fin, rudder and horizontal stabilizer skins and speed brake use boron/epoxy (B/EP)

(4) F-16 (U.S.A) — Uses graphite/epoxy (GR/EP) skins on the vertical fin box, fin leading edge, rudder and horizontal tail. The fin leading edge, rudder and horizontal tails [see Fig. 11.3.2(a)] are all full depth aluminum honeycomb sandwich structures with graphite/epoxy facesheets.

Horizontal stabilizer construction [a later design shown in see Fig. 11.3.2(b)]:
- Graphite/epoxy (GR/EP) used on both upper and lower skins
- Full-depth corrugated aluminum truss core replaces honeycomb core
- Multiple layers of protection separate metal and composite surfaces
  - Anodized aluminum
  - Epoxy primer
  - Liquid scrim (epoxy and chopped glass)
  - Sealant
- Glass/epoxy cloth (over entire skin) is cocured to inner surface of graphite/epoxy skin
- Corrosion-resistant steel fasteners with sealant

(5) F-18 (U.S.A) — Graphite/epoxy (GR/EP) is used on wing skin, horizontal and vertical tails, control surfaces, speed brake, leading edges, and doors, and account for 12.1% of aircraft structural weight.(see Fig. 11.3.3)

(6) AV-8B (U.S.A) — Fig. 11.3.4 shows graphite/epoxy (GR/EP) material is used in the wing box skins and substructure, forward fuselage, horizontal stabilizer, elevators, rudder, overwing fairing, engine bay door ailerons and flaps.
- One-piece solid laminate panel for both upper and lower wing skin (tip to tip), as shown in Fig. 11.3.5
- One of the unique features of the AV-8B aircraft is extensively use of sine-wave spars consisting of roll-formed webs with the caps attached, as shown in Fig. 11.3.6.
- The horizontal stabilizer box (tip to tip) is a carbon/epoxy multi-spar design with a one piece upper cover attached with special titanium blind fasteners, as shown in Fig. 11.3.7. The upper and lower covers are made from carbon/epoxy tape and the integral spars are channels made from woven cloth

(7) X-29A (U.S.A) — The X-29A, as shown in Fig. 1.1.8, uses graphite/epoxy (GR/EP) on the forward swept wing skins (156 plies at the thickest portion) and canard. Here "aeroelastic tailoring" is used to resist the characteristic nose-up twisting under load of the forward swept wing and to maintain the wing's structural, built-in twist throughout the flight envelope, delaying divergence.

(8) B-1B (U.S.A):
- Boron/epoxy was used to reinforce the dorsal longeron
- Weapons bay door, aft equipment bay doors and flaps are fabricated from full-depth aluminum honeycomb bonded with carbon/epoxy facesheets, except for the lower surface of the weapons bay door on which an Aramid fiber/phenolic outer layer provides damage resistance

Chapter 11.0

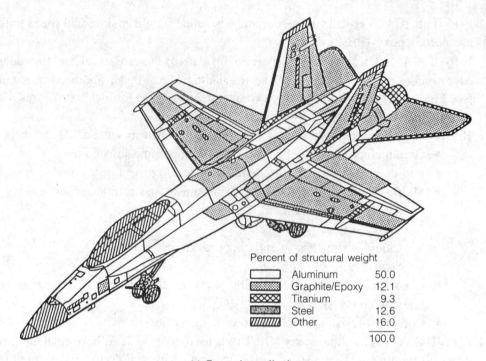

(a) Composite applications

By courtesy Northrop Corp.

(b) Graphite/epoxy composite vertical tail

Fig. 11.3.3 F-18 Composite Application

By courtesy of McDonnell-Douglas Corp.

Fig. 11.3.4 Composites Are Used Extensively On The AV-8B (26.3%)

By courtesy of McDonnell-Douglas Corp.

Fig. 11.3.5 AV-8B Lower Wing Skin Which is 28ft (8.53m) Across Wing Tips And 121 ft^2 (11.24m^2) in Area

Chapter 11.0

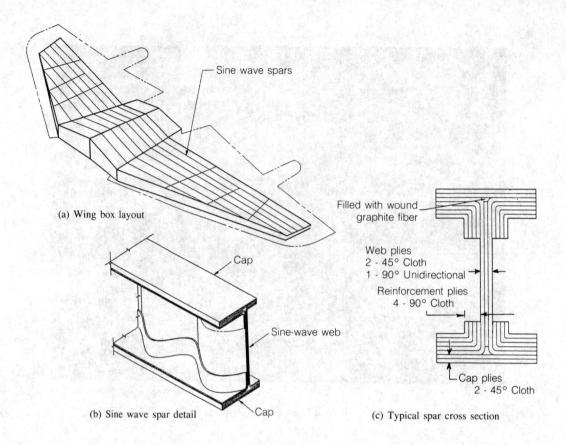

Fig. 11.3.6 AV-8B Composite Sine Wave Spar

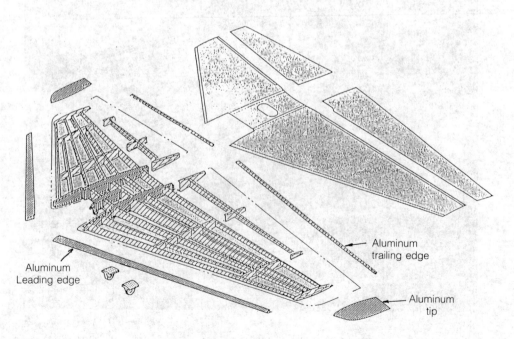

Fig. 11.3.7 AV-8B Composite Horizontal Tail

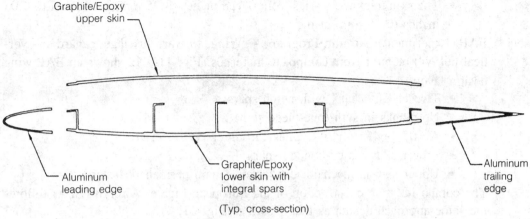

Fig. 11.3.7 AV-8B Composite Horigontal Tail (cont'd)

- Experimental composite horizontal and vertical stabilizers have been built for the B-1 bomber. Fig. 11.3.8 shows a vertical stabilizer constructed in a combination of graphite/epoxy (GR/EP) and boron/epoxy (B/EP) materials.

(9) V-22 V/STOL (U. S. A.):
  - The wing consists of integrally stiffened composite laminate cover panels, with composite ribs and spars
  - Fuselage is made of composite material which is 50% of its structural weight
  - Wing leading and trailing edges are also made of composite materials
  - Empennage is made of CFRP employing sine-wave spars

(10) A-6 Wing (U. S. A):
  - The carbon/epoxy composite replacement the A-6 wing box is shown in Fig. 11.3.9
  - The cover skins are fabricated as a one-piece panel using unidirectional tape
  - Composite intermediate ribs and spars
  - Titanium front and rear spars, and inboard and outboard tank end ribs.
  - All parts are assembled by mechanical fastening

(11) Alpha Jet (France and Germany) — Dornier uses composites on trimmable stabilizer box (see Fig. 11.3.10.) for the Alpha Jet.

(12) Jaguar (British) — Engine bay doors and Wing box structures. The wing box, as shown in Fig. 11.3.11, is a bolted multi-spar and rib configuration:
  - The front and rear spars are composite channel sections
  - Five intermediate spars are sine wave composite beams
  - The wing skins are predominantly $\pm 45°$ layup, which has low notch sensitivity (soft skin design approach)
  - Sandwiched between the wing skins and the spar flanges are booms of predominantly $0°$, layup
  - Wing bending is reacted mainly by the booms while shear and torsion are carried by the spar webs and the wing skins

(13) Tornado (British BAe and German MBB) — Composite taileron construction:
  - The main box has composite skins with full-depth aluminum honeycomb

Chapter 11.0

- The skin thickness varies from 0.128 inches (3.25 mm) at the root to 0.03 inches (0.75 mm) at the tip

(14) EAP (Experimental Aircraft Program) — Wing, forward fuselage, canard and vertical tail will be built from composite materials. Fig. 11.3.12 shows an EAP wing design layout:
- Two cell fuel tank with notch spars
- No planks in skin (one-piece skin)
- All 0°, ±45° and 90° plies carbon/epoxy material
- Lower skin has co-bonded spars
- Upper skin is mechanically fastened with permanent fasteners

The composite wing construction of the European Fighter Aircraft (EFA) follows the same approach design as that shown in Fig. 11.3.13.

By courtesy of Rockwell International Corp.

Fig. 11.3.8 B-1 Composite Vertical Stabilizer

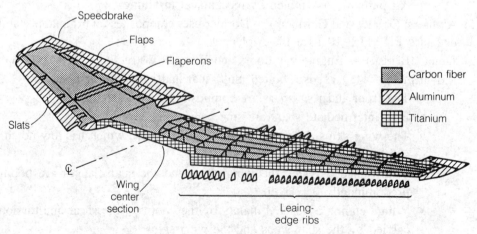

By courtesy of The Boeing Co.

Fig. 11.3.9 A-6 Composite Wing Box

Composite Applications

By courtesy of Dornier GMBH

Fig. 11.3.10 Alpha Jet Composite Tail

By courtesy of British Aerospace Military Aircraft Ltd.

Fig. 11.3.11 Jaguar Composite Wing Box

(a) EAP flight tests

Fig. 11.3.12 British Experimental Aircraft Program (EAP) Demonstration Aircraft

Chapter 11.0

(b) Wing box assembly uses Carbon/Epoxy; the multiple spars are cured and bonded to the lower wing surface in a single process

By courtesy of British Aerospace Ltd

Fig. 11.3.12 British Experimental Aircraft Program (EAP) Demonstration Aircraft (cont'd)

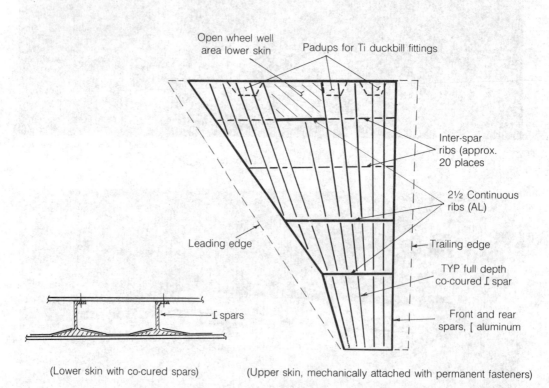

Fig. 11.3.13 European Fighter Aircraft (EFA) Wing Concept

586

Composite Applications

(a) Lavi aircraft

(b) Composite wing box (Fabricated by Grumman Corp.)

By courtesy of Israel Aircraft Industries Ltd.

Fig. 11.3.14 Lavi Composite Wing

Chapter 11.0

By courtesy of Aerospatiale

Fig. 11.3.15 Rafale Aircraft

By courtesy of SAAB-SCANIA AB, SAAB aircraft division

Fig. 11.3.16 SAAB JAS 39 Gripen

(15) Lavi (Israel — Grumman Corp. built the composite wing) — Composite material is used on wing skins (see Fig. 11.3.14), vertical tail, moving foreplane, control surfaces, various doors and panels, and fuselage.
(16) Rafale (France) — Composite material is used on the wing, fuselage (50%), control surfaces, canard, landing gear doors and access doors (see Fig. 11.3.15)
(17) Gripen (Sweden) — Gripen design (see Fig. 11.3.16) uses composite material on the wing, vertical tail, canard, intake duct and landing gear doors.

(18) MBB Fighter Fuselage Program (Germany) — Carbon material is used on the complete combat-aircraft forward fuselage (see Fig. 11.3.17) which is a two-piece fuselage shell featuring integrally cocured frames, stringers, and longerons.

(19) F-111 (U.S.A.):
- Boron doubler reinforces the overwing pivot joint structure
- Horizontal stabilizers (boron/epoxy skin over full-depth aluminum honeycomb) were fabricated for flight test.

(20) B-2 (U.S.A.) — The B-2 (see Fig. 11.3.18) is the biggest all-composite airframe structure ever built

(21) ATF (U.S.A.) — There is extensive use composite materials (nearly 40%) on the Advanced Tactical Fighter (see Fig. 1.1.2).

By courtesy of MBB GMBH

Fig. 11.3.17 MBB Composite Forward Fuselage (Fighter)

By courtesy of Northrop Corp.

Fig. 11.3.18 All-composite Airframe B-2 Flying Wing Aircraft

11.4 ALL-COMPOSITE UTILITY AIRCRAFT

Currently, all-composite aircraft are generally utility or smaller aircraft such as home built or kit aircraft (from one to four passengers). Other examples are shown in Fig. 1.1.13.

This section presents some unique designs used on various composite aircraft/programs that can be considered good examples for future composite airframe designs.

Lear Fan 2100

The Lear Fan 2100, shown in Fig. 11.4.1, is the first all-composite airframe aircraft in which Graphite/epoxy and Kevlar/epoxy composite materials account is for a saving of 40% of the aircraft structural weight. Titanium is used for all major fittings attached to graphite/epoxy structures to avoid galvanic corrosion.

(1) Fuselage construction:
- Large skin panels are adhesive-bonded to the fuselage shell. High load carrying members, such as windshield and door frames and pressure bulkheads, are both bonded and fastened. Skins are generally 0.0135 inches (0.343 mm) thick and made of 8-harness satin fabric
- The outer ply incorporates 0.004 inch (0.1 mm) aluminum wire for lightning protection
- Most of the pressure cabin skin consist of four plies of fabric laid up in a 45°, 0°, 0° and 45° sequence, which creates a nearly isotropic skin with strength and modulus comparable to aluminum

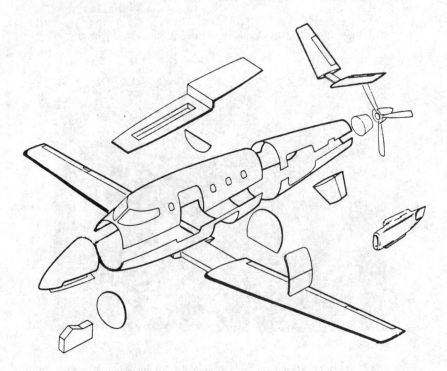

Fig. 11.4.1 Graphite/Epoxy Composites are Extensively Used on The Lear Fan 2100

- Each fuselage skin panel covers a quarter of the circumference with single-lap joints, as shown in Fig. 11.4.2, along the upper and lower shoulder lines
- The top and bottom skin panels each incorporate joggles to receive the side panels
- Skin panels continue fore-and-aft until terminated by openings such as the cabin door or engine bay.

(2) Wing Construction (see Fig. 11.4.3):
- The three-spar wing box is a continuous structure from tip to tip
- Wing skins are made of fabric with tape build-ups over the spar channels to carry bending loads. The entire wing structure is bonded
- Fasteners reinforce the bonded skin-to-spar attachment in areas of high shear loads. Internal T-shaped stiffeners help stabilize the wing skins against buckling under normal flight loads (compression)
- Monel or titanium fasteners and bonded joints are sealed to prevent fuel leakage from the fully wet wing (integral fuel tank).
- Aluminum wire mesh applied to the rear spar web helps shield metal control cables and hydraulic system components from lightning strike damage

(3) The rudder, elevators flaps and ailerons are constructed of Kevlar/epoxy sandwich skins with Nomex honeycomb cores. A mesh of very fine aluminum is applied over the rudder and elevator to protect areas subject to lightning strike.

(4) The lower tip of vertical stabilizer has an energy absorbing bumper to protect the propeller from ground strikes.

(5) Composites provide ample stiffness to the aft fuselage and tail cone to preclude propeller whirl-flutter problem.

(6) The carbon-reinforced plastic engine drive shafts are 74 inches (188 cm) long and approximately 5 inches (12.7 cm) in diameter.

Starship

The Starship, a twin pusher canard aircraft, is unconventional in typically configuration, material and design. The airframe (see Fig. 11.4.4) is constructed of graphite/epoxy facesheets on low density Nomex honeycomb.

(1) Wing Covers (see Fig. 11.4.5):

The Starship wing was designed to be bonded and have a minimum number of parts and associated tools. The wing panels are laid up on female tools by laying down the outermost skin first then a honeycomb core (HRH 3.0 and 6.0) and then the inner skin. The core is an inch (2.54 cm) thick at the root and steps down as it goes outboard to about 0.5 inch (1.27 cm). The steps are of 0.125 inch increments over an eight-to-ten inch ramp. Finally, the assembly is cocured.

There are three spars due to the nature of the assembly and the U. S. FAA requirement that the structure take limit load with failure of any single bond from wing root to tip. These spars are joined to the wing skins through corner clips and H-clips, as shown in Fig. 11.4.6. The clips have continuous fibers through the intersections to provide load path continuity.

Chapter 11.0

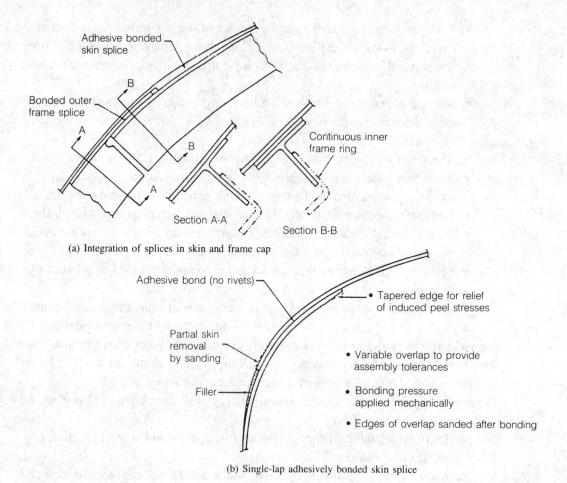

(a) Integration of splices in skin and frame cap

(b) Single-lap adhesively bonded skin splice

Fig. 11.4.2 Lear Fan Fuselage Longitudinal Splice

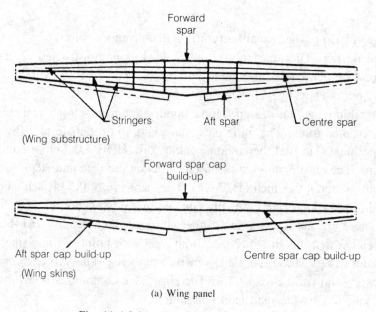

(a) Wing panel

Fig. 11.4.3 Lear Fan Composite Wing Box

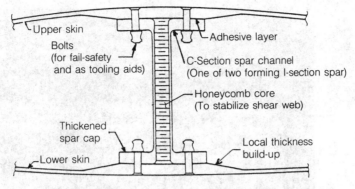

(b) Typical cross section of spars in No. 1 lear fan wing

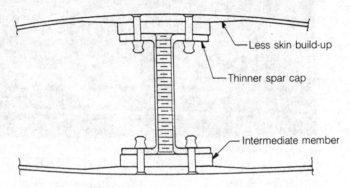

(c) Proposed manufacturing improvements in lear fan wing

Fig. 11.4.3 Lear Fan Composite Wing Box

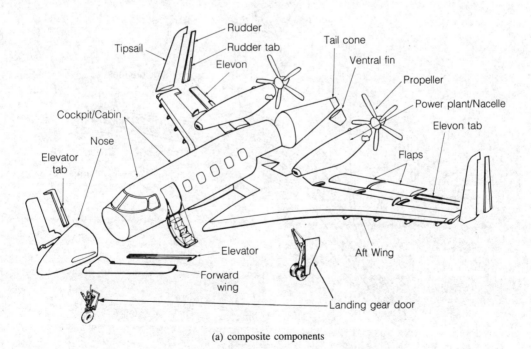

(a) composite components

Fig. 11.4.4 Starship All-Composite Airframe

Chapter 11.0

(b) Final assembly

By courtesy of Beech Aircraft Corp.

Fig. 11.4.4 Starship All-composite Airframe (cont'd)

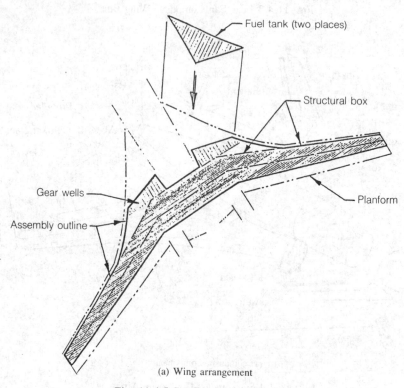

(a) Wing arrangement

Fig. 11.4.5 Starship Composite Wing

Composite Applications

(b) Wing skin built on tools

(c) Wing internal arrangement prior to final assembly

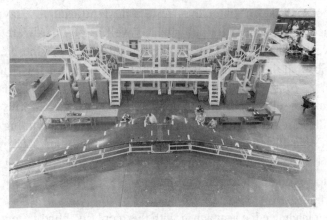

(d) Wing final assembly

By courtesy of Beech Aircraft Corp.

Fig. 11.4.5 Starship Composite Wing (cont'd)

Chapter 11.0

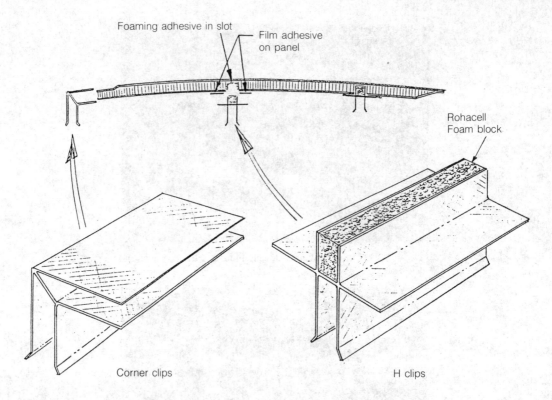

Corner clips and H clips are a woven product with continuous fibers through the intersections providing load path continuity.
H clip use Rohacell foam as a tooling aid and for shear strength in service.

Fig. 11.4.6 Starship Wing Cover Panel Details

Grooves are routed in the inner face and the corner clips and H clips are fitted up and adhesive is applied. A large fixture is nested over the assembly, as shown in Fig. 11.4.7, and clamped in place. Air supplied to the fixture provides the bonding pressure and the assembly is placed in an oven to provide the cure heat (oven cure on bonds, in lieu of autoclave, results in significant cost savings). The lower cover is then fitted with spars and ribs, as well as any hardware that will survive the next oven cycle [300°F (149° C)]. When all is ready, adhesive is applied to the clips and ribs and the assembly is returned to the oven for final closure bond cure.

(2) Fuel Tank:
The fuel tank is a self-supporting but non-structural box, and it forms the wing strakes. Internally they are heavily baffled to prevent surges. For lightning strike protection there are no metallic fasteners or parts inside the fuel tank. The outer ply of fuel tank has fine aluminum wires incorporated into the fabric itself (interwoven at 3 mm intervals) to dissipate strike energy.

(3) Wing Leading Edge:
The titanium leading edge is attached with thermoplastic blind fasteners. These special design fastener are heated in small electrical heaters, inserted in the fastener hole and squeezed by the metallic stem (see Fig. 5.2.33).

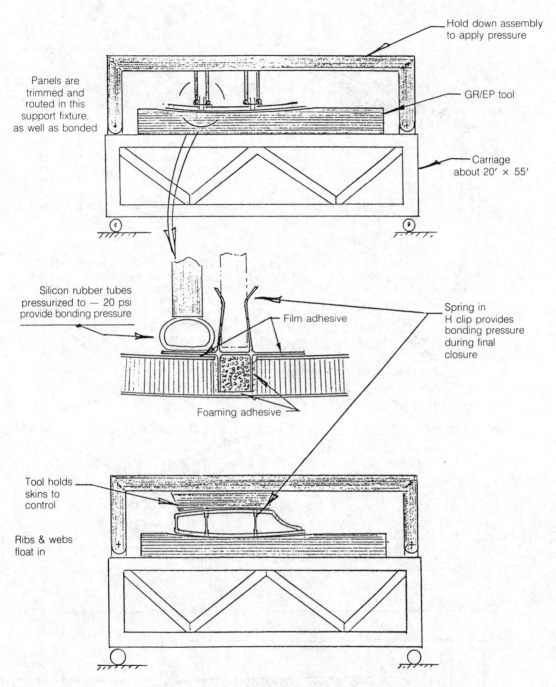

Fig. 11.4.7 Starship Wing Bonding Assembly

(4) Canards (forward wings):
   The canards swing about large trunnions and the configuration closely matches to that of a helicopter blade (see Fig. 11.4.8) The canards are built around two main beam caps that are wound around the root fitting. The caps are then built into the canards which are sandwich panel assemblies similar to the wing construction.

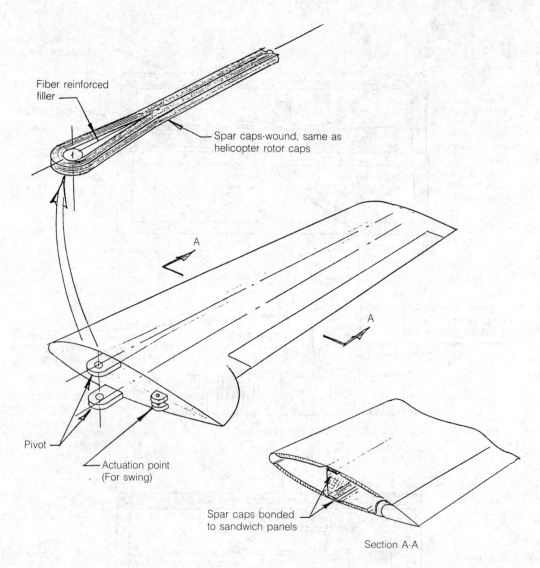

Fig. 11.4.8 Starship Canard Details

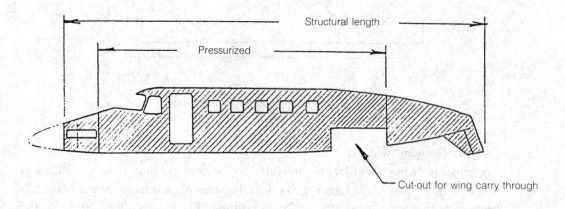

Fig. 11.4.9 Starship Composite Fuselage

Composite Applications

(b) Fuselage halve bagging prior to curing in autoclar

(c) Fuselage final assembly

By courtesy of Beech Aircraft Corp.

Fig. 11.4.9 Starship Composite Fuselage (cont'd)

(5) Fuselage:
   The fuselage of the prototypes have two different methods of construction:
   (a) Hand layup method — The fuselage, as shown in Fig. 11.4.9, is laid up in female half molds, starting at contour and working in. The inner facesheets are cocured on the Nomex core, as well as all doublers, window frames and necessary pad-ups. After cure, the halves are trimmed and fitted, then pulled apart to facilitate substructure installation. The two halves are then spliced together, top and bottom, using heat and pressure supplied by 250°F (121°C) bagging blankets.
   (b) Automatic filament wound method — This method involves hoop and helical winding around a large mandrel starting with the inner skin and working out. After winding, clamshell contour molds are clamped around the unit and pressure is applied to force the fuselage to contour for curing. After curing, the fuselage is cut and separated behind the cockpit to remove the mandrel. Once the mandrel is removed the fuselage substructure is then installed and the front and aft sections are spliced together using a heat and pressure technique. Doors and windows are then cut out.

Avtek 400

The primary consideration in the design of the Avtek 400 is the tandem-wing (canard) arrangement [see Fig. 11.4.10(a)] and the production method used is substantially different from that of other composite planes. The Avtek shells consist of a Nomex honeycomb and Kevlar cloth sandwich which is laid up in the dry state and at room temperature, and is impregnated with resin by hand in the female mold. The mold is then simply cured by heat at 250°F (121°C), without using an expensive autoclave. The Nomex and Kevlar, not being prepregs, do not require refrigerated storage or curing in costly heated metal molds or autoclaves. The structure design utilizes honeycomb to avoid the frame and stringer layout. The thickness and composition of the panel material is varied according to local stress levels.

The Avtek design includes the following features:
   (a) The entire airframe uses a modular design concept and contains approximately 51 parts [see Fig. 11.4.10(b)(c)]; the Nomex and Kevlar materials are laid up by hand for most of the parts [see Fig. 11.4.10(d)]
   (b) A thin aluminum coated glass-fiber mesh in the outer skin provides overall lightning resistance; metal straps are bonded on adjacent components to provide protection
   (c) Ultra-violet light resistance is provided by a Bostik paint on the outside and a resin paint containing carbon on the inside
   (d) A new resin is used along with a lower glass transition temperature to reduce moisture absorption, although at the cost of slightly poorer microcracking characteristics in the resin. It is easy to blend carbon with Kevlar when using lower temperature materials. Composites have been shown to recover strain characteristics when dried, after soaking at high humidity/low temperature.

Composite Applications

(a) Avtek aircraft

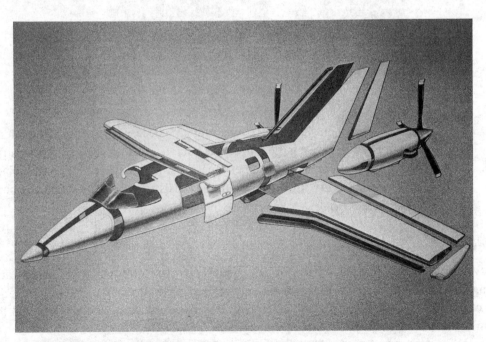

(b) Modular Design with approximately 51 parts

By courtesy of AVTEK Corp.

Fig. 11.4.10 AVTEK 400 All-Composite Airframe

Chapter 11.0

(c) Rear fuselage and fin section halve from the mold

(d) Some 72% of Avtek is hand-laid Nomex and Kevlar

By courtesy of AVTEK Corp.

Fig. 11.4.10 AVTEK 400 All-composite Airframe (cont'd)

(e) Some designers believe that fastening might be necessary to ensure the integrity of the bond between large half-shells such as fuselage halves. These structures use the reinforcement of "fiber ropes" (similar to that of Fig. 5.5.2) placed across the joint every few inches with the tails spread in a fan shape and bonded into the material.

Voyager

Voyager, shown in Fig. 11.4.11, is a large span [110 ft (36.09 m)], high aspect ratio, long range aircraft. It was a challenge not only of innovative design but also of fabrication to create this structural weight critical aircraft (the basic structural weight was 938 lbs, while takeoff weight was 9760 lbs).

(1) Wing Construction (see Fig. 11.4.11 and Fig. 11.4.12):
- The shell panels are 0.014 inch (0.36 mm) thick over 0.25 inch (6.35 mm) honeycomb, and are vacuum bagged and oven cured.
- Inner surfaces are squeeged with epoxy for sealing
- Ribs are profiled out of sandwich sheet stock and epoxy bonded at approximately 30 inch (76.2 cm) spacing

Composite Applications

By courtesy of Scaled Composites Inc.

(a) Voyager aircraft

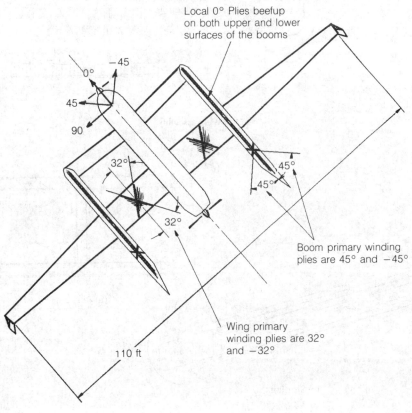

(b) Composite ply layup

Fig. 11.4.11 Voyager All-Composite Airframe

603

Chapter 11.0

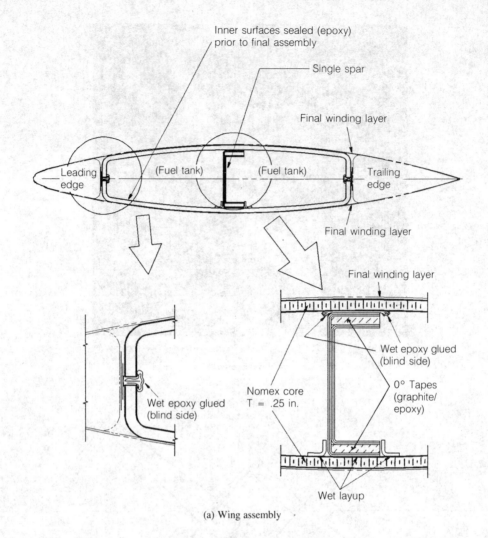

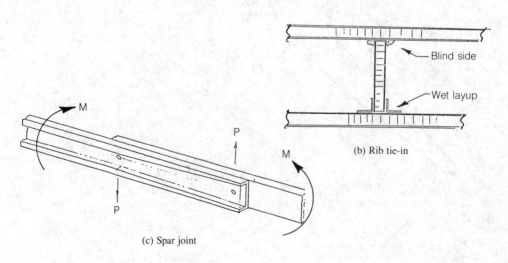

Fig. 11.4.12 Voyager Composite Wing Details

604

- No flanges are used around the ribs, as shown in Fig. 11.4.12(b)
- During assembly, the ribs and spars are bonded to the lower skin, all mating edges are epoxy painted and the top is pushed on, sealing the assembly and fixing the structure
- No access to the inside of the tank is possible after assembly, and since the entire airframe box stores fuel, all controls and electrical conduits are in the leading and trailing edges.
- The wing is made of graphite/epoxy tape (oriented at 32° to the chord), and Nomex honeycomb, laid up on female mold tooling.
- Laying the tape at 32° [see Fig. 11.4.11(b)] provides maximum torsional stiffness, but also produces a relatively weak wing with regard to bending.
- Wing spar is the only autoclave cured part and it is made up of graphite caps at 0° cocured with honeycomb stiffened webs [see Fig. 11.4.12.(a)]
- Only the wing spars go through the fuselage and they are bolted back to back, as shown in Fig. 11.4.12(c), for carrying through- wing bending loads

(2) Fuselage Construction:
- The fuselage is a typical sandwich shell, with the front and aft firewalls forming major bulkheads for wing and canard attachments
- Honeycomb core with facesheets of woven Aramid fiber or woven graphite fiber impregnated with epoxy
- Kevlar is used in lieu of carbon fibers over the cockpit area to provide electrical transparency for internal radio equipment
- Front and rear firewalls are fabricated from glass/polyimide honeycomb core with ceramic cloth impregnated with polyimide resin to withstand the heat [use FAA standard which requires the ability to withstand a 2000 °F flame for 5 minutes over a 5×5 ft^2, (1.52×1.52 m^2) area- without allowing anything behind them to ignite from the heat transfer].

(3) Booms:
- Booms support the canard and fins and allow for additional fuel capacity. The booms are generally a composite sandwich made of 2 plies of 0.007 carbon tape facesheets over 0.25 inches thick Nomex honeycomb core (1.8 lb/ft^3)
- The booms are about 30 inches (72.6 cm) in diameter with two ply facesheets laid at 45°
- The localized addition of a 0° ply, top and bottom, over the wing area enables the boom to carry bending loads
- Boom shells are laid up on female tooling, and then vacuum bagged and oven cured
- The entire shell is fabricated by a wet-layup process, with the inner surface squeegeed lightly with epoxy to seal any small pinholes.
- Wing beam carry through and boom fuel tank bulkheads provide a natural hard point for the main landing gear supports

Speed Canard

The Speed Canard, shown in Fig. 11.4.13, is based on the Rutan Vari Eze configuration. However, Rutan aircraft are only available as kits for home builders and cannot be put into production or used for training. Rutan aircraft are built from the inside out and start with a foam core on which layers of composite material are laid up wet. For an aircraft to be produced in large numbers and economically, it must be built from the outside in, using molds, to yield a high production rate.

The Speed Canard aircraft is constructed mainly of glass fiber and epoxy materials (although unidirectional carbonfiber spar caps are used) and it is cured using Low Pressure Cure (LCP) technology (discussed later). The design features of the Speed Canard are described below:

(a) All composite structures are made from PVC foam in combination with glass, Kevlar and/or a small percentage of carbon fibers and are cured per German composite design philosophy (see below)

(b) Carbon fibers are used on the compression side of the landing gear legs and Aramid is used on the tension side, as shown in Fig. 11.4.14.

(c) All composite curing takes place in a mold under vacuum and at 140°F (60°C) for eight hours, followed by post curing out of the mold for 15 hours at 176°F (80°C).

(d) Speed Canard loads were a little less severe in some areas and therefore only one spar was needed compared to the three spar design of the Starship aircraft.

(e) The single spar wing is a sandwich shell composed of glass/epoxy and PVC foam; it is double T-shaped and has carbon rovings in the flanges.

(f) Simple and quick mounting of the wing to the fuselage is shown in Fig. 11.4.15

Introduction of German Composite Design Philosophy:

This design philosophy (Ref. 11.21) involves the use of low pressure (used for Speed Canard) and high pressure cures, depending on which is the most cost effective. Both technologies will be described briefly. This discussion will close with an economic comparison of the two systems.

(1) Low Pressure Cured (LCP) Technology — LCP technology has essentially contributed to the worldwide success of a new generation of gliders and motor glider aircraft. This potential for unequalled performance is now being applied to general aviation and utility aircraft.

    (a) Characteristics and features of LCP include:
- For medium to high stress level applications
- Low pressure cure composites
- Room temperature cure
- Much lower cost
- Easily repaired

    (b) LCP technology is illustrated in [Fig. 11.4.16(a)]; numbers correspond to illustration:

      1 Single female mold of simple construction
      2 Fiber fabric sheets laminated wet resin (generally epoxy)

Composite Applications

By courtesy of FFT GMBH

Fig. 11.4.13 Speed Canard All-composite Airframe

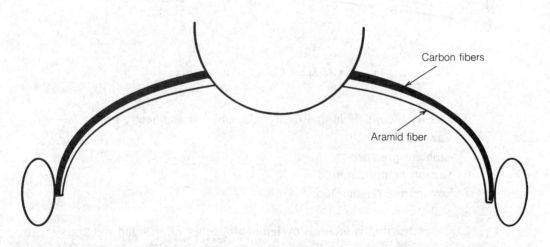

Fig. 11.4.14 Speed Canard Main Landing Gear Leg

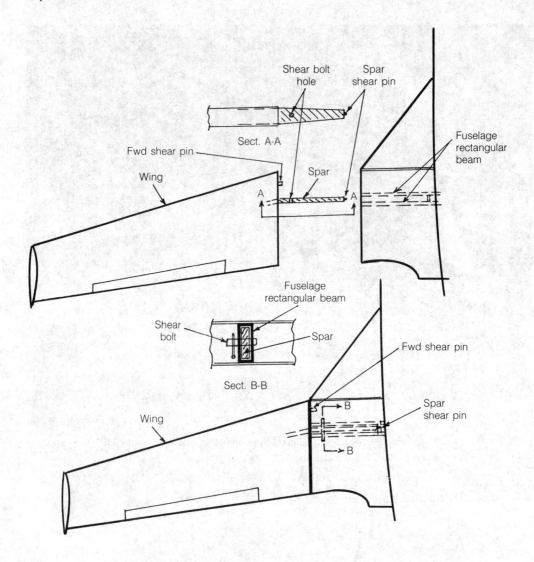

Fig. 11.4.15 Speed Canard Wing Mounting

    3 Sandwich core (light and easily adaptable) is frequently used
    4 Vacuum foil
    5 Ambient pressure
    6 Suction pump manifold
    7 Low curing temperatures
    8 Seal

LPC construction is attained by laminating fiber fabrics and wet epoxy resin together under low atmospheric pressure and medium heat. Curing of large airframe sections at 176°F (80°C) is accomplished outside the mold.

    This is in keeping with German composite design philosophy and is similar to the methods used for the Seastar Amphibious aircraft, shown in Fig. 11.4.17

Composite Applications

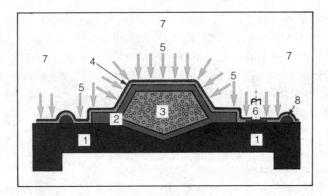

(a) LCP Technology

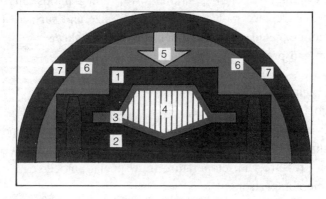

(b) HCP Technology

(Source: Ref. 11.22)

Fig. 11.4.16 German Composite Design Philosophy

By courtesy of Dornier Seastar GMBH & Co. KG (Claudius)

Fig. 11.4.17 Seastar Amphibious Aircraft

609

(2) High Pressure Cured (HPC) Technology — This construction method contrasts with the special characteristics of LPC. HPC is widely used in airframe components where high stress and temperature levels are required. Typical examples of HPC applications are primary structures of military and commercial aircraft.
   (a) Characteristics and features of HPC include:
   - For higher stress and temperature level applications
   - High pressure cure composites
   - High cost
   - Difficult to repair
   (b) HPC technology is illustrated in Fig. 11.4.16(b); numbers correspond to illustration:
   1 & 2 Two high strength upper and lower female molds (alternatively vacuum bag may replace molds, depending on size and surface specifications)
   3 Preimpregnated (prepreg) fiber/epoxy resin sheets
   4 Heat and pressure resistant honeycomb core
   5 High pressure
   6 High curing temperature
   7 Autoclave
   HPC is manufactured from semi-cured prepreg sheets incorporating fibers and the matrix material. The final chemical compound is achieved by application of high pressure and temperature in a special autoclave.
(3) The economic comparison:
   Since the field of general aviation aircraft design is at the threshold of increased structural stress levels, the proper choice of how to build the aircraft is of essential importance and is not only a technical problem, but also a cost issue.
   Certainly, LPC offers an attractive option with exceptionally low investment and construction costs, while providing the benefits of composite materials. As shown in Fig. 11.4.18, different construction methods correspond to the different stress level requirement of the airframe.

All-composite Kit Planes

The design and manufacturing philosophy behind the kit plane stresses both simplicity and low cost in what is a relatively complex high performance aircraft. Simplicity, economic, aerodynamic considerations and styling all point to composites as the only suitable materials for kit planes. On early models the composite airframe for the kit plane used room temperature, wet layup systems — a derivative from common "boat technology". However, because of the need for repeatable quality assurance and airframe longevity, these systems are not suitable for today's higher performance kit planes. The higher temperature matrix based composites are beginning to be used throughout the kit plane industry for airframe primary structures such as wing spars, longerons, etc. Kit plane structures must be sound; this requires analysis, testing, research and development work.

Composite Applications

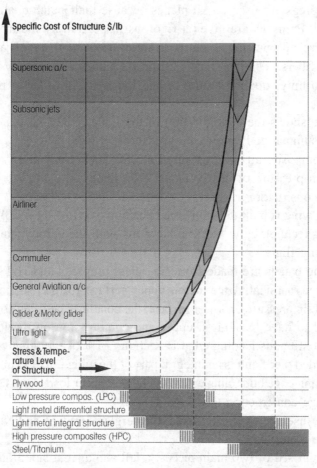

(Source: Ref. 11.22)

Fig. 11.4.18 Applied Construction Methods and Materials in Aircraft Primary Structures

By courtesy of Nieco Aviation Inc.

Fig. 11.4.19 All-Composite Laneair 320 Kit Plane (Seats 2)

611

Factory prepregs are used on kit planes because their batch quality and volume is always consistent. Prepregs are given a rigorous inspection by the material supplier to insure that the correct proportion of matrix (resin) is used and that the matrix is thoroughly blended with the fibers. The materials used on kit planes must follow the standards for traceability, uniformity, quality assurance and longevity in order to provide structural integrity.

Fig. 11.4.19 shows the configuration of the Lancair 320 kit plane; its design and construction are summarized below:

(a) The composite airframe is cured at 250°F (121°C) in an oven while under vacuum pressure of nearly 14 psi (103 kpa)

(b) Wing construction:
 - The wing is built in the inverted position (see Fig. 11.4.20) to allow the bonding technology to work in harmony with the aerodynamic load pressures during flight
 - Wing panels are made from fiberglass prepregs and H.T. Divinycell PVC
 - Aerodynamically, the airfoil upper surface formed suction pressure while push-in pressure on lower surface; the bondlines between cover and spars/ribs tend to have tension on upper and compression on the lower surface
 - A continuous leading edge D-section (25% of the wing chord) is one pre-molded piece and is an integral part of the wing upper skin itself to eliminate leading edge butt joints where the highest shear load occurs (see Fig. 11.4.21)

(c) Fuselage construction (see Fig. 11.4.22):
 - The fuselage shells are made of fiberglass prepreg and Nomex honeycomb core material
 - Comprised of two main pieces, a left and a right side; a third "belly pan" is added in conjunction with the center stub wing section

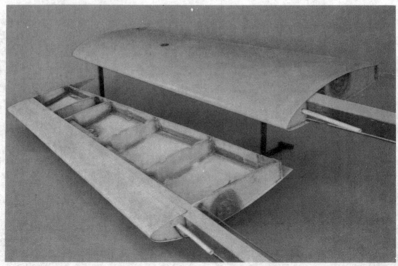

By courtesy of Nieco Aviation Inc.

Fig. 11.4.20 Lancair 320 Wing is Built in the Inverted Position

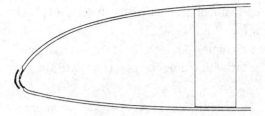
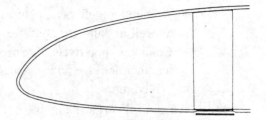

(a) Common leading edge "butt" joint on wing assembly is difficult and critical

(b) Lancair's unique "Continuous D section" offsets assured strength for safely

By courtesy of Nieco Aviation Inc.

Fig. 11.4.21 Lancair 320 Wing Leading Edge D-Section

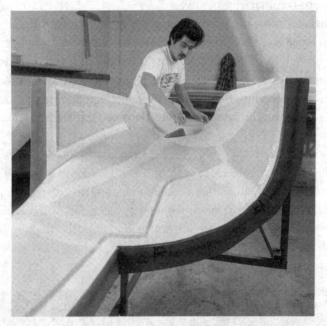

(a) Fuselage material laid up in female mold

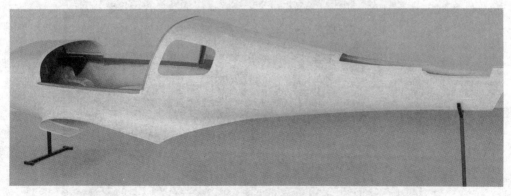

(b) Fuselage halves mated

By courtesy of Nieco Aviation Inc.

Fig. 11.4.22 Lancair 320 Fuselage

- A double joggle arrangement is used for alignment of the fuselage halves as well as fail-safe considerations (see Fig. 11.4.23)
- Common "pop rivets" are used as a clamping device to hold the seams tightly together until the adhesive cures, after which time the pop rivets are simply drilled out

(d) All control surfaces are 100% mass balanced; Fig. 11.4.24 shows an aileron cross-section with counterbalance weight mount and tail structure

Fig. 11.4.25 shows a Lancair IV kit plane which is a 4-seater all-composite plane. The basic construction is similar to that of the Lancair 320 but the airframe is predominantly carbon fiber combined with an epoxy matrix.

Fig. 11.4.26 shows the pusher configuration of the Cirrus VK30 kit plane; its design and construction are summarized below:

(a) The basic airframe is a vacuum bagged vinylester resin/fiberglass sandwich with polyurethane and PVC foam cores. Wings, stabilizer, fuselage, and control surfaces, as well as other selected parts, are built in female molds to ensure structural integrity and accurate shape. Highly loaded areas are reinforced with unidirectional carbon fibers. The horizontal tail and rudder use a Kevlar fabric for cost and weight savings. Wing flaps use carbon to increase stiffness and reduce weight.

(b) Wing construction (see Fig. 11.4.27):
- A high temperature resin system (Dow Derakane 470-36 vinyl easter resin) is used on the wing since it maintains good structural properties to 250 °F (121°C), insuring structural integrity. The Derakane resin also possesses outstanding resistance to solvents and chemicals, making it ideal for a fuel tank application

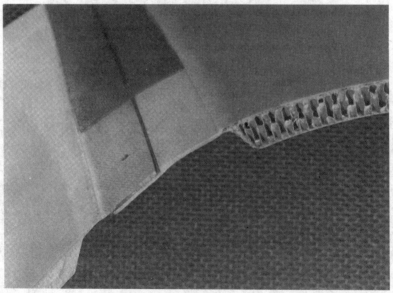

By courtesy of Nieco Viation Inc.

Fig. 11.4.23 Lancair 320 Fuselage Unique "Double Joggle" System

Composite Applications

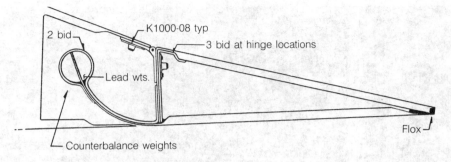

(a) Aileron cross-section

(b) Tail structures

By courtesy of Nieco Aviation Inc.

Fig. 11.4.24 Lancair 320 Aileron and Tail Structures

By courtesy of Nieco Aviation Inc.

Fig. 11.4.25 All-composite Lancair IV Kit Plane (Seats 4)

Chapter 11.0

(a) In flight (b) On ground display

By courtesy of Cirrus Design Corp.

Fig. 11.4.26 All-Composite Cirrus VK30 Kit Plane (Seats 4)

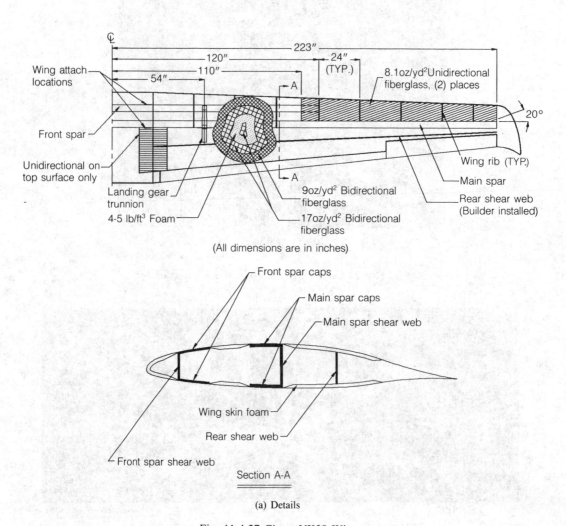

(a) Details

Fig. 11.4.27 Cirrus VK30 Wing

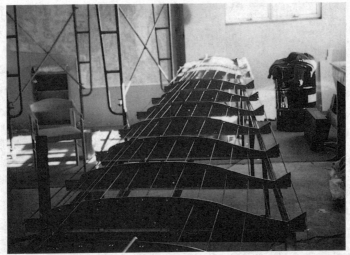

(b) Wing plus-machined rib templates (plaster is spread on the chicken wire between the templates)

By courtesy of Cirrus Design Corp

Fig. 11.4.27 Cirrus VK 30 Wing (cont'd)

- The wing box consists of two C-section spars facing each other with chordwise ribs every 24 inches
- The wing skin panels are a sandwich construction made of 4 and 5 lb/ft^3, polyurethance foam and 17 oz/yd^2 fiberglass fabric
- Unidirectional fiberglass is added to the wing skins from the end of the front spar to the wing tip, at a 20° angle, to aeroelastically tailor the wing
- Spars are made with unidirectional carbon fiber spar caps and fiberglass fabric shear webs
- The ribs are sandwich construction made with a 4 lb/ft^3 polyurethane foam and fiberglass fabric facesheets
- Spars and ribs are bonded onto the upper wing skin
- Next, the wing is turned over and bonded to the lower wing skin panel
- When cured, the wing is checked for proper twist, and fiberglass fabric is applied to the leading edge

(c) The shell of the fuselage to which all these parts and components are attached is a vacuum-bagged vinylester resin/fiberglass sandwich with a polyurethane foam core

- The fuselage consists of two half cones (see Fig. 11.4.28) mated at their wide ends and it so effectively distributes loads that little internal structure is needed
- Four plies of unidirectional fiberglass fabric cloth extend from nose to tail, creating an exceptionally strong, torsionally rigid fuselage shell structure
- A master bulkhead located just in front of the engine is the tie that binds the engine, wing, landing gear and cabin loads
- Carbon fiber, Kevlar and PVC foam are used in selected areas, such as the cabin door and the frame, where strength/weight is desirable

Chapter 11.0

(a) Fuselage halves mated

(b) Foam templates for fuselage half (plug)

By courtesy of Cirrus Design Corp.

Fig. 11.4.28 Cirrus VK30 Fuselage

References

11.1 Middleton, D. H.,"COMPOSITE MATERIALS IN AIRCRAFT STRUCTURES", Longman Scientific and Technical, Harlow, Essex, U.K. 1990.

11.2 Noton, B. R., "COMPOSITE MATERIALS, Volume 3-Engineering Applications of Composites", published by Academic Press, New York, NY. 1974.

11.3 Agnlin, J. M., "Aircraft Applications", ENGINEERING MATERIALS HANDBOOK, Vol 1 - Composites", ASM International, The ASM Composite Materials Collection, Metals Park, OH 44073. pp.801-809.

11.4 McMillan, J. C. and Quinlivan, J. T., "Commercial Aircraft Applications in Advanced Thermoset Composites", J. M. Margolis, Ed., Van Nostrand Reinhold, 1986.

11.5 Kuno, J. K., "Growth of the Advanced Composites Industry in the 1980's", Proceedings of the 31th International SAMPE Symposium, Society for the Advancement of Material and Process Engineering, 1986.

11.6 Zweben, C. "Advanced Composites — A Revolution for the Designer", AIAA 50th Anniversary Annual Meeting ans Technical Desplay, "Learn from the Masters" Series, May, 1981.

11.7 Anon. "Composites", FLIGHT INTERNATIONAL, 31 December 1983. pp.1735- 1748.

11.8 Vosteen, L. F. "Composite Aircraft Structures", Proceedings of the Fourth Conference on Fibrous Composites in Structural Design, Nov., 14-17, 1978. pp. 7-20.

11.9 McCarty, J. E. and Wilson, D. R. "Advanced Composite Stabilizer for Boeing 737 Aircraft", paper presented at 6th DOD/NASA Conference On Fibrous Composites in Structural Design, New Orleans, Louisiana. Jan. 24-27, 1983. pp. 57-77.

11.10 Sarh, B. "Principal Tests for the Automated Production of the Airbus Fin Assembly with Fiber Composite Materials", 29th National SAMPE Symposium, April 3-5, 1984. pp.1477-1488.

11.11 Anon. "Composite Materials in the Airbus", AIRCRAFT ENGINEERING. Dec., 1989. pp.20-29.

11.12 Anon. "Composites in Aircraft Construction", FLIGHT INTERNATIONAL, 23 May, 1981. pp. 1551-1555.

11.13 Parker, I. and Velupillai, D. "Lear Fan: Certification of A Plastic Aeroplane", FLIGHT INTERNATIONAL, 4 April, 1981. pp. 970-971.

11.14 Frisch, B. and Wigotsky, V. "Designer's Notebook — First Design Details of the All-composite Lear Fan", Astronautics & Aeronautics, May, 1983.

11.15 Noyes, J. v. "Composites in The Construction of The Lear Fan 2100 Aircraft", COMPOSITES, Vol. 14, No. 2. April, 1983. pp. 129-139.

11.16 Hart-Smith, L. J. "Design and Development of The First Lear Fan All-Composite Aircraft", paper presented to Institution of Mechanical Engineers, Conference on Design in Composite Materials. London, England. March 7-8, 1989.

11.17 Brewer, D. "Voyager's Composites 'Performed Superbly'", ADVANCED COMPOSITES, Jan./Feb., 1987. pp. 58-60.

11.18 Moulton, R. "Man Powered Flight Advances", FLIGHT INTERNATIONAL, 16 March, 1985. pp. 24-26.

11.19 Lambert, M. "Avtek 400, What is it?", INTERAVIA, March, 1986. pp. 275-277.

11.20 Postlethwaite, A. "Manufacturing: The Cutting Edge", FLIGHT INTERNATIONAL, 17 Sept., 1988. pp. 36-40.

11.21 Blech, R. "Turboprop Battle Hots Up", FLIGHT INTERNATIONAL, 13 July, 1985. pp. 29-31.

11.22 Anon. "General Aviation: German Composite Aircraft", publication of Dornier Seastar GMBH & Co. KG (Claudius), Germany.

11.23 Sharples, T. "Application of Carbon Fibre Composites to Military Aircraft Structures", THE AERONAUTICAL JOURNAL, Vol. 84, No. 834. July, 1980. p.177.'

11.24 Lambert, M. "Boeing Vertol's All-composite Helicopter", INTERAVIA, Aug., 1983. p. 837

Chapter 12.0

INNOVATIVE DESIGN APPROACHES

12.1 INTRODUCTION

Many current structural designs result in high manufacturing costs which limit preclude the application of advanced composites to other than subsidized hardware. Many of the unique problems of composite structural design have not been adequately addressed in either the current hardware programs or study efforts now underway. Early composite development programs utilized designs with structural configurations similar to those of metal counterparts. Hence, those designs embodied many of the undesirable characteristics of the past (e.g., excessive numbers of detail parts, complicated tooling and assembly techniques, etc.) with their only redeeming feature being lighter weight. Designs, materials, tooling concepts, and manufacturing methods are needed which take maximum advantage of relative low-cost processes such as filament (or tape) winding and resin transfer molding (RTM) techniques, and which can be single-stage cured at the required elevated temperature and pressure, without sacrificing structural integrity.

To obtain the utmost weight and cost savings potential of advanced composites in advanced aircraft structures, material forms and manufacturing methods must be taken into considerations during the development of innovative structural designs. This goal can be achieved by meeting the following objectives:
- Reduction of aircraft life cycles cost by advancing the state-of-the-art of composite innovative design concepts
- Paving the way for manufacturing and producibility methods that will find practical usage in airframe structures
- Incorporation of low cost manufacturing, tooling, assembly and repair methods

In order for a breakthrough to be made in composite structural technology, innovative composite design concepts are crucial. In general, innovative concepts can be categorized into two groups:
- Ideal or perfect design — impossible today (long term goal)
- Realistic or practical design — Possible but needs more work (near term goal)

Design Requirements

However, structural weight reduction and cost savings cannot be accomplished without input from the concurrent engineering teams (but not limited to these):
- Structural design
- Manufacturing and producibility

- Tooling
- Materials and processes
- Maintainability and repairability
- Stress engineering

Also, emphasis on modular monolithic types of structure, with reduced part counts, must be balanced against more complicated or exotic tooling (a cost consideration), minimization of cutouts and holes, and the use of cocured or co-consolidated composite structures. Innovative structural concepts should feature novel designs that not only exploit the unique characteristics of composite materials but also utilize structural mechanics to validate the concepts.

Rules of Innovative Design Approaches

The following concepts are basic components of innovative composite designs:
- Fiber orientation concept
- Modular concept
- Monolithic (integral) concept

The innovative composite design concepts discussed here are only a few of the many possible and all the concepts shown in this chapter are concepts only and may need more detailed work prior to practical application. Bear in mind that the philosophy of composite design should be completely different than that of conventional metal design which consists of many parts assembled together with numerous mechanical fasteners. In composite structures, elimination or most of all fasteners is critical not only to increase structural efficiency (eliminating composite material notched effect) and save structural weight, but also to reduce assembly cost. These innovative concepts should be considered a guide to composite engineers for the exploitation of true composite structural design and final achievement of the goal of more than 50% weight savings and more than 25% cost savings.

Incorporating manufacturing input during the creation of innovative concepts is strongly recommended to confine the design concept within reasonable costs as well as to insure technology availability within the near future. Another important aspect in developing new designs is to convince composite engineers to accept innovative concepts (e.g., by conducting structural tests which support the concept) and minimizing or eliminating structural conservatism, such as the widespread use of mechanical fasteners.

Patent Protection

It is obvious that composite design is not simply fabricating metal designs out of a new lighter material i.e., composites. Since the design of composite structures is a leading edge technology, many new concepts will be developed for weight reduction and cost-effective manufacturing processes. It is important to protect new ideas by filing by a patent as soon as possible so that priority rights to the invention may be secured.

12.2 FIBER ORIENTATION CONCEPT

In composite laminate structures, the fiber is the primary load-carrying material and should be used efficiently to avoid fiber interruption or termination. The fiber orientation concept accomplishes this by use of the following:

- Efficient usage of directional fibers, i.e., $0°$, $45°$, $-45°$, $90°$, etc.
- Convergent and divergent concept
- Straight-through concept
- Wrap-around concept

(1) Efficient usage of directional fibers:

Since fibers are the primary load-carrying materials, it is extremely important in composite design that the fibers, and therefore the load path, are continuous rather than being disrupted by fiber termination at mechanically fastened spliced joints. Splice joints cause structural deficiency, i.e., eccentricity, notched effect (reducing laminate strength), fuel sealing problems, etc. Fig. 12.2.1 and Fig. 12.2.2 show examples which illustrate this concept.

(2) Convergent and divergent concept:

The convergent concept involves converging all or part of the fibers of a laminate panel into one or several fittings, lugs, etc. which carry a very highly concentrated load. The fighter wing root fittings and lugs, shown in Fig. 12.2.1 and Fig. 12.2.2, illustrate the convergent concept.

In contrast, the divergent concept involves diverging or spreading a concentrated load evenly through built-in or integral trunnions or fittings in a panel to moderate the local notched effect or stress concentration. The landing gear trunnion mounted bulkhead panel, shown in Fig. 12.2.3, is a typical case; it uses the fiber-disc or fiber-sector to spread the trunnion concentrated load into the bulkhead panel.

(3) Straight-through (Geodesic panel) concept

From a structural standpoint, the fibers in composite geodesic structural grid (orthogrid or isogrid) panels should be kept as straight as possible to maintain mechanical strength. An orthogrid panel (see Fig. 12.2.4) and similarly, the isogrid panel (see Fig. 12.2.5) is constructed by using longitudinal and transverse grid walls which intersect orthogonally to form a grid system that provides strength and stiffness to the panel, allowing both axial and lateral loads.

The bar-grid panel design concept, shown in Fig. 12.2.6, consists of alternating layers of consolidated composite material bars (round, square, rectangular, etc.) and syntactic material to form the grids.

(4) Wrap-around concept:

The critical location on the bar-grid panel is where the grid terminates at the panel edge. Both Fig. 12.2.3 and Fig. 12.2.6 illustrate how the terminating bars can wrap around over the bars forming the grid at the edge of the panel.

Fig. 12.1.1(f) is another type of wrap-around concept which wraps fibers around a pre-located metal bushing and forms a wing panel lug or canard spar root lug able to carry concentrated loads.

Chapter 12.0

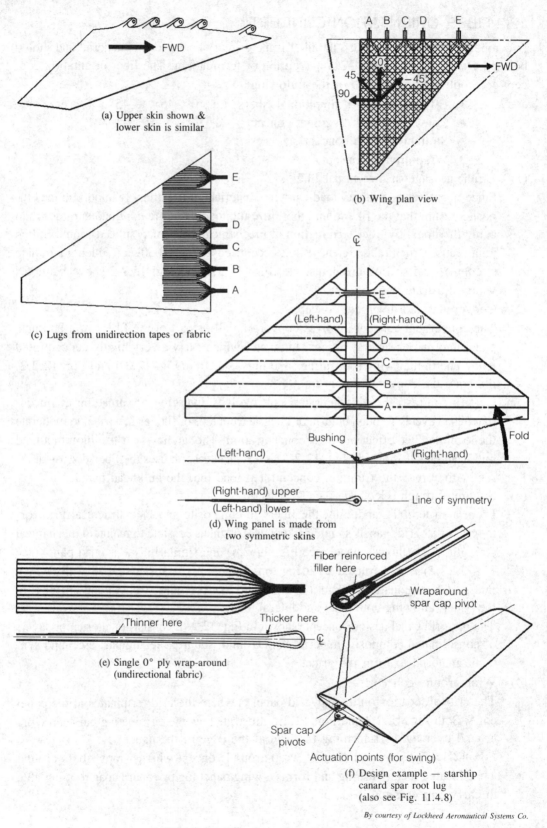

Fig. 12.2.1 Wing Root Integral Shear Lugs

Innovative Design Approaches

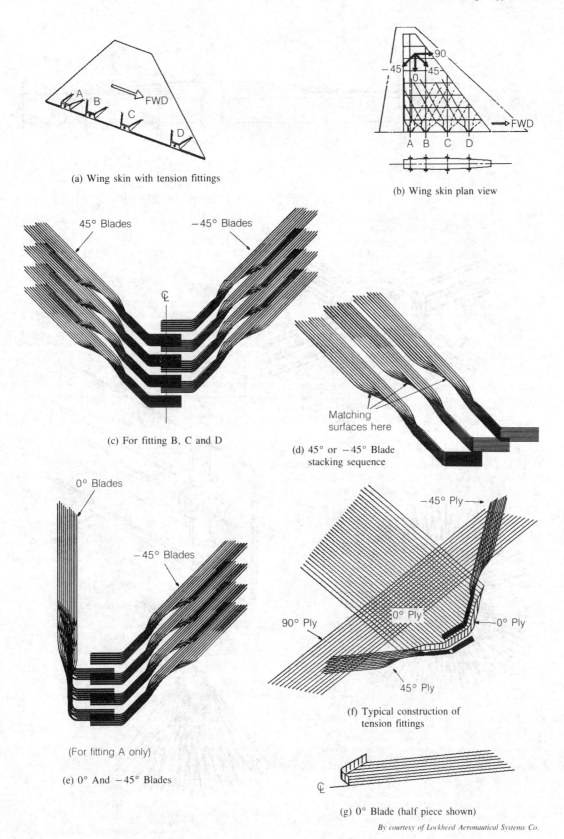

Fig. 12.2.2 Wing Root Integral Tension Fittings

625

Chapter 12.0

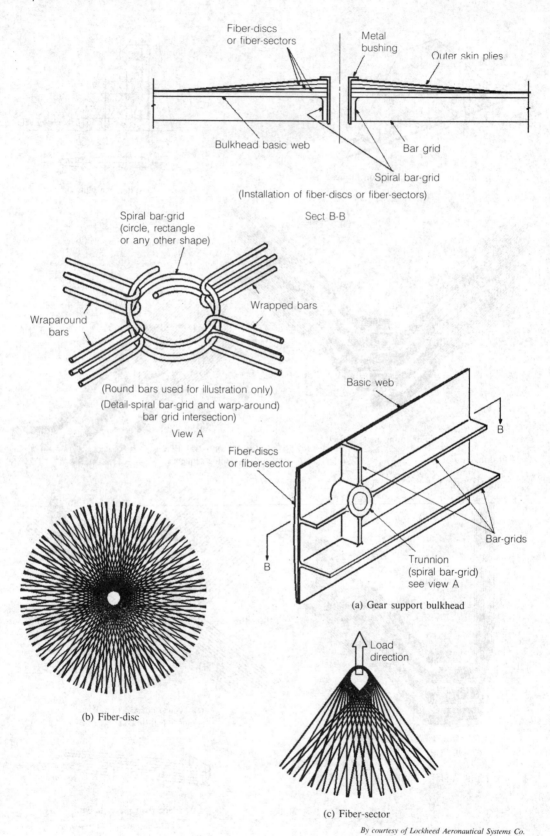

Fig. 12.2.3 Methods of Distributing Concentrated Load

Innovative Design Approaches

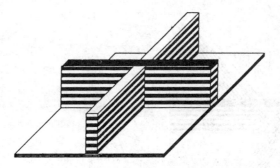

(a) Detail of orthogrid (see Fig. 7.5.2)

By courtesy of Lockheed Aeronautical Systems Co.

(b) Orthogrid panel and its tool

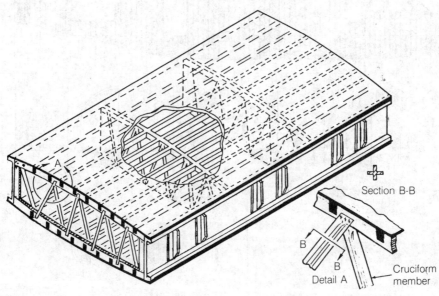

(c) Proposed orthogrid application (wing box)

Fig. 12.2.4 Geodesic Structure (Orthogrid)

Chapter 12.0

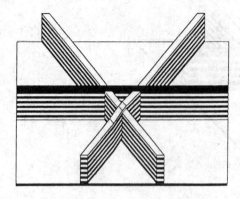

(a) Detail of isogrid (see Fig. 7.5.3)

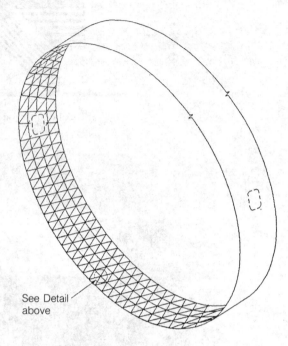

See Detail above

(b) Isogrid application (fuselage shell)

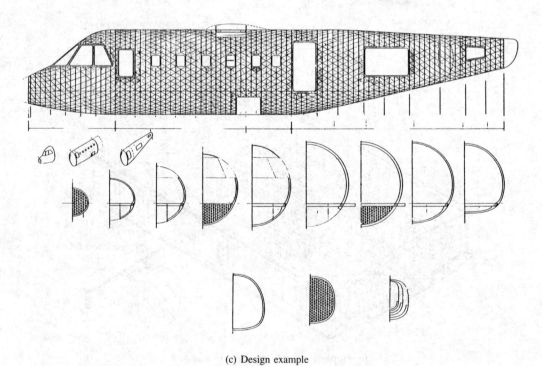

(c) Design example

Fig. 12.2.5 Geodesic Structure (Isogrid)

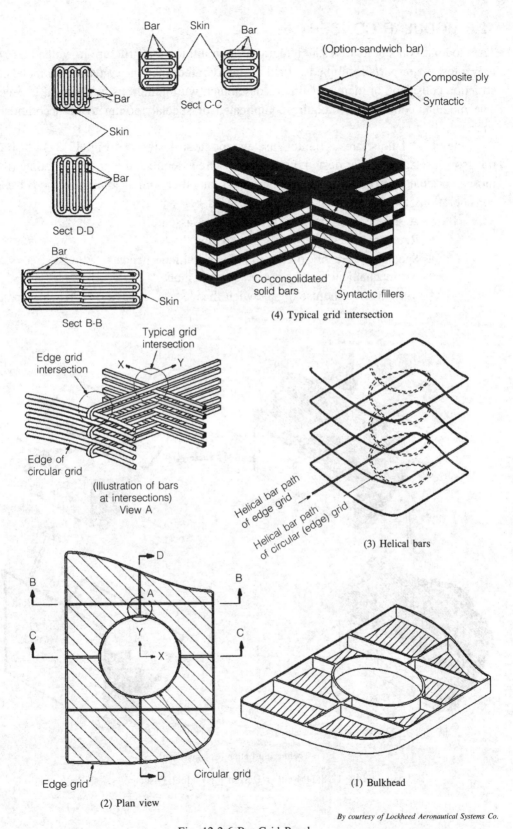

Fig. 12.2.6 Bar-Grid Panel

12.3 MODULAR CONCEPT

The modular concept involves an integral or monolithic type structure in which there are only a few major parts combined at final assembly rather than the conventional metal construction consisting of many small parts assembled with numerous mechanical fasteners. The modular concept may require complicated or special tooling to be accomplished successfully.

Fig. 12.3.1 illustrates a fighter fuselage modular design and Fig. 12.3.2 illustrates transport fuselage modular design; both concepts are assembled together by bonding (thermosets) or dual-resin bonding (thermoplastics) or another similar joining method. Design considerations for the modular concept are listed below:

- A simple component with little or no assembly
- Requires special mandrels and tooling
- Special cocure or co-consolidation technique needed
- No mechanical fasteners used for assembly
- May involve the use of sandwich, beaded, sine-wave or corrugated panels

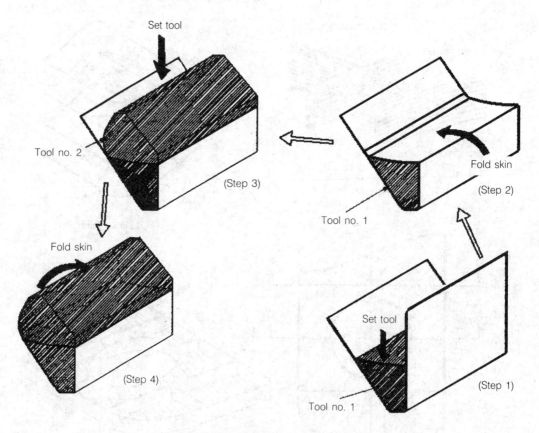

(a) Tooling illustration (example only)

Fig. 12.3.1 Modular Design for Fighter Fuselage

Innovative Design Approaches

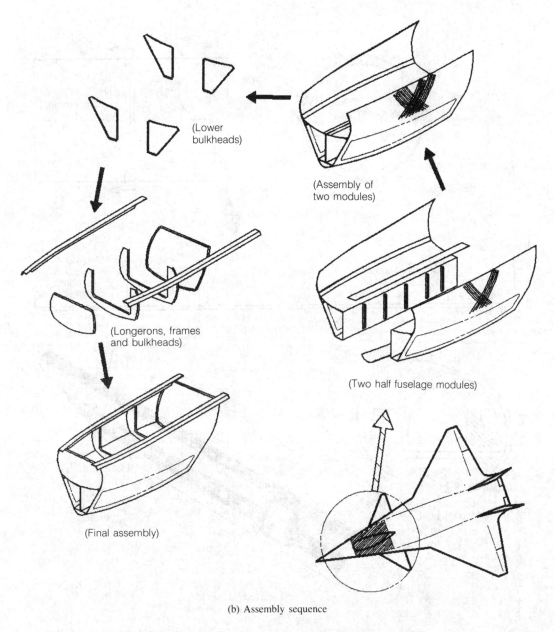

(b) Assembly sequence

Fig. 12.3.1 Modular Design for Fighter Fuselage (cont'd)

To produce a modular concept design, innovative and out-of-the ordinary tooling is a must. As mentioned previously, with this type of structure reduced part count must be balanced against more costly tooling, but increased tooling cost will become only a fraction of the total cost once it is divided by the hundreds of units produced. This modular design concept has been applied to all-composite kit planes (see Section 11.4 in Chapter 11) for ease of assembly by amateur aviation builders.

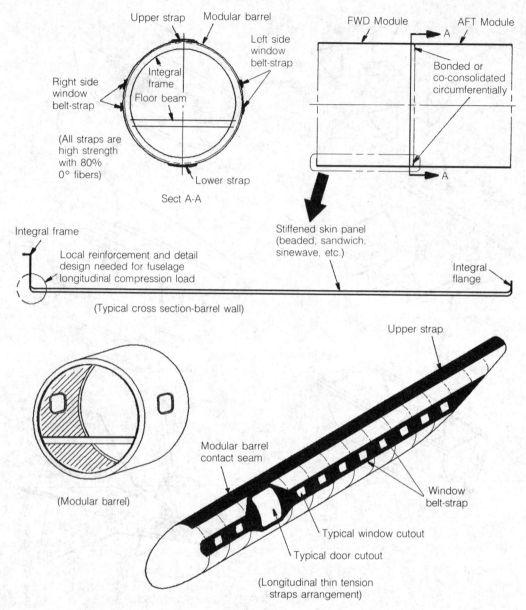

Fig. 12.3.2 Modular Barrel Fuselage

12.4 MONOLITHIC (INTEGRAL) CONCEPT

Fig. 12.4.1 shows a monolithic (one-piece) canopy frame which can be fabricated by low-cost filament winding using wash-out (disposable) mandrels (see Section 3.4 in Chapter 3) or other similar mandrels. This is a U-shaped frame with a constantly changing cross-section; all internal bulkheads, rear end hinges, etc. are built into the canopy frame by one cure operation. Both weight and cost savings are very substantial with this design concept.

Innovative Design Approaches

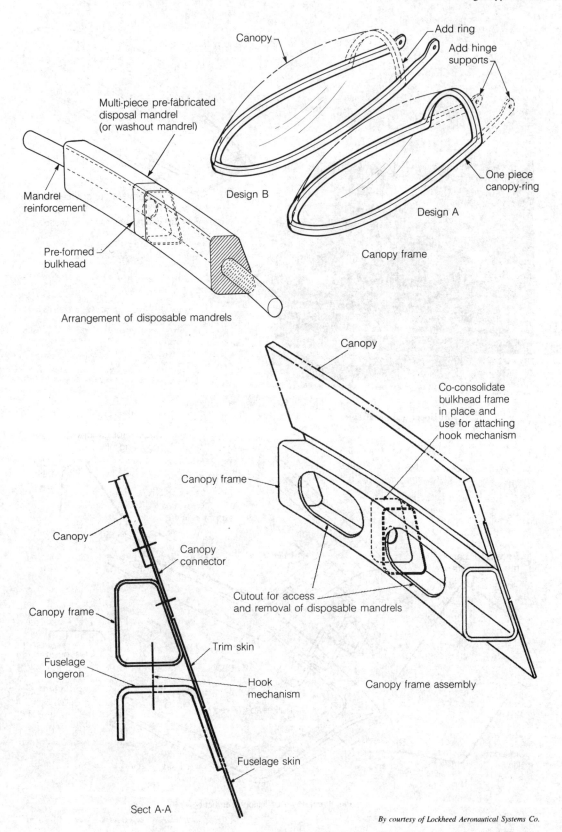

Fig. 12.4.1 One-Piece (Monolothic) Canopy Frame

Chapter 12.0

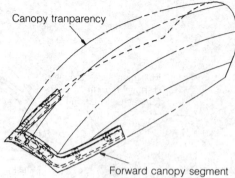

(a) Forward composite canopy segment

(b) Upper and lower skins

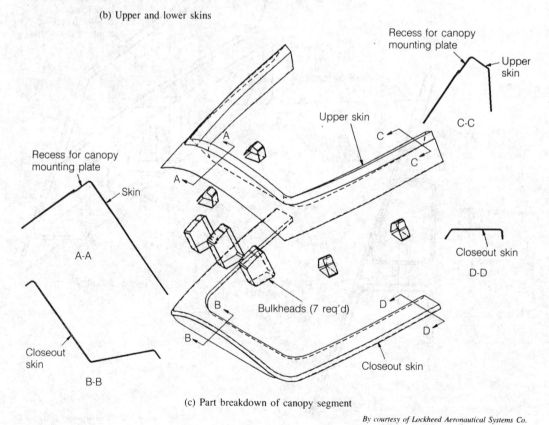

(c) Part breakdown of canopy segment

By courtesy of Lockheed Aeronautical Systems Co.

Fig. 12.4.2 Proof-of-the-Concept of Forward Composite Canopy Segment (Hand Layup Method)

634

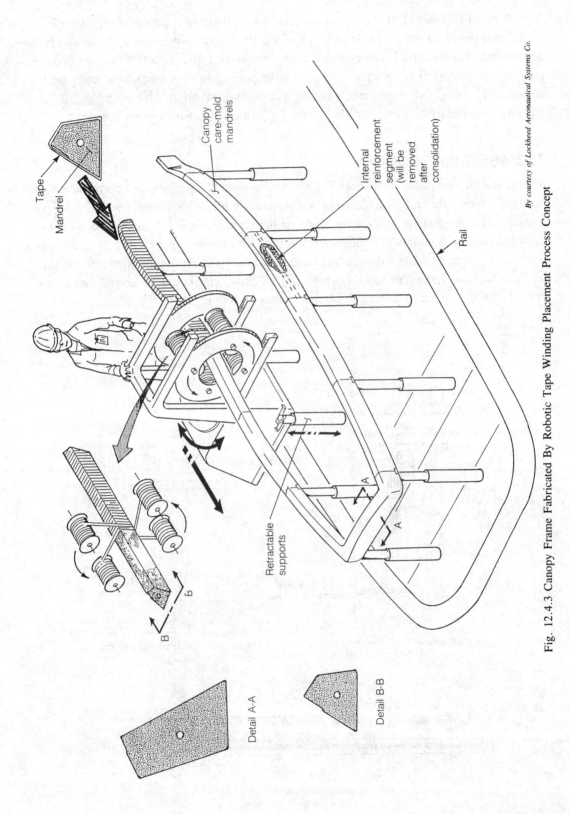

Fig. 12.4.3 Canopy Frame Fabricated By Robotic Tape Winding Placement Process Concept

A proof-of-concept of the full-scale forward segment of a canopy frame (see Fig. 12.4.2 and photo shown on the front page of this book) was fabricated by hand layup to demonstrate this concept. It achieved 50% weight reduction and manufacturing costs compatible with the metal counterpart. To attack the goal of lowering manufacturing costs, a proposed robotic/ automated tape winding placement technology, shown in Fig. 12.4.3, should be considered to produce the future full-scale composite canopy frame.

12.5 ASSEMBLY CONCEPT

Thermoplastic composite structures can be reconsolidated by applying appropriate heat and pressure onto the part. The mechanical tool shown in Fig. 12.5.1 allows out-of-autoclave joining and is designed to be used for applying both pressure and heat locally along structural joint areas. A compatible thermoplastic film with a lower melting temperature may need to be inserted between the contact surfaces of the joint. Many joining methods such as local heating, induction welding, resistant welding, etc. could be considered to co-consolidate the contact surfaces to obtain a sound bonding joint.

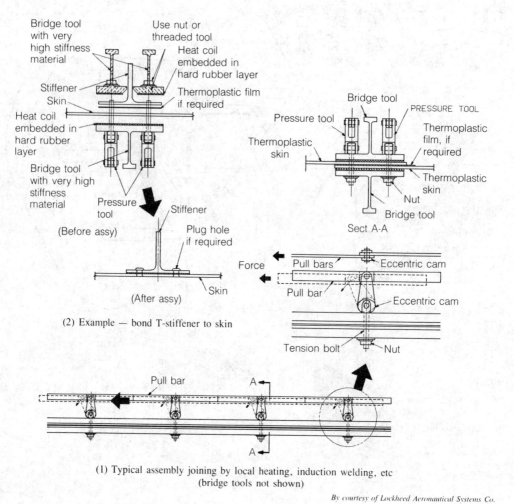

By courtesy of Lockheed Aeronautical Systems Co.

Fig. 12.5.1 Out-Of-Autoclave Assembly Joining

12.6 OTHER CONCEPTS

(1) Countersunk dimpled holes:

A minimum panel thickness at mechanical fastener joint locations is required due to local countersunk fastener heads. A method of dimpling a hole on a solid skin or SynCore sandwich panel of thermoplastic composites can be developed. During the dimpling operation, local heat and pressure are required. Fig. 12.6.1 illustrates this concept.

(2) Fits-in frame:

This concept utilizes stretchable thermoplastic material which is embedded into the frame at several selected locations. The frame is expanded by using clamp device and local heat source, as shown in Fig. 12.6.2.

(3) Step-hollow-grid:

This concept consists of a basic skin and a preformed step-over-grid-skin, which is superformed, as shown in Fig. 12.6.3. The depth and width of the grid-skin are either constant or variable as needed throughout the entire panel. Transverse and longitudinal reinforcements are cocured or bonded onto the skin-grid to provide additional strength and/or stiffness. This concept can be applied on door design as illustrated in Fig. 12.6.4.

(4) Even-hollow-grid:

This concept consists of a basic skin and a cross-hat-grid skin which are bonded (thermosets) or dual-resin bonded (thermoplastics) together to form a wing or fuselage panel (see Fig. 12.6.5). All grid caps (longitudinal and transverse directions) are continuous through grid intersections.

(5) Fastenerless construction:

This concept is an ideal method for use on control surface structures since it does not used a single fastener (see Fig. 12.6.6). This construction consists of a series of modular boxes which are wrapped around a pre-formed washout mandrel. The modular boxes are packed together and the outer skin is wrapped around the entire assembly. The assembly is cocured in one operation. Cutouts on the spar web are required not only to allow washout of the interior mandrel materials but to allow for inspection and future maintenance.

(6) Smart structures:

This concept represents a range of applications with one characteristic — the structure is more than just a framework designed to passively support the functional parts of the system. In airframe applications, the smart structure concept may be used at various levels of smartness. For example:

(a) Aircraft antenna systems have panels which incorporate built-in conformal electronics for transmitters, receivers and other sensors as "smart skins". The skin panel is designed to serve both as a structural member and as an electronic component (see Fig. 12.6.7).

(b) Structural components can be fabricated with an array of built-in fiber optic sensors to monitor processing conditions during manufacturing (see Fig. 12.6.8). In the finished part, changes in passive signals from the same embedded optical system can warn of either service or battle damage to prevent fatally damaged aircraft from continuing to fly.

Chapter 12.0

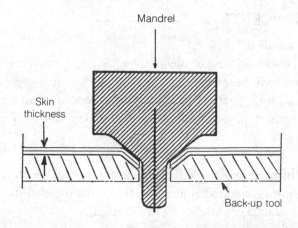

(a) Dimpled skin

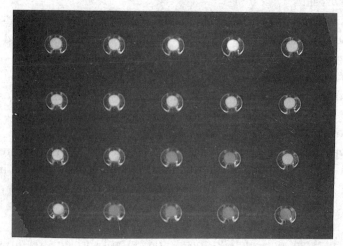

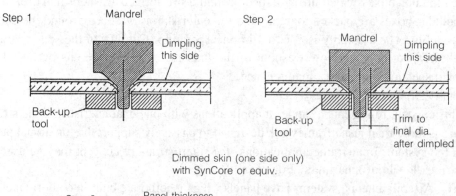

(b) Dimpled sandwich panel (SynCore)

By courtesy of Lockheed Aeronautical Systems Co.

Fig. 12.6.1 Dimpled Panel (Thermoplastics)

638

Innovative Design Approaches

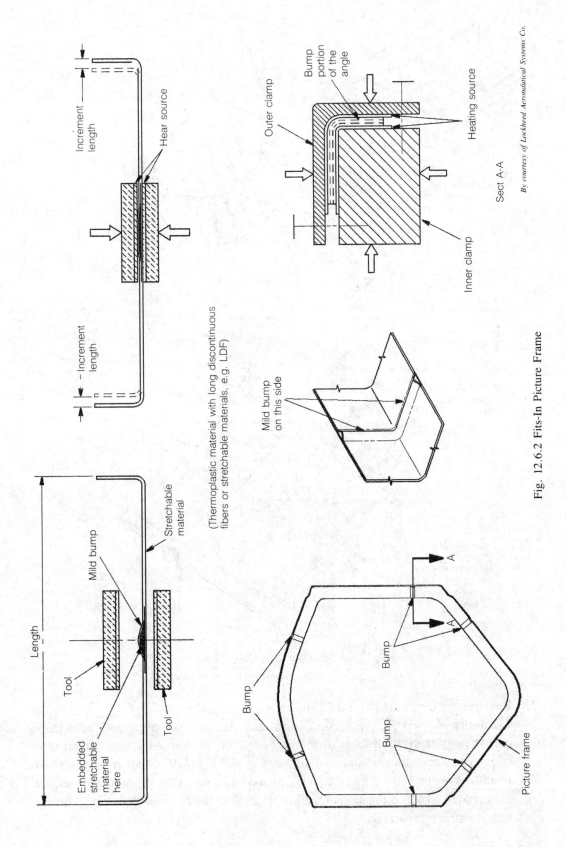

Fig. 12.6.2 Fits-In Picture Frame

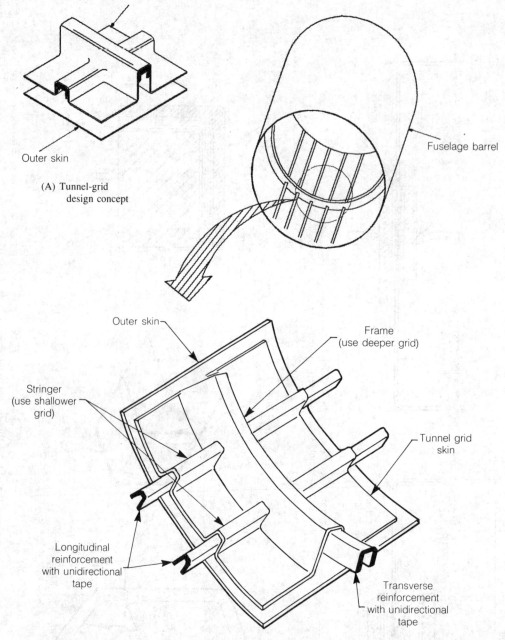

Fig. 12.6.3 Step-Over Grid Design Concept

(7) Pre-stressed integrally-stiffened panel:
The theory of composite pre-stressed structures is the same as conventional pre-stressed concrete construction. In composite structures, very high strength tensile fiber tows (similar to steel bars used in concrete) such as 700 ksi (4791 Mpa) are used. These fiber tows are embedded into the panel, as shown in Fig. 12.6.9, and this protection to ultimately allows utilization of the full strength of the tow materials which further reduces structural weight.

Innovative Design Approaches

(8) 3-D fiber-link composite structure

The marvelous handicraft weave method of rattan (see Fig. 2.5.25 in Chapter 2) is combined with resin transfer molding (RTM) to construct a sound composite structure which has all continuous fibers in three directions. This is a "dream composite structure" which provides the advantage of increasing not only load-carrying capability, but also impact resistance in composites and overcomes the weakness in today's composite technology.

Ultimately this method should rely on a robotic/automated technique to reduce the high costs which result from such a labor-intensive process. This concept should follows the same procedures as described for Very Complex Shapes (VCS) in Section 2.5 of Chapter 2).

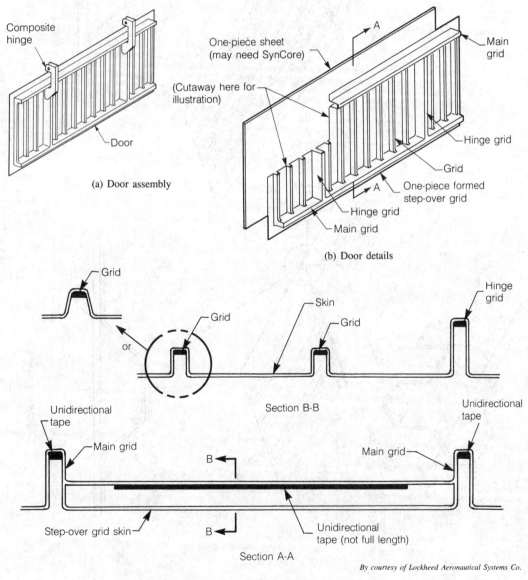

By courtesy of Lockheed Aeronautical Systems Co.

Fig. 12.6.4 Step-Hollow-Grid Door Design

Chapter 12.0

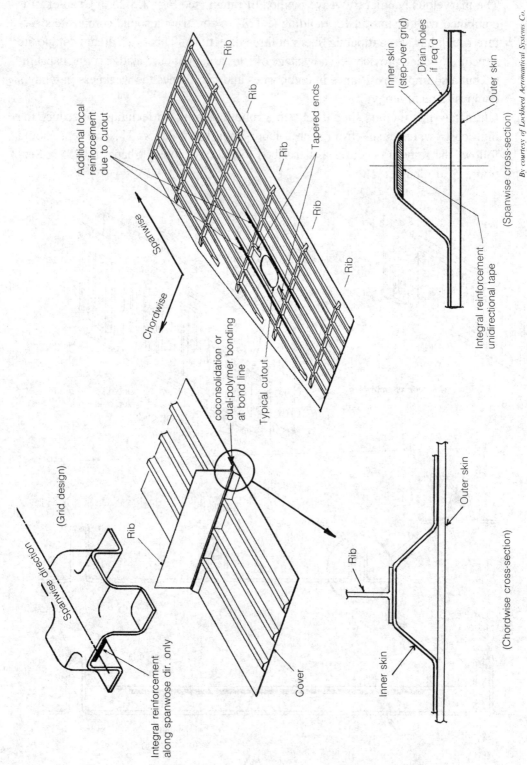

Fig. 12.6.5 Even-Hollow-Grid Wing Cover

Innovative Design Approaches

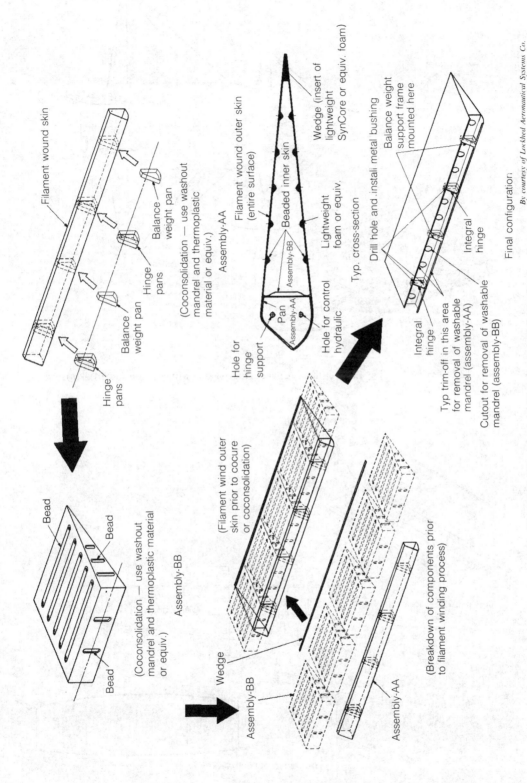

Fig. 12.6.6 Fastenerless Construction (Aileron Box)

By courtesy of Lockheed Aeronautical Systems Co.

Chapter 12.0

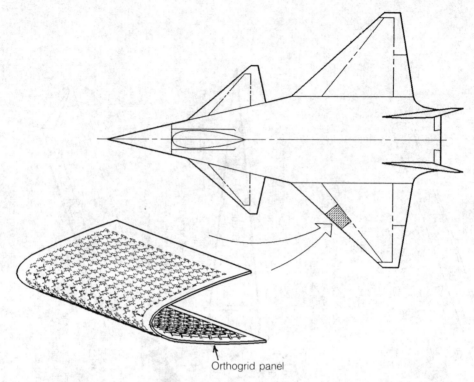

Fig. 12.6.7 Wing Leading Edge Orthogrid Panel And Avionics Integration (Smart skin)

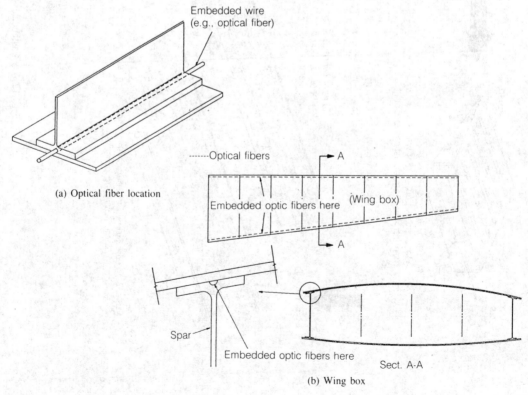

Fig. 12.6.8 Smart Structure Applications

Innovative Design Approaches

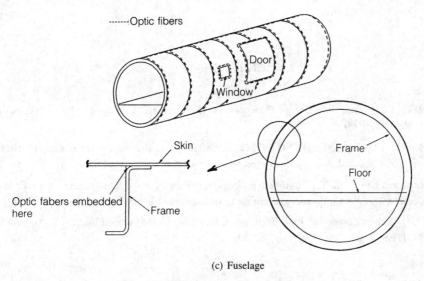

(c) Fuselage

Fig. 12.6.8 Smart Structure Application (cont'd)

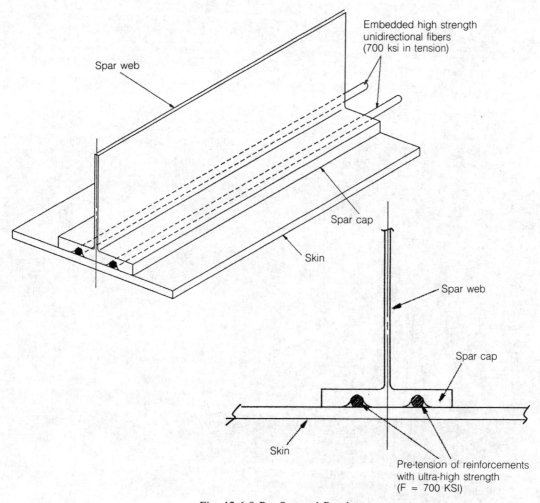

Fig. 12.6.9 Pre-Stressed Panel

References

12.1 Niu, M. C. Y., "Innovative Design Concepts for Thermoplastic Composite Materials", SAMPE Journal, Vol. 35, 1990.

12.2 Niu, M. C. Y., "ADVANCED COMPOSITE DESIGN — Innovative Design Concepts", Internal Publication of Lockheed Aeronautic Systems Co. 1990.

12.3 Chu, R. L. and Niu, M.C.Y., "Innovative Design and Fabrication of Composite Canopy Frames", Ninth Dod/NASA/FAA Conference on Fibrous Composites in Structural Design, November 4-7, 1991.

12.4 Leonard, L. "Smart Composite: Embedded optical Fibers Monitor Structural Integrity", ADVANCED COMPOSITES, Mar/April, 1989. pp. 47-50.

APPENDIX A

Commonly Used Conversion Factors
(English units Vs. SI units)

LENGTH:
    mil (0.001 inch) = 25.4 micron
    in = 2.54 cm
    ft = 0.3048 m
    ft = 0.333 yd

AREA:
    in^2 = 6.452 cm^2
    ft^2 = 0.0929 m^2

VOLUME:
    in^3 = 16.387 cm^3
    ft^3 = 0.02832 m^3

FORCE AND PRESSURE:
    psi = 6.8948 kpa
    ksi = 6.8948 Mpa
    Msi = 6.8948 Gpa
    Hg(32°F) = 3.3864 kpa

MASS:
    lb = 0.4536 kg

DENSITY:
    lb/in^3 = 27.68 g/cc
    lb/ft^3 = 0.6243 kg/m^3

TEMPERATURE:
    °F = (9/5) °C + 32

THERMAL EXPANSION:
    in/in °F × 10^{-6} = 1.8 K × 10^{-6}

THERMAL CONDUCTIVITY:
    Btu × in/hr × ft^2 × °F = 0.1442 (W/m)K

IMPACT ENERGY:
    ft-lb = 1.3558 J

COMMONLY USED SI PREFIXES:
    G — 10^9 μ — 10^{-6}
    M — 10^6
    k — 10^3

APPENDIX B

Composite Engineering Drafting Practice

This Appendix has been prepared to establishing the basic requirements for drafting practices to be following by engineers in the preparation of composite engineering drawings. It summarizes all relevant drafting requirements for composite structures; in addition, the following guidelines should be followed in order to create an "as manufactured" composite drawing system:

- Create an engineering drawing breakdown which follows the manufacturing sequence
- Provide a parts list which is compatible with manufacturing planning operations and methods
- Provide as many valid lower level drawings as possible
- Show views in the same direction in which manufacturing will work

General requirements for the preparation of engineering drawings for a part and/or assembly which is made of composite materials are described below:

(1) Requirements
   - Orientation symbol
   - Ply Identification number
   - Laminate ply view and ply tabulation
   - Material(s) identification
   - Tag-end piece locations
   - Definition of tool surface(s)

(2) Commonly used composite material designations (not limited to these)
   (a) Fibers:
      - GR — Graphite
      - FG — Fiberglass
      - B — Boron
      - KV — Kevlar
   (b) Matrices (resins):
      - EP   — Epoxy
      - BMI — Bismaleimide
      - PI   — Polyimide
      - PEEK — Polyetheretherketone
   (c) Prepreg material designation:
      e.g., GR/EP — Graphite fiber and epoxy resin
            GR/TP — Graphite fiber and thermoplastic resin

Appendix B

(3) Ply orientation and identification
- Typical ply orientation symbol is shown in Fig. 1.7.1 and Fig. B-1
- Ply orientation symbol shall be shown on the main view
- Additional ply orientation symbols may be used within the same laminate but an identification symbol shall be used to distinguish them (see Fig. B-2)
- 0° direction shall be the primary fiber direction fo the material used
- Each ply within a given part shall be assigned a ply number
- Ply material other than composite material such as plyset [see discussion in item (8)], lightning strike screen, adhesive film, etc. shall also occupy a space in both ply view and ply tabulation, but without an assigned ply number (see Fig. B-3)

(4) Material form identification (see Fig. 1.7.4)
- GR indicates a tape or unidirectional graphite tape only
- (GR) indicates a fabric or bidirectional graphite cloth only
- GR/EP Indicates a prepreg tape or unidirectional graphite tape with epoxy resin
- (GR/EP) indicates a prepreg fabric or bidirectional graphite cloth with epoxy resin

(5) Ply view and ply tabulation (see Fig. B-1)
- A ply-view is an appropriately selected cross-section of the laminate part with each ply delineated as a single line
- Each ply view shall be accompanied by an appropriate ply tabulation to identify each ply as to ply number, ply angle, material, rough size, etc.

(6) Laminate thickness (see Fig. B-1)
- Do not specify laminate thickness and call-out only as referenced thickness
- Laminate thickness may be specified if it is design critical for assembly

(7) Add-on (see Fig. B-3) or drop-off plies (see Fig. B-1)
- Use a ply view to define add-on or drop-off plies
- If ply drop-off occurs in more than one direction (x and y directions), a "plan view" may be used to define the shape
- Be careful in selecting the tool surface, since it will affect the assembly sequence (and tolerances)

(8) Plyset (see Fig. B-4)
- A plyset is a special case of add-on plies
- A plyset is a set of plies of composite material having the same or a different shape (considered as a standard, in general) which may be used at more than one place on the drawing
- Stacking sequence shall be defined
- Specify ply size, shape, and orientation for each ply

(9) Tool surface (see Fig. B-5)
- It is preferable to specify tool surface(s) only when required; generally leave the decision to manufacturing
- Tool surface is also mold surface and all laminate plies shall be cured against this surface
- Surface texture smoothness should be specified

Composite Airframe Structures

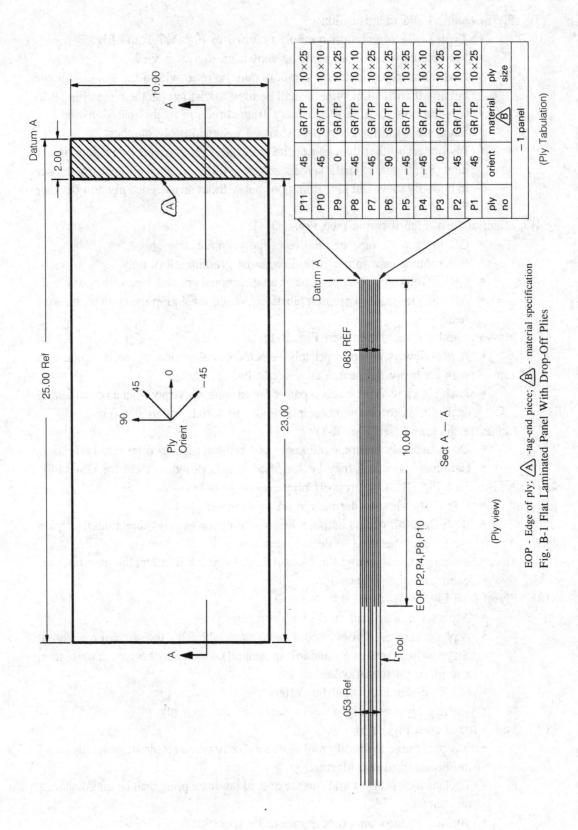

Fig. B-1 Flat Laminated Panel With Drop-Off Plies

Appendix B

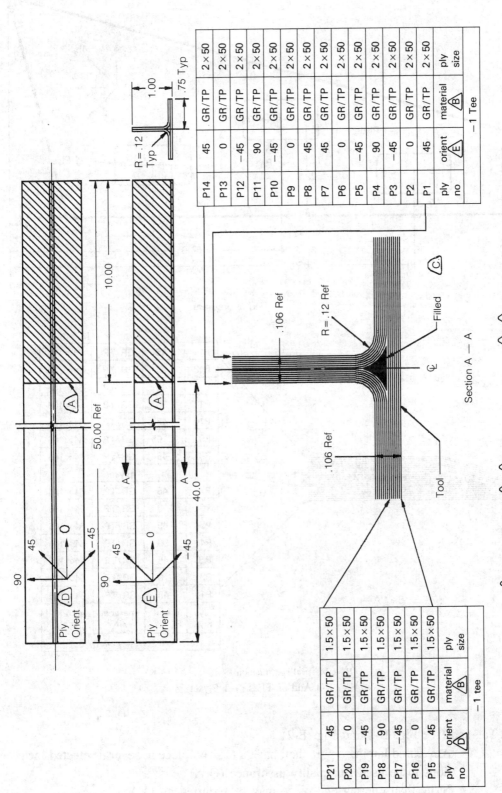

Fig. B-2 Constant Thickness Tee With Two Different Ply Orientations

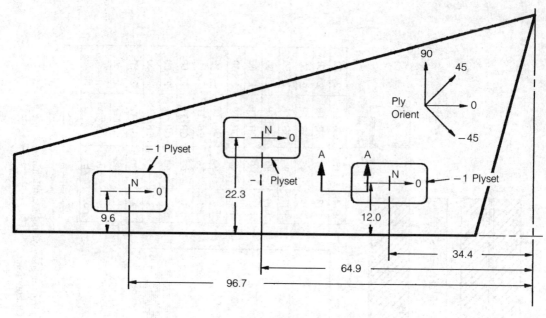

(filed of drawing)

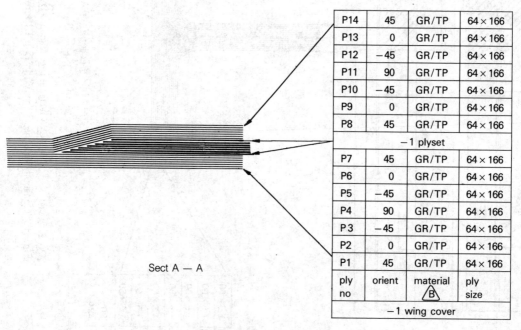

Sect A — A

ply no	orient	material /B\	ply size
P14	45	GR/TP	64 × 166
P13	0	GR/TP	64 × 166
P12	−45	GR/TP	64 × 166
P11	90	GR/TP	64 × 166
P10	−45	GR/TP	64 × 166
P9	0	GR/TP	64 × 166
P8	45	GR/TP	64 × 166
−1 plyset			
P7	45	GR/TP	64 × 166
P6	0	GR/TP	64 × 166
P5	−45	GR/TP	64 × 166
P4	90	GR/TP	64 × 166
P3	−45	GR/TP	64 × 166
P2	0	GR/TP	64 × 166
P1	45	GR/TP	64 × 166
−1 wing cover			

/B\ - Material specification
Fig. B-3 Add-on Plies (−1 Plyset)

(10) Tag-end (see Fig. B-1 and Fig. B-2)
- Any cured laminate part shall have a tag-end piece at several selected locations to be used for quality assurance (Q.A)
- More than one tag-end piece may be required by Q.A.

Appendix B

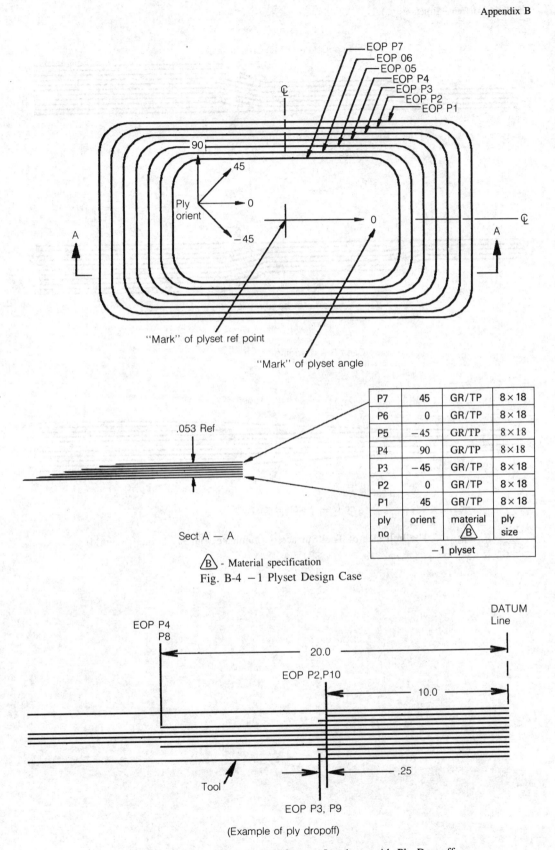

Fig. B-4 −1 Plyset Design Case

⚠B⃝ - Material specification

(Example of ply dropoff)

Fig. B-5 The Influence of Tool Surface on Laminate with Ply Dropoff

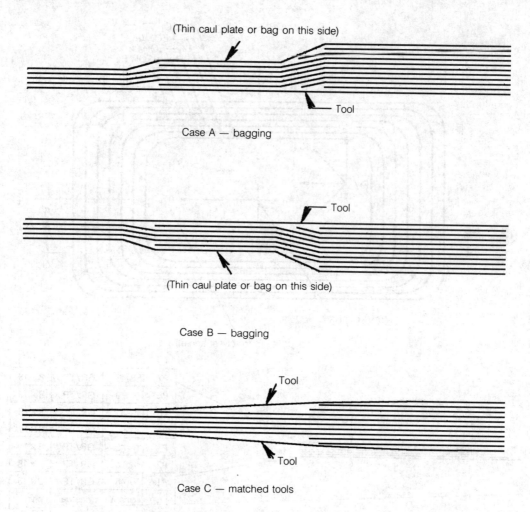

Fig. B-5 The Influence of Tool Surface on Laminate with Ply Dropoff (cont'd)

APPENDIX C

A List of Schematic Drawings Related to Composite Structures

The cutaway or schematic drawings shown here are those airplanes which have used composite materials on primary airframe structures such as control surfaces, empennage, fuselage, wing, helicopter blade, etc. The list given below is a good reference for the composite engineer who is interested in using them to exploit their composite design knowledge. A — Air International; F — Flight International; I — Interavia; AW — Aviation Week & Technology

Aircraft Name	Document	
	Title and Date	Page
Commercial Transports		
Boeing 747-400	A(May, 1988)	218
Boeing 747-400	F(12 Nov., 1988)	28
Airbus 320-200	A(May, 1987)	232
Airbus Industrie A340-300	A(Sept., 1988)	130
Utility Aircraft		
Claudious Dornier Seastar	A(Oct., 1988)	190
Learavia Lear Fan	F(26 Dec., 1981)	1898
Beech Starship1	F(3 may, 1986)	20
Piaggio P. 180 Avanti	F(23 July, 1988)	24
Piaggio P. 180 Avanti	A(April, 1988)	176
SAAB-Fairchild 340	F(30 Oct., 1982)	1276
Fighter Aircraft		
Grumman F-14A Tomcat	A(Jan., 1982)	26
McDonnell Douglas F-15 E Eagle	A(July, 1987)	8
McDonnell Douglas F-15 A Eagle	F(1 May, 1975)	708
General Dynamics F-16C	A(Nov., 1989)	222
General Dynamics F-16C	F(4 Sept., 1982)	696
McDonnell-Douglas F/A-18C Hornet	A(May, 1990)	224
Grumman X-29A	F(16 June, 1984)	1564
McDonnell-Dougals AV-8B	F(29 Dec., 1979)	2130
McDonnell-Douglas/		

Aircraft Name	Document	
	Title and Date	Page
British Aerospace AV-8B	A(Oct., 1986)	178
British Aerospace/ McDonnell-Douglas Harrier GR Mk 5	A(August, 1989)	64
Lockheed F-117A	A(Sept., 1990)	170
British Aerospace EAP	F(19 April, 1986)	28
British Aerospace EAP	A(June, 1986)	304
SAAB JAS39 Gripen	A(Nov, 1987)	228
Eurofighter EFA	A(Sept., 1988)	148
Dassault Rafale A	F(10 June, 1989)	84
Dassault-Breguet Rafale-A	A(June, 1986)	272
Helicopter		
Bell/Boeing V-22 Osprey	A(May, 1989)	216
Bell-Boeing V-22 Osprey	F(14 May, 1988)	26
Bell 214 ST	F(30 June, 1979)	2347
Boeing's Low-Profile Composite/Elastomeric Rotor Hub	F(23 July, 1983)	181
Boeing Vertol's All-composite Helicopter	I(Aug., 1983)	837
Bell and Sikorsky Advanced Composite Airframe Programme Designs	F(6 Nov., 1982)	1318
Westland's All-composite blade	F(10 April, 1982)	924
Bearingless Helicopter Main Rotor	AW(Aug. 23, 1982)	63
Vertol Tilt-wing VTOL Propeller	AW(Aug., 29, 1966)	41
MBB's New Rotor Bearingless Main Rotor	F(29 June, 1985)	32 — 35
Bell's Model 214 Composite Blade	I(May, 1979)	417

INDEX

A

A-basis 393, 469
Ablate 64, 67
Abrasion 62
Acoustic 380, 524, 563
Acoustic Microscopy 522
Adhesive 113, 265
Adhesive shear 331, 333, 339
Aging 493, 500
Airworthiness 30, 539
Aligned discontinuous fibers 88
All-composite airframe 10, 30, 590, 610
Allowables 19, 32, 34, 75-78, 81, 89, 116-123, 358, 390, 393
Alumina 75, 82
Aluminum 57, 131, 132, 141, 142
Aluminum flame spray 371, 376
AML plots 395, 446, 484, 485
Amorphous 54, 55, 56
Angle-ply 37, 407
Anisotropic 1, 15, 60, 385, 387, 403, 406
ARALL 61
Aramid 12, 74, 75, 77, 482, 579, 605
Aramid-epoxy 45, 482
Arcing 368, 369
A-scan 516
Assembly concept 636
A-stage 86
Autoclave 129, 159, 160, 162, 184
Autoclave forming 228
Automated systems 101, 107, 177
Automation 21, 27, 255, 262
Avamid-N 133, 141
Axes 383, 401

B

Bags 140, 223, 225
Bagging 217, 219, 222, 546, 599, 654
Balanced (laminate) 90, 438
Ballistic impact 77, 78, 430, 433
Bar-grid panels 623, 629
Battle damage repair 538
Bypass load (stress) 297, 298
B-basis 34, 393, 455, 469, 479
Bearing reduction factor 295
Bearing stress (or strength) 293, 298, 301, 443, 478
Bearing tests 478, 480
Bend radii (or radius) 101, 195
Binder fibers 94
Bismaleimide (BMI) 12, 49, 51
Bladder 369
Bleeder plies 220
Bleed system 221, 222
Blind fasteners 299, 302, 312, 313, 314, 315, 316, 317, 321, 541, 596
Boardy 53
Bolted joints 290
Bolted repairs 535, 541
Bonded joints 330, 337, 341
Bonded repairs 535, 544
Bonding 330, 335, 597, 636
Boron 1, 14, 57, 59, 60, 74, 75, 82, 380, 545, 561
Borsic 14
Braided pultrusion 265
Braided tube panel 270
Braiding 9, 96
Brakes 61
Breather plies 222
Bridging 162, 181, 200, 211, 220, 221
Brittle 52, 72, 82, 294
Broadgoods 43, 90
B-scan 516
B-stage 86, 551
Buckling 13, 418, 439, 494, 591
Building block testing 34, 453, 494
Bulk factor 139, 170

657

Burn-off 504
Butt joint 612
Bypass stress (or load) 297, 298

C

Carbon 1, 11, 75, 80, 81, 118
Carbon/carbon 61, 66
Carbon fiber reinforced plastic (CFRP) 571
Carbonization 67
CARE-MOLD 159, 257
Carpet plots 394, 395, 446, 482
Caul plate (sheet) 219, 221, 222, 654
Certification 29, 31, 458, 494, 497
Ceramic 11, 43, 67, 74, 81, 129, 131, 132, 141, 155
Chainstitch 95
Chamber systems 159
Chemical vapor deposition (CVD) 67, 69, 82
Chopped fibers 89
Clamp-up force 302
Cleavage 291
Cocuring 21, 195, 268, 346, 586, 622
Coefficient of thermal expansion (CTE) 46, 131, 141
Cold/dry 30, 343, 358
Collar 302, 304, 305, 309, 310
Commingled fabrics 56, 107, 110, 111, 228
Component tests 457, 486, 487
Composite mold tooling 147
Compression molding 242
Compression tests 469, 470, 471, 472, 487, 494
Compressive strength 13, 19, 77, 357
Computer tomography 525
Computerized techniques 400
Concentrated load 444, 494, 623, 626
Concurrent engineering 22
Constituents 400
Contamination 506
Convergent and divergent concept 623
Copper 58, 146
Core 102, 111, 112, 119
Corrosion 16, 58
Costs 6, 14, 16, 20, 24, 42, 49, 53, 72, 130, 141, 185, 622
Countersunk head fasteners 293
Coupling 385, 386, 407, 438
Coupon tests 28, 455, 457, 464, 482
Coweaving (cowoven) 107, 110

Cracking 13, 16, 72, 84, 149, 261, 422, 442, 496
Crashworthiness 30, 430, 432
Creel 261,
Creep 13, 16, 58
Creep forming 232, 236
Criteria 27, 392, 414, 537
Cross-plies 15, 265, 407
Crowfoot 93
Crystallinity 54, 55
C-scan 517, 524
C-stage 87, 536, 551
Cure
— cycle 27, 229
— pressure 49
— temperature 49
Cure-in-place repair 545, 550
Cutouts 490, 536, 541, 622, 637
Cutting 276

D

Dams 219, 221, 223
Damage 13, 16, 268, 359, 424, 429, 539, 540, 557
Damage tolerance 30, 34, 90, 255, 423, 427, 460, 462, 497
Debulk 203, 212, 226, 257
Deepdrawing forming 237
Defects 506, 525
Degradation 13, 18, 30, 201
Delamination 13, 16, 441, 442, 490, 496, 506
Densification 67
Density 46, 75, 77, 132, 647
Dents 501, 540
Diaphragm forming 243, 244
Dies 129
Dielectric 77, 78, 372
Diffusion bonding 59
Dimensional stability 71
Dimpled holes 637
Disbond 510
Discontinuous fibers 74, 88
Distortion 509, 525
Diverter 372, 373, 375
Double-lap joint 337
Drapability 86
Drape 49, 197
Drilling 189, 192, 326, 542, 543, 544
Dry 53, 54, 85, 111, 255

Dual-resin bonding   345, 346
Durability   34, 460, 496, 497, 538

E

Eccentricities   285, 292, 295, 297, 299, 332, 336, 341, 623
Eddy-current   529
Edge closeout   268
Edge distance   293, 544
Edge effects   416, 455, 490
Egg crating   135, 151
E-glass   75, 77, 78
Effector   212
Elastic constant   15, 405
Elastic trough   331
Elasticity   400
Elastormeric forming   232, 233, 234, 235
Elastomeric (rubber) tooling   129, 142, 168, 170, 171
Electrical conductivity   83
Electro-deposited nickel   145
Electroformed nickel tooling   133, 145
Electromagnetic interference (EMI)   44, 376
Electroplated nickel tooling   145
Element tests   457, 486
Elongation   72, 75, 77, 225
Encapsulate   57
Energy absorption   18, 77
Environment   12, 16, 30, 357, 380, 458, 494, 539
Environmental chamber   462, 468, 479, 493
Epoxy   11, 12, 49, 50, 52
Erosion   62, 360
Eutectic salt   157, 256
Even-hollow-grid   637, 642
Extensometer   468

F

Fabric   38, 74, 93, 201, 392, 551
Fabrication   16, 134, 272
Failure criteria   414
Failure modes   398
Failure strain   387, 395
Fail-safe design   79, 424, 614
Fastener (or bolt)   18, 21, 294, 318 380, 622
Fastener bending   299
Fastener hole   324
Fastenerless construction   637, 643
Fastener load transfer   297, 300
Fastener shear failure   295

Fatigue   13, 16, 17, 30, 33, 422, 423, 440, 460
Feather edge   325, 541
Female tooling   212
Fiber content   254
Fibers   24, 72, 74, 75, 81, 383
Fiber content (volume)   255, 466
Fiber-placement laying   204, 207
Fiberglass   74, 76, 117, 118, 367
Fiber orientation concept   623
Fiber tow placement   215
Fiber variations   508, 509
Fiber winding   215
Field repairs   538, 541, 544
Filament   2, 73
Filament winding (wound)   194, 206, 207, 212, 274, 600, 621, 632, 643
Fill direction   90
Finish   139, 220
Fits-in frame   637, 639
Fit-up   340
Flaws   500, 506, 519
Flexural tests   475, 476
Fluid heating   136
Flush repair   551, 552
Fluted-core   102
Foam   121, 122, 606, 617,
Foils   545, 551, 608
Forms   86, 111, 178
Forming   228
Fracture toughness   13, 53, 57
Fuel tank   299, 301, 369, 443, 591, 594, 596, 605
Full-scale testing   28, 33, 35, 457, 490

G

Galvanic corrosion   21, 296, 343, 379, 380, 440, 590
Gel coat   147, 149
Gelation   142
Geodesic   213, 214, 215, 623, 627, 628
Glass fibers   75, 77
Graphite   74, 80, 131, 152
Graphitization   67
Grind-down   504, 505
Grommet   327
Groove seal   299

H

Hail damage 359
Hand layup 201, 207, 600, 634
Harness satin 92, 104
Health 21, 185, 504
Heating blanket 136, 553
Heating pads 545, 547
Heat sink 152
Heat system 134, 159, 545
Helical winding 209
High pressure cure (HPC) 610
Holes 394, 475, 540, 622
Hollow microspheres 83
Holography 526
Honeycomb 112, 113, 221, 265, 440, 445, 510, 515, 519, 524, 527, 534, 554, 556, 563, 569, 600
Hooke's law 403
Hoop tension stress 299
Hot-roll 206
Hot-cured prepreg 149
Hot melt processing 157, 256
Hot-wire saw 270
Hot/wet 30, 52, 343, 358, 387
Humidity 357, 360
Hybrid-fabric (HF) 107, 109
Hybrids 44, 45, 181
Hybridization 45
Hybridized fabrication 272
Hydraulic ram 459
Hydroforming 247
Hydrocarbon 67
Hydroclave 161, 162
Hydroscopic 12, 13
Hydrostatic 161
Hydrotherm 16, 136

I

IITRI compression fixture 467, 472
Immersed inspection 516
Impact 13, 72, 74, 75, 79, 387, 391, 393, 422, 424, 428, 455, 463, 478, 562, 647
Impregnation 67, 600
Inclusion 507, 509
Induction heating 137
Induction welding 351, 352
Inflatable mandrel 171
Infrared (IR) heating 352, 544
Infiltration 64, 70
Injection mold 9, 251
Injection repair 557
Innovative design 6, 10, 21, 621
Inserts 270
Insulating blanket 136
Insulator 155
Integrally heated tooling 162, 167
Interference fit fasteners 301, 302, 324, 325, 326
Interlacing fabrics 107
Interlcok 106
Interply 45
Intraply 13, 45, 228
Invar 133, 141, 144
Iosipescu shear test 467, 475
Isogrid 214, 438, 623, 628
Isotropic 13, 15, 383, 403, 590

J

Joggle 222, 247, 591, 614
Joining 285, 286, 636
Joints 288, 291, 293, 336, 341, 489

K

Kapton 225
Kevlar 11, 13, 75, 77, 367, 600
Kit planes 590, 610, 631
Knits 203
Knitting 94
Knock-down factors 32, 493

L

Lamina 383
Laminae 19
Lamina interaction 399
Laminate design 438
Laminate thickness 189, 649
Lamination theory 397, 401, 416, 455
Laser 276, 468, 526
Laser cutting 276, 277
Layup 27, 201, 600, 604
Layup-over-foam method 270
Lightning
 — protection 30, 361, 370, 372, 375
 — strike 18, 301, 361, 440, 591
Liquid crystal 54
Liquid impregnation 67
Liquid shim 443
Lockstitch 95
Loss factor 79
Low pressure cure (LPC) 606

M

Machining 276
Macromechanics 397, 410
Magneforming 236, 237
Magnesium matrix 58
Magnetic 286, 351
Male tooling 212
Manufacturing 21, 176, 186
Master model (master plug) 147
Materials 41, 42, 46, 85, 111, 131, 141
    178, 377, 504, 649
Matched die forming 240, 241
Matched die (mold) tool 139, 173, 239
Matched metal mold 138
Material forms 52
Material safety data sheet (MSDS) 504
Matrix 2, 11, 52, 383
Mesh 155
Metal matrix composites (MMC)
    14, 43, 57, 59, 60
Microballons 83, 119
Microcracking 142, 358, 427, 440, 554, 600
Micrographs 505
Micromechanics 397, 400
Microwave heating 137
Modular concept (module concept)
    25, 572, 622, 630, 637
Modules 142, 572, 573, 632
Modulus 13, 42, 75, 77, 79, 80, 383, 401
Moen system 136, 545, 548
Moisture 358, 458
    — absorption 12, 30, 183, 458, 600
    — content 359, 458, 465
    — effects 458
    — resistance 16, 53
    — sensitivity 11, 12, 16, 53
Mold 128, 228
Moldable shim 328
Mold forming 228
Molding 251
Monitoring systems 225
Monolithic
    — assembly (concept) 622, 632
    — graphite 129, 132, 141, 152
Mosites rubber caul 222
Mylar 225

N

NDE 501
NDI 500, 501, 503, 511, 529
NDT 501
Near-net-shape 102, 255
Net tension failure 293
Nickel 131
Nickel coated fibers 370, 371
Nomex 116, 563, 568, 591, 600, 605, 612
Non-bleed bagging 221, 222
Non-metallic tooling 147
Notation 36, 37
Notch 16, 30, 391, 475, 537, 622, 623
Nozzles 62
Nuclear exposure 381
Nutplates 302
Nylon 139, 225, 546

O

Observable 71
Off-axis 544
Optic fibers 637, 644
Orbital winding 209
Organic
    — fibers 13, 74
    — matrices 47
Orientation 35, 38, 181, 189,
    255, 384, 510, 622, 623, 649, 651
Orthogonally reinforced weaving 105
Orthogrid 214, 437, 623, 627, 644
Othrotropic 1, 15, 383, 403
Outgassing 48
Oven 164, 185, 230
Oxidation 13, 62, 67

P

Padup (buildup) 295, 299, 334, 600
Paint stripper 380
PAN (polyacrylonitrile) 11, 73, 76, 80
Panel configurations 434
Particles 83
Patch 83, 536, 539, 541, 542, 551
Patent protection 622
Pattern 147, 544
Peel ply 139, 220, 341, 350, 351
Peel stress 332, 336, 341, 551, 553
Perforated 119, 220, 265
Phenolic 12, 49, 52, 118, 562
Picture frame shear tests 485, 488
Pigtail 367
Pilot part 183
Pitch 67, 76, 80
Placements 200, 203, 204, 635
Plain weave 90

Plaster 147, 157, 159, 256
Plastic film 162
Plied matrix 107
Ply level tests 465
Plywoods 1
PMR 49, 51, 257
Poisson's ratio 383, 386, 398, 399, 400, 416, 443
Polar weave 104
Polar winding 209
Polyarylenesulfide (PAS) 56
Polyester 12, 49, 51
Polyetheretherketone (PEEK) 56
Polyetherimide (PEI) 56
Polyethersulfone 56
Polyethylene (PE) 78
Polyimide 12, 49, 50, 52, 56
Polymers 43, 55, 56, 60
Polyphenylenesulfide 56
Polyurethane 360, 380, 617
Pop rivets 614
Porosity 181, 183, 220, 506, 513, 536
Porous 133, 220
Pot life 227, 253, 255
Potting compound 121, 123, 554
Powder Prepreg 110, 111
Precursor 11, 67, 80
Preforms 64, 86, 99, 254, 255
Preimpregnated 11
Pre-plies fabrics 86
Prepregs 11, 52, 56, 86, 87, 89, 553, 612
Press forming 185, 275
Pressure-bag forming 232
Pressure vessel 162
Pressure-vessel forming 230
Pre-stressed panel 640, 645
Process control testing 482
Processing 13, 16, 140, 179, 182, 184
Producibility 6, 24
Projectile 77, 78
Protruding head fasteners 293, 295
Pseudo-thermoplastics 53, 54
Pulforming 262, 264
Pull-through failure 291, 295, 478
Pull-off strength 486, 489, 544
Push-through tests 480, 481
Pulse-echo Ultrasonic 514
Pultrusion 194, 257, 258, 262, 265

Q

Quadrax 110, 111
Quality 502
Quality assurance (Q.A.) 28, 500, 502, 503, 652
Quartz 75, 82, 367
Quasi-isotropic 15, 19, 84, 228, 295, 344, 385, 446, 485, 568

R

Radiography 522
Radome 71, 79, 112
Rail (two-rail) shear tests 467, 469, 473
Rain erosion 360
Rattan 9, 107, 641
Rayon 11, 80
Refrigeration 48, 53
Release (separator) film 220
Repairs 31, 140, 533
Residual stresses 343
Resin
— content 261
— injection molding (RIM) 251
— rich 220, 255, 508
— starved 220, 221, 508, 513
— transfer molding (RTM) 194, 252, 621, 641
Resistance welding 349, 350
Retrofit testing 458
Reusable bags 225
Rivet-bonded joint 347, 348
Robot 107, 204, 212, 276, 281, 635
Roughness 220
Roll forming 249, 250
Rubber bagging 162
Rubber forming 248

S

S-glass 75, 77, 78
Safety 185
Salt 157, 360
Sand erosion 360
Sandwich 111, 265
Satin weave 90
Sawing 276, 277
Scarf joint 336
Scarf repair 534, 536, 551, 553

Scatter 469, 490, 513
Scrap 16, 24, 48, 53, 91, 129, 180, 207, 223, 281, 502
Scratch 510
Screen (non-woven) 377
Secondary curing 268
Self-contained tooling 138, 165
Semi-crystalline 55, 56
Sensors 225
Serrations 551, 552
Sewing 9, 95, 353
Shear head fastener 299
Shearography 526, 528
Shear stiffness 385, 386
Shear tests 469, 473, 474, 488
Shelf life 53, 210
Shim 189, 299, 327, 530
Short beam tests 475, 477
Shredder 281
Shrinkage 72
Silicon 75, 225
Silicon carbide 58, 67, 82
Silicon heating blanket 136
Silicon elastomer 164
Silicon rubber 132, 171
Sine wave beam (spar) 21, 234, 488, 568, 582, 583
Single-lap joint 298, 336
Slippage 198, 228
Sleeve fastener 301, 302
Smart structures 637, 644
Smoothness 189, 191, 649
Sodium-nitrate mixture 256
Solvent 13, 16, 48, 53, 220, 380
Sparking 368, 369
Specific heat 131
Specific gravity 504
Specifications 28, 29, 33, 44
Spectra fiber 78
Splice 195, 198, 592
Spray-metal tooling 155
Springback 195, 197
Stacking 1, 189, 384, 386, 416, 420, 649
Stamping 240
Standards 27, 455
Static tests 494, 497
Steel 131, 132, 141, 143
Step-hollow grid 637, 641
Stepped-lap joint 338, 339, 348, 349
Step-lap repair 536, 551, 553
Stiffness 385

Stitched fabrics 86
Stitching 9, 95
Striaght-through concept 623
Strain 13, 16, 18, 387, 389, 390, 414
Strain-to-failure 13
Strength 16, 42, 72, 75, 79, 80, 384, 391, 397
Stress concentration 287, 342, 381, 387, 442, 456, 494, 623
Superforming 243
Surface depression 510
Symmetrical balanced laminate 189, 193
Symmetrical laminate 181
SynCore 120, 637, 638
Syntactic foam 4, 119, 265, 554, 563, 629
Synthetic 13, 119

T

Tabs 466
Tack 48, 49
Tag-end tests 481, 482, 504, 505, 650, 651, 652
Tailor 11, 15, 44, 91, 102, 387, 388, 579, 617
Tape 2, 38, 87, 201
Tape laying 14, 25, 200, 203
Tape winding 209
Teflon 139, 220, 519, 546
Temperature 358, 458
— resistance 43, 75
— effects 458
Ten percent (10%) rule 440
Tensile tests 469, 470, 493
Tension head fastener 299
Tests 28, 30, 33, 34, 35, 400, 453, 456, 467
Textile 9, 73
Thermal conductivity 46, 131, 132, 141, 171, 647
Thermal contraction 142
Thermal distortion 343
Thermal expansion 18, 44, 46, 77, 126, 138, 181, 182, 343, 347, 440, 647
Thermal inertia 142, 230
Thermal resistance 11
Thermal shock 62, 72
Thermal stability 49, 51, 52
Thremal stress 343
Thermoclave 164
Thermocouple 183, 225, 546

Thermoforming (platen press) 238
Thermography 529
Thermoplastic fastener 320
Thermoplastics 12, 16, 47, 48,
    53, 54, 55, 56, 107, 111,
    179, 427, 544, 636, 637, 643
Thermoset sleeve fastener 319
Thermosets 11, 16, 47, 48, 49,
    52, 179, 239
Therm-X process (forming) 164, 231
Thread 95
Through-transmission 515
Titanium 58, 141
Tolerance 21, 139, 184, 188, 255, 299
Tool design 138
Tooling 32, 126, 129, 147, 152, 155,
    165, 167, 168, 173, 179
Tooling master 147
Tool life expectancy 139
Toughened material 8, 12, 49, 50, 52
Toughness 12, 48, 77, 90, 428
Tow 73
Toxic 185
Trial part 184
Triaxial braiding 96, 98
Twisted 73
Two-rail shear fixture 467

U

Ultrasonic cutting 276, 279
Ultrasonic inspection 512, 515
Ultrasonic testing 496
Utlrasonic welding 350
Ultraviolet (UV) 11, 30, 359, 600
Unbalanced bonded joint 334
Unidirectional fabrics 86, 90, 91
Unidirectional tape 11, 111
Unitape 86, 392
Upilex-R 225

V

Vacuum
    — bag 223, 225, 554, 617
    — forming 231
    — gauge 218, 226
    — lines 226, 227
    — oven cure 230
    — ports 218, 226, 227
Very-complex-shape (VCS) 107, 641
Vent 220
Vent lines 226

Vibration welding 352
Vinyl esters 52, 614, 617
Viscosity 13, 254
Visual inspection 512
Voids 49, 149, 195, 206, 465, 506
Void content 207, 255, 504
Volatiles 53, 183, 220

W

Warp direction 90
Warpage 181, 268, 407
Wash-out mandrel (disposable)
    157, 209, 633, 637, 643
Wash-out mandrel molding 256
Water heating 162
Water jet cutting 276, 277, 278
Wear 72
Weathering 30, 357
Weaving 9, 69, 90, 93
Welding 286
Wet
    — Layup 149, 179, 195, 270, 356, 533,
        541, 553, 605, 610
    — out 254, 255, 257
Wet-crush-rivet 353, 354
Whirling winding 209
Whiskers 84, 89
Winding 96, 204, 206, 209, 215
Wire 155, 165
Wire mesh 370, 376, 591
Woven fabric (cloth) 11, 86, 90
Wrap-around concept 623, 637
Wrapping 270
Wrinkle 181, 200, 220, 222, 225, 228, 509

X

X-axis 36, 41
X-ray 496, 522, 524, 525

Y

Yarns 73
Y-axis 36, 41

Z

Z-axis 36